Gravitational radiation has not yet been positively detected on Earth, although there is little doubt that astrophysical systems including neutron stars and black holes are extremely powerful sources of this radiation. In the opening chapters of the book the author gives a general introduction to the concepts and sources of gravitational waves, and the methods used to detect them.

David Blair has an extensive knowledge of the subject and has visited the sites of gravitational radiation experiments all over the world. He has compiled a book which will be of lasting value to undergraduates, postgraduates and researchers alike.

The detection of gravitational waves

The detection of gravitational waves

Edited by

DAVID G. BLAIR

University of Western Australia

CAMBRIDGE UNIVERSITY PRESS

Cambridge

New York   Port Chester

Melbourne   Sydney

CAMBRIDGE UNIVERSITY PRESS
Cambridge, New York, Melbourne, Madrid, Cape Town, Singapore, São Paulo

Cambridge University Press
The Edinburgh Building, Cambridge CB2 2RU, UK

Published in the United States of America by Cambridge University Press, New York

www.cambridge.org
Information on this title: www.cambridge.org/9780521352789

© Cambridge University Press 1991

First published 1991
This digitally printed first paperback version 2005

A catalogue record for this publication is available from the British Library

ISBN-13 978-0-521-35278-9 hardback
ISBN-10 0-521-35278-9 hardback

ISBN-13 978-0-521-02102-9 paperback
ISBN-10 0-521-02102-2 paperback

Dedication

This book is dedicated to my friends and colleagues at the University of Western Australia, and in Great Britain, China, France, Germany, Italy, Japan, USA and USSR, whose years of perseverance and effort will pay off in the end.

Contents

Contributors

H.-A. Bachor
Department of Physics
Australia National University
Canberra
A.C.T. 2601
Australia

David G. Blair
Department of Physics
University of Western Australia
Nedlands
WA 6009
Australia

A. Brillet
GROG
Bat 104
CNRS
91405 Orsay
France

R. W. P. Drever
LIGO Project
Astrophysics Department
California Institute of Technology
Pasadena
California
USA

J. Ferreirinho
Department of Physics
University of Western Australia
Nedlands
WA 6009
Australia

W. M. Folkner
Department of Physics and Astronomy
University of Maryland
College Park
Maryland
USA

J. Gea-Banacloche
Instituto de Optica Daza de Valdés
CSIC
Madrid
Spain

William O. Hamilton
Department of Physics and Astronomy
Louisiana State University
Baton Rouge
LA 70803
USA

Ronald W. Hellings
Jet Propulsion Laboratory
Pasadena
California
USA

J. Hough
Department of Physics
University of Glasgow
Glasgow
UK

G. A. Kerr
Department of Physics
University of Glasgow
Glasgow
UK

G. Leuchs
Max-Planck Institut für Quantenoptik
D-8046 Garching
and
Sektion Physik der Universität München
München
West Germany

D. E. McClelland
Department of Physics
Australia National University
Canberra
A.C.T. 2600
Australia

N. L. Mackenzie
Department of Physics
University of Glasgow
Glasgow
UK

C. N. Man
GROG
Bat 104
CNRS
91405 Orsay
France

B. J. Meers
Department of Physics
University of Glasgow
Glasgow
UK

G. P. Newton
Department of Physics
University of Glasgow
Glasgow
UK

G. V. Pallottino
Dipartimento di Fisica
Università 'La Sapienza'
P.le. A. Moro 2
00185 Rome
Italy

G. Pizzella
Dipartimento di Fisica
Università 'La Sapienza'
P.le. A. Moro 2
00185 Rome
Italy

J.-P. Richard
Department of Physics and Astronomy
University of Maryland
College Park
Maryland
USA

D. I. Robertson
Department of Physics
University of Glasgow
Glasgow
UK

N. A. Robertson
Department of Physics
University of Glasgow
Glasgow
UK

R. J. Sandeman
Department of Physics
Australia National University
Canberra
A.C.T. 2600
Australia

R. Schilling
Max-Planck Institut für Quantenoptik
D-8046 Garching
West Germany

Bernard F. Schutz
Department of Physics
University of Wales, College of Cardiff
Cardiff
UK

Kimio Tsubono
Department of Physics
The University of Tokyo
Bunkyo
Tokyo 113
Japan

P. J. Veitch
Department of Physics
University of Western Australia
Nedlands
WA 6009
Australia

J. Y. Vinet
GROG
Bat 104
CNRS
91405 Orsay
France

H. Ward
Department of Physics
University of Glasgow
Glasgow
UK

Walter Winkler
Max-Planck Institut für Quantenoptik
D-8046 Garching
West Germany

Preface

The detection of gravitational radiation will not only be a milestone in scientific achievement; it will also be of immense cultural and philosophical significance. It will perhaps complete the process by which Western culture has gradually been forced to let go of its absolutist heresy. The heresy goes back to Aristotle and beyond. It is intimately tied up with the Judeo–Christian prejudice of an unchanging homocentric universe. It is epitomised by the ancient belief in a heavenly crystalline celestial sphere rigidly rotating and unchanging above us.

This heretical edifice has been tumbling slowly under the onslaught of scientific investigation. Newton gave us absolute space, but contributed to the demolition of the geocentric universe brought about by Galileo, Tycho, Kepler and Copernicus. Darwin discovered the impermanence of species; the plate tectonic theory gave us impermanent continents. Einstein demolished Newtonian absolute space and time, and gave us both spacetime curvature and the theory of gravitational radiation. The *observation* of gravitational radiation will demonstrate that spacetime not only curves predictably in the presence of matter, but is also subject to unpredictable perturbations as gravitational waves ripple through the universe.

Absolutism is surely connected with prejudice. The absolutist prejudice has led to a lingering battle in the case of Darwinism, and most relativists suffer minor irritations from the Einstein-was-wrong brigade. Tycho Brahe wrote of 'his' supernova in 1572:

> During my walk contemplating the sky here and there, . . . behold, directly overhead a certain strange star was suddenly seen, flashing its light with a radiant gleam. Amazed, and as if astonished and stupefied I stood still . . . I was led into such perplexity by the unbelievability of the object that I began to doubt the faith of my own eyes.

His prejudice is transparent and seems naive. This shows how far we have gone today in giving up absolutism. Yet absolutism still exists in the world, and in social contexts such as issues of race and religion contributes much unhappiness. Ultimately society will absorb a world view that is free of absolutism. One aspect will include the fact that spacetime itself is stochastic. We will move closer to the

Buddhist world view of *anita,* impermanence. In absorbing the truth about our universe we will surely come to a deeper knowledge of our place in existence.

This book was conceived during a most pleasant visit to the Institute of Astronomy at Cambridge. I thank everyone there for their hospitality, especially Andy Fabian, Martin Rees, and the Fellows of Darwin College. Financial support by the University of Western Australia's Outside Studies Program is gratefully acknowledged, as is long term research support by the Australian Research Council. Finally, I wish to thank Thibault Damour for his careful checking of the manuscript and for his many invaluable suggestions, Raelene Selkirk for her brilliant work on the word processor, and all the co-authors who made this book possible.

David Blair   Perth

Introduction

This book is about gravitational radiation detectors. It is about experimental physics: the physics of extremely sensitive instruments designed to detect the infinitesimal time varying strains in spacetime which are gravitational waves.

For half a century most physicists considered the detection of gravitational waves to be an impossibility, but 30 years ago Joseph Weber first outlined possible means of detection, and followed this by a lonely pioneering decade of instrument development. About 20 years ago a range of new technologies appeared on the horizon, and we have now seen two decades of advance in a variety of areas, often driven by the needs of gravitational radiation detection. Looked at as a whole these represent a spectacular advance in technological capability, and now it is possible to look forward to a future when gravitational astronomy will plug a major gap in our knowledge of the universe.

The first area of intense effort was in the development of improved resonant bar antennas. This led to the development and understanding of systems and materials with ultralow acoustic loss, and ultralow electromagnetic loss. The development of low loss microwave cavities led to new technologies for vibration transducers and frequency standards. The need for sensitive amplifiers was met by the development of greatly improved superconducting quantum interference devices (SQUIDs) and cryogenic gallium arsenide field effect transistor amplifiers. The understanding of quantum mechanical limitations to measurement led to the development of techniques called variously squeezing, quantum non-demolition and back action evasion.

In the mid-1970s intense effort focussed on laser interferometer gravitational wave detectors. Methods were developed to enable lasers to be stabilised to unprecedented levels, optical components, especially mirrors, were dramatically improved, and methods were developed to lock optical cavities to stabilised lasers, and to isolate seismic noise to an extremely high degree.

Improved frequency standards allowed gravitational wave detectors to use the spacecraft Doppler ranging technique, and the discovery of high stability millisecond pulsars allowed the search for gravitational waves to be based on pulsar timing.

Gravitational radiation detection was a long shot in 1970. Today it appears inevitable. Gravitational observatories are now on the drawing boards; by the year 2000 we should be surprised and disappointed if the first gravitational signals have not been detected.

The purpose of this book is two-fold. The first is to provide an introduction to the field of gravitational wave detection at a general level. The second is to provide a state of the art account of the physics and technology of gravitational wave detection, and to provide a reference of lasting value to researchers in the field.

As an experimental text this book does not assume a detailed knowledge of general relativity. It introduces the concept of gravitational waves physically and intuitively, while only briefly sketching the theory. It uses SI units and not the gravitational units so dear to theoretical relativists but inappropriate when designing experiments.

PART I

An introduction to gravitational waves and methods for their detection

1
Gravitational waves in general relativity*

DAVID G. BLAIR

1.1 Introduction to general relativity

In Newtonian physics space is a rigid grid of Cartesian coordinates. All physical processes can be viewed and measured against this absolute framework. Einstein's general theory of relativity (Einstein, 1915a–d, 1916) refutes this view. Space is not an infinitely rigid conceptual grid, but a dynamical and deformable medium. Rather than considering three-dimensional space separately from time, general relativity creates a unified four-dimensional spacetime in which time is measured in light-travel distance, and all dimensions are lengths. This is possible because of the universality of the velocity of light, first demonstrated by the Michelson–Morley experiment (Michelson and Morley, 1887a, b).

The deformation, or curvature, of spacetime is described by the Einstein curvature tensor **G**. The magnitudes of the components of **G** express the magnitude of the spacetime curvature. The source of the spacetime curvature is expressed by the stress energy tensor **T**, which describes the distribution of mass, energy and momentum in the system. Einstein's field equations can be written

$$\mathbf{T} = \frac{c^4}{8\pi G}\mathbf{G}, \tag{1.1}$$

where the coupling constant is the very large number $c^4/8\pi G$, c is the speed of light and G is the gravitational constant.

Equation (1.1) has been deliberately written in a form which emphasises an analogy with Hooke's law, $F = kx$, or preferably $P = Eh$ where P is the applied pressure, E is the modulus of elasticity and h is the dimensionless strain or

---

* Some people will approach this book with a sound knowledge of general relativity, will know the properties of gravitational waves, and will want to skip or skim this chapter and possibly read a major review such as one of those listed at the end of the chapter. Many will come to it with a skimpy knowledge of general relativity. For these readers the chapter is intended to introduce the basic ideas and the most important formulae, but is not intended to be in any way rigorous or complete.

fractional deformation due to the imposed stress P. In the case of general elastic solids E is a tensor with 21 possible independent components, and the strain is similarly complex. However, in general relativity the 'elastic modulus' is a simple scalar: the complexity of general relativity arises because both the stress–energy $\mathbf{T}$ and the curvature $\mathbf{G}$ necessarily are tensor quantities. $\mathbf{T}$ is the source of the gravitational field. Thus $\mathbf{T}$ is the stress, and $\mathbf{G}$ is the strain, but measured as curvature. It is close enough to an equation of elasticity that we need not be surprised that it has a solution which is a wave equation, just as acoustic waves follow from Hooke's law. The analogy is not perfect, however, because the spring constant $c^4/8\pi G$ has dimensions of spring constant per unit length (force).

The form of equation (1.1) and the huge magnitude of c^4/G ($\sim 10^{43}$ N) leads us intuitively to two essential points:

(i)   Spacetime is an elastic medium: therefore it can sustain waves.
(ii) Spacetime has extremely high stiffness: hence extremely small amplitude waves have a very high energy density.

In some senses the step from Newtonian physics to general relativity is a small one; in others it is enormous. Spacetime goes from being infinitely rigid to being just extremely rigid. The effect on slow motion, weak field dynamics is so minute that it has required the ingenuity of several generations of physicists to measure it with any reasonable accuracy in the solar system. The relativistic corrections to Newtonian physics are seen in numerous phenomena such as the bending of starlight past the Sun, the precession of the perihelion of Mercury, and the gravitational redshift of clocks changing with altitude in the Earth's field. The effects are small because all gravitational fields in the solar system are weak (compared with that near a black hole) and all velocities are small (compared with the speed of light). Thus Mercury's orbit precesses by 43 arc seconds per century, starlight is deflected past the Sun by 1.7 arc seconds, and the rate of a clock varies by 3×10^{-16} in one metre of altitude.

The implication of these small effects is enormous, firstly because our fundamental concept of spacetime has to change radically, and secondly because the new concept results in new possibilities in the universe: black holes can exist; orbiting systems must always radiate away energy; and a new form of energy, gravitational radiation, can exist with intensities vastly greater than anything hitherto conceived.

1.2   Stress energy and curvature

Let us now look at equation (1.1) in a bit more detail. We might ask: why not use the mass density ρ for the source term as we do in Newtonian gravitation? Surely mass is the source of the gravitational field! The answer is that mass density is not invariant with respect to coordinate change. One of the beautiful elegances of general relativity is the introduction of the stress energy tensor as the source of gravity. The components of **T** have dimensions of mass–energy density (T^{00}), energy flux (T^{ij}), momentum density ($T^{0i} = T^{i0}$) and momentum flux ($T^{ij} = T^{ji}$)*. When **T** is used as the source the theory is independent of the choice of reference frame. Nature doesn't care about reference frames, so this property is essential. For slow motion the velocity dependent components of **T** are small compared to the rest mass energy; then **T** can be approximated by ρ. In general, the definition of **T** arises because internal energy is as much a source of gravity as is rest mass. However, the relative motion of external observers cannot alter the gravitational interaction, and **T** gives us the necessary invariance.

As we have seen, the curvature 'strain' in spacetime due to the stress **T** is described by the Einstein tensor **G**. Later we will see how true strain (in an engineering sense) is derived from **G**. First, another look at curvature. We will make the briefest contact with the tensor calculus, enough to define terms so as to be able to proceed further, but without any detail. The Einstein tensor is a contraction to second rank of the Riemann curvature tensor. The Riemann tensor is a fourth rank tensor $R_{\alpha\beta\mu\nu}$; it has 20 independent components in spacetime and satisfies a set of differential equations called the Bianchi identities. The Riemann tensor depends in general on the metric $g_{\alpha\beta}$, and on its first and second partial derivatives. The Einstein tensor is given by†

$$G^{\alpha\beta} = R^{\alpha\beta} - \frac{1}{2}g^{\alpha\beta}R, \tag{1.2}$$

where $R^{\alpha\beta}$ is the symmetric Ricci tensor which is contracted from $R_{\alpha\beta\mu\nu}$, and R is the Ricci scalar. The Einstein tensor is a symmetric tensor, with ten independent components, and therefore Einstein's field equations are a set of ten differential equations. This mathematical fact means that the matter and non-gravitational fields in a region of spacetime (as expressed by the energy–momentum tensor **T**) determine ten of the 20 independent components of the Riemann curvature tensor. The remaining ten independent components describe gravitational waves generated by distant matter–energy distributions.

The physical significance of curvature in four-dimensional spacetime is difficult to grasp. However, the meaning is simple when discussed in three-dimensional space. If space has a radius of curvature r, then the magnitude of a typical component of the Riemann tensor is simply $1/r^2$. But what is this curvature?

Imagine an instrument consisting of three laser beams, carefully configured so as to form a triangle. You make a device to automatically measure the included

---

* The coordinates have indices 0, 1, 2, 3, where 0 refers to the time axis and the indices i and j indicate the three spatial coordinates. Tensors are indicated by a bold font: their components (which are not tensors) are indicated by the indices.

† Greek indices conventionally indicate spacetime coordinates, and take values 0 (time) and 1, 2 and 3 (space).

angles where the beams cross. In Euclidean space we know that the sum of the angles Σ adds to π or 180°. In real space, near the Earth or the Sun for example, we know, from the fact that starlight is deflected as it passes the Sun, that Σ will generally be fractionally larger than π. We measure $(\Sigma - \pi)/A$, where A is the area of the triangle, and we are measuring a component of the Riemann tensor. A set of such instruments could measure the curvature in orthogonal planes, allowing us to define the spatial components of the curvature. The extrapolation to spacetime is a simple generalisation: Riemannian geometry tells us that to specify the curvature of spacetime we must measure ten independent components. The orthogonal planes can only be defined locally, however, in a local inertial frame.*

Riemannian spaces, of which spacetime is an example, are locally flat. On a small enough scale, in a local inertial frame, parallel lines are parallel! (That is, they never cross and they never diverge.) Einstein realised the connection between gravitational dynamics and Riemannian geometry. He showed that the geodesics in curved spacetime could be identified with spacetime trajectories of freely falling particles. Thus the laser beams map out the curvature of spacetime. Locally parallel lines cease to be parallel, and starlight curves past the Sun.

The problem of general relativity (perhaps I should say the problem of the universe!) is that it is non-linear. Unlike classical electromagnetism, we cannot assume superposition. We cannot rigorously separate the curvature of spacetime into the sum of two independent components. John Wheeler introduced the slogan 'matter tells space how to curve, space tells matter how to move' to describe general relativity. This hides the complexity, because in reality: *matter, the motion of matter, and radiation density, including propagating waves of curvature, tell spacetime how to curve.* Curvature creates curvature, and influences its propagation.

The non-linear complexity creates enormous mathematical difficulties. We cannot rigorously separate spacetime into the sum of a static curvature plus a time varying propagating curvature due to gravitational waves. Thus, although Einstein derived gravitational wave equations very early (Einstein, 1916, 1918) there followed a period of about 40 years during which the existence of gravitational waves was disputed. Eddington said that gravitational waves travel 'at the speed of thought'! Pirani (1957) and Bondi (1957) demonstrated that gravitational waves have a real existence, in that they can wiggle test masses, do work, and exchange energy. It was not until 1968 that the formal theory of gravitational wave propagation in curved space was fully developed (Isaacson, 1968a, b). Thus today there is no dispute: the wave solutions of Einstein's field equations exist, they represent a real flux of energy, and some of it (albeit a very small amount) may be absorbed by matter.

---

* A horizontal laser beam is deflected about one angstrom (1 Å) for every kilometre of path near the Earth's surface. What is the typical magnitude of the Riemann tensor components near the Earth?

1.3 Non-linearity and wave phenomena

Non-linearity means that in general wavefronts can interact with each other and with regions of static curvature. This can allow a variety of interesting phenomena, some of which we discuss below. There are probably plenty more such phenomena which will be among the surprises that await the birth of gravitational wave astronomy.

(a) Waves can be scattered by strong background curvature. For example, the gravitational waves emanating from the coalescence of a binary neutron star system, and its subsequent collapse to form a black hole, will be scattered in the intense curvature of the forming black hole (Thorne, 1987, and references therein).

(b) Waves can be focussed by background curvature. On a cosmological scale galaxies can focus gravitational waves from distant sources as they are already observed to do optically with quasars. On a more local scale, the Sun can act as a lens to focus plane gravitational waves to a focal point near to the orbit of Jupiter (about 20 AU from the Sun). There is a catch here, however, and it is probably not worth sending a gravitational wave detector out there. The problem is that the solar lens is not a lot larger than a gravitational wavelength. Thorne (1983) shows that strong focussing only occurs if the gravitational diameter of the Sun, $4GM/c^2$, is large compared to the wavelength; otherwise diffraction spreads the beam sufficiently that the focussed amplitude is not much increased. Thus only waves of frequency $> 10^4$ Hz will be strongly focussed, and, as we shall see later, these are not of great interest.

(c) Waves can be parametrically amplified in regions of time varying curvature. If the curvature is varying dynamically, for example in the early universe after the big bang, gravitational waves can be amplified as they interact with the background. This can increase the amplitude of gravitational waves emanating from fluctuations in the early universe. Similarly, the background curvature may be modulated by one source of waves (which an engineer might call the pump). A second incident wave (the signal) may be amplified and frequency shifted by this interaction exactly as a microwave or optical signal is amplified by a pumped non-linear medium. Thus a coalescing binary with orbital frequency ω_b will upconvert incident radiation to frequency $\omega_b + \omega_i$ while simultaneously generating radiation at twice the binary frequency. Similarly, one can imagine (probably very unlikely) processes whereby a massive black hole focusses intense gravitational waves causing non-linear harmonic generation.

(d) Waves interact with self generated curvature. This is epitomised by the geon, proposed by Wheeler (1962). A geon is a bundle of gravitational waves held together by its own self generated curvature. While most doubt that such objects could exist, one can conjecture about beams or pulses of self focussing gravitational waves, confining themselves in their own self generated valley of background curvature.

(e)   Colliding gravitational waves can focus each other so strongly that the wavefront may collapse to form a singularity. If the singularity forms only around one wavefront, a high velocity black hole will fly out from the collision, but if it forms about both wavefronts the resulting black hole will be nearly stationary (Yurtsever, 1988a–c).

These are some of the exotic aspects of the non-linearity. At a more immediately practical level, the problem of the non-linearity of general relativity affects above all the modelling of sources of gravitational waves. Thus the majority of source calculations can only be approximations. Even with the use of powerful numerical codes, the behaviour of matter under conditions of intense gravity and high densities is so uncertain that our confidence in their results cannot be high. (See, for example, Nakamura, 1987.) In spite of this, however, we will see in chapter 2 that good order of magnitude estimates exist for a range of wave generation processes.

1.4   Introduction to gravitational waves

In the case of the analysis for detection of gravitational waves it is a different matter. For even the strongest imaginable gravitational waves from astrophysical sources we need only consider plane, linear waves on an essentially flat space background. This is because the amplitude of waves crossing the solar system, even from relatively nearby strong sources, must always be exceedingly low, despite the fact that the energy flux may be enormous. We will come back to this point in chapter 2. As a result, in spite of non-linearity, we go ahead and make a separation, which is not fully precise. We assume superposition, which is essential to our concept of gravitational waves, which is not strictly rigorous, which is bothersome to some theorists, but which need bother experimentalists not at all. We separate spacetime curvature into a background term and a wave term by assuming that the background curvature varies slowly while the wave varies rapidly. That is

$$R_{\alpha\beta\mu\nu} = R^{B}_{\alpha\beta\mu\nu} + R^{W}_{\alpha\beta\mu\nu}. \tag{1.3}$$

It is possible to go on to further simplify things by assuming plane waves on a background of flat Minkowski spacetime, or alternatively to simplify Einstein's equations for weak fields and explicitly solve for a wave solution. We shall only sketch the barest details of the derivation here, as it involves considerable mathematical detail readily found elsewhere, and anyhow is outside the scope of this book.

The wave equation can be derived from the weak field approximation to general relativity, in which spacetime is nearly flat and we can use 'nearly Lorentz coordinates'. The metric $g_{\alpha\beta}$ is separated into the flat Minkowski metric $\eta_{\alpha\beta}$ and a perturbation term $h_{\alpha\beta}$.

$$g_{\alpha\beta} = \eta_{\alpha\beta} + h_{\alpha\beta}. \tag{1.4}$$

Here

$$|h_{\alpha\beta}| \ll 1; \quad \text{and} \quad \eta_{\alpha\beta} = \begin{pmatrix} -1 & 0 & 0 & 0 \\ 0 & 1 & 0 & 0 \\ 0 & 0 & 1 & 0 \\ 0 & 0 & 0 & 1 \end{pmatrix}. \tag{1.5}$$

In these coordinates, to first order the Riemann tensor is determined by a linear combination of second partial derivatives of the small metric perturbation term $h_{\alpha\beta}$. In weak field general relativity $h_{\alpha\beta}$ is replaced by a related pseudotensor $\bar{h}^{\alpha\beta}$, which is called the trace reverse of $h_{\alpha\beta}$. However, the wave solution in the transverse traceless gauge satisfies $\bar{h}^{\alpha\beta} = h_{\alpha\beta}$, so, at least in vacuum, this formality can be ignored. In terms of the perturbation tensor **h** (with components $h_{\alpha\beta}$), the wave equation can be written

$$\left(-\frac{1}{c^2}\frac{\partial^2}{\partial t^2} + \nabla^2 \right)h_{\alpha\beta} = -\frac{16\pi G}{c^4}T_{\alpha\beta}. \tag{1.6}$$

In the nearly flat spacetime of the solar system we can assume zero stress energy and go to the transverse traceless gauge where $\bar{h}_{\alpha\beta} = h_{\alpha\beta}$. Thus we go over to the vacuum weak field Einstein equations:

$$\left(-\frac{1}{c^2}\frac{\partial^2}{\partial t^2} + \nabla^2 \right)h_{\alpha\beta} = 0. \tag{1.7}$$

This is the three-dimensional wave equation, telling us immediately of the existence of waves with a velocity equal to c.

We can think of $h_{\alpha\beta}$ as the gravitational wave field. It is the second time integral of the curvature, and because the waves are pure transverse waves we can think of **h** as a time integrated spatial shear. The wave field is, by general relativity, *transverse* and *traceless*. Waves from distant sources will always be plane, so to describe them it is natural to use a specific reference frame in which they propagate in the z-direction. The transverse nature implies that all z-components of the wave are zero. That leaves only four components:

$$\mathbf{h} = \begin{pmatrix} h_{xx} & h_{xy} \\ h_{yx} & h_{yy} \end{pmatrix}. \tag{1.8}$$

The traceless property of the field means that $h_{xx} + h_{yy} = 0$. Hence $h_{xx} = -h_{yy}$. Because the Riemann tensor is symmetric, we have also $h_{xy} = h_{yx}$. Thus there can be only two independent polarisation states for gravitational waves, normally denoted h_+ and $h_\times$.

In different coordinate systems the form of equation (1.7) varies. It was this dependence of the solution on the choice of coordinate systems that caused years of doubts about the existence of gravitational waves. The problem is related to the indistinguishability between gravity and acceleration. In a local inertial frame the gravitational force may be eliminated, but the tidal force always remains.

In the transverse traceless coordinates gravitational waves are simple. Equation (1.8) describes the wave field components. Equally, we can go back to the Riemann tensor, which is simply the second time derivative of $h_{\alpha\beta}$. Because the wave amplitude is small we can construct any wavefront from a set of sinusoidal waves. Hence we do not need to restrict ourselves to sinusoids, and assume a general plane wave solution. Specifically

$$\frac{1}{2}\frac{\partial^2 h_{jk}}{\partial t^2} = -R_{j0k0}, \tag{1.9}$$

and in terms of $\mathbf{R}$ a plane gravitational wave can be described by

$$R_{\alpha\beta\mu\nu} = R_{\alpha\beta\mu\nu}(t - z/c) \tag{1.10}$$

where the waves propagate in the z-direction. The transverse traceless restriction on $\mathbf{h}$ applies also to $\mathbf{R}$, which likewise has only two degrees of freedom in general relativity. Thus the Riemann tensor of the wave reduces to the equivalent two pairs of independent components of $\mathbf{R}$.

$$R_{x0x0} = -R_{y0y0}, \tag{1.11}$$

and

$$R_{x0y0} = R_{y0x0}, \tag{1.12}$$

where 0 indicates a component along the time basis vector. Note, however, that components such as R_{x0yz} are not, in general, zero, but can be computed in terms of R_{x0x0} and R_{x0y0}. In figure 1.1 the propagation of a gravitational wave as a wave of spacetime curvature is illustrated.

1.5   The effects of gravitational waves

Consider a sinusoidal plane wave solution of equation (1.7):

$$h_+ = h_{xx} = -h_{yy} = \mathrm{Re}\{A_+ e^{-i\omega(t - z/c)}\} \tag{1.13a}$$

$$h_\times = h_{xy} = h_{yx} = \mathrm{Re}\{A_\times e^{-i\omega(t - z/c)}\}, \tag{1.13b}$$

where A_+ and $A_\times$ are the amplitudes of each polarisation.

As we can only measure relative accelerations, we ask what is the relative acceleration between two test masses in the x–y plane, due to the passage of a gravitational wave? Consider the relative acceleration between two particles subject to a Newtonian gravitational potential $\phi(x)$. A particle at coordinate x_1

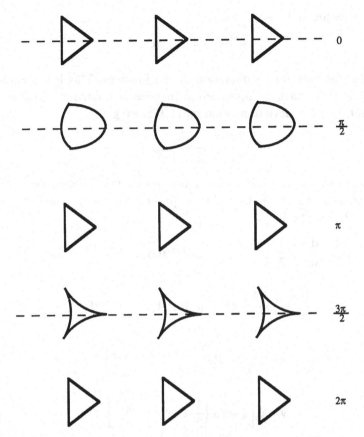

Figure 1.1. The propagation of a wave of spacetime curvature. The plane wavefront moves down the page and spacetime curvature, as determined by light ray triangles, oscillates from flat to convex to flat in the first half cycle, and then becomes concave during the second half cycle. The dotted lines indicate the wavefront. Adapted from Weber (1987).

suffers acceleration

$$\frac{d^2 x_1^i}{dt^2} = \frac{-\partial}{\partial x^i} \phi(x_1) \tag{1.14}$$

and likewise for a particle at x_2. The difference of acceleration is

$$\frac{d^2 x_1^i}{dt^2} - \frac{d^2 x_2^i}{dt^2} = \frac{-\partial}{\partial x^i} \phi(x_1) + \frac{\partial}{\partial x^i} \phi(x_2), \tag{1.15}$$

and if $x_2^i = x_1^i + \delta x^i$, then

$$\frac{-\partial^2 \delta x^i}{\partial t^2} = \frac{\partial^2 \phi}{\partial x^i \, \partial x^j} \delta x^j. \tag{1.16}$$

This is the deviation equation due to a tidal force of gravity gradient. In terms of the Riemann tensor

$$-\frac{d^2\,\delta x^i}{dt^2} = R^i_{\ 0j0}\,\delta x^j,$$

(1.17)

where the indices indicate a summation of the j-indices. This is the equation of geodesic deviation, which is fundamental to general relativity. If we substitute equation (1.9) into the deviation equation (1.17), we get

$$\frac{d^2\,\delta x^j}{dt^2} = \frac{1}{2}\frac{\partial^2 h^j_k}{\partial t^2}\,\delta x^k.$$

(1.18)

Now consider two test particles, one at the origin, the other at point (x, y, z). Assume that only the A_+ polarisation is present $(A_\times = 0)$. Then if h satisfies equation (1.13a), we find

$$\frac{d^2 x}{dt^2} = \frac{1}{2}\frac{\partial^2 h_{xx}}{\partial t^2} x = -\frac{1}{2}\omega^2\,\mathrm{Re}\{A_+e^{-i\omega(t-z/c)}\}x$$

(1.19a)

and

$$\frac{d^2 y}{dt^2} = \frac{1}{2}\frac{\partial^2 h_{yy}}{\partial t^2} y = \frac{1}{2}\omega^2\,\mathrm{Re}\{A_+e^{-i\omega(t-z/c)}\}y.$$

(1.19b)

The solution is

$$x = x_0\left[1 + \mathrm{Re}\left\{\frac{1}{2}A_+e^{-i\omega(t-z/c)}\right\}\right]$$

(1.20a)

and

$$y = y_0\left[1 - \mathrm{Re}\left\{\frac{1}{2}A_+e^{-i\omega(t-z/c)}\right\}\right].$$

(1.20b)

That is, the distance between two test masses is fractionally altered by amplitude $\frac{1}{2}A_+$. In the other polarisation the effects are similar. The sign of the displacement is the same for x and y. The displacements are given by $x = x_0 + \frac{1}{2}h_x y_0$ and $y = y_0 + \frac{1}{2}h_x x_0$.

To visualise the effects of plane polarised gravitational waves, imagine a set of test particles arranged in a ring in space. They will be deformed by the gravitational wave as shown in figure 1.2.

During the 'positive half cycle' of the wave, the ring is expanded along the x-axis, while the 'negative half cycle' contracts the x-axis. At any moment the deformation is invariant under 180° rotation, unlike electromagnetic waves which are invariant under 360° rotation. Thus the gravitational wave is a spin 2 transverse wave, while electromagnetic waves are spin 1 transverse waves.

The second polarisation state for gravitational waves has orthogonal displacement along the x- and y-axes, but maximal displacement for $x = y$. Deformation of the ring is along diagonal axes displaced 45° from the first. A linear combination of h_+ and $h_\times$ leads to either left or right circular polarised waves exactly analogous to the optical case.

We have seen that a gravitational wave produces a fractional, dimensionless

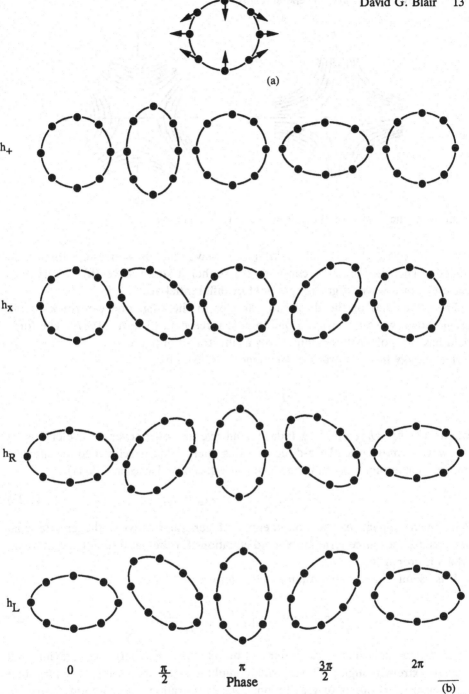

Figure 1.2. (a) The motion caused by a gravitational wave on a ring of test particles. (b) The deviation produced by a gravitational wave. The ring of test particles is successively distorted as shown, during one cycle of a plane gravitational wave. The diagram shows the two linear polarisations and the two linear combinations which give rise to left and right circular polarised waves.

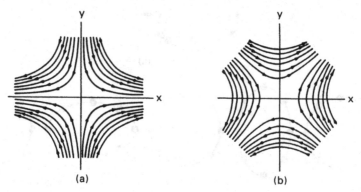

Figure 1.3. The quadrupole force field of a gravitational wave.

strain of space. Almost all gravitational wave antennas measure this strain directly. The wave may be considered as either a strain wave (**h**), a curvature wave (**R**) or a wave of gravity gradient or differential acceleration.

From the lines of the displaced particles in the ring, one can think of the gravitational wave as having a quadrupole force field (Thorne, 1987). The force field has two polarisation components as illustrated in figure 1.3.

The energy flux of a wave of frequency f is given by

$$F \approx \frac{\pi}{4} \frac{c^3}{G} f^2 \langle h^2 \rangle, \tag{1.21}$$

where $h^2 = h_+^2 = h_\times^2$, and $\langle \; \rangle$ indicates an average over several wavelengths. If the wave is moving in the z-direction, this energy flux is related to an effective energy momentum tensor of the wave (introduced by Isaacson, 1968a):

$$F = T_{00} = -T_{0z} = -T_{z0} = T_{zz}. \tag{1.22}$$

This again emphasises the non-linearity of general relativity: the gravitational wave is itself a source term in the field equations for the background curvature on which it propagates.

It is useful to remember a numerical figure for the energy flux:

$$F = 30 \; \text{Wm}^{-2} \left[\frac{f}{1 \; \text{kHz}} \right]^2 \frac{\langle h^2 \rangle}{(10^{-20})^2}. \tag{1.23}$$

Thus, as emphasised at the beginning of this chapter, the energy of a gravitational wave is extremely high for a very small amplitude. We shall see in chapter 2 that occasional short bursts of gravitational waves (duration $\sim 1 \; \text{ms}$) from sources in our galaxy could have a flux up to 10^4 times greater than the $30 \; \text{Wm}^{-2}$ scale factor of equation (1.23). Optimistic goals for detection of bursts down to $h \sim 10^{-23}$ represent energy fluxes as small as $30 \; \mu\text{Wm}^{-2}$, still a factor of 10^{23} larger than typical radio astronomical signals.

References

Bondi, H. (1957). *Nature* **179,** 1072.

Einstein, A. (1915a). 'Zür allgemeinen Relativitätstheorie', *Preuss. Akad. Wiss. Berlin Sitzber,* pp. 778–86 (Nov. 11).

Einstein, A. (1915b). 'Zür allgemeinen Relativitätstheorie (Nachtrag)', *Preuss. Akad. Wiss. Berlin Sitzber,* pp. 799–801 (Nov. 18).

Einstein, A. (1915c). 'Erklärung der Perihelbewegung des Merjur aus der allgemeinen Relativitätstheorie (Nachtrag)', *Preuss. Akad. Wiss. Berlin Sitzber,* **47,** 831–9 (Nov. 25).

Einstein, A. (1915d). 'Die Feldgleichungen der Gravitation', *Preuss. Akad. Wiss. Berlin Sitzber,* pp. 844–7 (Dec. 2).

Einstein, A. (1916). *Preuss. Akad. Wiss. Berlin,* Sitzungsberichte der physikalisch-mathematischen Klasse, p. 688.

Einstein, A. (1918). *Preuss. Akad. Wiss. Berlin,* Sitzungsberichte der physikalisch-mathematischen Klasse, p. 154.

Isaacson, R. A. (1968a). *Phys. Rev.* **166,** 1263.

Isaacson, R. A. (1968b). *Phys. Rev.* **166,** 1272.

Michelson, A. A. and Morley, E. W. (1887a). *Am. J. Sci.* **34,** 333.

Michelson, A. A. and Morley, E. W. (1887b). *Phil. Mag.* **24,** 449.

Nakamura, T. (1987). In *Proceedings of 1986 Kyoto Conference on Relativistic Astrophysics,* eds. T. Nakamura and H. Sato, World Scientific, Singapore.

Pirani, F. A. E. (1957). *Phys. Rev.,* **105,** 1089.

Thorne, K. S. (1983). In *Gravitational Radiation,* eds. N. Deruelle and T. Piran, pp. 1–57, North Holland, Amsterdam.

Thorne, K. S. (1987). In *300 Years of Gravitation,* eds. S. Hawking and W. Israel, pp. 330–458, Cambridge University Press.

Weber, J. (1987). In *Proceedings of the Sir Arthur Eddington Centenary Symposium,* Vol. 3, eds. J. Weber and T. M. Karade, World Scientific, Singapore.

Wheeler, J. A. (1962). *Geometrodynamics,* Academic Press, New York.

Yurtsever, U. (1988a). *Phys. Rev.,* D **38,** 1705.

Yurtsever, U. (1988b). *Phys. Rev.,* D **38,** 1721.

Yurtsever, U. (1988c). Caltech Preprint GRP-184.

The following textbooks may also be of interest.

Fang, L. Z. and Ruffini, R. (1983). *Basic Concepts in Relativistic Astrophysics,* World Scientific, Singapore.

Hawking, S. W. and Israel, W. (1979). *General Relativity – An Einstein Centenary Survey,* Cambridge University Press.

Misner, C. W., Thorne, K. S. and Wheeler, J. A. (1970). *Gravitation,* W. H. Freeman & Co., San Francisco.

Schutz, B. F. (1985). *A First Course in General Relativity,* Cambridge University Press.

Thorne, K. S. (1989). *Gravitational Radiation: A New Window on the Universe,* Cambridge University Press.

2

Sources of gravitational waves

DAVID G. BLAIR

2.1 Gravitational waves and the quadrupole formula

Gravitational waves are generated by the acceleration of matter. From the displacement pattern of figure 1.2 or the force field pattern of figure 1.3, we may assume that the sorts of motions which will efficiently generate gravitational waves will have a similar quadrupole form. The spherically symmetric collapse of a star for example will not generate gravitational waves.

As emphasised in chapter 1, for strong field sources the estimation of gravitational radiation power is difficult, partly because of the difficulty in modelling matter in the extreme conditions near to black hole formation, and partly because of the difficulty in calculating the gravitational redshift of the radiation at the time that it is in transition from the dynamical fields near to the event horizon to the region a few gravitational radii away where it can first be considered to be a wave. Thus for strong field sources we must be content with order of magnitude estimates.

In 1918 Einstein derived the quadrupole formula for gravitational radiation (Einstein, 1916, 1918). This formula states that the wave amplitude h_{jk} is proportional to the second time derivative of the quadrupole moment of the source:

$$h_{jk} = \frac{2}{r} \cdot \frac{G}{c^4} \frac{\partial^2}{\partial t^2} [D_{jk}(t - r/c)]^{\text{TT}}. \tag{2.1}$$

Here $[D_{jk}(t - r/c)]^{\text{TT}}$ is the transverse traceless projection of the quadrupole moment evaluated at retarded time $t - r/c$. While Einstein's derivation was for non-self gravitating systems in slow motion, Thorne (1987) emphasises that this result is accurate for all sources as long as the reduced wavelength $\lambdabar = \lambda/2\pi$ is longer than the source size L. This includes many astrophysical sources, but not all, and particularly does not include many of those that are most interesting and certain. See Damour (1987) for a full discussion of this point. For weak fields $\mathbf{D}$ is the second moment of the source mass density ρ:

$$D_{jk}(t) = \int \rho(t) \left[x^j x^k - \frac{1}{3} x^2 \, \delta^{jk} \right] \mathrm{d}^3 x, \tag{2.2}$$

where the second term in the square brackets which includes the Kronecker delta removes the trace, so that **D** is symmetric and trace free.

A rough estimate of the wave amplitude h can be obtained very simply from equation (2.1) because the second time derivative of D, $\ddot{D}$ is approximately the kinetic energy of that component of a source's internal motion which is non-spherical. Thus from equation (2.1)

$$h \sim \frac{G}{c^4} \cdot \frac{E^{ns}}{r}. \tag{2.3}$$

The quadrupole formula leads to a gravitational wave luminosity which depends on the third time derivative of D:

$$L_G = \frac{1}{5}\frac{G}{c^5}\sum_{jk}\left|\frac{d^3 D_{jk}}{dt^3}\right|^2. \tag{2.4}$$

Equation (2.4) is scaled by the extremely small number G/c^5. Its reciprocal c^5/G has a magnitude of 3.6×10^{52} W, or $2 \times 10^5 M_\odot$ per second. This is about the rate at which $\sim 10^{23}$ stars in the universe convert rest mass into electromagnetic radiation by nuclear burning, and is often called the luminosity of the universe.

We shall investigate the gravitational luminosity of various systems. First, let us consider the problem of making a laboratory source of gravitational waves. The simplest might be a mass spring dumb-bell as illustrated in figure 2.1, which oscillates sinusoidally, thus making a time varying quadrupole moment. The length L obeys $L = L_0 + a \sin \omega t$, and the quadrupole moment is seen from equation (2.2) simply ML^2. That is,

$$D = ML_0^2 + 2ML_0 a \sin \omega t + Ma^2 \sin^2 \omega t. \tag{2.5}$$

The last term will give rise to a 2ω component, but for $a \ll L$ this term is small compared with the second term. Differentiating, and neglecting the third term $\ddot{D}$ follows immediately:

$$|\ddot{D}| = 2ML_0 a\omega^3 \cos \omega t. \tag{2.6}$$

Thus by equation (2.4), L_G is

$$L_G = \frac{G}{5c^5} \cdot 4M^2 L_0^2 a^2 \omega^6 \cos^2 \omega t. \tag{2.7}$$

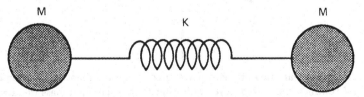

Figure 2.1. A mass quadrupole harmonic oscillator.

First note the form of equation (2.6): $|\ddot{D}|$ has the dimensions of power, so that L_G is of the form (power)^2/power. If the amplitude a was $\sim L_0$ we could think of L_G as simply the square of the circulating power in the system, scaled against the luminosity of the universe. By circulating power we mean the power represented by the rate of change of kinetic energy as it either changes with time (by exchange with potential energy) or as it changes in position, as it moves from one part of the system to another. This result is therefore closely connected to equation (2.3).

For a laboratory source we might make the dumb-bell into a solid bar, so as to maximise ω, on which the luminosity depends to the sixth power. Suppose $M = 10^3$ kg, $L = 1$ m, $a \sim 10^{-2}$ m (near to the elastic limit) and $\omega \sim 10^4$ rad s^{-1}. Then by equation (2.7), $L_G \sim 10^{-27}$ J s^{-1}. This is very small!

How could we obtain increased power? One way would be to effectively increase the amplitude to $a \sim L$ by replacing the oscillating bar with a rotating bar as illustrated in figure 2.2. In this case most power is emitted at the frequency 2ω. The quadrupole moment is reduced by a small numerical factor, and L_G is now:

$$L_G \approx \frac{2}{45} \cdot \frac{M^2 L^4 \omega^6}{c^5/G} \tag{2.8}$$

If we were ambitious we could construct a high tensile steel bar of 1000 tonnes, 100 m long, and rotate it to the limit of centrifugal destruction, as shown in figure 2.2. The top speed at the ends would be about 1 km s^{-1} ($\omega = 20$ rad s^{-1}). Put in the numbers and we find that this has not helped much: $L_G \sim 10^{-26}$ J s^{-1}.

Could the waves from this source be detected? We will see in chapter 3 that the best energy absorption cross section for a gravitational wave antenna is $\sim 10^{-20}$. Ignore the fact that the antenna has to be a distance $\sim \lambdabar = c/\omega$ from the source for the wave to be a wave and not a non-propagating near field fluctuation, and

Figure 2.2. A rotating bar gravitational wave source. From the plane of rotation it appears to expand and contract and is therefore equivalent to an oscillating mass quadrupole with an amplitude comparable to its length.

with wild optimism suppose we can absorb energy $E_a \sim 10^{-20}$ of L_G from the source: then $E_a \sim 10^{-46}$ J s^{-1}. This number is absurd. The energy of a quantum at frequency $\omega = 20$ rad s^{-1} is $\hbar\omega \sim 10^{-32}$ J. Thus even in our wildly optimistic scenario, and supposing we could detect the energy of a single graviton, the mean time between detections would be $10^{-32}/10^{-46} = 10^{14}$ s, about three million years.

At this point we abandon hope of laboratory generation and detection and go on to ask what astrophysical systems can do for us. Here we find a completely different picture: gravitational waves in astrophysics can radiate powers vastly greater than ever possible by electromagnetic waves. How can this be when our formula for generation scales with the ridiculously small number G/c^5?

The answer was explained to me in a hilarious seminar by Joe Weber at Louisiana State University in 1974. Never before and never since has mathematics on the blackboard brought tears of laughter to my eyes. Weber began with the quadrupole formula, equation (2.4), applied to the rotating bar that we have just analysed (equation (2.7)). He emphasised the terrible factor of G/c^5 which makes the output power so small. Weber said, 'wouldn't it be nice, if instead of scaling as G/c^5, this equation scaled as $c^5/G \ldots$ wouldn't it be nice if c^5/G could be moved to the numerator!'

'Well', he continued, 'let us see what we can do about it. The bar has mass M, and Schwarszchild radius $r_S = 2GM/c^2$, so we can substitute $c^2 r_S/2G$ for M. The surface velocity of its ends is $v = r\omega$, where $r = \frac{1}{2}L$. In relativity we should measure the velocity as a fraction of the light velocity. Therefore we should write $\omega = (v/c)(c/r)$'. The wiry and agile professor energetically made substitutions, along with numerous wry comments about academia, tenure and funding agencies. Try it yourself. With the substitutions Weber obtained:

$$L_G = \left[\frac{2}{45} \frac{c^4 r_S^2}{4G^2} 16 r^4 \left(\frac{v}{c} \cdot \frac{c}{r} \right)^6 \right] \bigg/ \left[\frac{c^5}{G} \right] \tag{2.9}$$

Lo and behold, he succeeded. The G's and c's cancelled and he was left with a factor c^5/G in front of two simple terms:

$$L_G = \frac{c^5}{G} \left(\frac{r_S}{r} \right)^2 \left(\frac{v}{c} \right)^6 \cdot \frac{8}{45}. \tag{2.10}$$

We enthusiastic young postdocs were most impressed. It is not a mathematical sleight of hand, but a means of emphasising the basic physics. Equation (2.10) tells us that in the relativistic limit, where internal velocities approach c and where the source size is comparable to its Schwarszchild radius, the luminosity can approach the luminosity of the universe. (This point was first emphasised by Freeman Dyson.) The two conditions are of course mutually compatible, because, if the source size is comparable to r_S, then the velocities must be comparable to c. (The escape velocity of a black hole is c. The orbital velocity for a particle orbiting near to r_S therefore approaches c.) Note also, however, that at this limit where L_G becomes comparable to c^5/G, the quadrupole formula is no longer accurate, as we described earlier. However, if we apply the formula to a pre-coalescing neutron star binary, we might have $r_s/r \sim (v/c)^2 \sim 3 \times 10^{-2}$, close to the breakdown of the quadrupole formula. Then $L_G \sim 10^{-9} c^5/G$, which is still a vast luminosity $\sim 10^{17} L_\odot$, comparable to the optical luminosity of 10^4 giant galaxies.

Thus there can be little doubt that astrophysical systems including neutron stars or black holes can be extremely powerful sources of gravitational radiation. Enormous luminosities cannot be sustained for long: the evolution must be dominated by the gravitational radiation emission. The emission will create a powerful radiation reaction on the system acting like viscosity as energy and angular momentum are radiated away. The lifetime of the above neutron star system will at this stage be less than one second, the orbital period about 10^{-2} s, and in much less than a second the orbital frequency will rise rapidly past 1 kHz. At the point of coalescence the physics is more complicated. There will be viscous heating and thermal neutrino emission. The system could become partially disrupted, but there is insufficient energy for more than a small part of the system to be ejected. The combined mass may well exceed the critical mass for neutron stars, so it will probably collapse to a black hole. Because of the huge angular momentum and the lack of spherical symmetry, this last stage of collapse may be close to the extreme relativistic limit of $v \sim c$ and $r \sim r_s$.

2.2   Strain amplitude, flux and luminosity

We have seen that high luminosity sources can exist in astrophysical systems. Any source can be characterised by an amplitude h and flux F detected at the Earth, or by a luminosity L_G which characterises the total rate of energy loss from the system. Generally we might presume isotropic radiation from a source, and therefore relate F to L_G by $L_G = 4\pi r^2 F$ where r is the distance of the detection from the source. However, since most sources probably have a dominant angular momentum axis along which the radiation will be weakly concentrated, as well as unknown polarisation, the factor 4π above must be seen as an indicative value only.

As we saw in chapter 1, the relationship between F and h is simple. The energy flux (in W m^{-2} or the equivalent cgs units) is given by

$$F = \frac{\pi}{4}\frac{c^3}{G}f^2 h^2 \qquad (2.11)$$

for a wave of amplitude h. In general h is the amplitude for two polarisations $h^2 = h_+^2 + h_\times^2$. Numerically, as given in equation (1.23) we can write

$$F \approx 30 \text{ W m}^{-2}\left(\frac{f}{10^3 \text{ Hz}}\right)^2\left(\frac{h}{10^{-20}}\right)^2. \qquad (2.12)$$

Note that the value of $h \sim 10^{-20}$, typical of experimental goals, still represents a very considerable energy flux, 3% of the solar intensity on Earth, although, as we saw above, such high flux densities can only be sustained in short bursts.

In the following sections we shall consider some of the main sources of gravitational waves. By its nature much of this work is somewhat speculative, in that it is not supported by observation. We shall concentrate on main features, and refer the reader to review articles for further reading.

2.3 Supernovae

Traditionally supernovae have been classified into two classes: Type I supernovae (SNI) and Type II supernovae (SNII). Type II supernovae represent the core collapse of a massive star and the shock-driven rebound expansion of an optically luminous shell. In a few instances it is certain that the collapsed core is a neutron star.

Supernova 1987A discovered on February 23, 1987, represented an historic landmark in astronomy. The supernova occurred in a nearby irregular dwarf galaxy, the Large Magellanic Cloud, at a distance of about 50 kpc. In this instance neutrinos from the inverse β-decay associated with the collapse were observed by several huge detectors originally designed to test for the radioactive decay of protons. This is the first instance of direct observation of core collapse, and as the first instance of extra solar neutrino detection it represents the opening of a completely new channel of astronomy – weak interaction astronomy.

So far there has been no confirming evidence of a neutron star having formed: no sign that the nebula is being energised by a rotating magnetised neutron star, and no sign of optical pulsations. There were suggestions that an unmagnetised or non-rotating neutron star had formed, or that the collapse had continued on to form a black hole. The supernova clearly demonstrated the inadequacies of existing supernova theory, in that the star which exploded was a blue supergiant and not a red supergiant that previous theory required. While attempts have been made to patch up the theory to fit the observations, the supernova has emphasised the need for theory and observation to go hand in hand. In early 1989 optical pulsations were reported from the supernova. This created great excitement. Unfortunately, in 1990 the pulses were found to be of spurious technical origin.

The supernova which has most defined the stereotype view of supernovae is SN1054, recorded by Chinese, Japanese and Korean astronomers in the year 1054. It was a Type II supernova, and modern observations have identified the energetic 30 Hz Crab pulsar which emits pulsed radiation across most of the electromagnetic spectrum. Here the collapse leading to neutron star formation is undisputed. The rotational kinetic energy of the neutron star energises the surrounding nebula, with a luminosity greater than $10^4 L_\odot$. The pulsed radiation contributes only a small fraction of the observed rate of rotational kinetic energy loss from the pulsar, determined by the spin down rate, but the total nebula luminosity is consistent with this figure, about 10^{31} W.

Of the half dozen supernovae to have occurred in our corner of the galaxy in the last 1000 years, SN1054 is the only one which fits the stereotype. Since we also observe a few black hole systems (such as Cygnus X-1) the most reasonable conclusion is that some SNII represent collapse to neutron stars, while others go over the edge and collapse to black holes. (Blair and Candy, 1988; Shklovsky, 1979).

The collapse to a black hole will occur if the shock wave which normally drives the explosive rebound of the surrounding material is weak. The envelope material will fall back, but neutrino emission and diffusion through the dense envelope will extend the timescale of collapse from milliseconds to seconds. In its final stages, however, the collapse through the event horizon will take place on a free fall timescale ensuring the presence of some high frequency components in the gravitational radiation pulse.

Type I supernovae are different. The traditional view is that a Type I supernova is the nuclear detonation of a white dwarf, after it has accreted matter from a companion. There appear to be many reasons why this may not be correct however. The white dwarf has the choice of collapsing or detonating. The choice is determined by detailed properties of degenerate matter. For example, depending on the detailed atomic composition, different atomic species such as neon and oxygen may undergo a phase separation. Models require knowledge of these details, which are not amenable to experimental investigation. The best guess today is that it is likely that at least a fraction of accreting white dwarfs will collapse, but only to a neutron star, since white dwarf mass is insufficient to allow collapse to a black hole.

Depending on the presence of surrounding matter at the time of collapse (accretion disc or photosphere) there may or may not be a strong optical display. Thus it is proposed that some neutron stars are born in low luminosity events, absurdly described as 'optically silent supernovae'.

The key question that concerns us, is: What is the magnitude of $\ddot{D}$ in such a collapse, or from equation (2.3), what is the kinetic energy of the non-spherical motion? There are many possibilities, very little astrophysical evidence to give guidance, and extreme technical difficulties in calculating the bulk properties of degenerate matter during collapse. Thus all calculations can only be considered, at best, order of magnitude estimates.

The main source of non-sphericity during collapse is angular momentum. Isolated supergiants may have relatively low angular momentum due to angular momentum loss by magnetic coupling to the strong stellar wind. However, supergiants in binary systems will be unable to dissipate their angular momentum, so perhaps in about half SNII we might expect reasonably high angular momentum.

During a high angular momentum collapse the angular frequency of the collapsing core will rise rapidly. It may pass through a succession of states, beginning by becoming an increasingly oblate spheroid (case (i)). In this case the time varying quadrupole moment is small and along the angular momentum axis. It may evolve to a bar mode instability (case (ii)) which, depending on the equation of state could continue to evolve briefly into a binary neutron star system (case (iii)). Alternatively it may develop a pancake instability, and break up into several 'neutron star' components (case (iv)): these again will evolve rapidly by gravitational wave emission, finally to coalesce.

The influence of neutrino and photon pressure during the collapse is uncertain. In cold collapse, the radiation pressure is negligible and the system collapses on a free fall timescale, so that gravitational wave frequencies are high. It is now more commonly believed that the collapse is hot and slow, so that for case (i) the radiation frequency is reduced from a peak of ~3 kHz to perhaps 500 Hz.

The conversion efficiency of rest mass to gravitational radiation is denoted by ε. Theoretical estimates, all of which represent a highly idealised situation, give a range of values for ε from a few per cent, to ~10^{-4}, and down to 10^{-7}. Thorne (1987) emphasises that there is no consensus on this value of ε. It clearly depends on two unknowns, (a) the physics of the collapse, and (b) the initial conditions, in the form of the range of angular momenta that exist in nature. It seems reasonable to suppose that there is a broad range of initial conditions, translating to a range of ε values. Where possible, detectors should be designed to detect events at least in the middle of this range.

If the collapsing nuclear matter core of a star is unable to drive off its surrounding envelope, and the core mass exceeds a critical value, the collapse will continue to form a black hole. The critical mass is not known for certain, but is certainly between $1.4 M_\odot$ and $3 M_\odot$.

Black holes can emit gravitational radiation through the excitation of their normal modes of oscillation. The normal modes can be excited by infalling matter, and in particular will be strongly excited if the high angular momentum components (corresponding to cases (ii)–(iv) above) coalesce to form a black hole. The initial wave burst as the black hole forms is difficult to calculate as it depends on complex initial conditions. However, once excited, the gravitational radiation from the normal modes can at least in principle be reasonably well determined since we have no internal structure to worry about.

Clearly the above possibilities will give very different results. There is a possibility of a large range of frequencies from below 100 Hz to ~10 kHz. The low frequencies will be generated by the formation of instabilities (cases (ii)–(iv)) and asymmetry during the early stage of collapse, whereas high frequencies will occur at the late stage; especially by the excitation of quadrupole modes in the neutron star.

Figure 2.3 shows some typical calculated waveforms. All show the general character of burst sources: extremely brief pulses with a duration of only a few

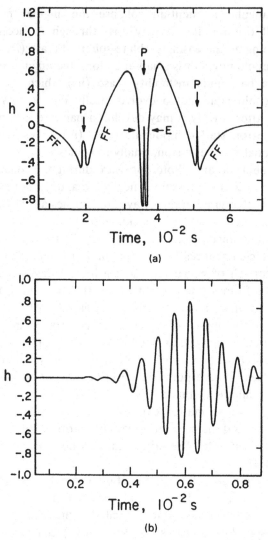

Figure 2.3. Typical gravitational waveforms calculated for idealised collapse scenarios. (a) Free fall collapse to a neutron star with three strong bounces. (b) Gravitational waves from the excited normal modes of a neutron star following gravitational collapse. (c) Waves from axisymmetric collapse to a black hole. (d) Waveform for a near miss interaction of a 'particle' with a Kerr black hole. (e) A particle plunging into the Kerr black hole.

cycles. Figure 2.3(a) by Saenz and Shapiro (1978) shows a waveform characteristic of a freely falling rotationally deformed oblate gravitational collapse to form a neutron star, with three strong bounces along the polar axis (narrow positive peaks) and a slower equatorial bounce (wide negative peak). The entire event lasts only 50 ms. Figure 2.3(b) by Saenz and Shapiro (1981) shows the waveform

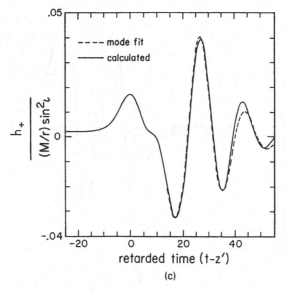

(c)

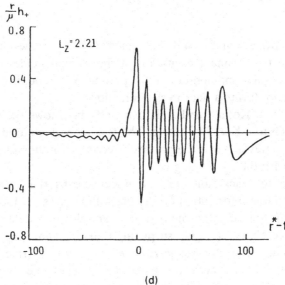

(d)

Figure 2.3 (*cont.*)

where the collapse has excited quadrupole oscillations of the neutron star. Near sinusoidal oscillations at 1.4 kHz are excited, and rapidly damped out.

Figure 2.3(c) from Stark and Piran (1985, 1987) shows a comparison of the waveforms obtained from a model of axisymmetric collapse of a star to a black hole. The waveform is dominated by the excitations of the quadrupole normal modes of black hole. The most weakly damped black hole modes dominate the emission. Even so, the damping by gravitational wave emission is so strong that the excitation lasts for only a few cycles.

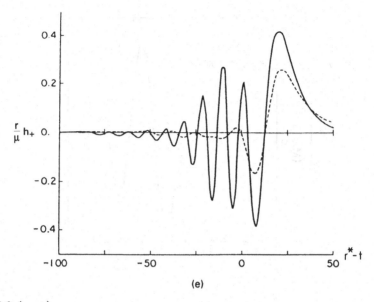

(e)

Figure 2.3 (*cont.*)

Figure 2.3(d) from Kojima and Nakamura (1984b) shows the gravitational waveform due to the resonant excitation of the normal modes of a Kerr black hole (one with angular momentum) by a particle with counter rotating orbital angular momentum during a near miss interaction.

Figure 2.3(e) from Kojima and Nakamura (1984a) shows the waveform for a particle plunging into a Kerr black hole. The total energy emitted as gravitational waves can be 15 times larger if the particle is counter rotating during infall (solid line) compared with the co-rotating case.

The excitation of black hole normal modes means that gravitational wave signals contain precise information on the black hole mass, and as emphasised by Detweiler (1980) the angular momentum can also be determined. Thus a gravitational wave signal can give surprisingly detailed information on forming black holes, and hence on the system from which they are formed.

As long as there is some angular momentum, it seems inevitable that collapse to form black holes must be more efficient than collapse to neutron stars. Efficiencies as high as 10% have been suggested, while for slowly rotating axisymmetric collapse ε is probably $\sim 10^{-3}$.

For collapse to form black holes, Thorne (1987) gives rough numerical estimates as follows:

$$f \approx \frac{c^3}{5\pi Gm} \approx (1.3 \times 10^4 \, \text{Hz}) \frac{M}{M_\odot} \tag{2.13}$$

$$h \approx \left[\frac{15\varepsilon}{2\pi}\right]^{1/2} \frac{G}{c^2} \frac{M}{r} \tag{2.14}$$

or

$$h \approx 7 \times 10^{-19} \left(\frac{\varepsilon}{0.01}\right)^{1/2} \left(\frac{M}{M_\odot}\right) \left(\frac{10\,\text{kpc}}{r}\right). \tag{2.15}$$

If we normalise to systems giving peak radiation at 1 kHz, corresponding to $13M_\odot$ black hole formation, using equation (2.13), we can rewrite equation (2.15) as

$$h \approx 10^{-17} \left(\frac{\varepsilon}{0.01}\right)^{1/2} \left(\frac{1\,\text{kHz}}{f}\right)^{1/2} \left(\frac{10\,\text{kpc}}{r}\right). \tag{2.16}$$

Such massive black holes are probably formed quite rarely in our galaxy. If one had formed in the late 1980s, existing detectors (when operating) should have been sensitive enough to detect it. If, however, we have to look to the Virgo cluster, a distance of about 10 Mpc, to have enough stars in our sample for an event to occur within a reasonable time interval, then we require $h \sim 10^{-20}$ which is beyond the reach of 1980s detectors.

2.4 Binary coalescence

The binary pulsar PSR 1913 + 16 is of special significance in this discussion. This pulsar is in a binary system with a neutron star companion, and has a $7\frac{3}{4}$ hour orbital period. The mass of each star is very close to $1.4M_\odot$. The pulsar has a fairly short rotational period (59 ms) and a very well behaved spindown, without the frequency glitches seen in younger pulsars. Simple application of Kepler's laws show that the system is significantly relativistic, with velocities $\sim 10^{-3}\,c$. The system has a high orbital eccentricity of 0.61, so that at periastron the stars move in strong background curvature and the relativistic effects are relatively large (about 10^{-6} of the Newtonian effects). We shall see below that this system is an ideal laboratory for testing general relativity and that it has provided an indirect proof of the existence of gravitational waves.

By the quadrupole formula the gravitational luminosity of a binary star system (Fang and Ruffini, 1983) is

$$L_G = \frac{32}{5} \frac{G^4}{c^5} \frac{(m_1 m_2)^2 (m_1 + m_2)}{a^5} f(e), \tag{2.17}$$

where a is the semi-major axis and $f(e)$ is the eccentricity function given by

$$f(e) = (1 + be^2 + ce^4) \cdot (1 - e^2)^{-7/2},$$

$$b = 73/24, \qquad c = 37/96. \tag{2.18}$$

In the simple case of a circular orbit and equal masses, $m_1 = m_2$, equation (2.17) reduces to

$$L_G = \frac{64}{5} \frac{G^4}{c^5} \cdot \frac{m^5}{r^5}. \tag{2.19}$$

A circular orbit is the minimum energy loss condition for a binary, and L_G increases steeply with eccentricity. For $e \sim 0.6$, $f(e) \sim 10$. As the eccentricity increases, the L_G is emitted at higher and higher harmonics of the orbital period. At $e = 0.6$, peak power is emitted at harmonics 6–10.

At present the PSR 1913 + 16 system has $L_G \sim 10^{25}$ W, and, as a result, the orbit decays by gravitational radiation emission. In 3.5×10^8 years the system will coalesce. Because the strongest radiation is produced near periastron, when the acceleration is greatest, the radiation reaction is also strongest at this time. This results in circularisation of the orbit.

When the orbit has become circular, the strain amplitudes for the two polarisations are given by

$$h_+ = \frac{2G^{5/3}}{c^4}(1 + \cos^2 i)(\mu/r)(\pi M f)^{2/3} \cos(2\pi f t)$$

$$h_\times = \pm 4 \frac{G^{5/3}}{c^4} \cos i (\mu/r)(\pi M f)^{2/3} \sin(2\pi f t).$$

(2.20)

The orbital inclination is i, and the projection of the orbit onto the sky looks elliptical. The axes which define the orientation of h_+ and $h_\times$ are simply the major and minor axes of the projected ellipse. The quantities M and μ are the total mass $M = m_1 + m_2$ and the reduced mass $\mu = m_1 m_2/m_1 + m_2$ for the system. The wave frequency f is twice the orbital frequency. The plus or minus sign is determined by the rotation direction of the orbit, and specifies whether the signal is left or right hand circularly polarised.

The increasing wave frequency as a function of time f is given by

$$f(t) = \frac{1}{\pi}\left(\frac{c^3}{G}\right)^{5/8} \cdot \left[\frac{5}{256} \cdot \frac{1}{\mu M^{2/3}} \frac{1}{t - t_0}\right]^{3/8},$$

(2.21)

where t_0 is the time of coalescence. Figure 2.4 shows the frequency evolution for the late stage of evolution, for two $1.4 M_\odot$ neutron stars. Note that the signal is a 'chirrup' rising rapidly in frequency, to a singularity at $t_s = t_0$. As it approaches this point the physics becomes that of a single coalesced system, and the frequency in reality reaches a maximum value. The frequency rises from 60 Hz to 140 Hz in about ten seconds, from 140 Hz to 330 Hz in the next second and finally reaches the coalescence in the last 0.1 seconds.

The entire data for PSR 1913 + 16 consists of pulse arrival times. The pulsar itself is a very accurate clock, and its motion causes Doppler shifts of the pulse period which can be related to the orbital parameters. This has enabled a complete solution of the orbital parameters of the system, with unprecedented accuracy (Taylor, Fowler and McCulloch, 1979; Taylor and Weisberg, 1989). It allows the orbital inclination and the individual stellar masses to be determined, as well as several relativistic parameters. The orbital periastron advance, whereby the orbital path instead of remaining fixed in space as it would in Newtonian

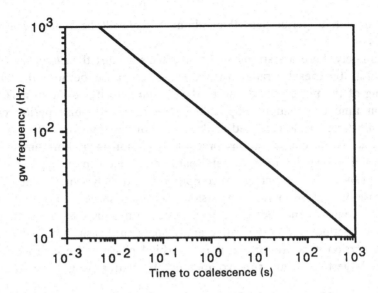

Figure 2.4. The frequency evolution for coalescence of a neutron star binary.

physics precesses around the star, is clearly observed. This anomalous property had first been recognised for the planet Mercury in the nineteenth century. The magnitude for Mercury is 43 arc seconds per century, and as we saw in chapter 1 the explanation of this effect was one of the first triumphs of general relativity. In PSR 1913 + 16 the effect is very large, about 4° per year, and is determined to about 10^{-4} degrees accuracy.

In relation to gravitational radiation, the most important observation is the spin-up of the binary (orbital) frequency.

The derivative of the binary period is found to be $\dot{p} = -2.4 \times 10^{-12}$ so the period decreases annually by an accurately measured amount $\sim 70\,\mu s$. Damour (1983, 1987) has analysed this system in detail. He concludes that the observations and theory 'provide the most sensitive available confirmation of the non-linear hyperbolic structure of Einstein's theory, and therefore also an indirect confirmation of the existence of gravitational radiation'.

PSR 1913 + 16 is also important in that it confirms the existence of one definite predictable source of intense high frequency gravitational radiation, although unfortunately it will not occur for 3.5×10^{8} years. Since this time is a small fraction of the age of the galaxy, and since we are only able to observe pulsars in a small fraction of the galaxy, it is fair to assume a reasonable population of such sources, existing presumably at random stages in their evolution. In section 2.6 we shall make a rough estimate of this population.

From equation (2.21) it is clear that only a small subset of such binaries will be 'interesting'. These are the systems able to coalesce within the age of the universe. The orbital period has only to increase by a factor of 2.5 for the lifetime to coalesce to increase by one decade. Thus for neutron star binaries only those born with a period less than about a day will be binary coalescence candidates.

Unfortunately there are strong selection effects against the discovery of short period pulsar binaries by radio astronomy. These arise because the Doppler broadening of the pulse period due to the orbital velocity variation makes long integration time observations impossible. The observed pulse period changes steadily with time so the expected peak in the Fourier transform of the signal is smeared out. Noise considerations preclude detection in a short time interval. Without prior knowledge of the orbital parameters one cannot apply a selective filter except by a massive increase in computation. Already searches must cover a two-dimensional (pulse period–dispersion measure) space. To include binary parameters would extend the search to a three- to five-dimensional search space (either pulse period–period derivative and dispersion measure, or at best, pulse period–dispersion measure–orbital period–orbital phase and eccentricity). Even with the largest existing supercomputers the latter would be practically impossible.

Thus our knowledge of the population of fast binary pulsars is poor, and probably we will have to make gravitational wave observations to find out how many really exist.

The waveform from a binary coalescence is special in that it is continuous, well defined and depends on only M, μ and f. Thus it is possible to design a powerful optimal filter if the detector has a sufficient bandwidth to observe a reasonable fraction of the chirrup signal. By Fourier transforming equations (2.20) for h_+ and $h_\times$, summing the squares of the polarisations and averaging over all directions we can get an expectation value for h. The optimal filter can be though as a device for summing the strains according to the known signature or template of the wave. The data is repeatedly run through the full set of templates until the one is found which coherently sums the signal while averaging out the noise over the time interval or bandwidth used. Therefore the signal is effectively summed from its instantaneous value, to a value which is about $n^{1/2}$ times larger, where n is the number of wave periods in the time interval the event is observed.

By assuming detector characteristics with peak sensitivity $\sim 10\,\text{Hz}$, Thorne shows that $n \sim 10^3$ for a neutron star binary, and obtains

$$h \approx 4 \times 10^{-18} \left(\frac{\mu}{M}\right)\left(\frac{M}{M_\odot}\right)^{1/3}\left(\frac{10\,\text{kpc}}{r}\right)\left(\frac{100\,\text{Hz}}{f_c}\right)^{1/6}, \tag{2.22}$$

where h is the typical summed strain obtained by an optimal filter and f_c is the characteristic frequency.

The instantaneous strain is typically 30 times smaller, but depends on the detector frequency and the binary system parameters. See Bernard Schutz's chapter 16 for more details.

Gravitational wave detectors can obtain powerful information about binary coalescence sources. An array of detectors on the globe (which has a gravitational wave travel time diameter of 40 ms) can measure differences in the signal phase and amplitude determined by the position and orientation of each detector relative to the incoming wave. This can give reasonably accurate directional information. The relative magnitude of each polarisation can be used to determine the system inclination using equations (2.20). A function of the masses can be determined, and the source distance can also be determined from the amplitudes. If the system could also be detected optically (a mini supernova) the redshift could be determined, and this would allow an accurate determination of the Hubble constant (Schutz, 1986). Detectors with strain sensitivity of about 10^{-21} could see binary coalescence events out to a distance of about 40 Mpc, and therefore monitor for coalescence events more than 10^4 galaxies.

2.5   Other sources of gravitational waves

We have discussed two types of source for which there is direct observational evidence. There are a few other sources in this category, and then many others which are of a more speculative variety. Almost all are variations of the sources discussed already: binary stars and binary black holes, stars infalling into massive black holes (a variant of stellar collapse) and rotating deformed neutron stars. In addition we will briefly mention primordial gravitational waves, and gravitational waves from cosmic strings.

2.5.1   Black holes

There is evidence that some galaxies and all quasars contain massive black holes $\sim 10^9$ solar masses, that our own galaxy has a black hole at the nucleus of $\sim 10^6 \, M_\odot$, and that some radio galaxies have a massive precessing object in their nucleus. Since all these systems formed at some stage, we need to consider possible events associated with their formation.

Figure 2.5 illustrates the range of possible ways supermassive black holes can form (Rees, 1983). Many of these are events that we may imagine occurring in the quasars we observe today, but are much less common in closer galaxies which do not show the same level of activity. Thus our primary interest might be in gravitational wave events which could be occurring at distances comparable to the Hubble distance.

The frequency of gravitational wave emission scales inversely as the mass. Thus for massive black holes the gravitational waves will be at very low frequencies. As black holes grow the gravity gradient at the event horizon reduces, so they become able to swallow larger and larger objects without tidal disruption. Thus a $10^9 M_\odot$ black hole can swallow whole main sequence stars, while $10^6 M_\odot$ black holes will disrupt main sequence stars strongly and be much weaker sources of gravitational waves. White dwarfs and neutron stars can be swallowed whole by

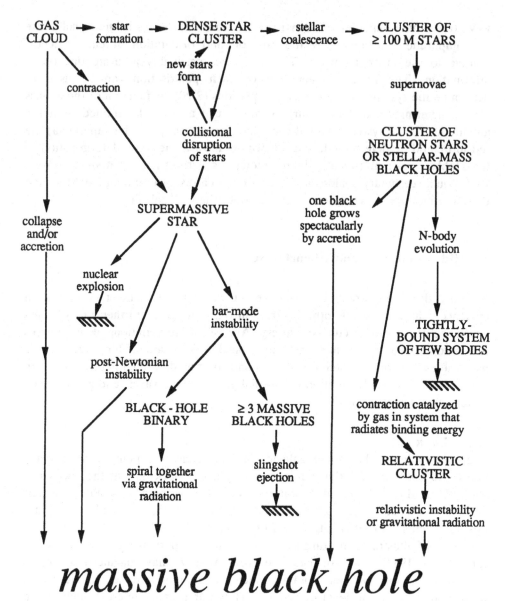

Figure 2.5. Possible pathways to the formation of massive black holes.

$10^3 M_\odot$ black holes, and of course stellar mass black holes are swallowed whole by all black holes.

In the case of infall of a star into a massive hole, order of magnitude numerical estimates for the frequency and strain amplitude are given by Thorne (1987). The characteristic frequency is determined by the light travel 'diameter' of the event horizon, so depends only on the black hole mass, M_H. The strain amplitude depends on the non-spherical component of the kinetic energy (equation 2.4) so depends only on the stellar mass, M_s:

$$f_c \approx \frac{1}{20} \frac{c^3}{G} \frac{1}{M_H} = 10^{-4} \text{Hz}\left(\frac{10^8 M_\odot}{M_H}\right) \qquad (2.23)$$

$$h_c \approx \frac{G}{c^2} \frac{M_s}{2r} = 2 \times 10^{-21} \left(\frac{M_s}{M_\odot}\right)\left(\frac{10\,\text{Mpc}}{r}\right). \qquad (2.24)$$

Another possibility is that an early population of stars (population III stars) were massive and evolved rapidly in the early galaxy, leaving a halo of massive black holes, some of which would be binaries (Bond and Carr, 1984). These could explain the missing mass, and could be coalescing today. If so, they could produce a relatively strong stochastic background of gravitational waves, as well as very powerful bursts at around 30 Hz if hole masses were $10^3 M_\odot$.

2.5.2 Pulsars

Pulsars and other rotating neutron stars can be sources of gravitational waves, due to non-axisymmetric distortions in their crust. The very low spindown rate of millisecond pulsars tells us that the non-axisymmetric distortions cannot be large. However, the presence of glitches in the rotation period of some pulsars is interpreted as the release of crustal distortion, while the non-axisymmetric magnetic field must also distort the star. Thus some radiation can be expected. While estimates for the amplitude on Earth from such systems is rather small, $h \sim 10^{-26}$–10^{-28}, the advantage of a known signal frequency from particular nearby and young pulsars, such as the Crab and Vela pulsars, makes detection a challenging goal pioneered by the late Professor Hirakawa's group in Tokyo (see chapter 9 by Kimio Tsubono).

Millisecond pulsars, which are observed with rotational frequencies up to 640 Hz, are produced either through the spinning up by accretion of old neutron stars, or by the collapse of one member of a white dwarf binary. From observations of X-ray pulsars, we know that spin up does sometimes occur, at moderately low periods. It has been suggested (Wagoner, 1984) that high frequency monochromatic waves could be generated when the spun up neutron star reaches a point of instability, where hydrodynamic waves on the surface are generated and emit gravitational radiation. In this case the gravitational luminosity counterbalances the spin up, and the gravitational flux on Earth will be proportional to the X-ray flux from the neutron star. Whether or not this happens depends critically on the viscosity of neutron stars as discussed by Lindblom (1988). If X-ray modulation could be detected to fix the signal frequency, narrow band detection of the gravitational waves could be possible.

2.5.3 Binary stars

Binary stars are a definite source of gravitational waves. The shortest binary known is a white dwarf–neutron star system which has a period of 11 minutes, and will produce gravitational waves at 3×10^{-3} Hz. There are large populations of various classical binary stars in the galaxy. These include the W Ursa Majoris type binaries which will emit at between 10^{-4} and 10^{-5} Hz, unevolved binaries in the 10^{-5}–10^{-7} Hz range, and cataclysmic variables (white dwarf–normal star systems) between 3×10^{-4} and 10^{-5} Hz. In addition our galaxy could contain up to a million binary neutron stars, with probable lifetimes to coalescence spread between 10^{10} and 10^{4} years (see section 2.6), and a substantially larger population of white dwarf binaries.

While the nearest sources may produce strain amplitudes $\sim 10^{-21}$ (see Douglas and Braginsky 1979), the dominant effect is to produce a stochastic background of gravitational waves due to the combined effect of all systems in the galaxy. This amplitude can be as high as 10^{-19} at 10^{-5} Hz, and depending on the number of close white dwarf binaries, could remain above 10^{-20} for frequencies up to 10^{-2} Hz. The spectra of these stochastic backgrounds have been analysed by Hils *et al.* (1987, unpublished manuscript), Lipunov and Postnov (1987) and Lipunov, Postnov and Prohorov (1987).

2.5.4 Cosmological sources

Whereas the microwave background radiation last scattered at the decoupling time about 10^{6} years after the big bang, at a redshift $Z \sim 10^{3}$, and neutrinos last scattered at about 0.1 seconds ($Z \sim 10^{10}$), gravitational waves decoupled from matter at the Planck time around 10^{-43} seconds, with $Z \sim 10^{30}$. Thus, analogous to the microwave background, we expect a background of gravitational waves to exist, but there are so many uncertainties that it is impossible to specify its amplitude. The initial conditions are not understood and there is the possibility of scattering or parametric amplification with dynamic spacetime. The amplitude is influenced by whether or not there was inflation and/or a QCD or electroweak phase transition.

We can be confident that the primordial background decoupled so early because matter is so extremely transparent to gravitational waves. Thus primordial waves represent an ideal probe of this earliest epoch in the universe. They may have been generated by quantum fluctuations in the Planck era, but also at various late epochs including the events that led to galaxy formation. Sazhin (1988) shows that the frequency of cosmological waves, which initially have a wavelength comparable to the scale of the horizon, is given by

$$f \sim 1.5 \times 10^{-8} \, N_G^{1/6} \frac{T_G}{1\,\mathrm{GeV}} \, \mathrm{Hz}, \qquad (2.25)$$

where N_G is the (small) number of degrees of freedom (or particle species) at the plasma temperature T_G, measured in GeV, characteristic of the epoch of formation. Thus terrestrial detectors in the frequency range 10 Hz–10 kHz can investigate background radiation that originated at a temperature $\sim 10^9$–10^{12} GeV.

The amplitude of the background waves also reduce as the universe expands, and those generated earlier (when T_G was larger) therefore have reduced amplitude. Primordial waves from the Planck era would today be in the microwave regime, and have amplitudes of perhaps $h \sim 10^{-35}$. Of greater interest are gravitational waves in the low frequency regime 10^{-8}–10^{-5}–10^{-3} Hz, which would have much larger amplitude, $h \sim 10^{-14}$–10^{-17}–10^{-21}, respectively. These could have been generated by the formation of domains during possible phase transitions in the quantum chromodynamic era (10^{-6} s, 100 MeV) or the electro weak era (10^{-12} s, 100 GeV–10 TeV). A phase transition in the proposed grand unified era (GUT phase transition at 10^{-36} s, 10^{14} GeV) would give rise to radiation in the MHz range.

Sazhin (1988) gives a numerical expression for the amplitude of cosmological waves of frequency f:

$$h \sim 10^{-24}\left(\frac{h_g}{10^{-4}}\right)N_G^{-1/6}f, \tag{2.26}$$

where h_g is the amplitude of the initial inhomogeneity, usually assumed to be in the range 10^{-2}–10^{-4}.

If cosmic strings exist as a result of a grand unified phase transition, the vibrations of closed cosmic string loops will produce a background (see Fang, 1988). In general the backgrounds can be expressed as an energy density per unit log frequency interval. This can be expressed as a fraction Ω of the closure energy density for the universe ($\sim 10^{-26}$ kgm^{-3}, or 10^{-9} Jm^{-3}). Cosmic string models for galaxy formation imply $\Omega \gtrsim 10^{-7}$ for all frequencies above about 10^{-8} Hz. If observations fail to observe this they will effectively be able to rule out these models. See figure 2.8 for a summary. The latest timing observations of the millisecond pulsar (Taylor and Weisberg, 1989) already place a limit on $\Omega < 10^{-1}$ for a narrow range of frequencies.

2.6 The rate of burst events

2.6.1 Galactic high frequency sources

Considerable effort has been spent in trying to determine, from our existing knowledge, the rate of gravitational collapse and binary coalescence in our galaxy. In the case of collapse to form neutron stars, we are able to observe the resulting population of pulsars and binary X-ray sources. In the case of supernovae, we have historical records for a few, extragalactic observations for several hundred, and supernova remnant observations in our galaxy for well over 100. In the case of black hole formation we have evidence only in the case of interacting black hole binaries such as Cygnus X-1, and the supernovae in which they might be born. However, we are unable to detect isolated black holes, and we cannot at present distinguish supernovae which form neutron stars from those which form black holes.

Here I do not wish to give a full analysis but simply to present main concepts, and the results. First it is important to know how our target population is distributed. Recent evidence tends to contradict the accepted beliefs, and finds that pulsars and supernovae have a surface density (projected onto the galactic plane) which is exponential in radius. There is no distinction between supernova type in this regard (Blair, 1988). Thus the population of supernovae and pulsars can be modelled as a two-dimensional disc with a surface density increasing exponentially towards the centre.

For any observed population in the vicinity of the Sun, we can use the exponential disc model to ask what fraction of the effective volume of the galaxy, V_{eff}, we are observing. V_{eff} is the volume of the galaxy if it was uniformly distributed with a stellar density equal to the mean stellar density in the solar neighbourhood. For historical supernovae and pulsars we find that we can see only about 2% of the galaxy. In the case of pulsars we must know their beaming factor: what fraction of the sky is covered by the beam of a pulsar. We need also to know how this evolves with time. In general we need to know the mean lifetime of the observed population.

Let us apply these arguments to obtain a rough estimate of the population of 'interesting' neutron star binaries. We start from our observation of a single system PSF 1913 + 16. We ignore all the uninteresting ones for which the time to coalescence is greater than the age of the universe. We know that the mean beaming factor for pulsars, f, is about 0.1 (Blair and Candy, 1988) and that the fractional effective volume of the galaxy observed, v_0, is about 0.02. We use the mean pulsar observable lifetime L_p of 5×10^6 years compared to a galactic lifetime L_g of 10^{10} years. We assume a constant formation rate. Then the total binary population is

$$N \sim \frac{N_{\text{obs}}}{v_0} \cdot \frac{L_g}{L_p} \cdot \frac{1}{f} \sim 10^6. \qquad (2.27)$$

If these are uniformly distributed in time we then would expect the most evolved one to be $\sim 10^4$ years away from coalescence, and the coalescence rate to be 1 per 10^4 years. Equation (2.21) tells us that it takes approximately five decades of time to evolve two decades of frequency. At 10 Hz the system is 10^3 s from coalescence, at 0.1 Hz it is a few years away and at 10^{-3} Hz it is 10^6 years away. We can expect there to be one binary with period ~ 100 s in the galaxy. This number would increase if selection effects had indeed prevented our detection of a larger population of interesting binary pulsars. In relation to this, the recently discovered (but still unconfirmed) binary pulsar in the bright southern hemisphere globular cluster 47 Tucanae may be one example of such a system

(Ables *et al.*, 1989). It has a period of $2000\,s$ and a lifetime to coalescence of 5×10^6 years. It appears to have overcome the selection effects through having an extremely improbable near optimum orbital inclination to our line of sight ($\sin i \sim 10^{-3}$). If this system is confirmed and is typical of the population of binary pulsars, then by equation (2.25), extended to allow random inclinations, we would have a population some 10^3 times higher.

If the historical data (on supernovae) is processed according to the exponential disc model of our galaxy, one finds an event rate of about one per ten years (Blair, 1988). In external galaxies the rate is more difficult to determine because of confusion near bright galactic nuclei, and recognition of the existence of a class of low luminosity Type II supernovae such as SN1987A, which are too dim to be detectable in most external galaxies. These effects are sufficient to raise the expected supernovae rate from the traditionally accepted value of one per 30 years, to one more consistent with the historical data. There is no consensus on the supernova rate, however, except that it is in the range one per 10–100 years.

The other key piece of evidence comes from the pulsar population. The birthrate of pulsars can in principle be determined rather accurately since we have a statistical sample of about 500 to work from. Unfortunately pulsar evolution is not fully understood. In particular the evolution of the beaming factor is uncertain, and this leads to a range of possible birthrates from about one per 50 years to one per ten years.

Except for the uncertainties, one might conclude from the above figures that pulsars and supernovae are clearly the same population, as implied by the Crab nebula–Crab pulsar association. Things are not so simple, however. The main problem is that pulsars are not found associated with enough supernova remnants, nor is there evidence for pulsar activity in the remnants. The association seems to be only about 30%, and one is forced to conclude that a significant fraction of supernovae do not form active neutron stars. That is, they either leave no gravitationally collapsed remnant, or else the remnant is an inactive (non-magnetised or non-rotating) neutron star, or a black hole. If the association is 30%, and if supernovae always represent gravitational collapse, then it is possible that the gravitational collapse rate is as high as one per three years in our galaxy. This is consistent with an estimate of the total death rate of stars in the galaxy (Bahcall and Piran, 1983) of up to one per four years. It will probably require gravitational astronomy to give us the answer to these questions.

We can conclude that binary coalescence is rare in our galaxy unless there is a large hidden population of binaries, and that gravitational collapse occurs in the range of one per few years to one per several decades. At the upper limit there are reasonable prospects for limited high frequency galactic gravitational astronomy, but at the lower limit we must extend the range of gravitational astronomy to a distance of at least $10\,Mpc$, where the stellar sample is increased by about 10^3.

2.6.2 Massive black hole events

It is possible that a large fraction of galactic nuclei at some stage in their evolution create a dense cluster of stellar mass black holes and neutron stars (see figure 2.5). If dense enough these will rapidly coalesce to form, ultimately, a supermassive black hole. Waves from coalescing binaries in the mass range $3M_\odot - 1000M_\odot$ could be detected at the Hubble distance, and at this distance the event rate could be of the order of one per year.

If a large fraction of the missing mass was due to black hole binaries (Bond and Carr, 1984) there could be several coalescences per year within the local group of galaxies.

Massive black hole binaries ($M \leq 10^8 M_\odot$) may sometimes form in the nuclei of galaxies (Rees, 1983). Rees estimates that this has occurred through galactic mergers for about 10% of galaxies. However, the timescale for coalescence is long, and the event rate is unlikely to be greater than one per century. If massive black hole binaries were more common, as perhaps the dominant outcome of dense cluster evolution, this rate could be significantly higher.

The capture of a low mass star by a massive black hole is a far less effective source of gravitational waves, and is only worth considering for nearer sources. Thorne suggests that there would be a 'reasonable' event rate for galaxies out to 10 Mpc, but in the absence of specific evidence from the centre of our own galaxy this must be seen as rather optimistic.

2.7 Thorne diagrams

Thorne (1987) has compiled diagrams which show our best knowledge, estimates, guesses and beliefs about possible gravitational wave amplitudes and frequencies. They span at least ten decades of frequency and 12 decades of amplitude (24 decades of flux). Figures 2.6, 2.7 and 2.8 present Thorne diagrams successively for burst sources, stochastic sources and periodic sources. They are adapted from Thorne (1987).

Figure 2.6 shows black hole coalescence, neutron star coalescence, supernovae and gravitational collapse and massive black holes swallowing stellar mass objects. For supernovae and gravitational collapse, the efficiency factor ε is given in the range 10^{-2} to 10^{-8}. The amplitude scale is the *characteristic amplitude*, which makes allowance for signal processing advantages for particular signals, such as binary coalescence events, for particular idealised detectors. (See Chapters 9 and 16.)

Figure 2.7 shows several binary stars, signals from asymmetric pulsar rotation, and steady state spun up pulsars in X-ray binaries. For continuous sources the detectors can integrate to obtain greatly reduced noise amplitude. Hence the experimental curves are much lower than in figure 2.6.

Figure 2.8 shows possible and certain stochastic sources. The possible sources are cosmological, while the certain sources are binary systems. Also shown are the signals from the possible population of black hole binaries.

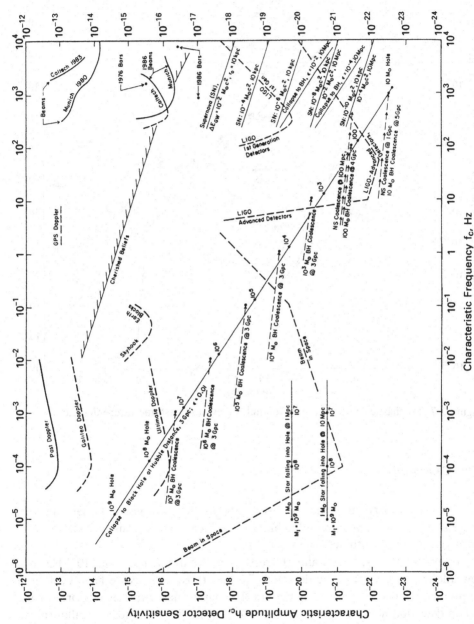

Figure 2.6. Burst sources of gravitational waves, showing the range of possible signals, and both past and future detector sensitivity.

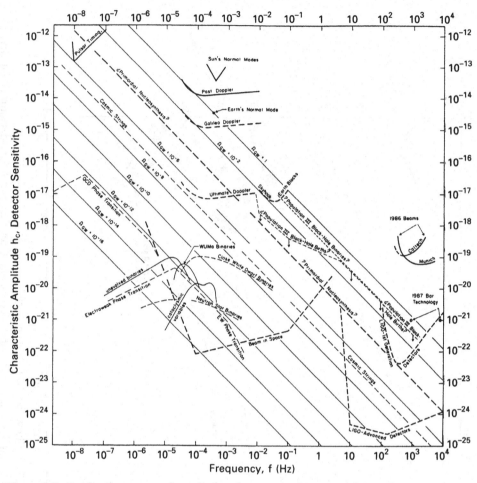

Figure 2.7. Stochastic sources of gravitational waves, and past and future detectors.

2.8   Conclusion

Now that the binary pulsar has confirmed the existence of gravitational waves, it is clear that at a low amplitude, but rather high energy density, spacetime is positively seething with activity.

Continuous waves, bursts and chirrups over the frequency range 10^{-6} Hz to a few kilohertz are there to be detected. Besides those sources we have been able to predict, we can be practically certain that nature has surprises in store for us. When detection does occur it will open a new window to the cosmos, allowing us to uncover the most energetic events to have occurred in the universe since the big bang itself, including the births of neutron stars, the vibrations of black holes as they swallow up stars, and the earliest signals from the young hot universe.

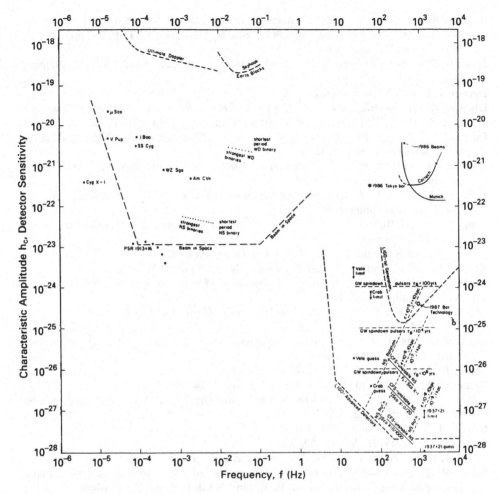

Figure 2.8. Periodic sources of gravitational waves. Detector sensitivity is greatly enhanced due to the ability to integrate over many cycles.

References

Ables, J. G., McConnell, D., Jacka, C. E., McCullough, P. M., Hall, P. J. and Hamilton, P. A. (1989). In *Proceedings of the Fifth Marcel Grossman Meeting on General Relativity*, eds. D. G. Blair and M. J. Buckingham. World Scientific, Singapore, p. 251.

Bahcall, J. N. and Piran, T. (1983). *Ap. J.* **267**, L77.

Blair, D. G. (1988). In *Supernovae, Pulsars, X-ray Binaries and the Rate of Gravitational Collapse in Experimental Gravitational Physics*, ed. P. Michelson, p. 36, World Scientific, Singapore.

Blair, D. G. and Candy, B. N. (1988). In *Timing Neutron Stars*, eds. H. Ögelman and E. van der Heuvel, Kluwer, Dordrecht.

Bond, J. R. and Carr, B. J. (1984). *Month. Not. Roy. Astron. Soc.* **207**, 585.

Damour, T. (1983). *Gravitational Radiation*, eds. N. Deruelle and T. Piran, pp. 59–144, North Holland, Amsterdam.

Damour, T. (1987). In *Gravitation in Astrophysics*, eds. B. Carter and J. Hartle, Plenum, New York.

Detweiler, S. L. (1980). *Astrophys. J.* **239**, 292.

Douglas, D. H. and Braginsky, V. B. (1979). In *General Relativity: An Einstein Centenary Survey*, eds. S. W. Hawking and W. Israel, Cambridge University Press.

Einstein, A. (1916). *Preuss. Akad. Wiss. Berlin*, Sitzungsberichte der physikalisch-mathematischen Klasse, p. 688.

Einstein, A. (1918). *Preuss. Akad. Wiss. Berlin*, Sitzungsberichte der physikalisch-mathematischen Klasse, p. 154.

Fang, L. Z. (1988). In *Experimental Gravitational Physics*, ed. P. F. Michelson, p. 172, World Scientific, Singapore.

Fang, L. Z. and Ruffini, R. (1983). *Basic Concepts in Relativistic Astrophysics*, World Scientific, Singapore.

Kojima, Y. and Nakamura, T. (1984a). *Prog. Theor. Phys.* **72**, 494.

Kojima, Y. and Nakamura, T. (1984b). *Prog. Theor. Phys.* **71**, 79.

Lindblom, L. (1988). In *Experimental Gravitational Physics*, ed. P. F. Michelson, p. 276, World Scientific, Singapore.

Lipunov, V. M. and Postnov, K. A. (1987). *Astron. Zhurn. (USSR)* **64**, no 2.

Lipunov, V. M., Postnov, K. A. and Prohorov (1987). *Astron. Astrophys.* **176**, L1.

Rees, M. J. (1983). In *Gravitational Radiation*, eds. N. Deruelle and T. Piran, p. 298, North Holland, Amsterdam.

Saenz, R. A. and Shapiro, S. L. (1978). *Astrophys. J.* **221**, 286.

Saenz, R. A. and Shapiro, S. L. (1981). *Astrophys. J.* **244**, 1033.

Sazhin, M. V. (1988). In *Experimental Gravitational Physics*, ed. P. F. Michelson, p. 179, World Scientific, Singapore.

Schutz, B. F. (1986). *Nature* **323**, 310.

Shklovsky, L. S. (1979). *Nature* **279**, 703.

Stark, R. F. and Piran, T. (1985). *Phys. Rev. Lett.* **55**, pp. 891, 56, 97.

Stark, R. F. and Piran, T. (1987). In *Proceedings of the Fourth Marcel Grossmann Meeting on General Relativity*, ed. R. Ruffini, North Holland, Amsterdam.

Taylor, J. H., Fowler, L. A. and McCalloch, P. M. (1979). *Nature* **277**, 436.

Taylor, J. H. and Weisberg, J. M. (1989). *Astrophys. J.*, in press.

Thorne, K. S. (1987). In *300 Years of Gravitation*, eds. S. Hawking and W. Israel, Cambridge University Press.

Wagoner, R. V. (1984). *Astrophys. J.* **278**, 345.

3

Gravitational wave detectors

DAVID G. BLAIR*, D. E. McCLELLAND†, H.-A. BACHOR† AND
R. J. SANDEMAN†

3.1 Introduction

The detection of gravitational radiation was pioneered by Joseph Weber in the early 1960s. He developed the first resonant mass detectors, and later investigated laser interferometer detectors. In this chapter we shall summarise these two main techniques. The extension of the latter technique to detection of very low frequency radiation using spacecraft and the signals from pulsars is described in the final chapter of this book.

We saw in chapter 1 that a gravitational wave acts to distort the ring of test particles, as shown in figure 3.1. In a 1960 *Physical Review* article Weber (1960) showed that a mass quadrupole harmonic oscillator will be excited by gravitational waves. The simplest such quadrupole oscillator is shown in figure 3.2(a). It simply represents a pair of the particles of figure 3.1, joined by a spring. The gravitational wave will do work on the oscillator. In practice the lumped parameter mass–spring oscillator is more easily replaced by a distributed system such as a bar or block or sphere of material. Weber suggested that the material could be piezoelectric. In this case the wave would be observed as the piezoelectric voltage across the oscillator. Alternatively the harmonic oscillator could be a metal bar, and a capacitive transducer could read out the motion, or the oscillator could be the Earth itself, and seismometers could read out signals at much lower frequencies. Later, Weber and Forward (Forward, 1971) considered the possibility that the spring of figure 3.2(a) could be replaced by a laser beam. In this case the motion would be detected by optical interferometry. See Weber (1987) for a full discussion of his early work. The optimum configuration for such a laser detection system is a laser interferometer as illustrated in figure 3.2(b).

The difference between the idealised quadrupole mass harmonic oscillator of figure 3.2(a), and the practical distributed-mass oscillator such as the Earth or a cylindrical bar, is that the distributed mass oscillator has a number of normal modes. The fundamental longitudinal mode is analogous to the mode of the idealised oscillator. All higher modes have reduced sensitivity to gravitational

* Sections 3.1–3.4.   † Sections 3.5–3.8.

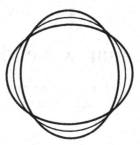

Figure 3.1. The distortion of a ring of test particles by a gravitational wave.

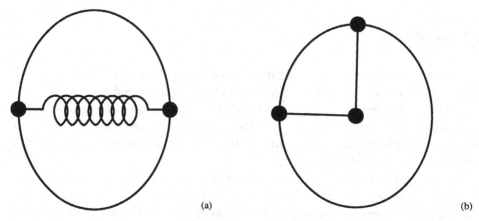

(a)

(b)

Figure 3.2. (a) A pair of test particles joined by a spring to make a mass quadrupole harmonic oscillator. (b) Two pairs of test particles connected via a laser interferometer. One arm of the interferometer expands, the other arm contracts, while laser beam fluctuations are balanced between the two arms, and cancel out.

waves owing to their reduced time-varying quadrupole moment: even modes are not excited at all, while the effective cross-section for odd modes falls as $1/n^2$, where n is the mode number. Thus follows the well known result that there is no advantage in making a resonant mass antenna longer unless it is desired to use it at a lower frequency.

In the case of the laser interferometer detector the mirrors would use a pendulum suspension against seismic noise, and would be effectively free masses. The detector closely approximates the ideal test particles, and the spring coupling, due to the pendulum and the laser beam, is very weak. In the resonant mass, the work done by the gravitational wave is stored in the oscillator and is available to be measured over a relatively long period, corresponding to a relatively narrow measurement bandwidth. In the laser interferometer the gravitational wave does work on the laser beam causing a small frequency shift, but in the simplest configuration there is negligible storage of this energy, and the detection process has a broad bandwidth.

In reality both of the descriptions are oversimplifications. The resonant bar has the convenient property that its sensitivity can be increased if operated as a narrow band detector, but broader bandwidths can be achieved. Techniques have recently been realised for enabling the laser interferometer to achieve a degree of resonant signal storage, allowing optimised operation in a narrower bandwidth. These points will be discussed further below and in greater detail throughout this book.

Let us now reconsider the problem of detecting gravitational waves, but now from an engineering viewpoint. Starting from Einstein's equations,

$$\mathbf{T} = (c^4/8\pi G)\mathbf{G},$$

we saw in chapter 1 that the coupling constant, $c^4/8\pi G$, where c is the speed of light and G is the gravitational constant, can be considered as a sort of metrical stiffness (see Sakharov, 1968) or modulus of elasticity for space-time, with the dimensions of energy density per unit curvature.

A gravitational wave is a wave in a medium with this extremely large stiffness. Since the propagation velocity is c, we can, by analogy with acoustic waves, identify the quantity c^3/G with the characteristic impedance of the medium. The material of a resonant antenna on the other hand has an impedance ρv_s per unit area, where ρ is the density and v_s is the velocity of sound. For an antenna of reasonable area made of a molecular solid such as steel this impedance is $\sim 10^8$ kg s^{-1}, whereas c^3/G is 4.5×10^{35} kg s^{-1}. Thus on entering a molecular solid the total impedance to a gravitational wave increases fractionally, by a multiplicative factor of $(1 + 4.5 \times 10^{-27})$. This solid therefore has a tiny effect on the wave motion and the interaction is very weak. The problem of detecting gravitational radiation can then be understood as an impedance-matching problem.

Objects made of nuclear matter (i.e. neutron stars) with v_s comparable to c are much more closely impedance-matched to gravitational waves so that significant absorption, refraction or diffraction can occur. However, the experimentalist, who has no nuclear matter available, is faced with a problem comparable to the problem of detecting electromagnetic waves if antennas could only be constructed from extremely dilute neutral gases with dielectric constant $\sim 1 + 10^{-27}$. (Individual nuclei are not useful either, because, although they have high density and high velocity of sound, they are infinitesimal compared with the wavelength, which at a frequency of 1 kHz is 300 km.)

In the case of the laser interferometer detector, the relevant mechanical impedance is that of the laser beam and the test masses. This again is extremely low, but by using the maximum possible laser intensity the stiffness is maximised. The impedance can therefore be varied, but conventional materials limit the optical power in the interferometer to about 10^6 W, leading to an impedance $\sim 10^5$–10^6 kg s^{-1}.

3.2 Resonant-bar antennas

During the 1960s, Weber constructed the first resonant-bar gravitational radiation antennas. (Weber, 1966.) These consisted of massive aluminium bars, suspended

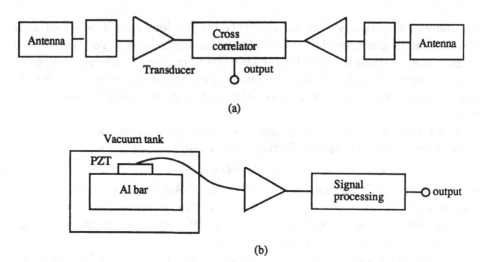

Figure 3.3. The basic resonant-bar antenna as designed by Weber. (a) Correlated detectors as proposed by Weber. (b) Schematic diagram of Weber's bar detector.

and vibration isolated in a vacuum container, as illustrated in figure 3.3. A gravitational wave passing through the bar induces a strain in the elongation of the bar. Since the bar has finite rigidity, work is done by the wave, and energy is deposited. In principle this can be detected. However, the detection is a very subtle process, and to understand this we return to an analogy.

A gravity wave is analogous to a water wave, a ripple in the curvature of space. But space is very stiff and normal matter is extraordinarily flimsy in comparison. If the gravity wave is like a water wave, then normal matter is like a piece of tissue paper floating on the water. It will move with the wave, and will absorb a negligible amount of its energy. If I tried to measure the motion of the tissue relative to another nearby floating piece of tissue, it too would move with the wave. There would be almost no relative displacement. Similarly the end of a massive metal bar and any device attached to or near the end of the bar which is intended to measure the motion of the bar, will move without any relative motion, so that the passage of the wave will go undetected. That is, a *local* measurement cannot detect the wave.

However, if the bar is resonant; that is, if it rings like a bell, then after the wave has passed the bar will continue to ring. This memory associated with resonance makes detection possible. A local measurement at the end of the bar can detect the effects of the wave. We will see that this memory is also related to noise reduction by compressing the noise into a very narrow bandwidth.

The impedance-matching problem means that only a very small fraction of the signal energy is coupled into a detector. As we saw in chapter 2, the amplitude of waves that we could expect from supernovae in our galaxy is about 10^{-18} m in a bar a few metres long. This is as small compared with an atom as an atom is compared with a page of this book. *It is 1000 times smaller than the typical intrinsic thermal vibration of an antenna cooled to a few degrees above absolute zero.*

Weber devised a very elegant method of solving this problem. This was again to use the resonant properties of the bar. The thermal energy of the fundamental mode of the antenna must have a mean value of kT. However, a bar with low acoustic losses, like a good tuning fork, will experience changes in its thermal vibration amplitude in times comparable to its 'ring down' or relaxation time τ_a. For example, if a bar rings for 100 s before its amplitude has dropped to $1/e$, then in 1 s the probable change in amplitude will be 1%, whether that vibration was caused by a hammer blow or by its internal heat. If the relaxation time is 1000 s, then in 1 s it will only have changed by 0.1%. Then if a 1 ms pulse of gravitational waves passes through the bar, changing its amplitude by, say, 0.2% this will be detectable above the thermal noise as long as the measurement can be made in less than 1 s, provided the ring down time is more than 1000 s.

Thus we can now define the properties needed for a resonant-bar gravitational wave antenna. It must be massive (so that the work done by the gravitational wave is as large as possible), it must ring well to reduce the effects of thermal noise, it must be at a very low temperature, and it must be equipped with a vibration sensor capable of detecting changes in vibration of less than 10^{-18} m. Moreover, outside vibrations from the environment must be filtered to an extremely high degree.

These criteria set a challenge which has kept a world-wide team of experimental physicists busy for the last 20 years. There has been a great deal of progress, but as in many other areas of science and technology, whether it be heavier-than-air flight or hydrogen-fusion reactors, the time from concept to completion is very long. Many problems have had to be solved: how to cool tons of metal near to absolute zero; how to make a vibration sensor with one million times the vibration-energy sensitivity of previous devices; how to make a bar that will ring 1000 times better than the best previously known acoustic resonator. Groups around the world have made impressive progress as they attempt to achieve these goals, as described in part II of this book.

3.3 Noise contributions to resonant bars

The measurement of gravitational radiation signals in a resonant-bar antenna involves the precision readout of the antenna motion, to a strain amplitude $h \sim 10^{-18}$ or 10^{-20}. For an antenna a few metres in length, this requires a transducer with sensitivity $\sim 10^{-18}$ to 10^{-20} m, and involves a careful optimisation of noise contributions. The receiver noise should allow energy sensitivity equivalent to the measurement of a small number of phonons of frequency ~ 1 kHz. Although piezoelectric transducers have been abandoned in favour of superconducting devices, the basic principles are the same as those developed for room-temperature antennas in the early 1970s. This involves identification of three fundamental noise sources, and optimisation of the sensitivity with respect to these (Michelson and Taber, 1981).

3.3.1   Brownian motion

The Brownian motion of an antenna is exactly analogous to the Nyquist noise of an LCR circuit, and represents the low-frequency part of the black-body spectrum of internal excitations. The mean energy of the fundamental mode of the antenna (determined from the low-frequency approximation to the Planck radiation formula) is simply equal to kT, but the effective noise is determined by fluctuations in this energy since the measurement of a gravitational wave is a measurement of the change in amplitude of the bar. Now the rate of fluctuation of the fundamental mode is related to coupling between the fundamental mode and the thermal reservoir, and this coupling itself determines the acoustic loss of the bar. Thus the effective noise energy reduces as the relaxation time increases, and is given by $kT(\tau_m/\tau_a)$ for short measurement integration times τ_m.

3.3.2   Series noise

The two additional sources of noise are due to the measurement process itself. The first is simple additive or series noise due to the transducer and amplifier which read out the motion of the antenna. This noise has the familiar Nyquist form as experienced in most areas of electronic instrumentation. We can model the measurement system as an antenna coupled to a noiseless transducer, coupled to a noisy amplifier, as shown in figure 3.4. Then this noise can be attributed wholly to the amplifier, which has a noise temperature T_A, and therefore a noise energy of kT_A per unit bandwidth. The narrower the measurement bandwidth, that is the longer the measurement integration time τ_m, the more this noise is reduced.

3.3.3   Back-action noise

The third source of noise is familiar to all students of quantum mechanics. It is noise due to the measurement process acting back onto the system being measured. In quantum mechanics back-action forces are used to illustrate how the uncertainty principle arises. However, the concept is not restricted purely to

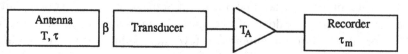

Figure 3.4. Model for a resonant-bar antenna measurement system. The antenna is coupled to a noiseless transducer coupled to a noisy amplifier. The thermal noise of the amplifier can be reduced by increasing the measurement integration time, but this increases the thermal noise contribution from Brownian motion.

quantum mechanics and, as we shall see, arises equally well in a classical system.

The back-action forces arise because the amplifier must always have a finite level of noise associated with its input. For example, the input resistance of the amplifier acts as a source of Nyquist noise which appears across the output of the transducer. It is fundamentally impossible for the transducer to have zero reverse transductance. (That is, there is no such thing as a perfect one-way valve.) Power can flow in reverse through the transducer, and Nyquist noise from the amplifier is therefore transduced to become a time–varying force noise which acts on the antenna. This force noise causes fluctuations in the antenna in much the same way that the thermal reservoir of modes in the bar couples to the fundamental mode causing the Brownian noise of the antenna.

Both the intrinsic thermal noise and the back-action noise increase as the measurement integration time increases. However, the series noise decreases with τ_m, so that there is clearly an optimum value of τ_m which minimises the noise of the system.

Figure 3.4 is useful for understanding another fundamental aspect of the detection process. Think of the antenna simply as an object at a temperature T. There is a coupling β between the antenna and the transducer. (In practice this is determined by the strength of an electric or magnetic field.) A gravitational wave causes a change in antenna temperature, causing energy to flow in or out of the transducer. If β is large the transducer will quickly come into equilibrium with the bar. However, if β is small the transducer will only slowly come into equilibrium; the measurement bandwidth will be very small. In the absence of intrinsic thermal noise in the antenna this condition is given by

$$\omega_a \tau_m \sim 1/\beta,$$

where ω_a is the antenna resonant frequency, typically $\sim 1\,\mathrm{kHz}$. If β is large (close to unity), then $\tau_m \sim 1/\omega_a$, and the antenna can have a bandwidth comparable with its resonant frequency.

In the past, technical problems have made it difficult to achieve $\beta \sim 1$, because (a) intrinsic losses in the transducer lead to a degradation of the antenna relaxation time when the coupling field is increased, and (b) it is fundamentally difficult to match an electromagnetic transducer to a massive bar without developing low-loss wideband mechanical impedance-matching networks.

3.4 Problems and progress with resonant bars

Figure 3.5 summarises the expected event rate-strain-energy relation for a typical resonant bar. It shows that the most likely events will cause a strain in a resonant bar equivalent to less than one phonon. Sensitivity at the single-phonon level is described as the quantum limit. In principle the quantum limit can be exceeded,

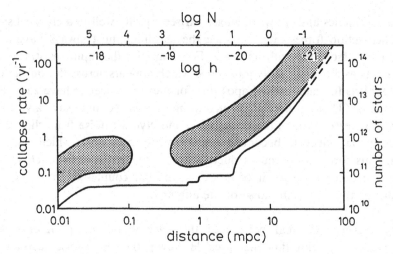

Figure 3.5. The shaded region shows that range of possible high-frequency gravitational wave event rates (based on optimistic data on the rate of gravitational collapse) as a function of possible strain amplitude h. The signal strength is also expressed as the effective number N of phonons induced in an optimally aligned 1 tonne resonant bar. That this number falls below unity as h falls towards 10^{-2} emphasises the quantum limit problem discussed in section 3.4.4. The lower curve shows the approximate number of stars as a function of distance, which can be monitored by such an antenna.

as discussed in section 3.4.4, but most efforts have been devoted to reducing noise levels from 10^8–10^{10} times the quantum limit (early 1970s' sensitivity) down to the quantum limit. We will look at various aspects of the noise and measurement problem below.

3.4.1   The acoustic-loss problem

As we saw above, the equivalent Brownian noise energy of an antenna is given by $kT(\tau_m/\tau_a)$. If we aim at a 1 kHz bandwidth, τ_m must be $\sim 10^{-3}$ s, and to reach the quantum limit we require the Brownian noise energy to be less than or equal to $\hbar\omega_a$ (ω_a is the antenna angular frequency). That requires τ_a to be $>10^5$ s if the antenna is to be operated at the boiling point of liquid helium, 4.2 K.

The traditional material used for antennas has been aluminium, for which τ_a has generally been less than 100 s. During the last ten years, materials with much lower losses at cryogenic temperatures have been discovered. These are sapphire, silicon, niobium and aluminium alloys containing 3–5% magnesium. They are best compared by their Q-factors, since this quantity is less dependent on frequency. (The Q-factor is the reciprocal of the fractional energy loss per cycle.) The Q-factor for conventional Al bars is 1–5×10^6. Sapphire has the highest Q ever observed, about 3×10^9 in the frequency range 20–40 kHz. This frequency range is determined by the maximum size of sapphire bars available. Here lies a major problem, however. There is no technical source for sapphire of sufficient

Figure 3.6. The 1.5 tonne niobium antenna at the University of Western Australia. The niobium cylinder is shown outside the cryogenic vessel in which it is suspended during measurements.

size or mass to make it a useful antenna material. Silicon has also been shown to have $Q > 10^9$, but large bars are also not available. The aluminium alloys 5056 and 5083 have been shown, under certain metallurgical conditions, to have a Q-factor between 4×10^7 and 1.4×10^8, whereas niobium has a Q-factor in excess of 2×10^8. Both these materials can be obtained in large ingots and both appear to be suitable materials for the present generation of antennas. See chapter 6 for a detailed discussion of acoustic losses.

Associated with the acoustic-loss problem is the antenna-suspension problem. It is difficult to suspend an antenna without degrading its acoustic properties, owing to the coupling of vibration to the outside world.

Various suspension cables from tungsten to silk have given excellent but often unreproducible results. Recently second and third order filters using intermediate masses and high-strength cantilever springs have given reproducible low losses. Such a mass–spring system for suspending the 1.5 tonne Nb bar at the University of Western Australia enables a Q-factor $Q \sim 2.3 \times 10^8$ to be achieved.

Magnetic levitation suspension has also been used. This uses superconducting coils to produce a field on which a Nb bar can float when it is below its superconducting transition temperature (about 9 K). This technique has been used successfully, but it is limited by the superconducting critical field to bars of diameter less than 200 mm.

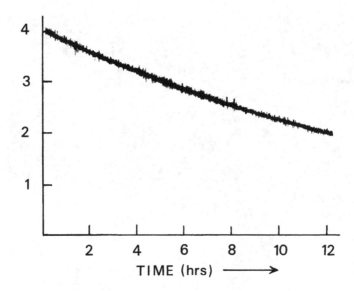

Figure 3.7. A ring-down curve for a 1.5 tonne niobium bar. Gas loading limits the Q-factor here to about 10^8. It is difficult to maintain such high Q-factors when a transducer is attached to the bar.

Figure 3.7 shows a ring-down curve for the 1.5 tonne Nb antenna. This Q-factor represents a rate of phonon exchange between the fundamental mode of the bar and the thermal reservoir, of about one phonon per millisecond. If τ_m could be reduced to 1 ms, then this noise would be at the quantum limit. To achieve full sensitivity at this limit would require the transducer to be coupled tightly. We discuss this problem below.

3.4.2   The impedance-matching problem

We saw earlier that the impedance of a metallic bar was extremely low compared with that of space. Unfortunately, however, the impedance is not low compared with that of a typically achievable electric or magnetic field (see chapter 8 for more details). The acoustic energy of the antenna is transduced into an electromagnetic signal either inductively or capacitively. A signal in the bar will couple well to the transducer if there is no impedance mismatch between them. That is, roughly, the mechanical stiffness of the bar should match the electro-mechanical stiffness of the coupling field of the transducer. In the simplest case the coupling of the transducer is determined by the ratio of the field energy of the transducer, compared with the elastic energy of that part of the antenna to which it is coupled. Because the field energy is low, the back-action forces are small, but the transductance is also small, so that series noise dominates the measurement. (The transducer is simply not very sensitive.) The sensitivity generally increases with field strength, but eventually either electric-field breakdown occurs (if it is a

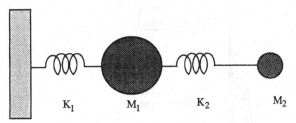

Figure 3.8. Schematic diagram of a coupled harmonic oscillator impedance-matching network. Energy is transferred from the high-mass resonator to the low-mass resonator, which oscillates at a larger amplitude.

capacitive device) or else magnetic breakdown occurs (for an inductive device) in the form of the superconducting critical current or critical field being exceeded.

Thus the properties of materials limit the transducer coupling that can be achieved, and then large values of τ_m are needed to overcome the series noise, so reducing the bandwidth and increasing the contribution of the Brownian noise.

The best solution to this problem is to design a mechanical impedance-matching network which transforms the vibration from being characteristic of a large mass and small amplitude, to one characteristic of small mass and large amplitude. In principle, various devices can achieve this, the simplest being a second resonator coupled to the first to form a coupled oscillator as illustrated in figure 3.8. As in all coupled harmonic oscillators, energy transfers at a beat frequency between each mode. The second oscillator can be a diaphragm, or a simple flap cut in the bar. The disadvantage of this system is that for half the time the energy resides in the bar and is inaccessible to measurement, leading to an inevitable loss of bandwidth.

The above scheme was first successfully implemented by the Stanford group (Michelson and Taber, 1981). Owing to the limitations, however, various other schemes are being pursued. All involve attempts to increase the transformer bandwidth. One approach is to increase the number of resonators in the coupled-oscillator system. A set of nested resonators of successively lower mass couple energy much faster from the first element to the last. At Maryland three- and five-mode resonators have been designed for this purpose (Richard, 1980). They show much greater bandwidth, but naturally are more complex to construct. See chapter 9 for details.

A second scheme consists of coupling the vibration of the antenna into a transverse wave in a tapered rod. Such a scheme is analogous to a whip, in which tapering increases the amplitude of a wave travelling along it. At the University of Western Australia we have demonstrated a 100-fold amplitude increase along such a whip attached to an aluminium bar (Blair, Giles and Zeng, 1987). This is shown schematically in figure 3.9.

Inspiration in designing devices such as these comes from the beautiful impedance-matching schemes designed for mechanical phonographs. The needle

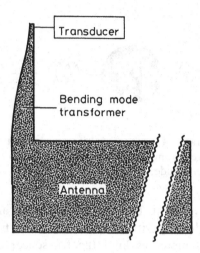

Figure 3.9. Schematic diagram of a 'whip' transverse-wave impedance-matching transformer. The vibration of the antenna is coupled with a tapered rod in which a transverse wave of much larger amplitude is induced.

in the record groove represents an extremely high impedance but low-amplitude signal source. By means of diaphragms and exponential horns good transformation to the low impedance of the air is achieved. For a resonant bar one can envisage a similar horn, but in this case tapering to a small diameter. Unfortunately such a horn must be large compared with the acoustic wavelength. For a resonant bar this means a length several times the bar's length (which is one-half a wavelength). To overcome this one is forced to transform to transverse modes: hence the choice of the whip configuration. Another non-resonant transformer has been designed by Paik of Maryland. This consists of a scissors-jack, which acts effectively as a lever to amplify the motion of an antenna.

3.4.3   The transducer problem

The transducer used on a resonant bar must have an intrinsic amplitude sensitivity which, as we have seen, is $\sim 10^{18}$ m. (This may vary depending on the impedance-matching scheme used.) The original use of piezoelectric crystals has been abandoned because of their high intrinsic losses. Two basic schemes are now in use. The first involves using a capacitor or inductor modulated by the motion of the antenna. The signal is directly coupled to an ultra-low-noise amplifier, for which the only suitable device is a SQUID (superconducting quantum-interface device), as shown in figure 3.10. The second involves using a high-frequency resonant LC circuit, of which the L or C is modulated by the motion of the antenna. A device of this type is called a parametric upconverter. The signal in the antenna is upconverted to a high frequency, and this high frequency is then

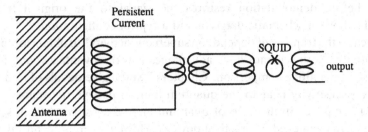

Figure 3.10. Schematic design of an inductively coupled superconducting transducer using a SQUID amplifier.

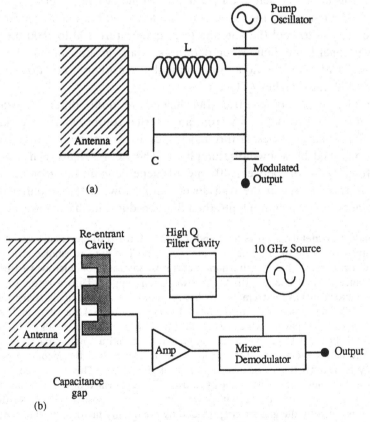

Figure 3.11. (a) Schematic diagram of a superconducting parametric upconverter using capacitive modulation. The high-frequency resonant LC circuit is coupled to the antenna, hence L or C is modulated by antenna vibration. (b) A parametric transducer can be practically implemented at microwave frequencies, using a re-entrant (capacitively loaded) cavity. The filter cavity is used to stabilise the microwave signal. See footnote on p. 56 for more details.

amplified before demodulation restores the signal to the original frequency. Figure 3.11 shows a schematic diagram and a typical configuration.

In both cases the transducer depends on superconductivity* to obtain low noise and high sensitivity. Both schemes have advantages and disadvantages, and it is difficult to make a definitive comparison. Chapters 7 and 8 give details of both types.

To obtain sensitivity near to the quantum limit in passive transducers requires the SQUID amplifier itself to be of quantum-limited sensitivity. Such sensitivity has not yet been achieved in practical devices. However, great technical progress in the development of low-noise SQUIDS (chiefly motivated by the need of gravitational radiation experiments) has occurred recently and there is hope for sufficient sensitivity before too long. See details and references in chapters 7 and 8.

The second more serious problem with passive transducers is the presence of a.c. losses in the superconducting circuit, especially as the current in the coils is increased. This limits the achievable Q in an antenna which is well coupled to the transducer. The reason for the losses is not understood, but one technical solution is possible. This is to cool the antenna to a temperature 100 to 1000 times lower than conventional liquid-helium temperatures. Many groups intend to cool their antennas to 50 mK; this should allow comparable sensitivity to that of an antenna at 4 K with 100 times higher Q-factor.

The problems with parametric transducers are quite different. Microwave power is introduced to the cavity from an external source, and the variations in phase of the signal as it passes through the cavity are observed after the signal has been amplified by a low-noise amplifier. The LC resonator can have a very high electrical Q-factor, between 10^6 and 10^8 depending on the geometry chosen. Thus the intrinsic losses in the transducer can be low. If the amplifier used has sensitivity near to the quantum limit, then the transducer itself can have similar sen-

---

* Superconducting metals, such as lead and niobium, have zero electrical resistance below some critical temperature T_c. For niobium T_c is 9 K. The zero resistance strictly applies only to direct current, but for alternating currents, say radio or microwave frequencies, the resistance is very small, $\sim 10^6$ times lower than copper. Thus a resonator made of superconductor can have extremely low losses, and have an extremely sharp resonance. At 10^{10} Hz (X-band microwave frequency) it is possible to make a very small resonator called a re-entrant cavity which is ideal for sensing displacements. The cavity is illustrated in figure 3.11. It has a tiny inductive post and a small capacitance gap. An intense alternating electric field is set up across the gap when the incoming microwave frequency is near to the resonant frequency of the cavity. This is altered by the gap separation. So when the cavity moves towards the end surface of the antenna there is a very sharp rise and fall of output power at a certain position. The width of this resonance is given by the gap spacing divided by the quality factor Q of the cavity, which is determined by the resistance of the superconductor. It is easy to attain a Q of 10^6 in a cavity like this, and d is chosen to be $\sim 10^{-5}$ m. Hence the resonance is 10^{-5} m/$10^6 =$ 10^{-11} m wide. In practice, using modern cryogenic low-noise electronics, it is possible to resolve this resonance to about one part in 10^8. This represents a distance sensitivity of 10^{19} m. This distance is measured between surfaces consisting of huge numbers of atoms, say 10^{14}. The distance is an average over these atoms, and so can be accurate to about one-billionth of the size of an individual atom.

sitivity. Fortunately three types of amplifiers are available which, at microwave frequencies, have a noise level no more than 20 times the quantum limit. These are ruby masers, parametric amplifiers and GaAs field-effect transistor amplifiers.

There are additional sources of noise, however. The chief one is noise in the incoming microwave signal. Fluctuations in the microwave signal are indistinguishable from fluctuations in the motion of the antenna. To overcome this problem it has been necessary to develop superconducting frequency sources, themselves very-high-Q resonators ($Q \sim 10^9$) to provide the pump signals. Sapphire-loaded superconducting cavities have been very successful for this purpose.

3.4.4 The quantum-limit problem

The technology described so far gives reasonable prospects of attaining an antenna sensitivity that is between 10 and 100 times the quantum limit. Compared with optics, say, this may not sound impressive, since at 10^{15} Hz one can measure individual photons, and therefore achieve quantum-limited performance. However, in energy terms this sensitivity is unprecedented. It represents an energy of $(10\text{--}100)\hbar\omega_a$, where $\omega_a \sim 5 \times 10^3$ rad s^{-1}. Thus it represents an energy of $10^{-29}\text{--}10^{-30}$ J.

Still, as we saw earlier, signal strength estimates lead us to expect to need sensitivity at or below the quantum limit. Recently it has been realised that the quantum limit is not a limit at all. It is a limit that arises because of the way in which we make our measurements. In fact it arises because when we measure using a *linear* position transducer with a *linear* amplifier we automatically measure a pair of quantum-mechanically conjugate observables. These are conventionally denoted X_1 and X_2; the real parts of X_1 and X_2 are given by $A \cos \phi$ and $A \sin \phi$ for antenna amplitude A and phase ϕ (see equations (4.1)). Because they are conjugate, maximum sensitivity is limited by the uncertainty principle

$$\Delta X_1 \Delta X_2 \gtrsim \hbar/m\omega_a.$$

Conventional linear devices lead to the quantum limit because they measure X_1 and X_2 symmetrically: the uncertainty in each variable is equal, and $\Delta X_1 = \Delta X_2 = (\hbar/m\omega_a)^{1/2}$. The problem is with the linear measuring system. If a phase-sensitive transducer were used, sensitive to X_1, but insensitive to X_2, then the uncertainty principle is satisfied for $\Delta X_1 \ll \Delta X_2$. A gravitational wave signal is then observed by making high-resolution measurements of X_1, and the resolution can enable the detection of strains which represent an energy less than $\hbar\omega_a$.

The transducer for such a device must allow no power flow which carries X_2 information. Back-action forces feed into the uncertain and unmeasured X_2 coordinate while one observes X_1. To achieve such a device one must use extremely narrow band (high-Q) filters to prevent power flow at unwanted frequencies. Thus the development of the so-called 'quantum non-demolition devices' involves a major technical challenge. The development of tunable ultra-high-Q filters and frequency sources, such as those under development for linear parametric transducers, will, it is hoped, eventually bring these ideas to

Table 3.1. *Resonant-bar gravitational radiation research groups.*

Group	Distinctive features	Frequency
University of Maryland J. Weber, J.-P. Richard, H. J. Paik *et al.*	Room-temperature Al bars. Massive cryogenic Al bars with three-mode SQUID transducers. Ultra-low temperatures planned.	1.6 kHz 800 Hz
Stanford University W. Fairbank, P. Michelson	Large-mass cryogenic Al bars. Two-mode transducer with SQUID readout. Ultra-low temperatures planned.	840 Hz
Louisiana State University W. O. Hamilton *et al.*	Large-mass cryogenic Al bars. Two-mode SQUID transducer. Ultra-low temperatures planned.	840 Hz

Moscow State University
V. Braginsky, Y. Rudenko
et al.
	Sapphire bars 10–100 kg, $Q \sim 3 \times$
	    10^9.
	'Horned' antenna with parametric
	    upconverter transducer.
		20–40 kHz

University of Rome
E. Amaldi, G. Pizzella, G. Pallotino
    et al.
	Various massive cryogenic Al bars.
	Ultra-low temperatures planned.
	Capacitive transducer using
	    SQUID.
		1.8 kHz
		840 Hz

University of Western Australia M. J. Buckingham, C. Edwards, D. G. Blair	Cryogenic niobium bar, 1.5 tonne. Parametric upconverter transducer.	1.7 kHz 700 Hz
Zhongshan University-Guangzhou Hu Enke *et al.*	Massive Al bar, tuned and low- frequency antenna 300 K. Cryogenic operation planned.	1.0 kHz 20–50 Hz
University of Tokyo H. Hirakawa (deceased), T. Tsubono *et al.*	Tuned antenna for Crab pulsar, 300 K. Cryogenic operation. Parametric transducers.	60 Hz

fruition (Caves *et al.*, 1980; Braginsky, Mitrofanov and Panov, 1985). Resonant-bar gravitational radiation detectors are in an advanced state of development at Stanford, Maryland and Louisiana State Universities, at CERN (University of Rome), and at the University of Western Australia. Other antennas are under development in Beijing and Guangzhou, Tokyo, and Moscow. The main research groups contributing to this field are listed in table 3.1.

3.5   Electromagnetic detectors

Weber's pioneering work spawned not only the resonant-bar type detectors introduced in the first part of this chapter, but also devices which use freely suspended test masses between which an electromagnetic signal is passed as a probe of the space-time curvature in the region between them. This is the fundamental idea behind four classes of detection:

(1) pulsar timing,
(2) spacecraft Doppler tracking,
(3) interferometry,
(4) planetary ranging.

In each of these techniques the gravitational wave can be considered as either affecting the propagation of the electromagnetic wave, or the motion of the test masses. In the absence of seismic noise, all these techniques can be used to explore the low- and ultra-low-frequency gravitational wave (gw) band – 10^{-8} Hz to 1 Hz, while laser interferometry can be sensitive up to 10 kHz. In the case of pulsar timing, the signals are simply received at mass 1 (the Earth), and accurately compared with a phase reference which must be derived from the most accurate available clock (which may be another pulsar). In the other cases the signal originates at mass 1, and is reflected or coherently transponded from mass 2 back to the transmitter. The signal at mass 1 appears Doppler shifted to $v + \delta v$ due (a) to the relative motions of the test masses induced by a gravitational wave, and (b) to gravitational, orbital and other classical perturbations which can be modelled and subtracted (for details refer to chapter 17). If the gravitational wave period τ_{gw} is much shorter than the light travel time τ_{tt}, then the electromagnetic signal will show a Doppler shift only if it was emitted, bounced or received at times when the gravitational wave amplitude h was non-zero during the event. This creates a unique signature which should appear in the time history of the signal, which can provide strong proof of the passage of a gravitational wave.

Ground based laser interferometers are restricted to detection in the range 10 Hz to 10 kHz, bounded from below by seismic noise and from above by acoustic modes in the mirrors. Typically, in this region τ_{gw} is of the order of or greater than achievable values of τ_{tt} and the signature alluded to above is washed out. Detection is now based on observing the phase difference imposed on the electromagnetic signal by the action of the gravitational waves on the test masses. The phase change is seen as a change in the apparent range between the masses leading to the familiar relationship between the strain and the wave amplitude

$$\delta L / L = h$$

for an optimally oriented detector, where L is the separation between the test masses.

This situation also applies to planetary ranging where the greatest sensitivity occurs when τ_{gw} is of the order of the orbital period, which is much greater than the light travel time.

For the remainder of the chapter we will discuss Earth based laser interferometers and examine fundamental limits to their performance.

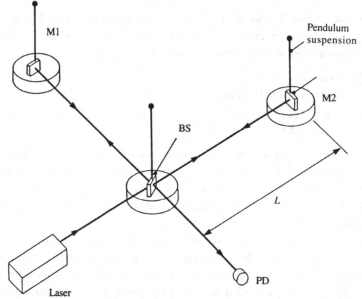

Figure 3.12. Standard Michelson interferometer layout. BS, beam splitter; M1 and M2, mirrors; PD photodiode.

3.6   The Michelson laser interferometer

The standard layout of a Michelson interferometer is shown in figure 3.12. The fundamental idea is strikingly simple: a partially reflecting mirror (beam splitter, BS) splits a beam of light into two beams of equal amplitude propagating in orthogonal directions. All components are mounted as pendula so that above the resonant frequency they respond as practically free masses. Mirrors M1 and M2 reflect the beams back on themselves to recombine at BS forming an interference pattern. Any relative change in the positions of the mirrors (or in the refractive index) will alter the optical path difference, resulting in a change in the interference pattern. It therefore responds to a *differential* rather than an absolute length change. The ultimate aim in gravity wave detection is to isolate the instrument from all other sources which can induce an optical path length change.

Indeed such an instrument is ideally suited for detecting gravitational waves since, due to their expected quadrupole nature, changes of opposite sign will be induced in the arms of a suitably oriented detector. Clearly, the difference in phase between the reflected beams can be increased by scaling up the arm length. However, beyond an optimum length L_{opt}, at which the storage time of the light within the interferometer arms is equal to half of the gravitational wave period, the wave changes sign and the induced strain reverses. Multipass techniques (described below) can be used so that the actual arm length can be considerably shorter than the optical path.

Typically L_{opt} is ~150 km. Two methods for achieving L_{opt} have been proposed and tested on prototypes (see table 3.2) – these are the Multi Pass Michelson

Table 3.2. *Overview of current prototypes, and of planned long baseline interferometers.*

Country Institute	USA MIT	FRG MPQ	GBR Glasgow	USA Caltech	FRA Orsay	PRC Guangzhou	PRC Beijing	JPN Tokyo	ITA Pisa	AUST UWA/ANU
Prototypes										
Start	1971	1975	1976	1980	1983	1985	1985	1986	1986	
Technique	MPM	MPM	FPM	FPM	(FPM)			(MPM)	(MPM)	
Armlength L	1.5 m (5 m)	(3 m) 30 m	(1.5 m) 10 m	40 m	5 m	3 m	0.5 m	10 m	2 m	
b or $2/\pi\,F$	50	100	3000	2000		(2)				
Strain sensitivity, $h(1/\sqrt{Hz})$	(4×10^{-17})	1×10^{-19}	1×10^{-19}	2×10^{-19}		$(>10^{-15})$				
Planning for large interferometers										
Planning	1982	1985	1985	1984	1986			1987	1986	1989
Armlength L	4 km	3 km	1 km	4 km	(3 km)				3 km	3 km
Technique	MPM	MPM	FPM	FPM					(MPM)	(FPM)

(MPM), proposed in this context originally by R. Weiss (1972), in which a delay line is inserted in each of the arms (Winkler, 1983), and the Fabry Perot Michelson (FPM), suggested by R. W. P. Drever *et al.* (1983a), where the delay lines are replaced with optical cavities.

Table 3.2 shows current proposals for long baseline instruments, with design sensitivities being (in general) of the order of 10^{-22}. Arm lengths vary from 1 km to 5 km. Here, we will examine fundamental limitations to the performance of a Michelson interferometer which could be important at the 10^{-22} sensitivity level.

3.6.1   Fundamental constraints

In practice there are many technical problems which may limit the performance of an interferometer. These include such things as: seismic noise; mechanical noise; thermal noise associated with the pendulum modes and internal resonances of the test masses; refractive index perturbations due to residual gas; scattered light mixing with the main beam, etc. Such problems can, at least in principle, be overcome. There are, however, three important fundamental limitations (Caves, 1981) which originate from manifestations of the Heisenberg uncertainty principle:

(i)   quantum uncertainty in the mirror position,
(ii)   photon counting error,
(iii)   radiation pressure fluctuations.

The standard quantum limit, which sets the 'single photon' sensitivity limit of a resonant bar also leads to a minimum uncertainty in the mirror position of an interferometer. This is also the smallest error obtainable from a balance between photon counting (pc) error and the radiation pressure (rp) error.

To obtain numerical estimates we will consider a standard interferometer of arm length $L = 3$ km, with mirrors of mass $m = 100$ kg; $b = 50$, so that $bL = L_{opt} = 150$ km; the optical wavelength $\lambda = 5 \times 10^{-7}$ m; and measurement integration time $\tau \sim 1$ ms so that the instrument bandwidth Δf is ~ 1 kHz.

(i)   Quantum uncertainty in test mass position

In a time τ the smallest measurable displacement of a mirror of mass m is restricted by the relationship $\Delta x \, \Delta p \geq \hbar$. Taking $\Delta p \sim m \, \Delta x / \tau$ then the minimum detectable strain $h_{min}(= \Delta x / L)$ is given by

$$h_{min} \sim (\hbar \tau / L^2 m)^{1/2}.$$

With the fiducial parameters given above, $h_{min} \sim 10^{-23}$.

(ii)   Photon counting error

Following Caves (1981) the output signal of an ideal photodetector placed at one of the output ports of the interferometer is given by

$$\langle N_{out} \rangle = N \sin^2(\phi/2) \quad \text{with} \quad \phi = 2b\omega x/c.$$

Here N is the number of photons passing through the device in a measurement time τ, given by $N = I_0 \tau / \hbar \omega$; x is the (static) arm length difference; and b is the effective number of bounces of the light in each arm. Assuming that the light is a coherent state, the noise on the signal due to photon statistics is

$$\Delta N_{out} = [N]^{1/2} \sin(\phi/2).$$

A change δx in the path difference is manifested as a change $\delta N_{out}(=(\partial \langle N_{out} \rangle / \partial x) \delta x)$ in $\langle N_{out} \rangle$:

$$\delta N_{out} \sim N(b\omega/c) \sin(\phi) \, \delta x.$$

It follows that the noise ΔN_{out} will appear as an error in x, known as the photon counting error, of the form

$$(\Delta x)_{pc} \sim c/(2b\omega N^{1/2} \cos(\phi/2)).$$

This error is minimised when the interferometer is operated around a dark fringe $(\phi = 0 = x)$ thereby making contributions from input power fluctuations negligibly small. In terms of the input power I_0, photon counting sets a limit $(\Delta x)_{pc}/L$ to the minimum detectable signal of the form

$$h_{min} = (1/4\pi bL)(2\pi \hbar c\lambda \, \Delta f/I_0)^{1/2}.$$

This equation shows the ideal dependence of h_{min} on the fundamental parameters of the interferometer indicating the need for low detection band-widths, high input intensities and optimal length baselines. Taking the fiducial parameters, a sensitivity of 10^{-22} can be achieved with an input laser power of the order of 1 kW.

Foreseeable developments in cw single mode stabilised laser power may achieve tens of watts. However, a technique known as recycling (Drever, 1983c; Meers, 1988; and Vinet et al., 1988) can be employed to increase the power onto the main beam splitter; and under the appropriate conditions (see next section) recent innovations in quantum optics (Loudon and Knight, 1987) can be used to reduce the power requirements. Also, more laser power can be obtained by coherent addition of several lasers (Man and Brillet, 1984). These methods will be addressed in detail in chapter 15.

(iii) Radiation pressure fluctuations

Vacuum fluctuations, which enter through the unused port of the beam splitter, when superposed with the laser light produce unequal fluctuating radiation pressure forces on the interferometer mirrors (Caves, 1980). The resulting difference in momentum of the end test masses (assuming that the central masses are much greater than the end masses) produces a momentum uncertainty Δp during a measurement time τ of the form

$$\Delta p \sim (2\hbar \omega b/c)N^{1/2},$$

with an associated position uncertainty due to radiation pressure of the form

$$(\Delta x)_{rp} \sim (\hbar \omega b/c)(\tau/m)N^{1/2}.$$

This limitation on the minimum detectable strain increases with the input laser power:

$$h_{min} \sim 2\pi \hbar (b/L)(\tau/m)(I_0\tau/\hbar c\lambda)^{1/2}.$$

Taking the fiducial parameters, $h_{min} \sim 10^{-25}$ for $I_0 = 100$ W.

Minimising the total error from photon counting and radiation pressure with respect to I_0 not only recovers the standard quantum limit but identifies an optimum power of the form (Caves, 1980) $mc^2/(2\omega\tau^2 b^2)$ which is of the order of 500 kW for the fiducial parameters. In the case of a photon counting limited interferometer, as we have here, manipulation of the vacuum fluctuations, entering through the unused port of the beam splitter can be used to effectively reduce the optimum power, thereby decreasing the photon counting error at the expense of an increase in the radiation pressure error. This can be achieved in principle using the squeezed states techniques described in chapter 15. Thus operation at the standard quantum limit (SQL) may be achievable at realistic power levels.

From these estimates of the fundamental limitations it is reasonable to expect that laser interferometers can be built that ultimately can reach the sensitivities shown in the Thorne diagrams in chapter 2, sufficient to be a powerful tool for gravitational wave astronomy.

Later, chapter 11 by Winkler and chapter 12 by Drever address in detail the characteristics of the two types of interferometer. A brief summary of the main features is presented below. In addition, we discuss another method using Fabry–Perot cavities (Newton *et al.*, 1986), referred to here as the locked double Fabry–Perot (LFP).

3.7   Michelson interferometer designs

3.7.1   Multi-pass Michelson (MPM)

In a multi-pass Michelson interferometer each arm contains a delay line made of spherical mirrors adjusted such that the beam traverses the arms b times before they are recombined on a beam splitter producing an interference fringe (figure 3.13). A phase change induced by a gravitational wave is thus sampled b times. Values of $b \sim 50$–100 have been achieved in a 30 m prototype (Shoemaker *et al.*, 1986), indicating that the effective armlength, bL, of a 3 km instrument can be made to approach L_{opt}.

Phase modulation techniques are employed to shift the signal to high frequencies, thereby minimising laser intensity noise.

By restricting the choice of mirrors to single, spherical surfaces at both ends of the interferometer the multi-pass Michelson can be figured to produce b equal diameter, non-overlapping spots on a circular perimeter of the mirror with the central part not illuminated. The size of the individual spots can be optimised by focussing the beam and using the mirrors in a confocal arrangement. Optimisation for minimum mirror diameter D results in (Maischberger *et al.*, 1987)

$$D = 2.5(b^2 L\lambda/\pi^3)^{1/2},$$

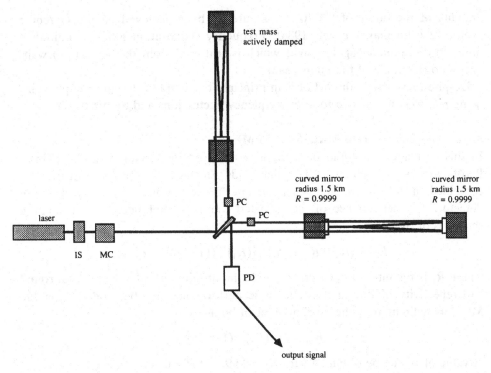

Figure 3.13. Schematic diagram of a multi-pass Michelson interferometer.

with the factor of 2.5 included to suppress beam truncation by the edge of the mirror etc. to less than 10^{-5}. With $L = 3$ km and $b = 50$, this diameter is 0.87 m for $\lambda = 0.5 \, \mu$m.

The larger the mirror, however, the lower is its resonance frequency. Setting the lowest permissible resonance frequency to 3 kHz would impose an upper limit of 0.6 m on the diameter. This limit can be met by removing the restriction of equal, non-overlapping spots at the expense of an increase in the noise level due to cross-talk between neighbouring beams. The main disadvantage of the MPM design lies in the cost of such large diameter mirrors and the corresponding large cross-section vacuum pipe.

Significant problems are also encountered with scattered light superimposing upon the main beam at the mirrors and being scattered back into the mode of the main beam, mixing with it to produce spurious signals (Shoemaker *et al.*, 1986).

In addition, in a practical instrument the pathlengths in each arm will not be exactly matched, due mainly to the different radii of curvature of the mirrors in the two delay lines, making the device responsive to laser frequency variations. The linear spectral density (square root of the spectral density) of the relative frequency fluctuations, $\delta v/v$, must then be kept below $hL/\delta L$,

$$\delta v/v < hL/\delta L.$$

Stability of the order of 5×10^{-18} has already been achieved on the Garching prototype (Shoemaker *et al.*, 1986). Frequency stabilisation is also required in order to discriminate against light which is scattered from the (vibrating) walls back into the mode of the main beam.

Despite these issues the MPM is in principle the simplest design to implement, being *relatively* insensitive to laser frequency fluctuations and to mirror tilt.

3.7.2 The Fabry–Perot Michelson (FPM)

In this design the multiple pathlength is achieved by placing resonant Fabry–Perot cavities in each arm. The lightfields reflected by the two cavities are combined in anti-phase, and again an interference minimum is formed at the detector. The dependence of the reflected complex amplitude E_r on the phase shift ϕ is of the form

$$E_r = ((R_1 - R) - i\phi R)/[(R_1)^{1/2}((1 - R) - i\phi R)],$$

where R_1 is the intensity reflectivity of the input mirrors and R is the total round trip reflectivity of the cavity. The basic performance can be modelled on the MPM by defining the 'effective' number of bounces as

$$b_{eff} = R/((R_1)^{1/2}(1 - R)).$$

A value of 50 can be obtained with $R_1 = 0.99$ and $R = 0.9805$.

In order to achieve good fringe visibility both cavities must have as near to identical properties as possible requiring matching of mirror substrates and coatings. Furthermore, intensities inside the cavity will be b_{eff} times higher than outside, increasing the absolute heat gain of the mirrors. Low loss mirror coatings ($<10^{-4}$) are therefore needed in order to reduce problems with thermal distortion of the mirror surfaces or coatings.

Analogous to the static pathlength difference that is likely in an MPM (see previous section) is the mismatch in cavity finesse arising mainly from variations in the radius of curvature and of the coatings of the mirrors. This places a somewhat more stringent condition on relative frequency fluctuations than required for the MPM though these higher stability levels have been achieved on the Glasgow prototype (Ward *et al.*, 1987). Techniques for achieving adequate cavity/laser lockings are discussed in the next section.

There have, however, been problems in achieving FPM operation in prototype detectors. The main reason for this is that the Pockels cell modulators in each arm distort the wavefronts to the extent that beam recombination gives poor fringe visibility (Ward *et al.*, 1987).

To eliminate this problem it is necessary for the locking of the Fabry–Perot resonators to be done without contamination of the input/output beams by Pockels cells. It may be possible to achieve this by injecting modulated light at 1% beam splitters before the optical cavities, and locking the cavities without interfering with the main beams. A similar system has been demonstrated on the bench by Brillet's group in Paris. More complex modulation schemes can be used involving three or more modulation frequencies, which may be especially useful in conjunction with recycling.

The FPM, however, has two significant advantages over an MPM. Firstly, the mirror size: the mode size supported in an optical cavity depends on the curvatures of the two mirrors and can be calculated using resonator theory. The mirror nearest the beam splitter (figure 3.14) should be flat in order to produce a minimum spot size through the transmission optics. The radius of curvature of the far mirror can then be chosen to be $1.5L$ in order to avoid radial mode overlap. In such a cavity the spot size on the close mirror is of the order of $(L\lambda)^{1/2}$. With $L = 3\,\mathrm{km}$ and $\lambda = 0.5\,\mu\mathrm{m}$ and multiplying by a factor of 2.5 to ensure that diffraction losses are less than 10^{-5}, the maximum mirror diameter would be $0.18\,\mathrm{m}$, approximately five times smaller than in the multi-pass Michelson.

Secondly, as all the light inside a cavity, both direct and scattered, travels the same path, problems with scattered light are greatly reduced. This was the initial motivation for the FPM design (Drever, 1983c).

3.7.3 The locked double Fabry–Perot interferometer (LFP)

This is a modification of the Fabry–Perot Michelson: it uses the same optical components. The major difference is that, rather than being optically superimposed, the lightfields reflected by the cavities are directed completely onto two separate photodiodes. The Glasgow group have achieved a quantum noise limited sensitivity of the order of 10^{-18} with this technique.

In detail, the two incident beams are sent through polarising beam splitters followed by quarter wave plates, all located between the beam splitter and the cavities (see figure 3.14). The light impinging on the cavities is circularly polarised and the cavities are optically isolated from each other. The locking scheme initially proposed by Drever et al. (1983b) can be employed: a Pockels cell in the input beam produces FM sidebands at a driving frequency w of several megahertz larger than the resonance profile of the cavities. Depending on the resonance condition of the cavities the light leaking back out of the cavity is phase shifted with respect to the reflected sidebands and the resulting Fourier components of the photocurrents at frequency w provide strongly dispersive error signals. One cavity is locked tightly to the laser by controlling the laser frequency as well as shifting the close mirror of this cavity. The second cavity is locked to the first cavity by shifting the close and the far mirrors to adjust the cavity length. The difference between the error signals is directly proportional to the differential strain h in the interferometer.

Not only is there a slight gain in sensitivity (~ 2 for the ideal device) with this operational mode, but wavefront distortion, which reduces fringe visibility and hence sensitivity in the Michelson systems, is not a critical issue here.

However, the requirements on the frequency stability of the laser are much

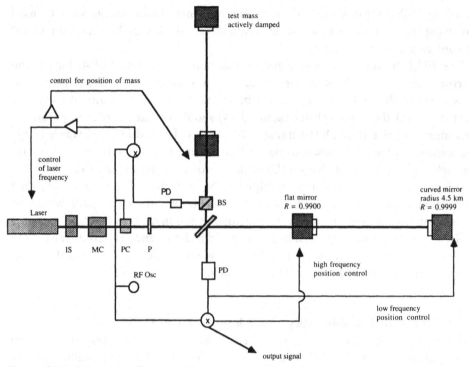

Figure 3.14. Schematic diagram of the double Fabry–Perot Michelson interferometer showing a possible locking arrangement. IS, intensity stabilisation; MC, mode cleaning cavity; PC, Pockels cell; P, polariser; BS, beam splitter; PD, photodiode and preamp.

more stringent since the phase shift due to a change in frequency of the laser is directly compared to differential phase shifts between the two arms of the interferometer. The other limitation of this design lies in the maximum permissible power on the photodiodes – typically only 1–50 mW for saturation, with maximum powers being under a watt. The photon counting error for the locking is correspondingly high. Operation with high power would require attenuation in front of the detectors, which does not lower the photon counting error since it depends on the intensity actually entering the detector. This is a serious limitation.

It is worth noting further that this design also precludes the use of 'recycling' and non-classical light techniques described in chapter 15.

3.8   Conclusion

We have examined resonant mass and laser interferometer gravitational wave detectors. The resonant bar detectors are relatively well developed, but are still undergoing steady improvement from their existing 10^{-18} sensitivity, towards

$\sim 10^{-20}$, where the quantum limit may hinder further development. Large scale laser interferometer detectors are expected to be constructed during the 1990s. They can be expected to achieve 10^{-22} strain sensitivity, sufficient to become powerful gravitational wave observatories.

Within the interferometer class there are major advantages and disadvantages with each of the proposed designs. Therefore the best choice of the operational mode of a laser interferometer observatory is debatable. Current proposals for long baseline instruments are evenly divided between the MPM and FPM. Only time and experience will provide the answers.

The rest of this book provides a state of the art description of gravitational wave detection. Besides the major focus on laser interferometers and resonant bars, one chapter describes the beautiful low frequency resonant detectors developed at Tokyo (chapter 9) and another (chapter 17) describes the spacecraft techniques that were introduced in section 3.5.

References

Blair, D. G., Giles, A. J. and Zeng, M. (1987). *J. Phys. D.* **20**, 162.

Braginsky, V. B., Mitrofanov, V. P. and Panov, V. I. (1985). *Systems with Small Dissipation*, University of Chicago Press.

Caves, C. M. (1980). *Phys. Rev. Lett.* **45**, 75–8.

Caves, C. M., Thorne, K. S., Drever, R. W. P., Sandberg, V. D. and Zimmerman, M. (1980). *Rev. Mod. Phys.* **52**, 341.

Caves, C. M. (1981). *Phys. Rev. D* **23**, 1693–708.

Drever, R. W. P. *et al.* (1983a). 'A gravity wave detector using optical cavity sensing', *Proc. Ninth Int. Conf. on Gen. Rel. and Gravit.*, (*GL9*), *Jena, 1980*, ed. E. Schmutzer, pp. 265–7, Cambridge University Press.

Drever, R. W. P., Hall, J. L., Kowalski, F. V., Hough, J., Ford, G. M., Munley, A. J. and Ward, H. (1983b). *Appl. Phys. B* **31**, 97.

Drever, R. W. P. (1983c). In *Gravitational Radiation*, eds. N. Deruelle and T. Piran, p. 321, North Holland, Amsterdam.

Forward, R. L. (1971). *Appl. Opt.* **10**, 2495.

Loudon, R. and Knight, P. L. (1987). *J.Mod. Opt.* **34**, (6/7), 709–59; see also references therein.

Maischberger, K., Rudiger, A., Schilling, R., Schnupp, L., Winkler, W. and Leuchs, G. (1987). 'Status of the Garching 30 Meter Prototype for a Large Gravitational Wave Detector', *International Symposium on Experimental Gravitational Physics*, ed. P. F. Michelson, pp. 316–21, World Scientific, Singapore.

Man, C. N. and Brillet, A. (1984). *Opt. Lett.* **9**, 333–4.

Meers, B. J. (1988). *Phys. Rev. D* **38**, 2317–26; see also references therein.

Michelson, P. F. and Taber, R. C. (1981). *J. Appl. Phys.* **52**, 4313.

Newton, G. P., Hough, J., Kerr, G. A., Meers, B. J., Robertson, N. A., Ward, H., Mangan, J. B., Hoggan, S. and Drever, R. W. P. (1986). *Proceedings of the 4th Marcel Grossman Meeting on General Relativity, Part A.*, Rome, 1985, ed. R. Ruffini, pp. 599–604, Elsevier, Amsterdam.

Richard, J. P. (1980). In *Gravitational Radiation, Collapsed Objects and Exact Solutions*, ed. C. Edwards, p. 370, Springer Verlag, Berlin.

Sakharov, A. D. (1968). *Sov. Phys. Doklady.* **12**, 1040.

Shoemaker, D. H., Winkler, W., Maischberger, K., Rüdiger, A., Schilling, R. and Schnupp, L. (1986). *Proceedings of the 4th Marcel Grossman Meeting on General Relativity, Part A*, Rome, 1985, ed. R. Ruffini, pp. 605–14, Elsevier, Amsterdam.

Vinet, J. Y., Meers, B., Man, C. N. and Brillet, A. (1988). *Phys. Rev. D* **38**, 433.

Ward, H., Hough, J., Kerr, G. A., MacKenzie, N. L., Mangan, J. B., Meers, B. J., Newton, G. P., Robertson, D. I. and Robertson, N. A. (1987). *International Symposium on Experimental Gravitational Physics*, ed. P. F. Michelson, pp. 322–7, World Scientific, Singapore.

Weber, J. (1960). *Phys. Rev.* **117**, 306.

Weber, J. (1966). *Phys. Rev.* **146**, 935.

Weber, J. (1987). In *Proceedings of the Sir Arthur Eddington Centenary Symposium*, Vol. 3, eds. J. Weber and T. M. Karade, World Scientific, Singapore.

Weiss, R. (1972). Electromagnetically coupled broadband gravitational antenna, Quarterly Progress Report, Research Laboratory of Electronics, MIT, **105**, 54–76.

Winkler, W., Dissertation, Munchen (1983), Internal Report MPQ 74 (unpublished).

PART II
Gravitational wave detectors

4

Resonant-bar detectors

DAVID G. BLAIR

4.1 Introduction

Resonant-bar detectors are designed to measure the acoustic signal induced in a massive bar due to its coupling to a gravitational wave. The large amplitude of thermal vibration in the bar normally considerably exceeds the amplitudes expected from astrophysical sources, and without methods to suppress this noise the principle of detection by resonant masses would be impossible. Weber's key contribution was the realisation that in a high Q antenna – one with a low acoustic loss – the effective noise energy is reduced by a factor $\sim \tau_i / \tau_a$, where τ_i is the effective measurement integration time, and τ_a is the antenna ring down time. The advantage from using a low acoustic loss antenna is a direct result of the fluctuation–dissipation theorem. A high Q antenna approaches an ideal harmonic oscillator, whose motion is exactly predictable at a time in the future from the observed amplitude, frequency and phase at an earlier time.

In this chapter we will examine the key concepts of resonant-bar detectors, and provide the framework for the following chapters on different aspects of resonant-bar technology.

4.2 Intrinsic noise in resonant-mass antennas

In 1971, Gibbons and Hawking gave an analysis of resonant-mass antennas which led to improved techniques and better understanding of the noise sources. They noted that Weber had monitored the *energy* or RMS amplitude of the fundamental mode of his antennas. Since the phase of an incoming gravity wave is random relative to the antenna, and since the energy deposited in the antenna is much smaller than the mean thermal energy of the mode, kT, it follows that the mode energy will only sometimes be increased by a gravitational wave. It is equally likely to be reduced in energy and the wave may simply cause a small phase shift.

It is convenient to describe the state of the antenna by the pair of symmetrical harmonic oscillator coordinates X_1 and X_2 given by

$$X_1 = A \cos \phi$$
$$X_2 = A \sin \phi,$$

(4.1)

where A is the antenna amplitude and ϕ is the phase. Experimentally X_1 and X_2 can be easily measured using two phase sensitive detectors in a configuration shown schematically in figure 4.1. The state of the antenna can be represented by a point P_1 in the (X_1, X_2) plane; the amplitude $A = |P| = (X_1^2 + X_2^2)^{1/2}$ and phase $\phi = \tan^{-1} X_2/X_1$. This is illustrated in figure 4.2. A gravitational wave causes the antenna to move from P_1 to P_2. The amplitude change is simply $|P_1| - |P_2|$. Clearly $|P_1| - |P_2| \leq |P_1 - P_2|$, and in general a measuring system that monitors $|P_1 - P_2| = [\Delta X_1^2 - \Delta X_2^2]^{1/2}$ will be superior to one sensitive to amplitude alone.

Thermal fluctuations cause the state vector P to execute a random walk in the $X_1 X_2$ plane. A high Q mode is weakly coupled to the thermal reservoir which is

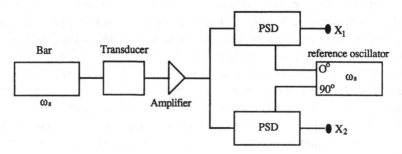

Figure 4.1. Antenna readout systems for obtaining harmonic oscillator coordinates X_1 and X_2.

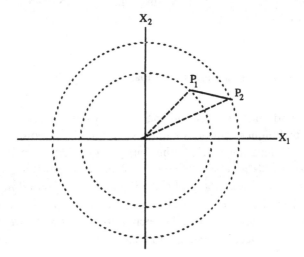

Figure 4.2. X_1–X_2 representation of the state of the antenna.

made up of all the higher modes of the system. The antenna loses energy slowly into the reservoir, and equally it is only weakly excited by the reservoir. The relaxation time $\tau_a = 2Q/\omega_a$ thus characterises both the rate of decay after a high energy excitation and the rate of amplitude change when the mode is in thermal equilibrium with the reservoir.

Clearly, if τ_a is large and the rate of fluctuation is low, the antenna becomes more deterministic on timescales that are short compared with τ_a. The mean energy is still kT, but the expected change in energy in a sampling time τ_i is $kT(\tau_i/\tau_a)$. The temperature $T(\tau_i/\tau_a)$ is the effective temperature or noise temperature of the resonator, and quite clearly can be made less than the actual temperature.

The above analysis describes a predictive filter. In this case our prediction is that the amplitude and phase of the detector will remain unchanged over the integration time. A more advanced type of filter – an optimum filter – would be based on the history of the antenna state vector. Such filters enable the antenna noise temperature to be improved somewhat, and are described in chapter 10.

To analyse a resonant-mass antenna Gibbons and Hawking used the electrical equivalent circuit shown in figure 4.3. The components L_1, C_1, R_1 represent the bar with mass L_1, spring constant $1/C_1$, and Q-factor $\omega_a L_1/R_1$. The bar is coupled to a piezoelectric transducer represented by capacitor C_2, and the resistor R_2 represents the losses in the transducer. (Gibbons and Hawking neglected amplifier noise). The resistor R_2 produces noise given by the Nyquist formula

$$V_{R_2}^2 = 4kTR_2 \cdot \frac{1}{\tau_i}, \tag{4.2}$$

where the bandwidth is the reciprocal of the sampling time τ_i.

The system noise with the addition of the transducer noise is then given by

$$V_n^2 = V_B^2 \tau_i/\tau_a + 4kTR_2/\tau_i, \tag{4.3}$$

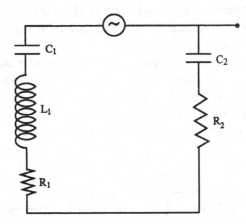

Figure 4.3. Gibbons and Hawking's equivalent circuit for a resonant antenna with a piezoelectric transducer.

where the first term is the antenna Brownian motion noise as discussed above. The presence of the 'series noise' contribution from R_2 completely alters the noise optimisation of the system. No longer can the noise be reduced arbitrarily by replacing τ_i because the series noise blows up as $\tau_i \to 0$. There is an optimum sampling time at which the two noise contributions are equal, and only by increasing τ_a (reducing the losses in the antenna) or decreasing the value of R_2 (improving the transducer) can improvements be made (at any given temperature).

Gibbons and Hawking also introduced a parameter β to characterise the coupling between the bar and the transducer. They define β as follows:

$$\beta = \begin{cases} \text{proportion of the elastic energy of the detector that can be} \\ \text{extracted electrically through the transducer in one cycle.} \end{cases}$$

A bar–transducer system with low β (weak coupling) requires more time for the signal energy to appear in the transducer. The longer the energy transfer takes, the more time there is for fluctuations in the antenna to dominate the noise. This point can be clarified by two alternative viewpoints. One is the thermodynamical model for the system discussed in chapter 3: the antenna is considered as a thermal bath at temperature $T_{\text{eff}} = T_a \tau_i / \tau_a$, coupled to a transducer with noise temperature T_T which itself is coupled to an amplifier of noise temperature T_N.

A gravitational wave causes slight 'heating' of the fundamental mode and energy flows through the coupling β shown schematically in figure 4.4. As long as $\beta > 0$ the transducer will eventually come into equilibrium with the bar, but for a rapid response β has to be large. This model emphasises an aspect of the analysis not included by Gibbons and Hawking. This is that the coupling is not unidirectional: thermal fluctuations in the amplifier or the transducer act back on the antenna producing *back-reaction noise*. Indeed, it is clear that the transducer is a source of thermal fluctuations comparable to those originating within the bar, and these will produce an additional noise contribution which will diminish as $\tau_i \to 0$ as does the Brownian noise.

The second viewpoint is that the antenna is effectively a transmission line which couples energy into the transducer. One can think in terms of phonons in the bar which may be absorbed by the transducer, with the emission of a photon into the amplifier, or they may be reflected back into the bar. Then β determines the *impedance match* between the output impedance of the bar, Z_{out}, and the

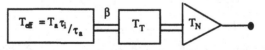

Figure 4.4. Antenna, transducer and amplifier: thermodynamical model.

transducer's mechanical input impedance Z_{11}. The ratio Z_{11}/Z_{out} is simply the coupling coefficient β.

Once we begin to think in terms of quanta we are led to ask: what happens if the induced strain in the antenna is equivalent to less than one quantum $\hbar\omega_a$? The profound significance of the quantum mechanical limit to macroscopic measurements was realised independently by several groups, particularly by Braginsky (1970) and Giffard (1976). Giffard used the much earlier result of Heffner (1962) who showed that, by the uncertainty principle, a linear amplifier has a fundamental limit to its sensitivity, given approximately by $\hbar\omega_a$. Similarly Giffard showed that a transducer used with a linear amplifier (an amplifier which preserves phase and changes the amplitude by a multiplicative factor) has a maximum sensitivity corresponding to a gravitational wave which produces an equivalent of two quanta in the bar. The term equivalent is used because the actual energy absorbed by the antenna depends on its instantaneous amplitude. For linear systems the signal-to-noise ratio is independent of the amplitude, and corresponds exactly to the signal produced in an ideal stationary antenna at absolute zero. Giffard's result meant that the maximum achievable sensitivity of an antenna would be limited to about the single phonon level (see figure 3.5) corresponding to a strain sensitivity between 10^{-20} and 10^{-21}. This is described as the standard quantum limit.

Meanwhile, at least as early as 1974, Braginsky and Vorontsov proposed that in principle it might be possible to devise *quantum non-demolition devices* which could read out the state of a system without disturbing it. Braginsky, Vorontsov and Kalili (1978), Caves *et al.* (1980), Unruh (1978) and others went on to define methods whereby gravitational waves of amplitude less than that required to induce one quantum can in principle be detected using *quantum non-demolition* or back-action evading techniques. These techniques are described in section 4.8 below, and in chapter 15 are applied to optical photons where they have so far been most successful.

4.3   The signal-to-noise ratio

A gravitational wave carries an energy flux $S(Jm^{-2}s^{-1})$ given by

$$S = \frac{c^3}{16\pi G}\langle \dot{h}_+^2 + \dot{h}_\times^2 \rangle, \tag{4.4}$$

where h_+ and $h_\times$ denote the dimensionless strain amplitudes of the two possible polarisations of the wave. Since the shape of expected gravitational wave pulses from gravitational collapse events is not accurately known, we cannot accurately determine the expected excitation of an antenna even knowing the total pulse energy. We need to know both the spectral distribution of the pulse energy, and the relationship between h and its time derivative. The details of the expected pulses depend not only on the dynamics of the gravitational collapse, but also on the mass of the collapsing object, both of which are uncertain.

If we assume only knowledge of the pulse duration τ_g (expected to be $\sim 10^{-3}$ seconds), and that it is predominantly a single cycle, it is sufficient to assume that $\dot{h} \sim 2h/\tau_g$. Then equation (4.4) can be rewritten

$$S = \frac{c^3}{16\pi G} \cdot \frac{4h^2}{\tau_g^2} \, \text{J m}^{-2} \, \text{s}^{-1}. \tag{4.5}$$

The total pulse energy E is then given by $S \cdot \tau_g$:

$$E \approx \frac{c^3}{16\pi G} \cdot \frac{4h^2}{\tau_g} \, \text{J m}^{-2}. \tag{4.6}$$

Now assuming that the spectral distribution of the pulse energy $F(\omega)$ is uniform over a bandwidth $\Delta\omega \sim 1/\tau_g$, it follows that

$$F(\omega) \approx E/\Delta\omega = E\tau_g \approx \frac{c^3 h^2}{4\pi G} \, \text{J m}^{-2} \, \text{Hz}^{-1}. \tag{4.7}$$

Numerically $F(\omega) \sim 20 \times 10^{34} h^2$. The assumption used in obtaining the result must be emphasised: the result is simply an order of magnitude estimate of the expected signal spectral densities. Moreover, variations in the pulse durations could make any chosen antenna frequency only suitable for a small proportion of actual events.

The energy deposited in an initially stationary antenna of mass M by a signal pulse $F(\omega)$ is given by

$$U_s \approx F(\omega_a) \sin^4 \theta \cos^2 2\phi \cdot \frac{8}{\pi} \left(\frac{G}{c}\right) \left(\frac{V_s}{c}\right)^2 M, \tag{4.8}$$

where θ and ϕ are coordinates describing the orientation of the bar relative to the incoming wave (as given in figure 4.5).

For a short pulse gravitational wave burst the bandwidth of the pulse is roughly the inverse of the pulse duration, and the frequency bandwidth is roughly equal to the peak frequency. Under these circumstances the strain amplitude $\delta l/l$ induced in the bar is roughly equal to the incoming wave amplitude h; there is no resonant excitation.

For the incoming gravitational wave to be detectable the signal U_s must be greater than the noise in the antenna U_n:

$$U_s \geq U_n. \tag{4.9}$$

To characterise the noise U_n we generalise the transducer to a two port device described by a 2×2 impedance matrix Z_{ij}. The transducer accepts force and velocity inputs F and v, gives current and voltage outputs I and V:

$$\begin{pmatrix} F \\ V \end{pmatrix} = \begin{pmatrix} Z_{11} & Z_{12} \\ Z_{21} & Z_{22} \end{pmatrix} \begin{pmatrix} v \\ I \end{pmatrix}. \tag{4.10}$$

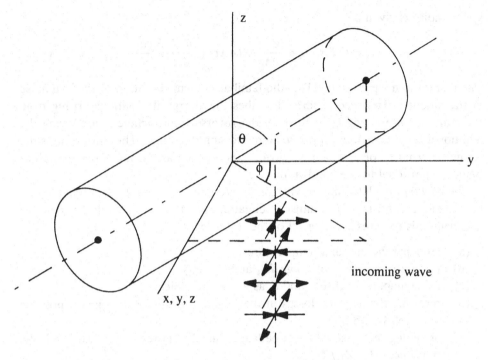

Figure 4.5. Coordinate system for resonant antenna.

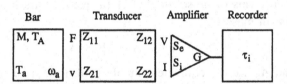

Figure 4.6. The various quantities used to characterise a gravitational wave antenna system.

The transducer has input impedance Z_{11}, measured in $\mathrm{kg\,s^{-1}}$, and output impedance Z_{22}, measured in ohms. The forward transductance Z_{21}, measured in $\mathrm{V/(m\,s^{-1})}$, determines the transducer sensitivity, whereas the reverse transductance Z_{12}, measured in $\mathrm{kg\,A^{-1}}$ determines the back-acting force on the antenna due to currents in the output circuit. See Blair (1980) and Veitch's chapter 8 in this volume for more details.

All the noise sources in the transducer and amplifier can be expressed as equivalent spectral densities of current and voltage noise at the input of the amplifier, denoted $S_i(\omega)$ and $S_e(\omega)$, respectively, as illustrated in figure 4.6.

The current noise S_i is the source of back-action noise in the antenna, whereas S_e describes the series noise contribution. In terms of these quantities the total system noise is given by

$$U_n = 2kT_a\tau_i/\tau_a + \frac{|Z_{12}|^2}{2M}S_i(\omega)\tau_i + \frac{2M}{|Z_{21}|^2}\frac{S_e(\omega)}{\tau_i}. \tag{4.11}$$

The first term in equation (4.11) is the familiar Brownian motion or thermal noise in the antenna. The second term describes the energy fluctuations arising from the current noise acting back through the reverse transductance, and giving the additional noise contribution proportional to sampling time. The third term is the series electronics noise, which for given S_e is reduced as Z_{21} increases, as well as being proportional to the bandwidth of τ_i^{-1}.

The problem of detecting gravitational waves with resonant-bar antennas to a large extent consists of minimising equation (4.11). The technical means of achieving this requires some or all of the following:

(a)   reducing the antenna temperature T_a,
(b)   using a transducer with high Z_{21} and low Z_{12},
(c)   using amplifiers with S_e and S_i as low as possible,
(d)   reducing the acoustic losses in the antenna to obtain the highest possible Q or relaxation time,
(e)   obtaining a reasonable impedance match between the bar and the transducer $\beta = Z_{11}/Z_{out} \rightarrow 1$.

It is convenient to scale the noise in the system relative to the 'standard quantum limit' of one equivalent quantum induced in the bar. To do this we rewrite the noise equation (4.11) in terms of noise number A (a quantity first used by Weber to characterise noise in masers):

$$A = U_n/\hbar\omega_a = A_T + A_B + A_S. \tag{4.12}$$

Here A_T, A_B and A_S are the equivalent numbers of noise quanta due to thermal noise, back-reaction noise and series noise in the measurement system. The experimentalists' goal is to achieve a total system noise number A approaching unity. To achieve this it is necessary not only to have a low noise transducer, but also to use a low acoustic loss antenna material, and to suspend and isolate the antenna so as not to increase the acoustic loss, nor to couple in excess noise.

4.4   Introduction to transducers

As we saw in chapter 3, there are two basic types of transducer for gravitational wave antennas, *passive* transducers and *parametric* transducers. Passive transducers have no external power source, and their power gain is less than unity. They must always be used with a high gain, low noise amplifier at the frequency of the antenna ω_a. Parametric transducers, on the other hand, have an external power source (a pump oscillator at frequency ω_p) and they have intrinsic power gain.

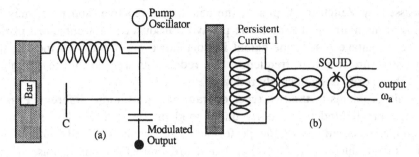

Figure 4.7. (a) Active or parametric transducer. (b) Passive transducer.

The output of a parametric transducer is generally upconverted to a frequency higher than ω_a. Most parametric transducers use high frequency resonant cavities combined with low noise high frequency amplifiers. Passive transducers use an inductive readout, coupled to a SQUID amplifier. The basic principles were described in section 3.4.3, and full details are given in chapters 7 and 8. Here we will discuss the major differences between them. Figure 4.7 illustrates their basic structure. Fundamentally the difference between passive and active transducers is not very great. Active transducers use a transduction process that is combined with amplification. (Additional amplification of the high frequency signal is still necessary.) Passive transducers have a complete separation between the transduction process and the amplification process. However, the amplifier itself (such as a SQUID) makes use of a parametric upconversion process. Thus the difference between passive and active transducers is simply in the choice of whether the parametric upconversion occurs during or after transduction. A laser interferometer is also a parametric upconverter. In this case the pump frequency is high enough that amplification is unnecessary: the entire power gain is realised through the upconversion of the signal frequency to the optical frequency.

One basic difference between passive and parametric transducers is in the transducer impedance mismatch ratio or coupling factor β. For the parametric transducer

$$\beta_{\text{para}} = \frac{\frac{1}{2}CV_p^2 Q_e}{m\omega_a^2 x^2}, \qquad (4.13)$$

but in the limit $Q_e \to \infty$, the factor Q_e is replaced by the frequency ratio ω_p/ω_a.

For the passive inductive transducer

$$\beta_{\text{pass}} = \frac{\frac{1}{2}LI^2}{m\omega_a^2 x^2}, \qquad (4.14)$$

the factor Q_e is replaced by unity. For a capacitive passive transducer $\frac{1}{2}LI^2$ would be replaced by $\frac{1}{2}CV^2$. The parametric transducer effectively samples the incoming signal up to ω_p/ω_a times, and therefore increases its coupling by the same factor.

As discussed by Veitch in chapter 8, the coupling is reactive, and there may be problems in maintaining stability. The passive transducer is not enhanced in this way. The advantage is to some extent illusory however, because by moving the coupling structure to a high frequency, one reduces its size, so that the absolute value of the L or C is significantly reduced.

The main problems of the two types of transducers are quite different. Passive transducers are limited by a poorly understood problem of AC losses in their superconducting circuits, and the performance of available SQUID amplifiers. Parametric transducers, on the other hand, are limited by phase noise in the pump oscillator and tuning difficulties. In principle both types of transducer can reach close to the quantum limit, and in practice it is not obvious which type will ultimately be the most successful.

All transducers, both active and passive, are limited in noise number by the noise number of the amplifier with which they are used. In chapter 8 Veitch summarises amplifier performance (figure 8.2). The data show that good noise number performance can be achieved either in the audiofrequency range, or at microwave frequencies. Thus the choice of a microwave pump frequency $\sim 10^{10}$ Hz for parametric transducers is clear. At this frequency ruby masers, GaAs FET amplifiers and parametric amplifiers all achieve noise numbers in the range 1–20. At the low frequency end of the spectrum superconducting quantum interference devices (SQUIDs) achieve a similar noise number. The SQUIDs are most easily matched to passive inductive or capacitive transducers.

Since SQUID devices can be operated at frequencies ~ 1 MHz, it is feasible to design a parametric transducer pumped at low radio frequencies. Such a transducer can use either inductive or capacitive sensing elements in a resonant LC circuit. Johnson and Bocko (1981, see also Bocko and Johnson, 1982, 1984) have shown that such a transducer pumped at less than 1 MHz and amplified by a DC SQUID can in principle approach the quantum limit, and can exceed it if back-action evasion techniques are used. See Braginsky (1970, 1983, 1985) and Caves *et al.* (1980) for further discussion of these issues.

4.5   Antenna materials

An ideal resonant bar would consist of a piece of nuclear matter, with high density and a velocity of sound comparable to the velocity of light! Since this is not available we must find a form of molecular matter which, to maximise coupling to gravitational waves, combines high velocity of sound v_s, high density ρ, and low acoustic loss Q^{-1}.

At a particular frequency the best antenna material will have the largest value of $Q\rho v_s^3$. This quantity is proportional to the ratio of energy absorbed ($\sim \rho v_s^3$) and the thermal noise in the antenna ($\sim Q^{-1}$). Of the three controlling parameters, only the Q-factor can be modified significantly in a particular material, depending

Table 4.1. *Comparison between antenna materials*

Material	ρ (g cm^{-3})	v_s (km s^{-1})	Q	ρv_s^3 (10^{13} kg s^{-3})	$Q\rho v_s^3$ (10^{20} kg s^{-3})
Aluminium 6061	2.7	5.1	5×10^6	36	18
Aluminium 5056	2.7	5.1	7×10^7	36	250
Niobium	8.57	3.4	2.3×10^8	34	800
Silicon	2.33	8.5	2×10^9	140	2.8×10^4
Sapphire	3.98	9.4	3×10^9	330	10^5
Lead	11.36	1.1		1.5	
Tungsten	18.8	4.3		150	

on its preparation and suspension. Table 4.1 lists the values of ρ, v_s and ρv_s^3 for various materials, along with the maximum achieved Q-value to date, and the figure of merit $Q\rho v_s^3$. The table shows that nearly one order of magnitude improvement is obtained (at a given frequency) in ρv_s^3 by going over from aluminium or niobium, to sapphire, and when the Q-factor is included the very low losses in sapphire make it about 500 times superior to Nb or Al (at a given operating temperature). Silicon is more than 100 times better than Nb and Al. Unfortunately, at present silicon and sapphire are not available in sufficiently large masses for these apparent advantages to be useful. Note also that a lower Q-factor can always be compensated for by sufficient cooling, so that fundamentally only the ρv_s^3 term need be considered.

For comparison, Table 4.1 also shows lead and tungsten. Lead is very poor, because of its low sound velocity, whereas tungsten is comparable to silicon. If massive high Q tungsten bars could be obtained, they would have the significant advantage that the cryogenic system necessary to house the antenna would be far smaller (and cheaper and simpler) than that needed for lower density materials.

4.6 Antenna suspension and isolation systems

Figure 4.8 illustrates a typical cryogenic resonant-bar antenna. The resonant bar is supported by a low loss suspension in an experimental chamber which is itself isolated from vibration. The suspension must satisfy two conditions:

(a) To achieve a noise temperature T_N, the extrinsic quality factor of the antenna (that is, the Q-factor loaded by the suspension and transducer and any other couplings) must satisfy $Q > (T_a/T_N)\omega_a\tau_i$, where T_a is the thermodynamic temperature, ω_a is the frequency of the fundamental resonant mode, and τ_i is the optimum integration time for the system.

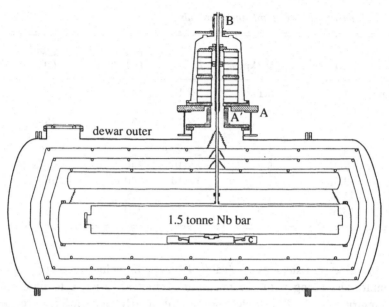

Figure 4.8. Schematic of the University of Western Australia cryogenic resonant-bar antenna. The antenna is supported by a cryogenic suspension and isolation stage indicated by C. The antenna is enclosed by an experimental chamber which is suspended from a nine-pole room-temperature isolation stack. This stack is supported by I beams (A) which rest on columns mounted in concrete footings separate from the laboratory floor. Soft bellows (A' and B) decouple the structure from the dewar outer.

(b)   To prevent noise degradation by ambient noise, the antenna must be isolated from the external environment. The isolation factor required for a given configuration can be determined (Veitch, 1989) by (i) calculating the force exerted on the antenna by the suspension (due to motion of the suspension base), (ii) determining the response of the antenna to the applied force (using the antenna effective mass appropriate to the suspension configuration), and (iii) multiplying the response by the appropriate gain factor to give the vibration of the antenna end-face. This can then be compared with the value due to Brownian motion. As an example, for the University of Western Australia (UWA) four-point suspension the vibration isolation must attenuate ground vibrations at the antenna frequency by $>10^{8.5}$ (170 dB).

Traditionally the isolation is achieved in two stages. The first stage is a room-temperature *isolation stack,* which is a mechanical analogue of a multistage LCR low pass filter. This consists of multiple layers of lead and rubber, or steel and rubber, which can in theory achieve very large isolation factors ($>10^{10}$). The full isolation has never been measured, however, since it becomes unmeasurable with conventional transducer systems. However, it is clear that 10^{10} suppression of noise will never in practice be achieved since self generated thermal noise in the final stages will be dominant. In addition such an isolator cannot be expected to be ideal, since instrumentation wires and residual gas can provide significant vibration paths. It is also worth commenting that laser interferometer groups have discovered anomalous vibration properties in some adhesives (Ward, 1987). There can be no expectation that the complex molecular structure of the rubber isolation material will not have excess noise properties, especially when placed under high compression. The well known problem of outgassing of volatile oils from these elements must also be seen as a possible source of vibration. A detailed high sensitivity experimental investigation of the noise properties of stressed elastic polymers would be of great value in defining the limits of these materials.

The second stage is a *cryogenic suspension and isolation stage*. This stage is designed (i) to isolate against the noise which bypasses the stack and the thermal noise of the stack, and (ii) to suspend the antenna using low loss elements so as to maintain a high antenna quality factor. It is essential that the cryogenic stage, at least, does not have any resonant modes near the antenna frequency. A variety of systems are being used, and it is worthwhile summarising each briefly.

4.6.1 Cable suspension

The first method of suspension, the belly cable, was pioneered by Weber. The bar is balanced on a high strength cable stretched around the longitudinal centre of the bar. Braginsky and Bagdasarov (see Braginsky, Mitrofanov and Panov, 1985) have achieved Q-factors of 3×10^9 in sapphire bars supported in this way by a fine polished tungsten wire. The cable is attached to an intermediate mass which is either supported by additional cryogenic suspension stages and then by a room-temperature isolation stack or directly by a multi-pole room-temperature isolation stack. The part of the cable in contact with the bar presumably is forced to participate in the 'breathing mode' of the bar. In general the motion of the bar will exert a time varying longitudinal force on the cable, which can lead to energy loss through longitudinal acoustic excitation. The energy may be lost in the cable itself, or into the cable suspension point. If the cable hangs exactly vertical, however, the breathing mode has no vertical component at the tangent point, and longitudinal excitation does not occur. Furthermore, if a cable is attached tangentially to a bar (not necessarily vertical) the longitudinal excitation of the cable is second order in the amplitude of the breathing mode, and may well be negligible. The breathing mode always causes a transverse motion at the tangent point of the cable. Thus a transverse acoustic wave always propagates into the cable. To obtain maximum Q-values the Moscow group carefully tuned the length of the suspension wire for their sapphire bar to one-quarter wavelength for the transverse wave at the fundamental longitudinal mode frequency of the bar.

At the lower Q-values associated with room-temperature Al bars, cable suspension appears to be reliable. It is easy to show, however, that a cable suspension can cause low frequency seismic noise to be upconverted via non-linear frictional coupling (Blair, 1982a). The process is analogous to the excitation of a violin string by low frequency bowing. In this case seismic noise at the natural pendulum or rocking frequency of the bar excites static/sliding frictional transitions near the cable tangent point. This gives rise to small strain changes in the bar, which can excite the fundamental mode to temperatures greatly exceeding the thermodynamic temperature. Moreover, occasional large seismic events can cause rare excitations which can easily mimic gravitational wave signals.

A simple analysis gives the following result for the mean acoustic power entering the low frequency longitudinal modes of the antenna as a result of a low frequency amplitude of oscillation of the bar–cable system:

$$P_\mathrm{a} = \frac{16 t^{5/2} \sigma^{1/2} a_\mathrm{r}^2 b \omega_\mathrm{r}}{\omega_\mathrm{a} m^2 l^3} (1/2\mu_\mathrm{s} - \mu_\mathrm{d}). \tag{4.15}$$

Here t is the tension in the cable, σ is the cable mass per unit length, a_r and ω_r are the amplitude and frequency of the rocking motion measured at the end of the bar, b is the effective offset of the cable from the centre of suspension, m and l are the antenna mass and length, respectively, and μ_s and μ_d are the static and dynamic coefficients of friction between the bar and the cable. For typical materials the last term is of magnitude ~ 0.1–0.3. Note that P_a increases quadratically with a_r; for typical antenna parameters one finds that vibration amplitudes ~ 1–$10\,\mu\mathrm{m}$ are sufficient to cause excitation of antennas to mode temperatures $\sim 10^3\,\mathrm{K}$. At the low rocking frequencies of antennas it is very difficult to prevent such motion.

Frictional transitions can be overcome by welding or other forms of rigid contact. All practical antennas take this precaution, but care must be taken to control all possible points of non-linear friction.

4.6.2   Magnetic levitation

Using superconducting materials it is possible to support an antenna by magnetic levitation. Magnetic levitation is a reliable means of obtaining a high Q suspension with good vibration isolation (Mann, 1982). However, it is restricted to bars made from, or coated with, superconductors having a reasonably high critical field Hc_1. The $100\,\mathrm{G}$ critical field of aluminium restricts magnetic levitation to bars less than $1\,\mathrm{cm}$ in diameter. Niobium, with a $1.9\,\mathrm{kG}$ critical field, is restricted to about $20\,\mathrm{cm}$ diameter.

In the early 1970s attempts were made at Stanford to plasma spray NbTi alloy (a high field type II superconductor) onto aluminium bars. Eventually the work was abandoned since high critical fields could not be reproducibly obtained. It is important that the superconducting materials have good stability to flux penetration. In niobium irreversible flux penetration occurs just above Hc_1. In type II superconductors flux penetration can occur far below the upper critical field Hc_2, thus high field superconductors cannot necessarily be used to coat non-superconducting bars.

A magnetically levitated antenna requires a large amount of instrumentation, both to levitate it and to enable its position in space to be accurately known. Superconducting 'pancake coils' do not automatically lead to a stable levitation configuration; thus the superconducting coil assembly generally needs additional trim coils to make adjustments to the position of the bar. Coils are also necessary to control the longitudinal position of the bar.

Much instrumentation, including precision radiofrequency position monitors (Mann and Blair, 1980) and flux pumps (Bernat and Blair, 1975), have been developed to enable accurate levitation control. The superconducting circuit diagram for a 67 kg prototype levitated antenna is shown in figure 4.8. Flux can be pumped into seven different superconducting coils, and transformer-coupled into several others. Small heaters (heat switches) are used to open and close superconducting circuits. A non-contacting transducer is levitated on a bifilar coil which produces a very soft linear bearing with a stiff transverse spring constant. A bifilar coil is also used to provide a stiff magnetic spring between the bar and the transducer, which helps reduce the low frequency motion of the transducer due to seismic noise. (The bar–transducer frequency can be raised to about 40 Hz). A superconducting magnetic drive assembly at the rear of the transducer allow it to be servo controlled and locked to the resonance of the re-entrant cavity.

Magnetic levitation can cause a degradation of the Q-factor of a levitated bar, since the vibration of the bar modulates the levitation field, giving rise to eddy current losses in the copper cladding of the NbTi superconductor and in the metallic structural elements. In practice this effect would limit the Q-value of a Nb bar to 10^8–10^9, depending on bar geometry and the height of levitation. In spite of its success, the critical field limit of magnetic levitation has caused this technique to be largely abandoned in favour of mechanical suspension.

4.6.3 4-Cables

The 4-cable suspension has been used successfully at Stanford. Cables are attached to welded lugs at four points in a horizontal plane, two near each end of the bar. The cables are attached to an intermediate mass which is itself suspended from a room-temperature stack by four more cables. Tuning is once again used to minimise the energy coupling.

The 4-cable system may seem at first sight to be less ideal because the cables couple to both the longitudinal and radial vibrations of the fundamental mode. However, excitation of the fundamental mode via its longitudinal vibration would require coherent excitation of the transverse modes of the cables. There is also less coupling to the radial vibrations of the antenna because the cables are near the ends of the bar.

4.6.4   Four-point suspension

Several groups have developed four-point suspensions, otherwise known as 'dead bugs', which were first developed by Richard at the University of Maryland. The suspension consists of a massive base which supports four flexible legs on which the bar sits. The legs participate in both the longitudinal and the breathing motion of the bar, so it is important that the legs have low acoustic loss, low stiffness in the longitudinal and vertical directions (relative to the bar), and small size and mass. For a specific example we will examine the Catherine wheel suspension developed at UWA, which is shown in figure 4.9.

The first stage of isolation is provided by a commercially pure titanium 'Catherine wheel' (cw) cantilever spring. Since the cw supports the bar at four relatively widely separated points, it must isolate against both longitudinal and radial vibrations. The cw arms are therefore designed to be relatively soft in all

Figure 4.9. Three-mode cryogenic suspension stage.

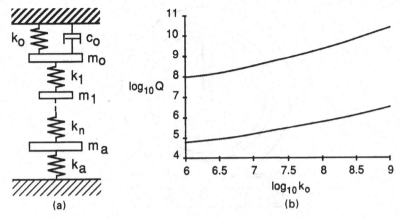

Figure 4.10. (a) Model used to calculate the Q limit imposed on the antenna by a lossy resonance. The quantities m_0, k_0 and c_0 represent the lossy resonance. It is assumed to be critically damped ($c_0^2 = 4m_0k_0$). $k_1 \ldots k_n$ and $m_1 \ldots m_{n-1}$ represent the cryogenic vibration isolation components. k_a and m_a are the equivalent spring constant and mass of the antenna. (b) Graph of the minimum antenna Q which can be obtained for any m_0, versus k_0. The lower and upper curves correspond to $n = 1$ and $n = 2$, respectively. The values used were $k_1 = 2 \times 10^8 \, \text{N m}^{-1}$, $m_1 = 50 \, \text{kg}$, $k_2 = 9 \times 10^6 \, \text{N m}^{-1}$, $m_a = 3.9 \times 10^6 \, \text{kg}$ and $k_a = 7.8 \times 10^{13} \, \text{N m}^{-1}$.

directions. This stage must have low acoustic losses since it is closely coupled to the antenna. If the lowest resonance of the cw arms is about 1 kHz then it must have a $Q > 10^3$ if the antenna Q is to be limited to 2×10^8 (assuming that there are no resonances at the antenna frequency in the isolation system). This stage by itself does not, however, provide sufficient isolation to guarantee the antenna Q since a critically damped loss in the experimental chamber could, as shown in figure 4.10, degrade the Q to $\sim 10^5$ (Veitch, 1989).

The second stage provides the additional isolation to prevent this Q degradation (see figure 4.10). It consists of a 50 kg mild steel mass which is supported by tapered aluminium cantilevers which are themselves supported by knife-edges. The cantilevers isolate the radial vibrations while the lateral flexibility of the knife-edges provides isolation in the longitudinal direction. The acoustic loss requirements of the second stage are much reduced over those of the first stage, but it is still important that this stage does not have any modes resonant at the antenna frequency. This stage, while providing sufficient isolation (105 dB) to enable an antenna $Q > 10^8$, does not provide sufficient isolation to prevent the antenna from being excited by ambient noise. To prevent this excitation a third stage is added to the suspension. The total isolation provided by this three-mode suspension is about 135 dB.

4.6.5 Nodal point suspension

The nodal point suspension for resonant bars has been proposed many times. This involves cutting a hole to the nodal point near the centre of mass of a

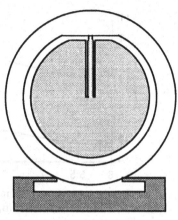

Figure 4.11. Schematic of a possible nodal point suspension. The antenna is supported from its central node by a cylindrical rod formed by boring an annular hole into the antenna. The rod is attached to a massive ring, which is supported by cantilevers.

resonant bar. A possible configuration is shown in figure 4.11. An annular hole decouples the surface of the bar from the fundamental longitudinal mode.

The difficulty is that strength limitations require a minimum area of contact; thus full decoupling is not achieved. Secondly, the attachment of a transducer displaces the nodal point from the centre of mass. This may necessitate mass loading both ends of the antenna to maintain symmetry. Thirdly, crystalline anisotropy may also displace the nodal point from the centre of mass. In spite of these problems the nodal point suspension has a simplicity and elegance which is very tempting, except for the fact that it is not reversible or easily modified. A nodal suspension is under active investigation by the Rome group (Pizzella, 1989).

4.6.6   Vibration isolation at room temperature

The suspension systems described above give a significant vibration isolation, since a system which minimises the degradation of the Q-factor of an antenna must also isolate against incoming noise at the same frequency. Further isolation is still essential, however, and traditionally, following Weber, bars have been isolated by stacks of steel and rubber that form a multi-pole low pass filter. The steel (or lead or concrete) itself is small enough (say 20 cm diameter × 4 cm thick) that its internal modes are well above the antenna frequency. Sheets of commercially available vibration isolation rubber are placed between the steel masses. Figure 4.8 shows a cross section of such a stack on the Perth antenna. Three stacks using 10 kg lead discs support a 50 cm diameter steel disc from which hangs the experimental chamber which contains the 1.5 tonne Nb antenna. Figure 4.12 shows a typical transfer function for the isolation system.

In a cryogenic system it is essential to isolate the antenna from the vibration caused by boiling cryogenic fluids. This leads to complex cryostat design as

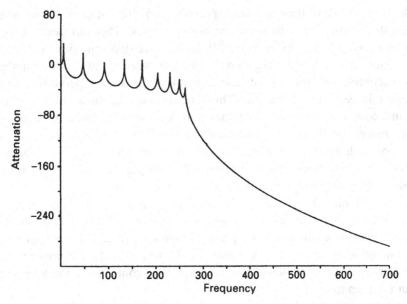

Figure 4.12. The transfer function for the lead and rubber anti-vibration stack used to isolate the 1.5 tonne antenna at the University of Western Australia (attenuation in dB).

described in chapter 5. In most cryostats the bar suspension system is carefully isolated from the cryogenics. However, electrical leads must be thermally grounded to prevent excess heat from entering the antenna or transducer. Thus some noise in the leads is inevitable. Most groups have used 'Tabor isolator' filters, first developed at Stanford, consisting of chains of masses hanging from each other on fine wires to attenuate noise in the cables (see chapter 5 for more details).

Possibly the most serious vibration source is *low frequency* seismic or environmental noise. Most anti-vibration stacks have a roll off frequency of 10–100 Hz. Below this frequency seismic noise can be enhanced. The low frequency noise can excite non-linear processes, which, as we saw earlier, can lead to strong excitation at the signal frequency. Stress-induced stress relaxation in the antenna itself is one such process which could cause excitations correlated with low frequency accelerations of the antenna. The Guangzhou group has used a massive concrete platform supported by air-springs to obtain roll-off frequencies as low as 0.5 Hz. Even lower frequency isolation may be necessary to obtain sufficient environmental immunity.

4.7 Excess noise and multiple antenna correlation

Besides seismic noise excitation, all cryogenic antennas have shown evidence of excess noise of indeterminate origin. Low noise performance may be achieved for considerable periods of time; but interspersed are periods of excess noise which is not identifiably correlated with known noise sources. This can degrade performance by many orders of magnitude. While there are opportunities for non-linear upconversion driven by low frequency pendular modes, and for thermal stress driven excitation as regions of the cryostat vary in temperature, no firm correlation is generally apparent. Thus the seismic, acoustic, electromagnetic pulse and cosmic ray shower detectors that are generally used to discriminate possible noise signals are not sufficient to eliminate excess noise, and much careful work still needs to be done. It has been shown that at a noise temperature of ~1 mK about 60 muon shower events will occur per day, and this increases to more than 10^4 events per day at the $1\,\mu$K noise sensitivity.

Multiple antenna correlations can minimise the effects of excess noise, as demonstrated by the coincidence analysis recently reported by Hamilton (1987), which sets new lower limits on the gravitational wave flux. However, as we shall see below, greatly improved results will be achieved when four or more antennas of the highest possible bandwidth are operated in coincidence for long periods without interruption.

Candidate gravitational wave events consist of either unknown environmental perturbations, occasional rare Gaussian high energy excursions, and possibly gravitational wave signals. These may be idealised as an independent set of background events, occurring at a constant rate R per unit time. There is evidence that the background events are not entirely independent but to some extent are clustered. However, as antenna suspension and isolation is improved this will be a better and better approximation.

The probability of a background event in one antenna during the antenna resolving time τ_i (which is generally the optimum sampling time) is given by

$$P_1 = R\tau_i. \tag{4.16}$$

Now if there are N independent antennas, the probability of accidental coincident excitation of all N antennas in the same sampling time interval is

$$P_N = \tau_i^N \cdot \prod_{i=1,N} R_i, \tag{4.17}$$

and if all antennas experience the same background at the rate R,

$$P_N = R^N \tau_i^N. \tag{4.18}$$

That is, the probability per unit time P_N of an accidental coincidence is

$$P_N = R^N \tau_i^{N-1}. \tag{4.19}$$

It is useful to express this result numerically. Consider a range of optimum integration times from 1 to 10^{-2} s, and $N = 1$–4.

A realistic value for R is one event per 100 seconds (10^3 events per day). Table 4.2 summarises the resulting probabilities. To detect rare events we should be looking for $P_N < 10^{-8}$ (one accidental event per three years).

Table 4.2. *Probability of accidencal coincidence for $R = 10^{-2}\,s^{-1}$.*

Number of antennas	τ_i	1 s	0.3s	0.1s	0.01 s
1	P_1	10^{-2}	10^{-2}	10^{-2}	10^{-2}
2	P_2	10^{-4}	3×10^{-5}	10^{-5}	10^{-6}
3	P_3	10^{-6}	10^{-7}	10^{-8}	10^{-10}
4	P_4	10^{-8}	3×10^{-10}	10^{-11}	10^{-14}

It is clear that $P_N < 10^{-8}$ can be achieved by four narrow band detectors ($\tau_i = 1\,\text{s}$), or three detectors with $\tau_i = 0.1\,\text{s}$, but it cannot be achieved by a pair of detectors even if τ_i is reduced to $10^{-2}\,\text{s}$, unless R is reduced to less than $10^{-3}\,\text{s}^{-1}$. Table 4.2 emphasises that multiple antenna operation is essential to reduce the background coincidence rate.

Note that if the antennas have different resolving times, the τ_i used in equation (4.19) is generally the longest resolving time. If the candidate events are non-Gaussian (in energy distribution) and non-Poisson (in time distribution) it is likely that accidental coincidences will have a higher probability than estimated here.

To assess the probability of multiple antenna coincidences we must investigate the antenna pattern of a set of antennas on the globe. We consider the case for the four antennas at Perth, CERN, Louisiana and Stanford. In their present orientations all are placed in a north–south (N–S) direction (within about 1°), except for the antenna at CERN which is oriented east–west (E–W). Since all antennas are horizontal, their orientation with respect to the sky is largely determined by their various locations on the Earth, and considered as a whole this leads to a complex antenna pattern. Moreover the antenna pattern is also dependent on the data processing and detection criterion used, which itself depends on background noise. Antenna patterns for one and two resonant-bar antennas have been analysed previously by Frasca (1980) and Nitti (1980), while Schutz and Tinto (1987) have analysed antenna patterns for pairs of laser interferometers.

The presence of local sources of background noise lead to the minimum detection requirement that a two antenna zero time delay coincidence be observed. With unknown source direction there can be no constraint based on the relative signal sizes. However, with four antennas it is possible to consider much stronger detection criteria, the strongest of which is the presence of four antenna coincident events with relative amplitudes consistent with a plane gravitational wave of a particular polarisation originating from some source direction. Unfortunately the present generation of resonant-bar detectors have insufficient time resolution to be able to use phase information.

The angular dependence of the signal S observed in a single resonant bar antenna is given by

$$S(\theta, \phi, \varepsilon) = (0.5(1 - \varepsilon) + \varepsilon \cos^2 2\phi) \sin^4 \theta, \qquad (4.20)$$

where θ is the angle of the incoming plane wave relative to the cylinder axis of the antenna, and ϕ is the polarisation angle of the wave measured relative to the plane of the antenna and the source. The polarisation fraction ε measures the fraction of linear polarisation of the wave. For $\varepsilon = 0$ the wave is circularly polarised, whereas for $\varepsilon = 1$ the wave is 100% linearly polarised, with polarisation angle ϕ.

The antenna sensitivity can be expressed in geodetic coordinates relative to a source of given hour angle α and declination d, using trigonometric expressions given by Tyson and Douglass (1972). Two separate criteria can be used for analysing the four-antenna array. The first consists of a measure of the signal product for all antennas in the array, or for a subset of them. The four-antenna signal product function, P_4, has the advantage of being analytic at the expense of imprecision in specification of the individual antenna signals:

$$P_4 = \prod_{i=1,4} S_i. \qquad (4.21)$$

To measure three- and two-way coincidences we define the functions P_3 and P_2 as follows:

$$P_3 = P_4 / S_{m1} \qquad (4.22)$$

$$P_2 = P_4 / S_{m1} \cdot S_{m2}, \qquad (4.23)$$

where S_{m1} is the smallest and S_{m2} is the second smallest of $\{S_i\}$.

Since the antennas considered here are expected to have roughly equal sensitivity, at about the same frequency, 700–900 Hz, we make the assumption that all S_i have a maximum value of one when the antenna is optimally aligned relative to a source direction. Thus P_N all have a maximum value of 1.0. A value of $P_4 = 1$ implies all antennas optimally aligned, while $P_4 = 0.1$ implies a range of S_i such as (0.1, 1.0, 1.0, 1.0) to (0.6, 0.6, 0.6, 0.6). Since signals can only be expected to appear marginally above the noise (otherwise unambiguous detection would have already occurred) reasonable thresholds for four-way coincidences vary from 10^{-1} to 10^{-4} corresponding to equal signals in each antenna varying from 0.6 to 0.1.

Since the product functions allow S_i to occupy a range of values (some of which would be buried unmeasurably in the noise), it is useful to define a second criterion, defined by the locus of point L_N for which four, three or two antennas are above a specified threshold.

The loci L_N are useful in determining the fractional sky coverage by four, three or two antenna coincidences, and the dependence of sky coverage on threshold. The locus criterion is useful when antenna signals are analysed independently to obtain lists of possible events. However, they do not make maximal use of the data available. An optimal filter for multiple antennas would take into account both Gaussian and non-Gaussian noise components, and would be expected to see significantly below the noise thresholds implied by use of the locus criterion. In practice L_N and P_N give very similar results for fractional sky coverage, although the antenna patterns generated differ qualitatively in structure.

The antenna patterns for P_N and L_N can be plotted as a map, and because the four-antenna array is symmetrical with respect to wave propagation direction, a half-sky map suffices. Figure 4.13 shows an example of P_4 and L_4 contour maps in the case of circularly polarised radiation. The map is most usefully expressed in sky coordinates for a given universal time, thus expressing the areas of sky to which the antenna is sensitive. The antenna pattern clearly rotates on the sky with a sidereal day period.

The fractional sky coverage C_N can be obtained for P_N as follows:

$$C_{N,P} = \frac{1}{4\pi} \iint_{\alpha\beta} L_N \, ds \qquad N = 2,4. \tag{4.24}$$

Similarly $C_{N,L}$ can be defined for the locus criterion L_N.

An analysis of antenna patterns for the present geographical locations of antennas (Blair, Frasca and Pizzella, 1988) shows that four antennas are sufficient to obtain near 100% sky coverage for two-way coincidences. That is, if we are content with only two antennas being suitably aligned for a random source, we can observe practically 100% of the sky. On the other hand, if we are to demand four-antenna coincidences, then we require the antenna orientations to be adjusted such that they optimally search the same part of the sky, and the sky coverage is reduced to about 50%. With eight operating antennas one can achieve near 100% sky coverage and a minimum of four antennas in coincidence for any one event.

4.8   Quantum non-demolition and back-action evasion

We have seen that for a sufficiently low acoustic loss a resonant bar can have negligible thermal noise, but the measurement noise is limited directly by the uncertainty principle. The coordinates X_1 and X_2 are conjugate variables, so that an ideal measurement is limited by an uncertainty relation $\Delta X_1 \Delta X_2 \geq \hbar/2m\omega_a$.

The standard quantum limit arises because the assumed linearity of the transducer means that it measures the X_1, X_2 coordinates symmetrically. This symmetry is characteristic of almost all linear measurement devices, and as a consequence inflicts a minimum measurement uncertainty of the order of one quantum on whatever system is measured. Thus $\Delta X_1 = \Delta X_2 \geq (\hbar/2m\omega_a)^{1/2}$. Most experimentalists are quite familiar with a device that does not have this linear property, however. This is the lock-in amplifier or phase sensitive detector. It has strongly phase sensitive gain and thus will measure only the chosen quadrature on

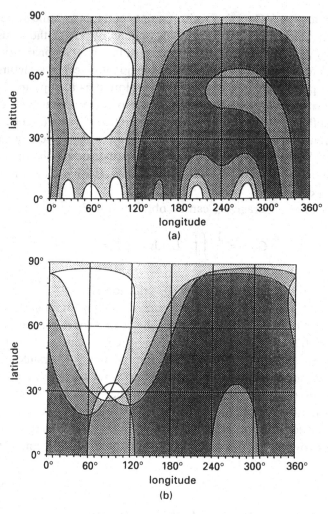

Figure 4.13. Typical antenna patterns for four existing antennas on the Earth's surface. (a) Contour map for product P_4 with circularly polarised radiation. Antenna orientations: LSU and Stanford, N–S; UWA and CERN, E–W. Shading density indicates higher values of P_4. Highest density $P_4 > 10^{-2}$, intermediate density $10^{-3} < P_4 < 10^{-2}$, light density $10^{-4} < P_4 < 10^{-3}$, unshaded $P_4 < 10^{-4}$. (b) Contour map for loci L_2, L_3, L_4, for circularly polarised radiation, and threshold 0.05. Orientations: Stanford, LSU and UWA, N–S; CERN, E–W. Highest density in the L_4 locus. Intermediate density in the L_3 locus (which includes the L_4 locus), and low density in the locus for two-antenna coincidences L_2 (which includes L_3 and L_4).

the $X_1 X_2$ plane. But it is not a back-action evading device, since the phase sensitivity is not intrinsic to the measurement (which may be performed by a linear transducer or linear amplifier) but is simply part of the post measurement signal processing.

A back-action evasion device transfers this phase sensitivity to the front end. A lossless transducer with phase sensitive transductance (obtained by modulating the electric field that couples the transducer to the antenna) will have sensitivity to one quadrature of the signal only, while being insensitive to the other. With appropriate design back-action noise can feed back only into the quadrature to which the transducer is not sensitive. Since quantum mechanics only sets a limit on the product $\Delta X_1 \Delta X_2$, one can obtain arbitrarily high sensitivity to X_1 at the expense of X_2. Because the gravitational wave is a classical wave, one can build two identical detectors and arrange in one to measure X_1 to high precision, while in the other X_2 is measured. Then the product $(\Delta X_1)_A (\Delta X_2)_B$ can be arbitrarily small.

For antennas with thermal noise number $A_T > 1$ there is very little advantage to such schemes since all one achieves is more sensitivity to the limiting thermal noise. It is possible that future high Q low temperature antennas will benefit from BAE techniques, however. Meanwhile experimentalists are offered the opportunity to exercise their minds in devising practical designs for the proposed devices. This could at least lead to improved designs for conventional transducers.

A typical equivalent circuit is illustrated in figure 4.14. A balanced and lossless transformer couples a modulated pump signal to the transducer. The device is described as a double pumped parameter upconverter. The two pump sidebands are obtained by modulating the carrier ω_p at frequency ω_a, and removing the carrier to obtain a signal $A \sin \omega_p t \cos \omega_a t$, which consists simply of two sidebands $\omega_p \pm \omega_a$. This pump signal is coupled across a split capacitor, which is part of a high Q resonator. The transformer which couples in the pump must avoid losses from the resonator. The importance of this resonator is that it acts as a filter to allow power flow at the frequency ω_p which now contains signal information X_1, but prevents power flow at frequencies $\omega_p \pm 2\omega_a$ which contains unwanted X_2 information. One of the first approaches to a practical design was proposed by Johnson and Bocko (1981).

Due to the difficulties in designing a lossless transformer, a simpler approach to

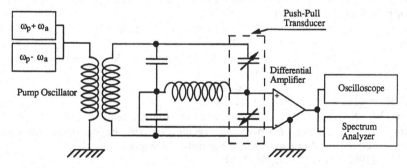

Figure 4.14. The balanced variable capacitors are modulated in opposition by the motion of a transducer element. The other two capacitors balance the bridge. (From Onofrio, Rapagnani and Ricci, 1986.)

BAE transducers might consist of using an existing microwave cavity transducer and modulating the pump signal. This suggestion exposes some of the pitfalls, however. For the transducer to work, the cavity Q must be high enough to significantly reject unwanted power flow at $\omega_p \pm 2\omega_a$. This in turn means that phase and amplitude noise in the pump signal will be significantly degraded, since the cavity is modulated off-resonance. The Lorentzian response of the cavity enhances pump noise at the signal frequency ω_p, while attenuating the pump power at frequency $\omega_p \pm \omega_a$ (see Blair, 1982b). It follows that pump noise sets severe limits on the maximum achievable sensitivity in single cavity BAE transducers. There is an optimum Q-factor above which sensitivity is degraded due to the pump noise. As a result the single cavity approach is unlikely to enable large increases in sensitivity. Hamilton (1982) has demonstrated that the single cavity BAE scheme does indeed work in a high noise classical limit. However, much work and new ideas are still needed before a transducer on a resonant bar exceeds the standard quantum limit.

References

Bernat, T. P. and Blair, D. G. (1975). *Rev. Sci. Inst.* **46**, 582.

Blair, D. G. (1980). In *Gravitational Radiation, Collapsed Objects and Exact Solutions*, ed. C. Edwards, p. 299, 314, Springer-Verlag, Berlin.

Blair, D. G. (1982a). In *Proceedings of 2nd Marcel Grossmann Meeting on General Relativity*, ed. R. Ruffini, p. 1125, North Holland, Amsterdam.

Blair, D. G. (1982b). *Phys. Lett. A* **91**, 197.

Blair, D. G., Frasca, S. and Pizzella, G. (1988). *II Nuovo Cimento* **11**, 185.

Bocko, M. and Johnson, W. W. (1982). *Phys. Rev. Lett.* **48**, 1371.

Bocko, M. and Johnson, W. W. (1984). *Phys. Rev. A* **30**, 2135.

Braginsky, V. B. (1970). *Physical Experiments with Test Bodies*, NASA Tech. Trans. F672 NTIS, Springfield.

Braginsky, V. B., Vorontsov, Y. I. and Kalili, F. Y. (1978). *Sov. Phys. JETP, Lett.* **27**, 276.

Braginsky, V. B. (1983) In *Gravitational Radiation*, eds. N. Deruelle and T. Piran, p. 387, North Holland, Amsterdam.

Braginsky, V. B., Mitrofanov, V. P. and Panov, V. I. (1985). *Systems with Small Dissipation*, University of Chicago Press.

Caves, C. M., Thorne, K. S., Drever, R. W. P., Sandberg, V. D. and Zimmerman, M. (1980). *Rev. Mod. Phys.* **52**, 341.

Frasca, S. (1980). *Il Nuovo Cimento* **3C**, 237.

Giffard, R. P. (1976). *Phys. Rev. D.* **14**, 2478.

Hamilton, W. O. (1982). Lecture presented at Les Houches Summer School.

Hamilton, W. O. (1987). *International Symposium on Experimental Gravitational Physics*, ed. P. F. Michelson, World Scientific, Singapore.

Heffner, H. (1962). *Proc. I.R.E.* **50**, 1604.

Johnson, W. W. and Bocko, M. (1981). *Phys. Rev. Lett.* **47**, 1184.

Mann, A. G. and Blair, D. G. (1980). *Cryogenics* **20**, 645.

Mann, A. G. (1982). Ph.D. Thesis, University of Western Australia.

Nitti, G. (1980). *Il Nuovo Cimento* **3C,** 420.

Onofrio, R., Rapagnani, R. and Ricci, F. (1986). Nota Interna no. 872, Universita di Roma Dip. Fisica.

Pizzella, G. (1989). *Proceedings of the 5th Marcel Grossman Meeting,* eds. D. G. Blair and M. J. Buckingham. World Scientific, Singapore.

Schutz, B. F. and Tinto, M. (1987). *Mon. Not. R. Astr. Soc.* **224,** 131.

Tyson, J. A. and Douglass, D. H. (1972). *Phys. Rev. Lett.* **28,** 991.

Unruh, W. G. (1978). *Phys. Rev. D* **18,** 1764.

Veitch, P. J. (1989). Preprint, University of Western Australia.

Ward, H., Hough, J., Kerr, G. A., MacKenzie, N. L., Mangan, J. B., Meers, B. J., Newton, G. P., Robertson, D. I. and Robertson, N. A. (1987). *International Symposium on Experimental Gravitational Physics,* ed. P. F. Michelson, pp. 322–7, World Scientific, Singapore.

5
Gravity wave dewars

WILLIAM O. HAMILTON

5.1 Introduction

The design and construction of the cryogenic apparatus for a cooled resonant bar gravity wave detector is one of the more interesting and challenging tasks in low temperature physics. The designer is confronted with the requirement that a large mass must be cooled economically, that it must be kept at low temperature for an extended period of time, typically a major fraction of a year, and that the mass must be extraordinarily well isolated from external vibration, including the vibrations of the cryogenic system itself. The second generation detectors, of which we will speak here, characteristically have been operated at temperatures of 4 K. The third generation detectors are operated with the gravity wave antenna at 50 or 100 mK. Since these detectors are just being built as this article is written, their design will not be included. We will use the dewars constructed for the LSU and Stanford experiments as our design example in what follows (Hamilton, 1982). The joint problem of vibration isolation and thermal stability is the central design challenge. Figure 5.1 shows the cryogenic layout. Details of the support system have not been included. Figure 5.2 is the schematic of the antenna mechanical support. A perspective drawing of the support is shown in figure 5.3.

It must also be noted that all of the design considerations must be considered as a package, that is, the thermodynamic considerations are also driven by some of the factors that affect the mechanical design. Thus in our discussion we will separate the mechanical and thermodynamic considerations for convenience of organization. This is a purely arbitrary separation that must be ignored when design of a similar package is undertaken.

5.2 Thermodynamic considerations

The enthalpy of aluminium (Johnson, 1960) is shown in figure 5.4. The graph shows the enthalpy for a gram of material, whereas the usual antenna is an aluminium bar with a mass of over 2000 kg. The LSU antenna, for example, has a mass of 2296 kg. That means that the cryogenic system must be designed to

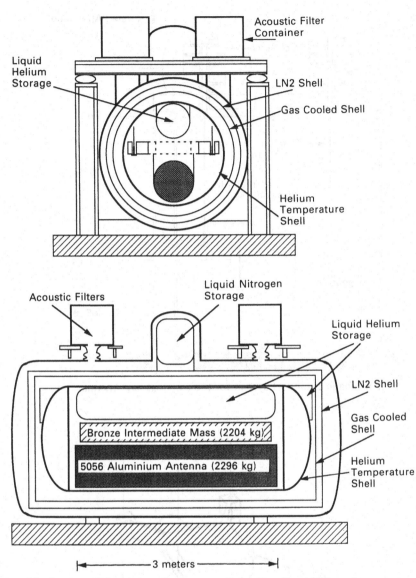

Figure 5.1. Layout of the cryogenic system at Louisiana State University.

remove more than 4×10^8 J in a reasonable time. This is no small requirement. We show the theoretical time constant for the LSU antenna in figure 5.5(a). This is the longest time constant that arises in the solution of the equations for the temperature of an aluminium bar originally of uniform temperature subjected to a sudden change of temperature on its surface (Carslaw and Jaeger, 1959). This graph shows the time constant for attainment of internal equilibrium in the antenna. We have calculated this graph using the measured properties of the aluminium antenna only.

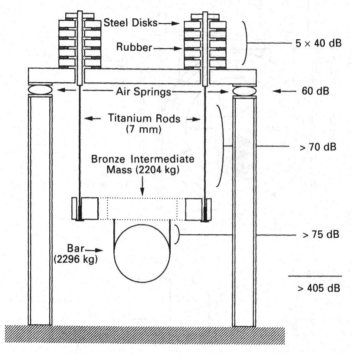

Figure 5.2. The mechanical support system for the resonant bar antenna.

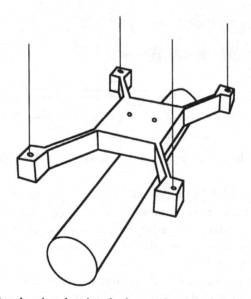

Figure 5.3. Perspective drawing showing the bronze intermediate mass and the aluminium bar.

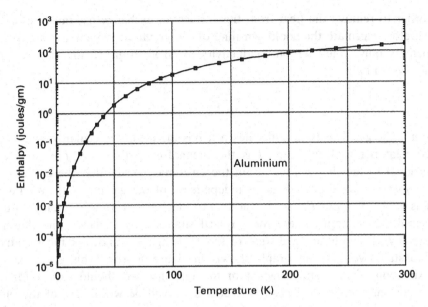

Figure 5.4. The enthalpy of aluminium.

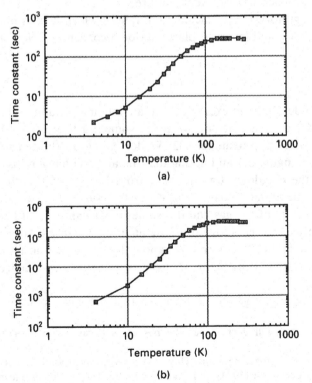

Figure 5.5. (a) The intrinsic thermal time constant of the aluminium antenna at Louisiana State University. (b) The thermal time constant of the antenna limited by exchange gas conduction.

In order to remove the heat from the surface an exchange gas must be used. We therefore calculate the cooldown time of the gravity wave antenna using gas conductivity only. The thermal conductivity of an ideal gas in the viscous flow region is given by

$$\lambda = \frac{n\bar{v}cl}{3},$$

where n is the gas density in molecules/m^3, $\bar{v}$ is the mean molecular speed, c the specific heat per molecule, and l is the mean free path. Since l is inversely proportional to the density, we attain the customary result that the amount of heat removed by an exchange gas is independent of the gas pressure. We know that $\bar{v}$ is proportional to $T^{1/2}$ so the thermal conductivity drops as the temperature is lowered. If we approximate the physical situation by replacing the thermal conductivity of aluminium with that of the exchange gas and plot the resultant time constant we get the graph shown in figure 5.5(b). This gives us an approximation of the time constant of the response we should expect for an antenna during cooldown. Recalling that a body will be within 2% of the final temperature after a time equal to five time constants we estimate that it will take about 17 days to reach nitrogen temperature.

It is also straightforward to calculate the rate at which heat is removed from a cylinder of radius r_1 and length L placed inside a concentric cylinder of radius r_2:

$$\dot{q} = \frac{2\pi L\lambda}{\ln(r_2/r_1)}\,\Delta T.$$

Using the LSU dewar as an example, with an antenna of radius 30 cm suspended inside a cylinder of radius 75 cm, we may estimate a heat removal rate, due to conduction only, of approximately 320 W depending on the temperature of the gas. Referring to figure 5.1 and noting that the antenna has a mass of 2296 kg we estimate that the cooldown time to 77 K from room temperature for only the antenna should be about 13 days. This estimate assumes that we depend only on thermal conductivity of the gas and it assumes a constant rate of heat removal.[*]

We can improve the cooldown time by adding enough gas so that convection currents can aid the heat transfer.[†] We know that the heat transfer coefficient for convection is pressure dependent, though the pressure dependence is relatively weak.

Assuming the helium gas to be ideal, so that the gas expansion coefficient is

---

[*] Of course the rate will decrease as the bar cools because the temperature gradient decreases.

[†] I have chosen an engineering approach to the convection problem. Many examples are worked out in Chapman (1974) and Thomas (1980). The reader is referred to these or other texts to get the correct units on the natural convection equations. Thomas's chapter 7 has the correct form for the average Grashof and Nusselt numbers which combine to give the heat transfer coefficient.

Table 5.1. *Convection contribution to heat transfer[a] (last column calculated for the LSU dewar).*

Pressure (torr)	$T_s - T_f$ (K)	ΔT (K)	$\bar{h}$ $\mathrm{Wm^{-2}\,K^{-1}}$	$\dot{q}$ (W)
12	100	100	0.26	150
12	200	200	0.4	400
760	200	200	2.8	3200

[a] We use these figures for helium: $C_p = 5.2\,\mathrm{Jg^{-1}\,K^{-1}}$, $\mu = 1.566 \times 10^{-5}\,\mathrm{kg\,m^{-1}\,s^{-1}}$, $\lambda = 0.1177\,\mathrm{W\,m^{-1}\,K^{-1}}$ and $Pr = 0.694$.

simply the inverse of the temperature, we calculate first the Grashof number for flow around a horizontal cylinder:

$$Gr = \frac{g\beta D^3 \rho^2}{\mu^2}(T_s - T_f),$$

where g is the acceleration of gravity, β is the expansion coefficient, D is the diameter of the cylinder, ρ is the density of the helium, μ is the helium dynamic viscosity, and T_s and T_f are the temperature of the convecting surface and the helium, respectively. The viscosity μ is that usually calculated in a kinetic theory course and is approximately pressure independent.

We next calculate or look up the Prandtl number for helium:

$$Pr = \frac{\mu C_p}{\lambda},$$

and we take the product of the Grashof number and the Prandtl number to obtain the Rayleigh number for the flow: $Ra = (Gr)(Pr)$.

Finally we assume laminar flow and use the recommended empirical relation for the heat transfer coefficient:

$$\bar{h} = 0.53\frac{\lambda}{L}(Ra)^{1/4},$$

where the units for $\bar{h}$ are $\mathrm{W\,m^{-2}\,K^{-1}}$ and L is the antenna length. The actual heat transferred by convection is obtained by multiplying $\bar{h}$ by the area of the antenna and by the temperature difference between the hot and cold surfaces.

The range of convection transfer rates may be seen in table 5.1. We have used the pressures from the 1988 LSU run for the lower of the figures and have demonstrated the entire range by also assuming an exchange gas pressure of 1 atmosphere when starting the cooldown. (In most experiments the density of the exchange gas is a constant because gas is not added after the cooldown starts.)

For the exchange gas pressure used by LSU in 1988 the cooldown time was reduced to approximately 11 days.

We can further promote convection in the region surrounding the antenna by arranging to have a cold container above the antenna. The additional internal storage container in the LSU and Stanford dewars serves that purpose and makes a marked improvement in the cooldown rate.

We must not ignore the effects of radiation heat transfer. While the antenna is near room temperature radiation heat transfer can result in an additional 100 W of cooling power. Because of the dependence on the fourth power of the temperature, however, this mechanism soon becomes ineffective, but it is important while the antenna is hot and has a lot of enthalpy to lose.

The dewar for a gravity wave detector must have a relatively long hold time or must be designed so that replenishment of the cryogenic coolants does not excite any vibrational modes of the antenna or its support system. All groups that have constructed cryogenic detectors have reported that the vibrations associated with filling have had a tendency to excite some of the support modes of the system. To minimize this the LSU and Stanford detectors have been constructed with enough capacity that they need to be refilled only once every three or four weeks.

A large system can have an excessively high helium consumption if the insulation is at all lacking in efficiency. From the diagram of the dewars shown in figure 5.1 we see that the LSU and Stanford systems have a liquid nitrogen cooled shell, a gas cooled shell and a liquid helium cooled shell. Each of these shells is surrounded with a number of layers of aluminized mylar superinsulation.* The LSU antenna has 125 layers of insulation between the room temperature shell and the nitrogen cooled (77 K) shell, 30 layers between the nitrogen shell and the gas cooled (20 K) shell, and five layers covering the liquid helium cooled (4 K) shell. Using the figures given in White (1987) it is easy to see that this will reduce the radiation heat leak to a very small level (see also Scott, 1959; Timmerhaus, 1960). For engineering estimates one may estimate that superinsulation between room temperature and nitrogen temperature has a thermal conductivity of approximately 10^{-4} W m^{-1} °C^{-1}.

The insulating vacuum for a dewar of the LSU–Stanford design should be at least as good as 1×10^{-6} torr when the dewar is in operation. This figure is purely empirical because with superinsulation, as noted earlier, it is difficult to calculate performance accurately. To the low temperature physicist this pressure requirement does not seem terribly stringent but the size of the dewar leads to a tendency for one's intuition to be incorrect.

Because of the large area of the shells (the area of the LSU helium temperature shell is approximately 19.5 m^2) the helium consumption is very sensitive to the residual gas pressure in the insulating vacuum. The superinsulation represents a

---

* NRC-2 type crumpled mylar insulation, aluminized on one side, produced by King Seeley Thermos Co., Winchester, Mass.

Table 5.2. *Partial pressures in the insulating vacuum (torr).*

	Ion gauge	RGA total	N_2	O_2	H_2O
77 K	6.6×10^{-6}	5.4×10^{-6}	2.4×10^{-6}	4.9×10^{-7}	1.7×10^{-7}
4.2 K	1.5×10^{-6}	4.6×10^{-7}	1.6×10^{-7}	3.4×10^{-8}	1.2×10^{-7}

huge surface area which can trap a large amount of gas which is subsequently released into the insulating vacuum as the experiment runs. The vacuum must be made as good as possible before any coolant is added to the system or the resultant virtual leak will cause a large helium boiloff. In the LSU 1988 run the insulating vacuum was maintained at 1.6×10^{-6} torr as measured by an ionization gauge on the top of the dewar.

The helium boiloff was $1.2 \, l \, h^{-1}$. A residual gas analyzer[†] showed that the gas in the insulating vacuum had all of the characteristics of ordinary atmospheric air. Table 5.2 shows the partial pressures of the gas measured early in the run. If, however, the partial pressure of helium was increased slightly by allowing gas to diffuse into the space around the 4 K shell (from a small leak in a line that was ordinarily kept pumped out) the boiloff increased dramatically. The increase was much more than the indicated increase in the helium partial pressure. The boiloff could be doubled with only a small increase in the indicated partial pressure. The lesson to be learned from these figures is that simple cryopumping is relatively ineffective in reducing gas in the insulating vacuum if superinsulation is present. With the advent of relatively inexpensive cryogenic refrigerators it would be well to reconsider the effectiveness of multi-layer insulation. Fewer layers to reduce the trapped gas might result in less expensive operation, with cryogenic refrigerators being used to keep the shells cold.

Another consideration for a gravity wave detector is that the temperature should be as constant as possible while the apparatus is in operation. Since a gravity wave appears as a strain in the antenna, and since temperature changes result in strains due to differential thermal expansion, the lowest noise detector will also be that detector which is the most stable. The design of the LSU and Stanford detectors was predicated on keeping the antenna at as stable a temperature as possible. The cryogenic storage was kept as high on the shell as possible so that there would be little change in the environment of the antenna when the helium level changed. Both detectors eventually included an internal tank above the antenna so that the surface facing the antenna would always be at 4 K as long as helium was in the dewar. The counter example is the Rome antenna at CERN which is built so that the thermal environment around the antenna changes considerably as the helium level changes. The noise performance

[†] VG ARGA 1-80 amu; VG Instruments, 1515 Worthington Ave., Clairton, PA 15025.

of the Rome antenna shows a periodicity which is well correlated with the dewar's helium level.

5.3   Mechanical considerations

The Stanford and LSU antennas are constructed, as much as possible, from non-magnetic materials, primarily aluminium. At the time of original design it was felt, probably incorrectly, that it was crucial for the apparatus to be non-magnetic. This is an expensive choice because of the difficulty in obtaining good vacuum welding of aluminium. A better choice would have been to use non-magnetic stainless steel. The Rome dewar is constructed with stainless steel in the low temperature environment but the vacuum container is made of iron.

If there is one bit of advice to be given to any prospective builder of a gravity wave dewar, that advice would be that it is almost impossible to leave too much clearance between the various vacuum shells. There is always the cost constraint which makes it important to save material by making the dewar and vacuum shells as small as possible. It is very difficult and expensive to construct large apparatus to the same degree of precision as a precision machine shop constructs small apparatus. Manufacturing tolerances will add up at some point to make clearances much smaller than the design specifies. The LSU and Stanford dewars were designed with a minimum clearance between any two shells of 10 cm. It should have been at least 50% greater. The extra cost in manufacturing would have been more than recovered in the cost of putting the equipment into service.

There have been several approaches taken with respect to vibration isolation of the antenna. Initially workers in the field assumed that if a support was used which kept the Q of the antenna high that support would be good for vibration isolation. The early room temperature detectors all used some type of support similar to that first used by Weber whereby the bar was supported by a cable around its center. This support has the advantage of coupling poorly to the fundamental longitudinal mode of the antenna so that it does keep the Q high. It has been most thoroughly examined by Fulgini and Ricci (1981) and has turned out to be the most successful antenna support.

Several experimenters have tried to use the 'dead bug' or 4 point support first pioneered by Richard (1976). The Italian group used the 4 point support on their first small antenna but were unable to eliminate occasional large mechanical excursions from thermal equilibrium and so switched to the cable support on their later models (Coccia, 1982). The LSU group worked with several configurations of 4 point support but was never able to get consistent results for the Q of the antenna or for good isolation when there was a varying low frequency excitation applied to the support. The work of the Rochester group (Karim, 1984) and of the Perth group (Veitch, 1987; Veitch et al., 1987) has demonstrated the need for a compound support, utilizing an intermediate mass, to get good isolation with a 4 point support, but no group has been able to attain excitation at the Brownian motion level by using a 4 point support only. The most promising 4 point support is the 'Catherine wheel' support used by the Perth group, which has demonstrated that it does allow the antenna to have high Q, ~10 K excess noise is observed, and this could be caused by noise originating in the suspension.

The direct support of the antenna is not the whole story for vibration isolation. Hirakawa points out that the average power fluctuation of the antenna may be obtained by considering the vibrational energy of the normal mode of the antenna to be fluctuating about its average, kT, over its effective relaxation time Q/ω_0. This results in thermal noise power of approximately 2×10^{-24} W. Thus we must attenuate all external disturbances to transfer less than this at the antenna frequency. Steel and rubber isolation stacks have been used as the first stage of defense by all experimenters since Weber (Narihara and Hirakawa, 1976). The direct antenna support then provides the final isolation.

Because the noise power must be so low, it is important that the antenna support system be separate from that of the cryogenic system so that boiling cryogens do not cause antenna excitation. The schematic figures of the LSU system show the antenna supported, through the various filters, from a table that is separate from the cryogenic dewar. The table in turn rests on air springs for high frequency isolation from the earth. On the table rest stacks of steel and silicone rubber. The electrical analog of such an isolator is a ladder of π section low pass filters. It is therefore straightforward to determine the isolation. On the LSU system each of the support rods hangs from a stack of five 90 kg steel masses, each separated from the other by silicone rubber vibration isolation material.* The steel masses are discs 7.5 cm thick and 80 cm in diameter. The measured isolation at the antenna frequency for each stage of steel and rubber is greater than 50 dB and the attenuation for the entire stack is unmeasurable at room temperature. The estimated attenuations are shown on figure 5.2.

Mechanical stability is also an important consideration in design of the support system. Large excursions in a soft support system should be eliminated if at all possible. The first design of the LSU antenna used 11 stages of steel and rubber. Such a filter was very unstable against horizontal motion. Narihara and Hirakawa point out that all such stacked filters are much softer in the horizontal modes and hence should be better isolators against horizontal motion at the antenna frequency. Isolation is not the whole story, however. A very soft support allows a large excursion in the horizontal direction. This large displacement then makes the system much more susceptible to any non-linearities which may exist. As an example if there are any surfaces which might slide over each other they are more likely to do so if the relative motion can be large. In turn this sliding motion can introduce, through friction, vibrational frequencies at the antenna frequency.

---

* Vib-χ™ pad, 45 durometer silicone, purchased from VibraSciences Inc., 234 Front Ave., West Haven, CT 06516.

This gives up-conversion of the low frequency energy to energy at the frequencies of interest. It is, incidentally, this type of sliding which has probably resulted in the poor results seen by all groups which have attempted to use only a 4 point support for the antenna. Most such supports were made so that they would slide on the surface of the antenna if the antenna were moved vertically.

To gain additional mechanical stability the Italian group and the LSU group introduced a very large intermediate mass between the low pass filters and the antenna itself. Both groups have supported the antenna directly from the intermediate mass. The Italian intermediate mass consists of a bronze ring which is supported on a dead bug support resting on the bottom of the cryogenic shell. The LSU intermediate mass is a bronze cruciform shape supported by cables at the arms with the antenna hung from the center of the mass. The LSU intermediate mass is 2204 kg. Its purpose is to provide a stable mechanical ground for the antenna. To that end it has been made so that all known vibrational modes are at least 40 Hz away from the antenna frequency. By making the supported mass so large, the amplitude of any excitation is kept small, thus avoiding up-conversion.

We again emphasize the central difficulty with all such designs. How does one support the antenna and intermediate mass *with vibration isolation* and also provide a low heat leak to the antenna? One approach is to attempt to put the springs represented by the rubber in the isolators at low temperatures. Weber tried to use felt at low temperature in his low pass vibration filter but the felt gradually lost its elasticity under a constant load and the filtering efficiency decreased. The Rome group uses no rubber but mounts its intermediate mass directly on a dead bug support supported on the bottom of the helium shell. The Rome antenna thus is better coupled to the vibrations of the cryogenic fluids. The LSU and Stanford antennas put most of the filters at room temperature and use as long a support rod as possible. Both groups construct the support rod from Ti-6Al-4V alloy. This alloy has a lower thermal conductivity than stainless steel so the conduction heat leak can be kept to an acceptably small value while keeping an acceptable mechanical safety factor on the support. The LSU antenna for instance is designed with a safety factor of more than two on each of the support rods. The rods are then blackened to enhance heat transfer by radiation exchange with the surrounding tube. The calculated heat leak, assuming conduction only, is less than 0.1 W for all four supports. The Stanford group puts additional masses between the support rod and the antenna to provide an acoustic impedance mismatch. They then weld the support system directly to the antenna to eliminate possible up-conversion.

An additional source of vibration to a gravity wave antenna is conduction of vibration along wires connected to the transducer or to thermometers or other sensors. Most groups use some variant of the Taber isolator to eliminate vibrational coupling through the wires. A diagram of a typical Taber isolator is shown in figure 5.6. The isolator consists of five or more masses connected together

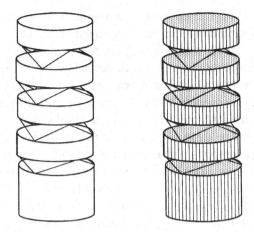

Figure 5.6. Diagram of a Taber isolator, as developed at Stanford.

as shown by piano wire or other similar thin strong wire. Each brass mass is approximately 1.4 kg. There is provision for strongly fastening the sensor wires to the individual masses. The bottom mass is approximately 4.5 kg so as to keep all of the support wires under high tension. A Taber isolator provides a series of mechanical grounds for vibrations conducted along the sensor wires. Analysis of such a coupled mass system demonstrates that it has no normal modes in the low frequency range of most gravity wave antennas. A Taber isolator may be mounted directly on the antenna to mechanically ground the wires before they attach to the transducer. Such mounting, surprisingly, does not degrade the Q of the antenna, at least at the level of $Q = 10^7$.

5.4   Practical aspects

We delineate in this section some of the practical aspects of operation of the LSU antenna. We hope that some of these details will enable operators and designers of large cryogenic experiments to proceed to test their apparatus with some degree of confidence and have some numbers from which to scale their experiments.

5.4.1   Pump out time

The LSU dewar is first evacuated with a 10 cfm Alcatel mechanical pump. The pump is throttled until the pressure in the dewar drops below 1 torr. This throttling is necessary to protect the mylar superinsulation. Pumping too rapidly will cause the superinsulation to tear itself up due to the stress of the escaping air trapped in the insulation. In our experience it takes approximately 12 hours until the pump can be opened fully.

A 4 inch diffusion pump is then used to continue the evacuation. The diffusion pump will begin to contribute to the pumping speed when the pressure at its input is approximately 0.5 torr. An additional 12 hours is required to get from 1 torr to 0.5 torr and an additional day is required to get below 10^{-3} torr. From that point on the time depends very strongly on the weather.

The weather is an important factor because of the influence of the local humidity on the amount of water vapor trapped on the internal surfaces and in the superinsulation. There are a number of stories in the vacuum industry about space simulators built for NASA that would work well when built and tested in California where the climate is dry but which would not meet specifications when taken apart and reassembled in Houston or Florida where the humidity is much higher. The experience at LSU should be used realizing that Baton Rouge is in a humid semi-tropical climate where the summer dew-point is usually 24°C. Even though the laboratory is air conditioned the adsorbed water vapor is a major factor in the vacuum performance of the dewar. Incidentally, in a humid climate where large volumes are to be pumped by mechanical pumps, it is wise to use single stage rather than two stage mechanical pumps. The two stage pump is more susceptible to water condensation in the oil with a subsequent decrease in the pumping efficiency.

We continue to pump on the insulating vacuum until the pressure, measured with an ion gauge, is approximately 5×10^{-5} torr. This can take an additional week of continuous pumping. We have attempted to speed up the evacuation procedure by first pumping and then backfilling the dewar with dry N_2 gas in the hope that the N_2 molecules will displace some of the adsorbed water vapor and allow the attainment of low pressure more quickly. There is mixed opinion as to the effectiveness of backfilling. The system returns very quickly to the pressure at which the backfilling occurred but sometimes does not seem to pump down more quickly after reaching that pressure.

5.4.2   Cooldown time

Helium exchange gas is added to the 4 K shell prior to starting the cooling of the experiment. We use as much pressure as we can get for the exchange gas pressure. Our experiment is limited by the fact that the large flanges on our dewar are warped and have a tendency to leak if the internal pressure is too great.* We have run with initial pressures of 150 torr (measured at room temperature), but during the 1988 run our initial pressure was limited to 12 torr. The Rome experiment has used an initial pressure of 300 torr, thus allowing the exchange gas pressure to act as a crude gas thermometer. The cooling rate is proportional to the total amount of gas circulating in convective currents so it is important to use as much as possible.

---

* The flanges were not stress relieved prior to welding them on the shell. This was a serious error. If aluminium is used to construct a large flange that flange *must* be stress relieved.

The nitrogen shell is cooled first and cools quite quickly. The insulating vacuum improves to at least 10^{-5} torr when the 77 K shell gets completely cold. We have had occasional difficulty cooling the 77 K shell because of vapor locks. The shell is cooled from the top and the LN2 is contained in pipes which define the 77 K shell. If simple gravity feed is used to cool the shell a vapor lock can prevent the liquid from filling the tubes and thus retards cooling. Vapor locking can be prevented by forcing the liquid to flow through the tubes on the shell while the cooldown is taking place. The 77 K shell reaches liquid nitrogen temperature at the bottom in less than one day if there is no problem with vapor locking.

A phenomenon that has been noted at both LSU and Rome is the occasional sudden increase in pressure in the insulating vacuum as the shells cool. This is due to cryopumped gases being released as cold surfaces warm when the temperature distribution changes during cooldown. The size of the pressure bursts can be quite surprising. It was not uncommon to see the pressure change from 5×10^{-6} torr to 10^{-4} torr or so during such a pressure excursion. The bursts seem to become larger and less frequent as the temperature gets lower. The Rome group has stated that on several occasions, when they were using molecular sieve material on some of the cold surfaces, they saw pressure increases to as large a value as 5×10^{-2} torr with a slow recovery back to the original pressure over the period of a half hour or so. Large pressure changes can be expected when changing liquid nitrogen dewars during cooldown or during any other activity that might reasonably result in a redistribution of the temperature inside the dewar.

As soon as the LN2 shell is cold we begin to add LN2 to the helium storage space. The initial cooling rate is controlled because, in our configuration with warped flanges, too rapid an initial cooling rate will cause the flanges to warp further and open, thus allowing helium exchange gas into the insulating vacuum. We bypass the gas cooled shell while we are cooling the 4 K shell and its contents. This allows a more efficient transfer and a more rapid one. Our experience is to transfer approximately 500 l of liquid nitrogen in a 24 hour period. At this rate we approach nitrogen temperatures in 11–14 days.

The enthalpy removal is quite efficient while the antenna and isolation mass are warm because of the large temperature difference. However, as the temperature of the antenna approaches 100 K the cooling rate slows tremendously. The helium storage space fills with liquid nitrogen but there is little enthalpy removal from the contents of the 4 K shell. Progress is painfully slow after the antenna cools below 90 K. We begin to pump at that point on the liquid nitrogen to reduce its temperature and the temperature of the 4 K shell and thus promote more rapid cooling of the antenna. While we pump we pull gas through the gas cooled shell and bring its temperature down to 77 K or below. About five days of pumping, being careful to never let the pressure above the liquid nitrogen to fall below the LN2 triple point pressure, are required to get the antenna temperature below 71 K. This last pumping is very important from the standpoint of total helium consumption since the enthalpy of aluminium is a factor of two lower at 70 K than at 77 K.

As soon as approximately 70 K is reached we are ready to cool to helium temperatures. At this point we force the liquid nitrogen out of the dewar. We do this by putting helium gas pressure over the liquid and forcing it out of a tube which goes down to the lowest part of the dewar. An important design consideration is that one should be careful to avoid traps for the liquid nitrogen in the piping. The helium storage space is then pumped to evaporate any remaining liquid. Helium liquid is then transferred into the liquid helium storage region. The boiloff gas is again bypassed around the gas cooled shield for more efficient transfer. In 1988 we took about four days to get the entire 4 K shell and its contents to liquid helium temperature. We used about 1800 l of helium in 1988 in the initial cooling. About 4500 l of liquid nitrogen were required.

5.4.3   Recovery from accidents

A large system can have a long time constant in its recovery from a perturbation in any system parameter. As an example we use a mishap that occurred on the cooldown of the LSU detector. With the 77 K shell at liquid nitrogen temperature and the other shells between 200 K and 250 K, a leak was accidentally opened from the atmosphere into the insulating vacuum. By the time the leak was closed the indicated pressure in the insulating vacuum was above 0.3 torr. The 4 inch diffusion pump was still pumping on the space at the time of the accident. In one hour the pressure had recovered to 6.5×10^{-2} torr, and in 2.5 hours it had recovered to 9.4×10^{-6} torr. It was 7.7×10^{-6} torr prior to the accident. This recovery was due to a combination of cryopumping by the low temperature shells and the continuous operation of the diffusion pump.

Acknowledgements

The construction and operation of a gravitational wave experiment is an effort involving a great many people over a very long time span. All of the groups have had contributions from individuals who have never been recognized for the importance of the work they did. The LSU group is no exception. We have had major contributions from faculty, postdoctoral, graduate and undergraduate students. Administrators at the local and national levels have given us a helpful hand and wise counsel. This chapter is not the place to make a list of everyone that has helped. The people who did the work know who they are. Their efforts are appreciated.

References

Carslaw, H. S. and Jaeger, J. C. (1959). *Conduction of Heat in Solids*, 2nd edn., p. 194, Oxford University Press.

Chapman, A. J. (1974). *Heat Transfer*, Macmillan, New York.

Coccia, E. (1982). *Rev. Sci. Instrum.* **53**, 148.

Fulgini, F. and Ricci, F. (1981). *Il Nuovo Cimento* **4C**, 93.

Hamilton, W. O. (1982). *Cryogenics* **22**, 107.

Johnson, V. J. (gen. ed.) (1960). Wright Air Development Division Technical Report 60-56, ASTIA Publication AD249786.

Karim, M. (1984). *Rev. Sci. Instrum.* **55**, 103.

Narihara, K. and Hirakawa, H. (1976). *Jap. J. Appl. Phys.* **15**, 833–42.

Richard, J. P. (1976). *Rev. Sci. Instrum.* **47**, 423.

Scott, R. B. (1959). *Cryogenic Engineering*, Van Nostrand, Princeton.

Thomas, L. C. (1980). *Fundamentals of Heat Transfer*, Prentice Hall, New Jersey.

Timmerhaus, K. D. (1960). *Advances in Cryogenic Engineering*, vol. 8, Plenum Press, New York.

Veitch, P. J. (1987). Ph. D. Thesis, University of Western Australia.

Veitch, P. J., Ferreirinho, J., Blair, D. G. and Linthorne, N. (1987). *Cryogenics* **27**, 586.

White, G. K. (1987). *Experimental Techniques in Low-Temperature Physics*, 3rd edn., p. 303, Oxford University Press.

6

Internal friction in high Q materials

J. FERREIRINHO

6.1 Introduction

The subject of low temperature acoustic loss mechanisms in crystalline solids has assumed some practical importance due to the requirement for very high mechanical quality factors in resonant gravitational antennae of the type first proposed and built by Weber (Weber, 1960). This requirement arises from the effects of the thermal noise or Brownian motion which limits the sensitivity of resonant bar antennae and other high Q mechanical resonators used in sensitive electromechanical transducers (Braginsky, Mitrofanov and Panov, 1985; Paik, 1976; Thorne, 1980).

The equation of motion of an elastic body such as a gravitational radiation antenna may be written generally as (Landau and Lifshitz, 1970a)

$$\rho \partial^2 w_i / \partial t^2 + \partial \sigma_{ik} / \partial x_k = \rho f_i. \qquad (6.1)$$

Here $\mathbf{w}$ is the elastic displacement and σ_{ik} is the stress tensor, related to the linearised strain tensor $\varepsilon_{ik} = \frac{1}{2}(\partial w_i / \partial x_k + \partial w_k / \partial x_i)$ by

$$\sigma_{ik} = C_{iklm} \varepsilon_{lm} + \eta_{iklm} \, \partial \varepsilon_{lm} / \partial t, \qquad (6.2)$$

ρ is the material density, and $\mathbf{f}$ is the force per unit mass acting on the body.

In an isotropic material, symmetry considerations reduce the number of independent components of the elasticity tensor C_{iklm} and the viscosity tensor η_{iklm} to two. The viscosity tensor is introduced into the theory of elasticity in order to take account of the dissipative properties of the elastic medium. The divergence $\partial \sigma_{ik} / \partial x_k$ is therefore the sum of the elastic restoring force (per unit volume) and a small frictional force. For an elastic body with a force per unit area $\mathbf{p}$ applied to its surface, the boundary condition for equation (6.1) is

$$\sigma_{ik} n_k = p_i, \qquad (6.3)$$

where the vector $\mathbf{n}$ is normal to the surface. In the case of the bar antenna, the stress $\mathbf{p}$ is zero everywhere except at the suspension points and those surface regions which are coupled to an electromechanical transducer.

For a gravitational radiation antenna, $\mathbf{f}$ may be expressed as a summation

$\mathbf{f} = \mathbf{f}_g + \mathbf{f}_N$ of a gravitational force $\mathbf{f}_g$ due to the gravitational radiation and a noise force $\mathbf{f}_N$. If the suspension system is assumed to isolate completely against external seismic vibrations, and if the effects of the transducer system are also ignored, then $\mathbf{f}_N$ is primarily due to intrinsic thermal fluctuations in the antenna, although $\mathbf{f}_g$ due to extragalactic sources, may in fact be so small that quantum effects such as the zero point motion of the antenna may also have to be considered.

It should be noted here that the theory of elasticity is inadequate for the proper description of mechanical dissipation in solid bodies, since Hooke's law requires that the elastic strain be determined by the instantaneous value of the stress, and vice versa. In a real solid the response is not instantaneous, even in a small sample where the finiteness of the speed of sound may be neglected. The strain response at any instant depends to a small extent on the previous stress history. The more general theory of anelastic solids which takes this into account (Zener, 1948) is reviewed in the next section. It will be seen there that the anelastic behaviour (including the dissipation) can be represented by complex, frequency-dependent elastic constants. While the frequency dependence of the real part is very slight, perhaps one part in 10^6 over the entire audio frequency range in a high Q material, that of the imaginary part is much stronger, and may involve order of magnitude changes. On the other hand, Hooke's law can at most accommodate complex elastic constants which are frequency independent, and do not therefore represent accurately the dissipative properties of the material.

The partial generalisation of Hooke's law by the addition of viscous terms proportional to $\partial w_{ik}/\partial t$ to the stress tensor in equation (6.2) is also inadequate, since it requires similarly that the viscosity coefficients η_{iklm} be frequency independent. Although the viscous part of the response is not instantaneous, the assumption of equation (6.2) leads to a viscous stress which is proportional to the frequency ω for a harmonically varying strain, so that $Q^{-1} \to \infty$ at high frequencies. This is not in agreement with the observed behaviour of crystalline solids. A further generalisation of equation (6.2) to be discussed in the next section gives the standard anelastic model, which approximates well and over a wide frequency range the dissipative properties due to many of the loss mechanisms in crystalline solids.

It will also be seen in the next section that the simple viscosity model, equation (6.2), gives a correct description of dissipative processes at frequencies which are sufficiently low for the mechanical response to be approaching quasi-static conditions. At the typical frequencies of the antenna modes, the response is usually quasi-static, depending on the characteristic time scale involved in the loss mechanisms which determine the dissipation in the material in question.

If the dissipation is small, a normal mode expansion

$$\mathbf{w}(\mathbf{r}, t) = \sum b_n(t)\mathbf{u}_n(\mathbf{r}) \tag{6.4}$$

allows the equation of motion (6.1) to be transformed into a set of independent, one dimensional harmonic oscillator equations

$$b_n + b_n/\tau_n + \omega_n^2 b_n = F_n/M_n,$$ (6.5)

where

$$F_n(t) = F_{ng}(t) + F_{nN}(t) = \rho \int d^3r\, \mathbf{f}_n(\mathbf{r}, t) \cdot \mathbf{u}_n(\mathbf{r}),$$ (6.6)

and the effective mass M_n for each normal mode is given by

$$M_n = \rho \int d^3r\, u_n^2$$ (6.7)

The gravitational force $F_{ng}(t)$ vanishes for even n, and the fundamental mode $n = 1$ is of greatest interest, since this mode has the largest absorption cross-section for gravitational radiation (Misner, Thorne and Wheeler, 1973; Rees, Ruffini and Wheeler, 1974).

If the length L of a cylindrical bar antenna greatly exceeds its diameter d, the vibrations may be approximated as being purely longitudinal and the eigenfunctions $\mathbf{u}_n(\mathbf{r})$ for the antisymmetric modes have the form

$$u_n^\zeta(\mathbf{r}) = \sin(n\pi\zeta/L), \quad n \text{ odd,}$$

$$u_n^\rho(\mathbf{r}) = 0$$ (6.8)

$$u_n^\theta(\mathbf{r}) = 0$$

in cylindrical polar coordinates (ρ, θ, ζ) with origin at the centre of mass and ζ-axis coincident with the longitudinal axis of the bar. For these modes, the effective mass $M_n = M/2$, where M is the bar mass, and the eigenfrequencies ω_n are given by $\omega_n = n\pi v_s/L$, where $v_s = (E/\rho)^{1/2}$ is the speed of extensional sound waves in a material with Young's modulus E. The relaxation times τ_n of the modes are related to the extensional viscosity coefficient H by

$$\tau_n = E/\omega_n^2 H = Q_n/\omega_n,$$ (6.9)

where Q_n is the quality factor of the nth mode. As mentioned previously, E and H are to be regarded as frequency dependent quantities. The small three dimensional corrections of order $(d/L)^2$ which must be applied to some of the above results are discussed by Paik (1974).

The stationary and random Brownian noise force $F_{nN}(t)$ due to thermal fluctuations in the antenna is related to the damping coefficient H by a fluctuation–dissipation relation of the form

$$\langle F_{nN}^2 \rangle = (2k_B T/\pi) \int d\omega H(\omega) \approx 4k_B T H(\omega_n)\, \Delta f,$$ (6.10)

where Δf is the frequency bandwidth of interest. For the gravitational radiation antenna, Δf is of order τ_g^{-1}, where $\tau_g \ll \tau_n$ is the typical duration of the force F_{ng} due to a burst of gravitational radiation. Relation (6.10) leads to the well-known detectability condition

$$|F_{ng}|^2 > 4k_B TH(\omega_n)/\tau_n = 4k_B TM_n/(\tau_g \tau_n) \tag{6.11}$$

for the minimum impulsive force detectable by the antenna (Braginsky and Manukin, 1977). The antenna sensitivity is proportional to the relaxation time $\tau_n = Q_n/\omega_n$, and the desirability of large mechanical quality factors Q_n for the antenna modes is thus apparent. A qualitatively similar conclusion obtains for a resonant antenna designed to detect continuous monochromatic gravitational radiation.

For the purposes of this chapter, it will be assumed that the acoustic loss Q^{-1} of a gravitational radiation antenna is the sum of contributions from several nominally distinct loss mechanisms, so that

$$Q^{-1} = \sum Q_i^{-1}, \tag{6.12}$$

where the sum (6.12) includes contributions which are not 'intrinsic' to the antenna material, such as surface and suspension losses. These 'extrinsic' contributions will not be discussed in any detail in this chapter, which is concerned only with the intrinsic internal friction mechanisms. A brief discussion of surface and suspension losses is given by Braginsky et al. (1985).

Particular emphasis will be placed on experimental internal friction results relating to Nb, since the most recent results for this metal have not been published elsewhere at the time of writing. An extensive investigation of the effects of various pre-treatments on the mechanical Q of Nb has been carried out in conjunction with the gravitational radiation detection project at the University of Western Australia (UWA). A 3 m long, 1.5 tonne Nb antenna with a very low acoustic loss level $Q^{-1} = 4.3 \times 10^{-9}$ is currently being operated at UWA, and this is by far the highest Q factor obtained with any antenna of this size. Niobium also has the lowest internal friction of any metal reported to date both at liquid helium temperature and at room temperature (Veitch et al., 1984). The technical ease of machining polycrystalline metals in comparison with brittle, covalently bonded high Q materials such as sapphire (Braginsky et al., 1985) is also well known.

A further justification for this emphasis is given by the excellent superconducting properties of Nb, which render this material generally suitable for the construction of smaller, high Q mechanical resonators such as those used in some antenna transducer designs (Paik, 1974). Experimental results relating to other promising materials with low Q^{-1} such as sapphire and quartz (Braginsky et al., 1985), silicon (McGuigan et al., 1978) and 5000 series aluminium alloys (Suzuki, Tsubono and Hirakawa, 1978) will also be reviewed briefly.

6.2 Anelastic relaxation

6.2.1 The anelastic model

It was mentioned in the previous section that Hooke's law could be generalised to provide a more accurate macroscopic model for the elastic and dissipative response of a solid, known as the anelastic model (Zener, 1948). A review of this model will be given in this section, following closely the approach of Nowick and Berry (1972). Some of the limitations of this model will then be considered, and a more detailed analysis of the internal friction in terms of microscopic and macroscopic loss mechanisms will be given in section 3.

The anelastic model takes account of the relaxation in the linear elastic response which occurs in solids as a result of the finite time required for the microscopic degrees of freedom of the solid to attain a new thermal equilibrium when the macroscopic stress or strain parameters are varied. There is thus an analogy between this model and the models of dielectric and magnetic relaxation which were developed by Debye (Daniel, 1967; Debye, 1929) and others for the linear electric and magnetic response of material media.

It is assumed that the relaxation processes are characterised by finite relaxation times so that the anelastic model is naturally consistent with the results which are to be obtained from qualitative kinetic arguments at the microscopic level, which also rely on a relaxation time approximation. The 'anelastic solid' is therefore a qualitative model which should not be expected to yield quantitatively accurate results. It is nevertheless useful as a qualitatively correct model with wide applicability in the interpretation of internal friction measurements. The degree of accuracy which is achieved in some specific cases will be discussed in the next section.

The finiteness of the characteristic relaxation times τ (not to be confused with the antenna mode relaxation times τ_n) requires that the deformation should be reversible both in the quasi-static limit $\omega \to 0$ and in the instantaneous limit $\omega \to \infty$. In the quasi-static limit, the model is thus consistent with the simpler viscosity model of equation (6.2), in which the internal friction is proportional to ω. The physically unreasonable behaviour of the latter model in the instantaneous limit is however eliminated.

It is clear from the above considerations that the stress–strain relationship for the anelastic solid must be of a more general nature than equation (6.2). The most general linear relationship possible is of the form

$$a_0 \sigma + a_1 \, \partial \sigma / \partial t + a_2 \, \partial^2 \sigma / \partial t^2 + \cdots = b_0 \varepsilon + b_1 \, \partial \varepsilon / \partial t + b_2 \, \partial^2 \varepsilon / \partial t^2 + \cdots. \quad (6.13)$$

Since the stress σ and the strain ε are second-rank tensors, the coefficients a_n and b_n should in general be tensors also. The essential features of the model are, however, most clearly illustrated in a simplified scalar form which may then be generalised where necessary to take account of the three-dimensional and anisotropic tensor character of elasticity in crystalline solids.

The particular case in which a_0 and b_0 are finite and all other coefficients are zero clearly corresponds to ideal elasticity (Hooke's law) with an elastic modulus $M = b_0/a_0$, while the case

$$a_0\sigma = b_0\varepsilon + b_1\,\partial\varepsilon/\partial t \qquad (6.14)$$

corresponds to a simplified scalar form of the viscosity model (6.2) with $\eta_1 = b_1/a_0$. It is helpful to represent these models in terms of simple mechanical equivalents (Nowick and Berry, 1972), in which strain corresponds to displacement and stress to force, so that elasticity is represented by a spring with stiffness $K = M$ or compliance $K^{-1} = J = M^{-1}$, and the resistive element is a 'Newtonian dashpot' with a viscous resistance $\eta = \tau M$, where τ is the characteristic relaxation time of the model. It is again emphasised that τ is not to be confused with the relaxation times of the modes of a resonator such as a gravitational radiation antenna. The viscosity model (6.2) is thus equivalent to the parallel spring–dashpot combination shown in figure 6.1, and is obviously incapable of instantaneous response.

The simplest mechanical model which has the physical properties of anelasticity as outlined above is shown in figure 6.2. The differential stress–strain relationship corresponding to this model is

$$\sigma + \tau\,\partial\sigma/\partial t = M\varepsilon + (M + \delta M)\tau\,\partial\varepsilon/\partial t, \qquad (6.15)$$

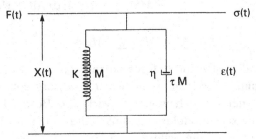

Figure 6.1. Simple mechanical equivalent of 'viscosity' model with stress–strain relationship $\sigma = M\varepsilon + \eta\,\partial\varepsilon/\partial t$. Force $F(t)$, displacement $X(t)$, spring constant K and dashpot viscosity η correspond to $\sigma(t)$, $\varepsilon(t)$, M and τM, respectively.

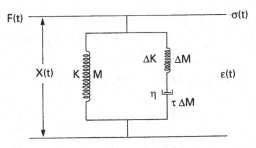

Figure 6.2. Simple mechanical equivalent for the standard anelastic solid.

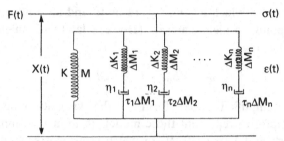

Figure 6.3. Mechanical equivalent corresponding to a discrete spectrum of relaxation processes.

which is an equation of the form (6.13) with $a_k = 0$, $b_k = 0$, $k > 1$. It is clear from figure 6.2 that this model is characterised by a relaxed or quasi-static elastic modulus $M_r = M$, an unrelaxed or instantaneous modulus $M_u = M + \delta M$, and a relaxation time τ at constant strain. The model which obeys the stress–strain relation (6.15) is known as the standard anelastic solid (Zener, 1948). A more general anelastic model may be obtained by connecting n relaxing elements in parallel with the ideal spring as shown in figure 6.3.

For $n = 2$, the stress–strain relation is readily found to be (Nowick and Berry, 1972)

$$d^2\sigma/dt^2 + (1/\tau_1 + 1/\tau_2)\,d\sigma/dt + \sigma/(\tau_1\tau_2) = M_u\,d^2\varepsilon/dt^2 + [(\delta M_1/\tau_1 + \delta M_2/\tau_2)$$
$$+ M(1/\tau_1 + 1/\tau_2)]\,d\varepsilon/dt + M\varepsilon/(\tau_1\tau_2), \quad (6.16)$$

where

$$M_u = M + \sum \delta M_i. \tag{6.17}$$

For an arbitrary value of n the differential relation (6.13) will be of the same order n in both σ and ε, and this is in fact a necessary condition for relation (6.13) to represent anelastic behaviour (Nowick and Berry, 1972). Expression (6.17) for the unrelaxed modulus remains valid for arbitrary n, with the summation extending over the values $i = 1, 2, \ldots, n$. The set $\tau_1, \ldots, \tau_n$ of relaxation times is known as the relaxation spectrum of the anelastic solid.

For a harmonically varying strain $\varepsilon = \varepsilon_0 e^{i\omega t}$, the stress will be of the form

$$\sigma = (\sigma_1 + i\sigma_2)e^{i\omega t}, \tag{6.18}$$

as a consequence of the linearity of relation (6.13) and the time independence of the coefficients. It is therefore possible to define a complex, frequency dependent elastic modulus

$$M(\omega) = M_1(\omega) + iM_2(\omega) = \sigma(\omega)/\varepsilon, \tag{6.19}$$

so that

$$M_1(\omega) = \sigma_1(\omega)/\varepsilon_0 \tag{6.20a}$$

and

$$M_2(\omega) = \sigma_2(\omega)/\varepsilon_0. \tag{6.20b}$$

Substitution of the above expressions for σ and ε into the relation (6.15) for the standard anelastic solid (case $n = 1$) gives the results

$$M_1(\omega) = M + \delta M(\omega\tau)^2/(1 + \omega^2\tau^2) \tag{6.21a}$$

and

$$M_2(\omega) = \delta M\omega\tau/(1 + \omega^2\tau^2). \tag{6.21b}$$

The internal friction of the standard anelastic solid may be defined in terms of the quality factor, Q, as

$$Q^{-1} = (2\pi)^{-1} \int \sigma \, d\varepsilon/(M\varepsilon_0^2/2), \tag{6.22}$$

where the integration is over one complete cycle of the alternating strain. Substitution of equations (6.20) and (6.21) in (6.22) then gives

$$Q^{-1} = \Delta\omega\tau/(1 + \omega^2\tau^2)$$
$$= (\Delta/2)\,\text{sech}[\log(\omega\tau)], \tag{6.23}$$

where $\Delta = \delta M/M$ is known as the relaxation strength (Nowick and Berry, 1972; Zener, 1947). It may be seen from the plot of equation (6.23) as a function of the variable $\log(\omega\tau)$ in figure 6.4 that Q^{-1} attains its maximum value $\Delta/2$ at $\omega\tau = 1$ and that the curve is a symmetrical function of $\log(\omega\tau)$ about this peak value. This symmetry is made explicit by the expression of Q^{-1} in terms of the symmetrical hyperbolic function sech. The curve, known as a Debye peak, is of

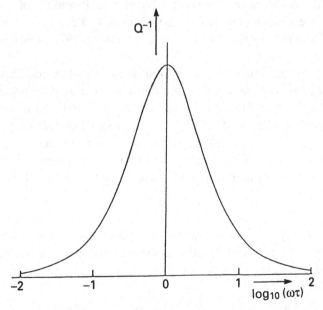

Figure 6.4. Classical Debye relaxation curve showing symmetric frequency dependence of Q^{-1} for the standard anelastic solid.

great importance in the theory of relaxation processes. In the low internal friction limit which is of interest here, $\delta M \ll M_u$ and $M_2 \ll M_1$, so that the result (6.23) for the internal friction may also be expressed as

$$Q^{-1} \approx M_2/M_1 = \tan\phi \approx \phi, \tag{6.24}$$

where ϕ is the phase angle between the stress and the strain.

In the more general model represented by figure 6.3, it is clear from the additivity of the stresses in the relaxing elements that the frequency dependent moduli will be given by

$$M_1(\omega) = M + \sum_i \delta M_i \omega^2 \tau_i^2/(1 + \omega^2 \tau_i^2) \tag{6.25a}$$

and

$$M_2(\omega) = \sum \delta M_i \omega \tau_i/(1 + \omega^2 \tau_i^2) \tag{6.25b}$$

while the internal friction is given in the limit $Q^{-1} \ll 1$ by

$$Q^{-1} = \phi = M_2/M_1 = \sum (\delta M_i/M)(\omega \tau_i)/(1 + \omega^2 \tau_i^2)$$

$$= \frac{1}{2} \sum \Delta_i \operatorname{sech} \log(\omega\tau_i). \tag{6.26}$$

An important simplification which therefore obtains in this limit of low internal friction is the simple additivity of the contributions from each of the relaxation processes in the spectrum.

As it has been outlined above, the anelastic model represents only a plausible generalisation of ideal elastic behaviour. Some understanding of the physical significance of the model parameters may, however, be gained from thermodynamic arguments (de Batist, 1972; Nowick and Berry, 1972) which will now be reviewed briefly.

It is assumed that the small departure from thermodynamic equilibrium arising from the deformation can be described by the introduction of thermodynamic or internal variables ξ_p in addition to the variables ε, σ and T which appear in the equilibrium equation of state. For small deformations this latter equation is of course Hooke's law, in a generalised form which takes account of phenomena such as thermal expansion and the temperature dependence of the elastic constants. The first law of thermodynamics may then be expressed in the form

$$dU = T\,dS + \sigma\,d\varepsilon - \sum_p A_p \xi_p, \tag{6.27}$$

where $A_p = (\partial U/\partial \xi_p)_{\varepsilon, S, \xi}$, and thermodynamic functions such as the free energy may be expanded in terms quadratic in these variables. The Helmholtz free energy F, for example, then has the form

$$F(\varepsilon, T, \xi_p) = F(0, T_0, 0) - \frac{1}{2}(c_\varepsilon/T)(\Delta T)^2 + \frac{1}{2}M_u\varepsilon^2 - \varepsilon\sum_p \lambda_p \xi_p$$

$$- \alpha M_u \varepsilon \Delta T - \Delta T \sum_p \chi_p \xi_p + \frac{1}{2}\sum_{pq} \beta_{pq} \xi_p \xi_q \tag{6.28}$$

for small deformations ε. Here, the coefficients are the various derivatives of F with respect to the corresponding variables, e.g. $c_\varepsilon/T = (\partial^2 F/\partial T^2)_{\varepsilon,\xi}$, where c_ε is the specific heat at constant strain. The Maxwell relation $(\partial S/\partial \varepsilon)_{T,\xi} = (\partial s/\partial T)_{\varepsilon,\xi}$ has also been used to transform the fifth term. The equilibrium values $\bar{\xi}_p$ of the internal variables ξ_p for a particular value of ε may be obtained from equation (6.28) by minimising F with respect to these variables.

For small deformations, the departure of the internal variables from thermodynamic equilibrium is sufficiently small for the internal variables to obey first order kinetic equations characterised by a set of relaxation times τ_r corresponding to relaxation rates τ_r^{-1} which are linearly related to the kinetic coefficients (Landau and Lifshitz, 1970b; Nowick and Berry, 1972). In the particular case where there is only one internal variable ξ, the single kinetic equation has the form

$$d\xi/dt = -(\xi - \bar{\xi})/\tau, \tag{6.29}$$

with a single relaxation time τ, and it is readily shown that the stress–strain relationship may be expressed by the equation

$$\sigma + \tau\dot{\sigma} = (M_u - \lambda\mu)\varepsilon + \tau M_u \dot{\varepsilon}, \tag{6.30}$$

which may be identified with the relationship (6.15) for the standard linear or anelastic solid if $\delta M = -\lambda\mu = \lambda^2/\beta$. It may also be shown more generally that, in the case of n variables, the stress–strain relationship satisfies the nth order differential relation (6.13) which defines anelastic behaviour, so that the existence of such variables would provide some theoretical justification for the anelastic model.

Since the internal variables must satisfy a kinetic equation of the form (6.29), it is reasonable to expect that such variables may be derivable from the microscopic theory of the solid state, which relies heavily on kinetic theory for the interpretation of the dynamical behaviour of solids. For most crystalline solids it is in fact possible to identify several such variables and to associate with each a nominally distinct microscopic relaxation mechanism. Some specific examples of anelastic relaxation mechanisms in crystalline solids will be considered in the next section.

6.2.2   Thermal activation

In concluding this section it is important to note that the anelastic internal friction is usually very sensitive to the presence of crystal defects, and also that the presence of defects in the crystal structure may greatly increase the number of these mechanisms. Defect-induced relaxation mechanisms in fact often dominate the internal friction in real metallic crystals, and the experimental results to be presented in section 6.4 suggest that this is the case also for niobium at liquid helium temperatures.

An important feature of many such mechanisms is a 'thermally activated' relaxation process with a relaxation time of the Arrhenius form

$$\tau = \tau_0 \exp(H_a/kT) \tag{6.31}$$

due to the presence of an energy barrier H_a (Vineyard, 1957), over which the defects must be thermally excited in order to be relaxed under the influence of the applied stress or strain.

In the simple case where there are unique, well-defined values for H_a and τ_0, the variable $\log(\omega\tau)$ which defines the symmetric form of the Debye peak, equation (6.23), is linearly related to the inverse temperature $1/T$ by

$$\log(\omega\tau) = \log(\omega\tau_0) + H_a/kT. \tag{6.32}$$

Expression (6.23), for Q^{-1}, then becomes

$$Q^{-1} = \frac{\Delta_M}{2} \operatorname{sech}\left[\log(\omega\tau_0) + \frac{H_a}{kT}\right], \tag{6.33}$$

so that a plot of Q^{-1} versus $1/T$ also produces a symmetric peak similar to the Debye peak, but with the scale of the abscissa changed by a factor H_a/k and a displacement of the peak position from the origin $1/T = 0$ by an amount $-k\log(\omega\tau_0)/H_a$.

Many examples of such internal friction peaks associated with thermally activated relaxation processes are observed experimentally, particularly in metals (Bordoni, 1949, 1954; Chambers, 1965; Snoek, 1939; Zener, 1943). Some of these processes will be discussed briefly in section 6.3.3, and the problem of the interpretation of the temperature dependence of the internal friction in terms of continuous distributions of values for H_a and τ_0 will be considered very briefly in section 6.4. In the case of such distributions, Q^{-1} is obtained by integration of the contribution (6.33) over the 'range' or spectrum of values of H_a and τ_0 (Nowick and Berry, 1972).

6.3   Anelastic relaxation mechanisms in crystalline solids

6.3.1   Outline

In order to obtain theoretical estimates of the anelastic internal friction which are readily comparable with experiment, it is generally necessary to take into account the three-dimensional and anisotropic tensor character of the strain ε_{lm} and the stress σ_{ik}. For the sake of simplicity, the approximation of an isotropically elastic solid characterised by a bulk modulus K and an isotropic shear modulus μ (or equivalently by an isotropic Young's modulus E and Poisson's ratio σ) will be retained wherever possible, with the elaboration of specific results to account for the effects of crystalline anisotropy where necessary.

Since it will be seen in section 6.4 that most of the samples used in the Nb experiments were discs vibrating in flexure, the results for the internal friction in thin plates undergoing flexural vibration will be of particular interest, as well as those for the relaxation Δ_E in Young's modulus and the internal friction associated with extensional vibrations, which are relevant to gravitational radiation antennae. For the above reason, it is convenient to define a 'plate flexure modulus' $G = E/(1 - \sigma^2)$, which characterises the plane bending of a thin plate (flexure with vanishing intrinsic curvature: (Blair and Ferreirinho, 1982; Ferreirinho, 1986). This modulus is thus another practical elastic modulus in addition to K, E and μ, defined by the condition of bilateral (planar) constraint of the 'Poisson contraction'. Although the flexural vibration modes of a disc with free edge have a non-vanishing, negative intrinsic curvature (Rayleigh, 1894), this may for some mechanisms be accounted for by multiplying the relaxation strength and the internal friction by a numerical factor depending on the mode geometry (Blair and Ferreirinho, 1982).

6.3.2 Relaxation mechanisms in a perfect crystal

At the temperatures and energy scale of interest for gravitational radiation antennae, the only microscopic degrees of freedom of a perfectly crystalline solid which are dynamically relevant and which contribute significantly to the internal friction are the phonons or lattice vibrations and, for metals, the conduction electrons also. These two nominally separable sets of degrees of freedom give rise to three different anelastic relaxation mechanisms. Two of these are microscopic mechanisms due to the relaxation of the respective dynamical (energy and wavevector) distributions of the phonons and the conduction electrons, and one is a general macroscopic mechanism of irreversible heat flow which occurs in any thermodynamic system whose entropy and free energy depend on the state of deformation, if the system is subjected to a time-varying and spatially in-homogeneous deformation.

In solids, this mechanism is known as the thermoelastic effect (Nowick and Berry, 1972; Zener, 1937, 1947) and may be regarded as the thermodynamic reciprocal of thermal expansion. Since no thermoelastic temperature gradients are induced by a homogeneous deformation, it is clear that the thermoelastic dissipation vanishes for such deformations, while the relaxation of the micro-scopic phonon and conduction electron distributions by contrast give rise to dissipation for homogeneous deformations also.

(i) The thermoelastic effect

As mentioned above, this is a general mechanism through which the energy of a non-uniform and time-varying deformation such as an acoustic wave is dissipated by thermal conduction or diffusion due to the temperature gradients induced by the deformation. In the thermodynamic description, this mechanism is considered to arise from the coupling of the strain (or stress) and the temperature expressed in thermodynamic functions such as the free energy, equation (6.28), where this coupling is represented by the fifth term. The thermoelastic effect may be regarded as a relaxation mechanism if the temperature (or the entropy) is adopted as the internal variable. The temperature shift ΔT relaxes towards a uniform equilibrium value $\Delta T = 0$ as a result of the diffusion of heat from 'heated' (compressed) to 'cooled' (dilated) regions of the material. As mentioned previously, the thermoelastic effect is simply the thermodynamic reciprocal of thermal expansion, generally giving rise to dissipation of energy by irreversible heat flow in time-varying, inhomogeneous deformations.

Although the thermoelastic effect may be classified as a relaxation mechanism, it has the atypical feature of a macroscopic rather than a microscopic relaxation time, which is moreover dependent on the sample geometry. It will suffice here to note that the thermoelastic effect is a general mechanism for the dissipation of acoustic energy in material media when the frequency is sufficiently low for the temperature to be well-defined throughout the cycle of the deformation. It is therefore a mechanism which is limited to frequencies which are low compared with the important characteristic relaxation times in the medium on the microscopic scale (e.g. the electron and phonon relaxation times), although the frequency may be high compared with the thermal diffusion time, which is the (macroscopic) characteristic relaxation time of the relaxation process. In the latter limit the deformation is nearly adiabatic and the dissipation is relatively small.

If the temperature is taken as the relaxing internal variable, the Helmholtz free energy may be written (Landau and Lifshitz, 1970a)

$$F(\varepsilon_{ik}, T) = F(0, T_0) - (c_\varepsilon/2T_0)\,\Delta T^2 - 3K\alpha\varepsilon_{jj}\,\Delta T + (K/2)\varepsilon_{jj}^2 + \mu(\varepsilon_{ik} - \delta_{ik}\varepsilon_{jj}/3),$$

$$(6.34)$$

where the various quantities have been defined previously. Expression (6.34) differs from equation (6.28) only in the explicit tensor representation of the strain ε_{ik} and the neglect of all internal variables other than T. From this expression, it is clear that, in an isotropic material, the thermoelastic coupling occurs only through the 'dilatational' or 'hydrostatic compression' component ε_{jj} of ε_{ik}, so that the relaxation strength Δ and the internal friction Q^{-1} vanish for a pure shear. This important and simplifying feature is preserved in anisotropic crystalline materials of cubic symmetry, for which the thermoelastic coupling is similarly characterised by a scalar thermal expansion coefficient α (Landau and Lifshitz, 1970a).

It is readily shown from expression (6.34) for the free energy that the thermoelastic relaxation strength $\Delta_M = \Delta M/M$ for the two cases of a 'Young's modulus' deformation without lateral constraint and plane flexure of a thin plate are given, respectively, by (Ferreirinho, 1986; Nowick and Berry, 1972; Landau and Lifshitz, 1970a)

$$\Delta_E = \Delta E/E = E\alpha^2 T/c_\sigma \tag{6.35}$$

and

$$\Delta_G = \Delta G/G = [(1+\sigma)/(1-\sigma)]E\alpha^2 T/c_\sigma, \tag{6.36}$$

where G is the 'plate flexure modulus' defined earlier in this section, and

$$c_\sigma = c_\varepsilon + 9K\alpha^2 T \tag{6.37}$$

is the specific heat at constant stress. For solids at low temperatures, the second term on the right hand side of equation (6.37) is negligible, so that $c_\sigma \approx c_\varepsilon$ for the purposes of this chapter.

In order to calculate the thermoelastic internal friction in terms of the parameters of the anelastic model, it is also necessary to determine the characteristic relaxation times τ_{th} for the thermoelastic temperature gradients. These may be obtained by solving the heat conduction equation for a material with thermal conductivity κ,

$$\partial T/\partial t = (\kappa/c_\varepsilon)\nabla^2 T - (K\alpha T_0\,\partial\varepsilon_{jj}/\partial t)/c_\varepsilon, \tag{6.38}$$

which may be derived from expression (6.34) for the free energy F (Ferreirinho, 1986; Nowick and Berry, 1972; Zener, 1948). As mentioned previously, the thermoelastic relaxation times τ_{th} are macroscopic relaxation times obtained from the macroscopic heat equation (6.37). These relaxation times moreover depend on the associated boundary conditions, and are therefore strongly dependent on sample geometry.

The two sample geometries which are of practical interest in this chapter are the extensional vibration modes of cylindrical bars, which for the present purposes may be approximated to sufficient accuracy as thin rods, and also the flexural vibration modes of thin discs with free edge. In the former case, the temperature distribution associated with the approximate normal modes, equation (6.8), is sinusoidal in form, and a Fourier decomposition of the heat equation (6.38) shows that the thermoelastic relaxation of the nth normal mode of a bar antenna is characterised by a single relaxation time given by

$$\tau_{th}^{-1} = D_{th}(n\pi v_s/L)^2, \tag{6.39}$$

where $D_{th} = \kappa/c_\varepsilon$ is the thermal diffusivity of the material, and the quantities on the right hand side have been defined in section 6.1. The relaxation time (6.39) is just the characteristic time required for the diffusion of heat over the half-wavelength separation $\lambda/2 = L/n$ between the regions of maximum compression and extension.

The thermoelastic relaxation time of the low frequency flexural modes of a thin disc is by contrast independent of the mode number n for small n (Blair and Ferreirinho, 1982; see also section 6.4), and is given to good approximation by (Nowick and Berry, 1972; Zener, 1948)

$$\tau_{th}^{-1} = \pi^2 D_{th}/h^2, \tag{6.40}$$

which is the characteristic time for thermal diffusion across the plate thickness h. This result is again physically reasonable, since one side of the plate is dilated and cooled during flexure, while the other side is compressed and heated.

Table 6.1. *Estimated thermoelastic internal friction Q_{th}^{-1} for the University of Western Australia GR antenna fundamental longitudinal mode with $\omega = 4.4 \times 10^3\,s^{-1}$.*

T (K)	κ (W m^{-1} K^{-1})	Q_{th}^{-1}
300	5.3×10^{-6}	1.5×10^{-11}
70	5.0×10^{-7}	3.7×10^{-12}
9.2 (n)	3.5×10^{-7}	5.4×10^{-13}
9.2 (s)	8.5×10^{-7}	9×10^{-14}
4.2	50	2×10^{-15}
2.0	20	4×10^{-15}

With the aid of equation (6.23), it may be seen from the above results that the thermoelastic internal friction Q_{th}^{-1} for the low frequency longitudinal modes of a bar antenna is given to good approximation by

$$Q_{th}^{-1} = (E\alpha^2 T/c_\sigma)\omega\tau_{th}/(1 + \omega^2\tau_{th}^2), \tag{6.41}$$

with τ_{th} given by equation (6.39). The corresponding result for the flexural modes of a thin disc is

$$Q_{th}^{-1} = A_n[(1 + \sigma)/(1 - \sigma)](E\alpha^2 T/c_\sigma)\omega\tau_{th}/(1 + \omega^2\tau_{th}^2), \tag{6.42}$$

where $A_n \leq 1$ is a dimensionless numerical constant characteristic of the nth flexural mode and $A_n = 1$ corresponds to plane bending (Blair and Ferreirinho, 1982; see also section 6.4 for the definition of the mode index n). The thermoelastic relaxation time τ_{th} is of course given by equation (6.40) in this case.

For various high Q materials, estimates of the numerical values at a given temperature of the material properties appearing on the right-hand side in equations (6.39) to (6.42) may be obtained from published tables (e.g. Touloukian *et al.*, 1970), or from the general literature. In estimating the values of highly microstructure-dependent transport properties, such as the thermal conductivity κ, published data for samples of a comparable purity and microstructure to the actual antenna or test sample of interest should be used in the estimation. Where sufficiently accurate published data are not available, existing data may be appropriately scaled, although direct measurement of κ in the antenna or test sample or in an offcut from similar material is preferable in that case (Blair and Ferreirinho, 1982).

Numerical estimates of Q_{th}^{-1} at different temperatures for the 3 m long, 1.5 tonne polycrystalline Nb GR antenna at UWA are given in table 6.1, and estimates for a 3 mm thick polycrystalline Nb plate undergoing plane flexure at 1 kHz are given in table 6.2. It is clear from table 6.1 that the thermoelastic relaxation contributes negligibly to the internal friction for the low frequency

Table 6.2. *Estimated thermoelastic internal friction in an annealed Nb plate of thickness 3 mm, assuming plane bending and $\omega = 6.3 \times 10^3 \, s^{-1}$.*

T (K)	τ_{th} (s)	Q_{th}^{-1}
300	4×10^{-6}	6.2×10^{-6}
77	2.4×10^{-2}	1.7×10^{-6}
9.2 (n)	1.5×10^{-4}	1.0×10^{-7}
9.2 (s)	2.7×10^{-4}	2.6×10^{-8}
4.2	5.5×10^{-5}	5×10^{-10}
2.0	5.6×10^{-6}	1×10^{-11}

longitudinal modes of bar antennae, which are currently limited to internal friction levels of $Q^{-1} \geq 10^{-9}$ by other mechanisms (Braginsky et al., 1985). The smallness of Q_{th}^{-1} in this case is due essentially to the long wavelength of the elastic deformation which characterises the fundamental longitudinal mode. This results in a very long thermal diffusion time τ_{th}, of the order of several seconds or longer, so that the audiofrequency longitudinal vibrations are very nearly adiabatic, with $\omega\tau_{th} \gg 1$ and very little irreversible heat flow.

It is also clear from table 6.2 that the thermoelastic contribution Q_{th}^{-1} is much more significant in the case of a thin plate, where the thermal diffusion time (6.40) is several orders of magnitude shorter than the corresponding bar antenna result, equation (6.39), and the large difference between the results in tables 6.1 and 6.2 clearly underlines the strong dependence of Q_{th}^{-1} on sample geometry. In the case of a superconductor such as Nb, Q_{th}^{-1} is readily seen to change abruptly at the superconducting transition temperature T_c due to 'discontinuities' in α and c_σ at this temperature. Clear evidence of this behaviour in Nb is shown in table 6.1 (Blair and Ferreirinho, 1982).

More importantly, this disparity indicates a general need for caution in extrapolating results obtained with small test samples such as discs to a much larger bar antenna. Although the microscopic relaxation mechanisms which are to be discussed in this section are normally much less geometry dependent, reflecting 'bulk' material properties, there are nevertheless some further and significant geometry-dependent effects, arising from sample surface treatments such as machining, which will be discussed briefly in section 6.4.

For antenna materials other than Nb, it may readily be seen (Braginsky et al., 1985; Ferreirinho, 1986) that the magnitude of the longitudinal thermoelastic contribution Q_{th}^{-1} for the fundamental mode of a gravitational radiation (GR) antenna of similar dimensions to the UWA Nb antenna is comparable with or smaller than the values shown for Nb in table 6.1. In the cases of silicon and sapphire, this is essentially due to their very small thermal expansion coefficients

α at liquid helium temperatures, and, in the case of the 5000 series aluminium alloys, to their poor thermal conductivity. No useful purpose would therefore be served by the similarly detailed presentation here of numerical estimates of Q_{th}^{-1} for these materials. In those cases where an estimate is required for small test samples, where Q_{th}^{-1} may be of practical significance, it may readily be obtained from the results (6.39) to (6.42) and the appropriate thermophysical and mechanical data. A table of typical numerical values of some of the thermoelastic parameters appearing in equations (6.39) to (6.42) is given by Braginsky *et al.* (1985).

A slight complication arises for the covalently bonded insulators and semiconducting Si, which all have a crystal symmetry which is lower than cubic and which are therefore anisotropic in their thermal expansion and thermal conduction (Landau and Lifshitz, 1970a). It should also be remembered of course that the elastic moduli such as E and G are anisotropic even in cubic crystals, varying by as much as 50% with direction relative to the crystallographic axes in bcc Nb (Hayes and Brotzen, 1974). An accurate estimate of Q_{th}^{-1} for a single crystal sample then requires a knowledge of the orientation of the crystallographic axes relative to the antenna or sample axes. A rough estimate obtained from a crude averaging of α, κ, E, etc. over the 4π solid angle of possible orientations would, however, suffice for the practical purposes of experimental GR research.

(ii) Thermoelastic loss due to intercrystalline thermal currents

Although Nb and Al have bcc and fcc structure, respectively, so that α and κ are both scalars, the necessarily polycrystalline nature of the large metal ingots from which all the currently operating, metallic antennae are fabricated gives rise to a further anisotropic effect which must be considered. In a polycrystalline sample, the crystalline anisotropy gives rise to inhomogeneity in the elastic deformation and thus also the temperature of neighbouring crystal grains, and therefore to an additional thermoelastic contribution Q_{ic}^{-1} due to intercrystalline thermal currents (Landau and Lifshitz, 1970a; Nowick and Berry, 1972; Zener, 1948). Strictly speaking, the intercrystalline grain boundaries are classified as crystal defects, and this intercrystalline thermoelastic relaxation is therefore not a perfect crystal mechanism. It will nevertheless be discussed here rather than in the section on defect-induced mechanisms, because of its close relationship and similarity to the intracrystalline thermoelastic relaxation discussed above.

The magnitude of Q_{ic}^{-1} depends on the size and relative orientation of the grains and, in general, only a rough estimate of it can be obtained for a given polycrystalline specimen. A crude, order of magnitude estimate is easily obtained in the case where the grains are assumed to be randomly oriented (Zener, 1947). A rough measure of the inhomogeneity or relative mean fluctuation R in the strain and the stress in neighbouring crystal grains is then given by the relative mean-square variation in a characteristic elastic constant such as Young's modulus E or the reciprocal compliance E^{-1}. If the latter quantity is used, the anisotropy and inhomogeneity factor R is to be defined as (Zener, 1947)

$$R = (\langle E^{-2} \rangle - \langle E^{-1} \rangle^2)/\langle E^{-1} \rangle^2, \qquad (6.43)$$

where the average is to be taken over the 4π solid angle of possible orientations relative to the crystallographic axes. Substitution of the relationship between E^{-1} and the elasticity tensor c_{ijkl} (Ferreirinho, 1986; Landau and Lifshitz, 1970a) into equation (6.43) with numerical values appropriate for Nb (Hayes and Brotzen, 1974) gives $R \sim 0.025$ for this metal.

If the wavelength λ of the acoustic wave is long compared with the typical linear dimension d of the crystal grains, then the condition $\omega\tau_{th} \ll 1$ is also satisfied, where $\tau_{th} = c_\sigma d^2/\kappa = d^2/D_{th}$ is the time required for heat to diffuse across the typical crystal grain diameter d. The vibrations are then very nearly isothermal and simple dimensional arguments (Landau and Lifshitz, 1970a) show that

$$Q_{ic}^{-1} = R(E\alpha^2 T/c_\sigma)\omega\tau_{th} = R\, \Delta_E \omega\tau_{th}. \qquad (6.44)$$

In the opposite limit where the acoustic wavelength is short compared with the grain size, so that $\omega\tau_{th} \gg 1$, the deformation is nearly adiabatic in the interior of each grain. Heat conduction is then confined to the grain boundaries, with the propagation of 'temperature waves' over a characteristic distance of order $\delta = (D_{th}/\omega)^{1/2}$ from the boundaries. Since only a fraction δ/d of the volume of each grain is thus involved in the dissipation of energy by thermal conduction, Q_{ic}^{-1} is approximated in this limit by

$$Q_{ic}^{-1} = R(E\alpha^2 T/c_\sigma)(\omega\tau_{th})^{-1/2} = R\, \Delta_E(\omega\tau_{th})^{-1/2}. \qquad (6.45)$$

As the Nb bar antenna at UWA is polycrystalline with a typical grain size $d = 50$ mm, it is of interest to estimate the numerical value of Q_{ic}^{-1} for vibrations at the fundamental longitudinal mode frequency $\omega = 4 \times 10^3\,s^{-1}$ at various temperatures, and these estimates are given in table 6.3. Finally, it may be noted that in a non-cubic material such as a sapphire polycrystal, there is an additional source of intercrystalline thermal damping due to the anisotropy of the thermal expansion coefficient.

It will be seen in section 6.4 that the measured value of Q^{-1} for the fundamental mode of the UWA antenna at 300 K (Veitch et al., 1987) is about a factor of two larger than the value of Q_{ic}^{-1} shown for this temperature in table 6.3. This suggests that Q_{ic}^{-1} is an important loss mechanism contributing about 50% of the observed room temperature internal friction of the antenna. It must be remembered, however, that the results in table 6.3 are much more uncertain than those in tables 6.1 and 6.2, so that any agreement to within a factor smaller than two would in any case have to be regarded as fortuitous.

(iii) Phonon relaxation
A more typical example of a relaxation mechanism occurring in crystalline solids is that of phonon relaxation (Akhieser, 1939; Bommel and Dransfeld, 1960;

Table 6.3. *Estimated numerical values of intercrystalline thermoelastic damping for the University of Western Australia GR antenna fundamental longitudinal mode with* $\omega = 4.4 \times 10^3\,s^{-1}$.

T (K)	Q_{ic}^{-1}
300	9×10^{-8}
77	2×10^{-8}
9.2 (n)	2×10^{-10}
9.2 (s)	4×10^{-11}
4.2	2×10^{-12}
2.0	2×10^{-12}

Maris, 1971; Woodruff and Ehrenreich, 1961). This is in fact just the low-frequency limit of the 'phonon–phonon interaction' arising from the anharmonicity of the crystal forces, and the relaxation time τ_{ph} for the mechanism is the typical or mean lifetime of the thermal phonons. This interaction generally provides a significant mechanism for acoustic attenuation in insulators, especially at ultrasonic frequencies (Bommel and Dransfeld, 1960; Braginsky et al., 1985).

The relaxation formalism of this section is only applicable to this mechanism when $\omega\tau_{ph} < 1$, i.e. when the period of the acoustic wave is longer than τ_{ph}. In this frequency range the mean free path of the thermal phonons is shorter than the wavelength of the sound and a 'local relaxation' or 'hydrodynamic' limit obtains, in which the strain-modulated phonon distribution relaxes towards a local equilibrium determined by the local value of the strain as a result of the anharmonic coupling. The necessity that such a local limit should prevail in order that a given dissipative mechanism may be classified as a relaxation mechanism may be seen from the fact that the anelastic relaxation model is itself local in character, like the theory of ideal elasticity which it is intended to generalise.

Since the thermal phonons propagate through the crystal with typical velocity $v_s = \omega/q$ and mean free path $l_{ph} = v_s\tau_{ph}$, the condition $ql_{ph} \approx 1$ where $\mathbf{q}$ is the acoustic wavevector therefore marks a transition between the local or 'relaxation' regime $ql_{ph} < 1$ and non-local or 'acoustic phonon scattering' regime $ql_{ph} > 1$. In the former regime, the entire phonon distribution is modulated by the acoustic wave, while, in the latter, the individual acoustic phonons are scattered by the interaction (Landau and Rumer, 1937; Morse, 1955; Pippard, 1955). Only the relaxation regime $ql_{ph} < 1$ is of practical interest in this chapter, since this condition is always well satisfied in practice for the low frequency modes of mechanical resonators, even in near perfect crystals at the lowest attainable temperatures where l_{ph} is greatest.

Although the high frequency limit $\omega\tau_{ph} > 1$ in which the relaxation formalism fails is not of practical interest in this chapter, it is generally of great importance in solid-state physics, (Pippard, 1960; Ziman, 1960). Here, it is sufficient to note that, because the relaxation formalism usually fails in the later regime, the high-frequency side of the 'Debye' relaxation peak shown in figure 6.1 is unattainable physically for those microscopic mechanisms involving thermal excitations or microscopic degrees of freedom which propagate through the crystal at speeds comparable with or greater than the speed of sound, such as electrons and phonons.

Returning to the specific mechanism of phonon relaxation, it is clear that, as a result of the anharmonicity, the frequencies of the thermal phonons will be modulated by the acoustic strain. If all of the phonon frequencies were shifted by the same relative amount $\Delta\Omega/\Omega$, a phonon distribution which was initially in thermal equilibrium would remain in equilibrium and no relaxation process would occur. However, the three-dimensional, anisotropic and dispersive nature of phonon propagation ensures that different parts of the phonon spectrum will experience different relative frequency shifts characterised by generalised Grüneisen parameters $\gamma_{k\alpha}$ given by

$$\Delta\Omega_{k\alpha}/\Omega_{k\alpha} = \gamma_{k\alpha}\varepsilon. \tag{6.46}$$

The thermal phonon wavevector $\mathbf{k}$ and the polarisation vector α specify a mode or 'branch' of the phonon spectrum, and the tensor character of the strain ε has again been ignored for the present. Since the equilibrium distribution of the phonons is the Planck distribution

$$N_{k\alpha} = [\exp(h\omega_{k\alpha}/kT) - 1]^{-1}, \tag{6.47}$$

it may be seen that the 'effective temperature' of the various branches is shifted by differing amounts by the acoustic strain and that the interactions between the thermal phonons will tend to restore thermal equilibrium. Crudely therefore, the phonon relaxation mechanism may be regarded as irreversible heat flow between different branches of the phonon spectrum, induced by the deformation associated with the acoustic wave or vibration.

The internal variable which corresponds to this mechanism in the present macroscopic description must be a parameter which measures the mean deviation of the phonon distribution from equilibrium. In principle, such a parameter can be obtained by appropriately averaging the parameters $\gamma_{k\alpha}$ over the branches of the phonon spectrum. The complexity of anharmonic lattice dynamics is such, however, (Leibfried and Ludwig, 1961; Maris, 1971) that such a derivation of a single parameter which describes the complex kinetic processes occurring in phonon relaxation and whose value may be compared with experimental results is not possible without a gross oversimplification of the theory. Similar remarks apply to the mean phonon lifetime τ_{ph}. In any case, the values of the parameters $\gamma_{k\alpha}$ from which these parameters could be calculated are not known for most materials, and crude estimates only can be made.

In writing the relation (6.46) which specifies the Grüneisen or anharmonicity parameter $\gamma_{k\alpha}$ as a specified scalar equation, the tensor nature of the strain ε_{ik} was of course unspecified. The usual definition of the Grüneisen parameter involves the dilatational component ε_{jj} (Ashcroft and Mermin, 1976; Bommel and Dransfeld, 1960), so that

$$\Delta\Omega_{k\alpha}/\Omega_{k\alpha} = \gamma_{k\alpha}\varepsilon_{jj} = \gamma_{k\alpha}\,\Delta V/V, \tag{6.48}$$

where V is the sample volume. Strictly, therefore, any estimate of the phonon relaxation contribution Q_{ph}^{-1} to the acoustic loss which is based on equation (6.48) is limited to the 'pure dilation' component of the acoustic strain.

More general treatments which relate the attenuation of an acoustic wave with arbitrary polarisation β or acoustic strain ε_{ij} to a generalised Grüneisen tensor $\gamma_{k\alpha}^{ij}$ are given by Maris (1971) and Woodruff and Ehrenreich (1961). For the purposes of this chapter, it will however suffice to give a rough estimate of Q_{ph}^{-1} for a pure dilation based on a highly simplified treatment (Braginsky et al., 1985), and to note that the shear attenuation is generally of comparable magnitude (Bommel and Dransfeld, 1960). The estimate to be given here is moreover specialised to the relaxation limit $ql_{ph} = \omega\tau_{ph} \ll 1$, which is of practical interest here. In this limit, the spatial variation of the acoustic strain is neglected on the microscopic scale of l_{ph}, and the deformation is effectively regarded as homogeneous $(ql = 0)$.

With the above simplifications, an estimate of Q_{ph}^{-1} is readily obtained by application of the relaxation formalism reviewed in section 6.2 to the phonon relaxation process (Braginsky et al., 1985; Ferreirinho, 1986; Nowick and Berry, 1972). The resulting expression for Q_{ph}^{-1} is the very crude approximation

$$Q_{ph}^{-1} = (K\alpha_g^2 T/c_g)\omega\tau_{ph}, \tag{6.49}$$

valid for $\omega\tau_{ph} \ll 1$, where α_g and c_g are the respective lattice contributions to the thermal expansion α and the specific heat c. This result may at best provide an order of magnitude estimate of Q_{ph}^{-1} in a crystalline insulator at temperatures below the Debye temperature (Ferreirinho, 1986). As mentioned previously, the difference in Q_{ph}^{-1} between dilations and shear deformations or between longitudinal and shear waves may reasonably be ignored at this level of accuracy.

In order to apply the result (6.49) to metals also, the further approximation $\alpha_g/c_g \approx \alpha_{el}/c_{el}$ is also required, with the right hand side denoting electronic contributions to the subscripted quantities, so that there should be little net heat exchange between the phonons and the electrons during the deformation process (Ferreirinho, 1986). In practice, this is certainly a good approximation at the present level of accuracy (Ashcroft and Mermin, 1976), so that equation (6.49) may be used to estimate very roughly the magnitude of Q_{ph}^{-1} in metals such as Nb also, provided that the lattice contributions α_g and c_g to the thermal expansion and specific heat and the typical phonon lifetime τ_{ph} are known at a given temperature.

Estimated numerical values of Q_{ph}^{-1} for sapphire and Nb are given in tables 6.4

Table 6.4. *Estimated numerical values of the phonon relaxation contribution Q_{ph}^{-1} for sapphire monocrystal at various temperatures for a frequency of $\omega = 6.3 \times 10^3 \, s^{-1}$.*

T (K)	Q_{ph}^{-1}
300	5×10^{-12}
4.2	4×10^{-15}
2.0	2×10^{-15}

Table 6.5. *Estimated numerical values of Q_{ph}^{-1} in annealed Nb at various temperatures for a frequency $\omega = 6.3 \times 10^3 \, s^{-1}$.*

T (K)	τ_{ph} (s)	Q_{ph}^{-1}
9.2 (n)	2.5×10^{-11}	1×10^{-15}
9.2 (s)	2.5×10^{-11}	1×10^{-15}
4.2	2.5×10^{-10}	1×10^{-14}
2.0	2.5×10^{-8}	8×10^{-14}

and 6.5, respectively. Since the microscopic relaxation time or thermal phonon lifetime τ_{ph} is usually short compared with the acoustic vibration period, Q_{ph}^{-1} is proportional to ω.

(iv) Conduction electron relaxation

The final example of an anelastic relaxation mechanism to be considered here is conduction electron relaxation in metals, which is analogous to the phonon relaxation mechanism just discussed. The electron relaxation mechanism is the low-frequency limit of the electron–phonon interaction, which is responsible for the well-known ultrasonic attenuation in normal metals (Pippard, 1960) and in superconductors (Morse, 1959, 1962). It is the dominant attenuation mechanism in perfect or very pure normal metal crystals at low temperatures and in perfect superconducting crystals near the transition temperature. It is of major importance in real metals of high purity, and may dominate the internal friction at low temperatures in high-purity specimens at frequencies as low as 100 kHz (Lax, 1959).

As in the case of phonon relaxation, the electron–phonon interaction gives rise to a relaxation mechanism only in the limit where the electron mean free path l of the conduction electrons is shorter than the characteristic wavelength of the acoustic wave. Except in very pure metals, this condition is well satisfied at all practically attainable cryogenic temperatures and at frequencies of up to 10 MHz or more. In a highly simplified picture in this limit (Mason, 1955; Morse, 1955; Pippard, 1955), the acoustic strain produces non-equilibrium modulation of the conduction electron distribution as a result of the electron–phonon interaction. This modulated distribution relaxes through collisions with thermal phonons and crystal defects towards a local equilibrium distribution determined by the local values of the strain and the ionic velocity (Holstein, 1959; Pippard, 1955).

The conduction electrons may thus be regarded as exerting both viscous and elastic forces on the ionic lattice, giving rise to acoustic attenuation. These forces are characterised in a simple approximation by the Fermi energy or 'Fermi pressure', so that the latter is an appropriate internal variable in a simplified macroscopic description of the electronic contribution within the formalism of the anelastic model. Clear experimental evidence of a significant electronic relaxation contribution to Q^{-1} in Nb at temperatures near the superconducting transition temperature T_c will be presented in section 6.4.

The form of the relaxed and unrelaxed Fermi–Dirac distributions of the conduction electrons in the presence of an acoustic wave is readily obtained in the highly oversimplified, spherically symmetric free electron approximation (Ferreirinho, 1986; Nowick and Berry, 1972). It is found that, in this approximation, no departure from equilibrium occurs for a pure dilation, so that the electronic contribution Q_{el}^{-1} to the internal friction is proportional to the shear component of the acoustic deformation.

The free electron approximation naturally neglects the large anisotropy of the electron distribution characterised by a non-spherical Fermi surface and band structure in transition metals such as Nb. The free electron approximation nevertheless gives a useful order of magnitude estimate of Q_{el}^{-1} in such metals.

For a pure shear deformation, the conduction electron contribution to the internal friction is given by (Nowick and Berry, 1972)

$$Q_{el}^{-1} = nmv_F^2 \omega \tau_{el}/5\rho v_s^2, \qquad (6.50)$$

where n is the conduction electron density, m is the effective electron mass, v_F is the free electron Fermi velocity, ρ is the mass density and v_s is the speed of the acoustic shear wave.

The corresponding result for a Young's modulus' extension without lateral constraints on the Poisson contraction in a metal with Young's modulus E may similarly be obtained as (Filson, 1959; Lax, 1959)

$$Q_{el}^{-1} = \Delta_E \omega \tau_{el} = (4/15)(1 + \sigma)^2 nmv_F^2 \omega \tau_{el}/\rho v_E^2, \qquad (6.51)$$

where σ is Poisson's ratio and $v_E = (E/\rho)^{1/2}$. The result for an acoustic deformation with unilateral constraint characteristic of the flexure of a thin plate is found to be (Ferreirinho, 1986)

$$Q_{el}^{-1} = \Delta_G \omega \tau_{el} = (4/15)[(1 + \sigma^3)/(1 - \sigma)]nmv_F^2\omega\tau_{el}/\rho v_E^2, \qquad (6.52)$$

where the subscript G denotes the 'plate flexure modulus' $G = E/(1 - \sigma^2)$. As in the case of the thermoelastic result (6.42) this result applies to simple or plane bending of a thin plate with zero intrinsic curvature, and must be multiplied by some numerical factor less than unity representing the proportion or degree of shear involved in a particular flexural mode of a real disc sample. Such modes always have negative intrinsic curvature due to the unilateral Poisson contraction and expansion. The corresponding coefficients for the degree of dilation are discussed by Blair and Ferreirinho (1982).

It is worth noting here that the three results (6.50), (6.51) and (6.52) differ only by a factor of order unity, which is not significant in the context of the drastic oversimplification and approximation involved in the use of a free electron model to describe the complex electronic structure of a transition metal such as Nb. It is also interesting to note here that the typical magnitude of the electronic relaxation strength Δ_{el} in equation (6.50) to (6.52) is of order 10^{-1} and therefore several orders of magnitude larger than the corresponding relaxation strengths for the thermoelastic effect and phonon relaxation, especially at low temperatures. This result suggests that acoustic vibrations in a very pure, near perfect normal metal crystal will be very highly damped in the limit $T \rightarrow 0$, where the kinetic factor $\omega\tau_{el}$ attains its maximum value. This is in fact consistent with experiment (Lax, 1959), although τ_{el} is in practice always limited by impurities and other defects.

It may be shown (Ferreirinho, 1986) that the internal friction of a hypothetical gravitational radiation antenna fabricated as a large, extremely pure, perfect normal metal single crystal would approach $Q_{el}^{-1} = 10^{-3}$ in the limit $T \rightarrow 0$. Such an antenna material would be a very poor choice therefore, even if it were entirely free of dislocations. The presence of dislocation induced internal friction mechanisms makes matters even worse, so that a real normal metal antenna certainly needs to be constructed from a relatively impure metal (Suzuki et al., 1978).

Since Nb undergoes a superconducting transition at $T_c = 9.2$ K, it is important to consider the effect of superconductivity on Q_{el}^{-1}. A discussion of the theory of superconductivity is, however, beyond the scope of this chapter. It must suffice here to note that the presence of an energy gap $\Delta(T)$ and a modified density of states at the Fermi surface in the superconducting electronic excitation spectrum (Bardeen, Cooper and Schrieffer, 1957) lead to the well-known result (Kadanoff and Pippard, 1966)

$$Q_{el,s}^{-1} = Q_{el,n}^{-1}2/[\exp(\Delta(T)/kT) + 1] = Q_{el,n}^{-1}[2f_0(\Delta)] \qquad (6.53)$$

for Q_{el}^{-1} in the superconducting state in the relaxation regime. Here, f_0 is the well-known Fermi–Dirac function.

Table 6.6. *Estimated numerical values of electronic contribution Q_{el}^{-1} for the $\omega = 4.4 \times 10^3 \, s^{-1}$ fundamental mode of the UWA GR antenna and for plane flexure of a thin Nb plate with $\omega = 6.3 \times 10^3 \, s^{-1}$.*

T (K)	Antenna $Q_{el,a}^{-1}$	Plate $Q_{el,p}^{-1}$
9.2	5.1×10^{-9}	4.5×10^{-9}
4.2 (s)	2.6×10^{-10}	2.3×10^{-10}
2.0 (s)	3.1×10^{-12}	2.7×10^{-12}

Estimates of Q_{el}^{-1} for the fundamental mode of the UWA antenna and for a 1 kHz flexural mode of a thin plate sample in the plane bending approximation are given in table 6.6 for various temperatures. Experimental evidence for a significant electronic contribution to the internal friction at higher frequencies of order 10 kHz will be presented in section 6.4.

(v) Limitations of the anelastic model

The examples of relaxation mechanisms which have just been discussed suggest that the anelastic model has several limitations. The requirement that a local limit should prevail on the microscopic level is not restrictive within the scope of this chapter, since it is well satisfied at audiofrequencies even in very pure metals and at very low temperature. However, the evaluation of model parameters such as the internal variables and coupling coefficients is in general very difficult, as indicated by the phonon and electron relaxation examples. The model nevertheless provides a unifying description and a useful guide to the identification and qualitative analysis of a variety of important loss mechanisms in crystalline solids. Some of these mechanisms are of particular importance in the interpretation of experimental results to be presented in section 6.4.

A more serious and practical limitation of the anelastic model from the point of view of this chapter, however, is the disagreement of the model with experiment at very low frequencies, especially in metals, since it predicts vanishingly small internal friction in the quasi-static limit $\omega\tau \to 0$ (see equation 6.23). This prediction is reasonable for perfect crystals, where the existence of an equation of state should indeed ensure thermodynamic reversibility of small deformations in the quasi-static limit. It has however long been known (Love, 1944; Zwikker, 1954) that there is a significant 'background' component of internal friction in metals that is essentially independent of frequency from audiofrequencies to very low frequencies below 1 Hz. Such internal friction is observed in many metals at room temperature (Routbort and Sack, 1966) and also at low temperatures (Bruner, 1960; Suzuki et al., 1978).

It is clear that the low frequency background is due to the presence of crystal defects and there is evidence that some of the mechanisms are hysteretic in nature and involve the motion of dislocations. These mechanisms are not yet understood in detail (Granato and Lücke, 1966; Koehler, 1952; Nowick and Berry, 1972).

It will be seen in section 6.4 that deviations from the 'classical' anelastic behaviour predicted by equation (6.33) for thermally activated relaxation mechanisms also occur at liquid helium temperatures as a result of a transition to quantum-mechanical tunnelling as the dominant relaxation mode. In this 'quantum relaxation' regime, the kinetics of the defect relaxation are characterised by a quantum-mechanical transition rate, τ_t^{-1}, rather than a 'classical' thermally activated relaxation time, $\tau = \tau_0 \exp(H_a/k_B T)$. It is not clear at present whether the symmetrical 'Debye' frequency dependence, equations (6.23) and (6.26), for Q^{-1} on frequency ω remains generally valid in the tunnelling regime.

6.3.3   Defect relaxation mechanisms

(i) Scope of section
In contrast with the three basic 'perfect crystal' mechanisms just reviewed, there are many more defect-induced mechanisms, with several new relaxation peaks being discovered each year in recent years (Benoit and Gremaud, 1981; Fantozzi and Vincent, 1983). Although many mechanisms due to point defects such as impurities are well understood (Nowick and Berry, 1972), the general understanding of the intrinsic relaxation processes involving dislocations or the interaction of dislocations with other defects is far from complete (Fantozzi and Ritchie, 1981; Groh and Schultz, 1981; Seeger and Wüthrich, 1976). For reasons which will be outlined later, the dislocation-induced internal friction spectrum in bcc metals such as Nb is usually more complex than that of fcc or hcp metals.

Considerable progress has however been made recently, particularly in the understanding of the effect of residual hydrogen content on the low-temperature internal friction spectrum of the bcc transition metals, including Nb (Cannelli, Cantelli and Vertechi, 1982; Funk, Maul and Schultz, 1983; Funk and Schultz, 1985; Groh and Schultz, 1981; Maul and Schultz, 1981; Poker et al., 1979). The extent and the rapid development of these areas of research place a comprehensive review of the defect-induced internal friction mechanisms beyond the scope of this chapter.

Those results which are relevant to this chapter will be reviewed in this section, though in much less detail than the 'simpler' perfect lattice mechanisms reviewed earlier. In most instances, the review will be no more than a summary of recently reported experimental results and a brief outline of models for their interpretation which are discussed in the literature. All of these results relate to anelastic relaxation mechanisms.

(ii) Point defect relaxation

The long-range elastic strain fields surrounding point defects such as interstitial or substitutional impurities or vacancies provide an obvious coupling of the defect system to an external stress or strain applied to the crystal. This coupling can give rise to a variety of anelastic relaxation processes. In the cases of interstitials, for example, the equilibrium distribution of impurity atoms among crystallographically equivalent interstitial sites of a particular symmetry may be altered by the applied stress. A necessary condition for this anelastic coupling of the defect system is that the symmetry of the defect in its interstitial site should be lower than that of the crystal, since the crystallographically equivalent sites would otherwise remain equivalent even in the presence of the applied stress (Nowick and Berry, 1972). If the temperature is sufficiently high for thermally activated transitions between sites, a 'stress-induced ordering' occurs which is analogous in some respects to the ordering of magnetic dipoles by an external magnetic field.

For interstitial impurities other than hydrogen, the activation energy H_a for transitions between sites is of the order of 1 eV, while the characteristic frequency τ_0^{-1} of the defect vibrations is of the order of the Debye frequency, i.e. $\tau_0 \approx 10^{-14}$ s (Nowick and Berry, 1972). Substitution of these approximate values into the expression (6.31) for the thermally activated relaxation time τ shows that for audiofrequencies ω the internal friction maximum at $\omega\tau = 1$ is attained at temperatures of order 500 K, i.e. above room temperature. Such thermally activated anelastic processes due to stress-induced ordering were first observed by Richter (1938) in bcc iron containing carbon, and related to the presence of the carbon impurity by Snoek (Nowick and Berry, 1972; Snoek, 1939). Relaxation processes of this type are therefore called Snoek relaxations.

In estimating the strength of the Snoek relaxation, it is reasonable to expect that for low impurity concentrations the interaction between defects will be slight, so that the relaxation strength Δ_{SR} is proportional to the concentration C. Simple kinetic and thermodynamic arguments (Nowick and Berry, 1972) suggest that the internal friction Q_{SR}^{-1} due to the Snoek relaxation is given by

$$Q_{SR}^{-1} = (\lambda\rho v_s^2 Cv_0/nkT)[\omega\tau/(1+\omega^2\tau^2)] \qquad (6.54)$$

for temperatures T which are sufficiently high so that $kT \gg \sigma v_0$, where σ is the applied stress. Here ρ is the density, v_s is the speed of sound, C is the defect concentration as a molar or atomic fraction, n is the number of crystallographically equivalent sites, the coefficient λ is a generalised size factor of order unity which characterises the local elastic strain produced by the defect, and v_0 is the atomic or molecular 'volume of formation' of the defect.

The factor $\lambda\rho v_s^2 v_0$ is usually termed the elastic dipole strength, because of the lower symmetry of the defect strain field which forms an 'elastic dipole' within the more symmetric crystal structure. In a bcc crystal, the defects usually retain the tetragonal symmetry of the octahedral and tetrahedral interstices in the bcc structure (Fast, 1976; Nowick and Berry, 1972). A rough numerical evaluation of equation (6.54) at the peak temperatures shows that the proportionality between the relaxation strength and the defect concentration is typically $\Delta_{SR}/C \approx 0.1$ (at. %)$^{-1}$, in agreement with experiment (Nowick and Berry, 1972; Powers and Doyle, 1959).

Many other point defect relaxation processes besides the Snoek effect have been observed and analysed (Nowick and Berry, 1972), such as the Zener relaxation due to the stress-induced reordering of substitutional solute pairs in dilute alloys (Zener, 1947). These processes also have activation energies of order 1 eV (deBatist, 1972; Nowick and Berry, 1972) and therefore become insignificant at low temperatures. From the point of view of this chapter, the basic theory of the Snoek relaxation is however of some interest, since it is relevant to the behaviour of interstitial hydrogen in Nb at low temperatures (Baker and Birnbaum, 1972; Cannelli et al., 1982; Huang, Granato and Birnbaum, 1985; Schiller, 1976; Zapp and Birnbaum, 1980a,b), which influences the internal friction spectrum. It should be noted here also although the Snoek peaks for impurities such as oxygen and nitrogen occur at temperatures of order 400°C and are thus completely 'frozen out' at cryogenic temperatures, the low temperature 'tail' of such peaks can contribute significantly to Q^{-1} at room temperature. This indicates a need to use metals with low residual oxygen and nitrogen concentrations in the construction of high Q resonators for room temperature applications.

It is clear that even a very small concentration in the p.p.m. range can contribute significantly to the internal friction in the temperature range in which stress-induced reorientation can occur. Unlike oxygen and nitrogen, which produce Snoek peaks at 370°C and 600°C with activation energies of 1.2 eV and 1.5 eV, respectively (Powers and Doyle, 1959), interstitial hydrogen remains mobile in solid solution at much lower temperatures because of the smaller defect volume and activation energy. The Snoek effect due to hydrogen in Nb has not however been observed, because the equilibrium concentration of the dilute α phase of interstitial hydrogen which could give rise to a hydrogen Snoek relaxation in Nb is too small at temperatures of order 50 K where the peak would be expected to occur (Ferreirinho, 1986; Schober and Wenzl, 1978). Estimates of the degree of hydrogen contamination in the various samples used in the experimental investigation will be given in later chapters.

Since the activation energy for defect reorientation by the Snoek relaxation is in effect the energy barrier to 'jumps' between neighbouring interstitial sites, it is the same as the activation energy for the long-range migration or diffusion of the solute in the crystal (Nowick and Berry, 1972). The value of H_a for hydrogen in Nb can therefore be estimated from diffusivity measurements on the macroscopic scale.

Some understanding has been gained in recent years of the importance of quantum-mechanical tunneling in determining the mobility of crystal defects at low temperatures (Flynn and Stoneham, 1970; Kehr, 1978). The importance of quantum effects is essentially due to the small mass of the H atoms, whose wavefunctions are therefore not quite localised within their interstitial sites by the energy barrier of order 0.1 eV, in conformity with the requirements of the uncertainty principle.

It is clear that any 'residual' mobility of H at very low temperatures must be entirely due to the tunnelling process. In the intermediate temperature range, a combination of thermal activation and tunnelling is believed to operate, and measurements between 30 K and 200 K indicated that the diffusion coefficient in Nb obeys an Arrhenius law with an activation energy of 0.068 eV (Engelhard, 1979; Kehr, 1978). No measurements at lower temperatures have been reported to date, and it can only be stated here that the intermediate temperature regime would therefore appear to extend down to 30 K at least, where the diffusion coefficient is of order 10^{-14} cm^2 s^{-1} (Engelhard, 1979). Above 250 K, the experimental results for the diffusion coefficient obey an Arrhenius law with activation energy $H_a = 0.11$ eV, suggesting that this is in fact the magnitude of the energy barrier to interstitial migration, and that this barrier is overcome solely by thermal activation at these high temperatures.

From the present point of view, the significance of the above results is that the mobility of H in Nb would appear to obey an Arrhenius law down to temperatures below 50 K where the internal friction maximum due to the Snoek relaxation is expected to occur at audiofrequencies. In the absence of an observed Snoek relaxation even at relatively low internal friction levels of order 10^{-6} in a sample containing 2000 at. p.p.m. H (Cannelli and Cantelli, 1982; Cannelli et al., 1982), it may be surmised that stress-induced reordering of single interstitial hydrogen atoms does not contribute significantly to the internal friction in Nb at liquid helium temperatures. Since this chapter is concerned with much lower internal friction levels of order 10^{-8}, however, further experimental investigation of the influence of hydrogen on the internal friction spectrum in Nb at liquid helium temperatures is desirable. In this connection, it should be stated here that no such experiments were carried out in the experimental investigation at UWA.

Although the hydrogen Snoek relaxation has not been observed in bcc transition metals, at least four different relaxation peaks related to the presence of interstitial H are observed in the internal friction spectrum of Nb below room temperature. The peak temperatures and typical peak heights are shown in figure 6.5, which is only a composite summary of the basic features of the low-temperature relaxation processes in Nb which have been reported in the literature, and should not be interpreted as a quantitative or even qualitative description of the experimental results which may be obtained for a particular sample. Some of these processes cannot strictly be regarded as point defect relaxations, since they involve the interaction of interstitial H with dislocations (Bruner, 1960; Funk and Schultz, 1985).

A broad internal friction peak which is not shown in figure 6.5 has been observed in the temperature range from 200 K to 300 K in Nb with significant hydrogen concentrations ($C_H > 200$ at. p.p.m.) and has conclusively been demonstr-

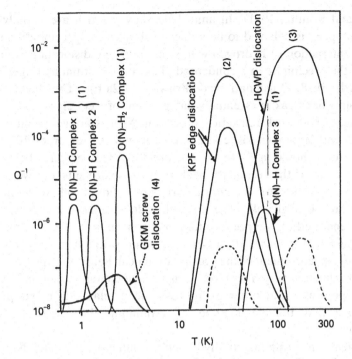

Figure 6.5. Survey of some low-temperature relaxation processes observed or expected in Nb at frequencies of order 1 kHz. (1) Hydrogen point defect relaxation peaks: O(N)–H Complex 1, extrapolated from 10 MHz results of Huang *et al.* (1985); O(N)–H Complexes 2 & 3, extrapolated from results of Cannelli and Cantelli (1982); O(N)–H$_2$ Complex, Huang *et al.* (1985). (2) Intrinsic dislocation relaxation peak due to thermally activated kink-pair formation (KPF) on edge dislocations. The heights of the continuous curves are typical of cold worked material, while the dashed curve indicates the expected height in the annealed state (Funk and Schultz, 1985). (3) Hydrogen cold work peak (HCWP), due to interaction of non-screw dislocations with interstitial hydrogen. Height of continuous curve is typical of cold worked material, while the dashed curve indicates the expected height in the annealed state (Funk and Schultz, 1985; Verdini and Baci, 1980). (4) Relaxation peak observed in recrystallised Nb disc samples and attributed in section 6.4 to the geometrical kink migration (GKM) mechanism proposed by Seeger and Wüthrich (1976). This peak can also be identified with the similar peak observed by Kramer and Bauer (1967).

ated to be due to stress-assisted precipitation of the condensed Nb–H phases (Yoshinari and Koiwa, 1984). The peak height decreases with increasing frequency, and the peak is always accompanied by a secondary peak in the range 120 K to 150 K.

The internal friction peak at about 140 K which accompanies the precipitation peak is believed to be the 'hydrogen – cold work peak' (Fantozzi and Ritchie, 1981; Funk *et al.*, 1983; Verdini and Baci, 1980) which is observed in cold worked bcc metals containing hydrogen impurity. It has recently been demonstrated conclusively that this latter peak arises from an interaction between the dislocations generated during cold working and the interstitial hydrogen (Funk and Schultz,

1985; Maul and Schultz, 1981). In annealed samples which are initially free of cold work, the peak is believed to be induced solely by the precipitation process, by this same interaction of hydrogen with stress-relieving dislocation loops which form around the precipitate of a condensed phase of interstitial hydrogen (Ferron and Quintard, 1978; Yoshinari and Koiwa, 1982a,b). The height of the hydrogen–cold work peak is essentially independent of frequency.

The hydrogen–cold work relaxation process in Nb has an activation energy of about 0.28 eV and a characteristic attempt frequency τ_0^{-1} of order $10^{-12} \, \mathrm{s}^{-1}$, so that the relaxation should again freeze out rapidly at temperatures below 100 K. If the identification of the secondary peak with the hydrogen–cold work peak is correct, the contribution to the internal friction at liquid helium temperatures should therefore be negligible. The possibility does remain, however, that the dislocations generated by the precipitation may contribute to the internal friction at liquid helium temperatures through some other mechanism. It is also worth noting here that measurement of the precipitation peak height may provide a useful technique for detecting very high levels of hydrogen contamination in samples. It will be seen presently that other hydrogen-induced relaxation peaks provide a more sensitive and reliable method of estimating the level of hydrogen contamination.

Some evidence for trapping of hydrogen by interstitial O and N in Nb is provided by the two low-temperature internal friction peaks at 2.6 K and 90 K shown in figure 6.5. These peaks have been observed only in samples in which significant levels of oxygen (or nitrogen) and hydrogen are present together (Cannelli and Cantelli, 1982; Huang *et al.*, 1985; Poker *et al.*, 1979). The height of the 100 K peak is proportional to the hydrogen concentration C at low concentrations and saturates when C becomes comparable to the oxygen or nitrogen concentration in doped samples (Baker and Birnbaum, 1973). For these reasons, the relaxation peak at 100 K is shown as the O(N)–H peak. The ratio Q_p^{-1}/C of the peak height to hydrogen concentration in polycrystalline samples is of order 5×10^{-4} (at. %)$^{-1}$ in the unsaturated region for hydrogen concentrations of order 2000 at. p.p.m. (Cannelli and Cantelli, 1982).

This behaviour of the 100 K relaxation peak may be interpreted as a saturation of the O or N trap sites or as a reduction in the effective dipole strengths of the O(N)–H complexes, or as a combination of these two effects (Qi, Volkl and Wipf, 1982). In polycrystalline samples, the O(N)–H relaxation process is thermally activated with a spectrum of activation energies around 0.2 eV in polycrystalline samples (Cannelli and Cantelli, 1982). This is consistent with the existence of different species of O(N)–H complexes, each of which has several different modes or pathways for stress-induced reorientation which contribute to the relaxation with different activation energies (Zapp and Birnbaum, 1980a,b).

In contrast with the O(N)–H peak at 100 K, the smaller peak at liquid helium temperatures has been discovered more recently and has been less extensively investigated, with only three observations of helium temperature peaks which are definitely associated with the presence of H in Nb having been reported to date (Cannelli and Cantelli, 1982; Huang *et al.*, 1985; Poker *et al.*, 1979). Measurements of the sound velocity or the real part of the elastic modulus in samples doped with O and H have also been made in this region by Bellessa (1983).

Cannelli and Cantelli have observed a thermally activated relaxation peak at low ultrasonic frequencies ($Q_p^{-1} = 2.6 \times 10^{-5}$, peak temperature $T_p = 2.6\,\text{K}$ at 20 kHz and $Q_p^{-1} = 4 \times 10^{-5}$, $T_p = 2.9\,\text{K}$ at 74 kHz) in a polycrystalline Nb disc sample containing 2000 at. p.p.m. H and a combined O and N concentration of about 1500 at. p.p.m. The peak observed by Poker *et al.* was in the ultrasonic attenuation of C' shear waves (i.e. propagating in the $\langle 110 \rangle$ direction with $[\bar{1}10]$ polarisation) at a much higher ultrasonic frequency ($Q_p^{-1} = 3 \times 10^{-5}$, $T_p = 2.3\,\text{K}$ at 10 MHz) in a single crystal sample containing 200 at. p.p.m. H.

The mechanisms of the relaxation peaks observed by Poker *et al.* and Huang *et al.* at ultrasonic frequencies have been identified clearly as relaxations of thermally excited states of O(N)–H complexes (Huang *et al.*, 1985). The mechanism of the peak observed by Cannelli at 20 kHz is known to involve interstitial hydrogen since the peak is not observed in UHV annealed samples in the hydrogen-free state. The nature of this mechanism is not however known at present (Cannelli and Cantelli, 1982). It is also not clear at present whether there is any relationship between this latter peak and those observed by Poker *et al.* and Huang *et al.* in the MHz region.

It will be seen in section 6.4 that the height of the O(N)–H peak observed in the 1.5 tonne gravitational radiation antenna at the University of Western Australia is indicative of a trapped hydrogen concentration of order 20 at. p.p.m. The presence of hydrogen contamination at these levels is also expected to result in a significant contribution to Q^{-1} in the antenna by this associated helium-temperature relaxation process. Further systematic study of the kinetics and relaxation mechanisms of defect complexes in Nb which involve interstitial hydrogen is therefore desirable from the viewpoint of this chapter.

(iii) Dislocation relaxations

As in the case of point defects, the long-range elastic strain associated with dislocations (Read, 1953), provides a strong coupling of these defects to an externally applied stress or strain. The various internal friction mechanisms including anelastic relaxation processes which arise from this coupling, (Fantozzi and Ritchie, 1981; Lenz and Lücke, 1975; Nowick and Berry, 1972; Seeger, 1981), are even more complex and less well understood than the point defect relaxation processes and the discussion of this and other dislocation-induced mechanisms for internal friction will be very brief, concentrating on the most important qualitative features of those mechanisms which are most relevant for this chapter.

If interaction between dislocations and other defects such as impurities is ignored, the internal friction mechanisms arising from the 'intrinsic' mobility of dislocations in an otherwise perfect crystal may be separated into two categories. The first of these involves the motion of dislocation lines or segments whose direction differs significantly (by more than about 5°) from a close-packed crystallographic direction (e.g. $\langle 111 \rangle$ in the bcc Nb lattice). It is readily shown that such dislocations are in general free to move under vanishingly small external stresses (Read, 1953), and that such motion is describable in terms of (perhaps anisotropic) continuum mechanics (Granato and Lücke, 1966; Nowick and Berry, 1972). These features of the dislocation motion are due to the 'averaging out' over the length of the dislocation line of effects due to the discrete nature of the crystal structure when the direction cosines with respect to the crystallographic axes are not in the ratio of integers.

It may also be shown (Granato and Lücke, 1966; Nowick and Berry, 1972) that, under these conditions, each dislocation segment should behave like a damped vibrating string with constant tension and with an inertia or effective mass which is negligible at audiofrequencies, but which gives rise to 'resonant' behaviour in the MHz frequency region if the damping or drag coefficient of the dislocation motion is sufficiently small. In the low-frequency region of interest here, this 'vibrating string' model of dislocation motion in any case gives rise to a contribution to Q^{-1} given by (Nowick and Berry, 1972),

$$Q_d^{-1} = \frac{\Lambda B l^4 \omega}{36 G b^2},\tag{6.55}$$

where Λ is the dislocation density, B is the dislocation drag coefficient, l is the typical length of dislocation segments between pinning points, G is the shear modulus, and b is the magnitude of the Burgers vector of the dislocation.

It maybe be seen from equation (6.55) that the vibrating string model predicts a very strong dependence of Q_d^{-1} on the typical segment length l in the low-frequency limit. While this predicted l^4 dependence appears to be in good agreement with experiment (Thompson and Holmes, 1956), the experimental results also show that Q_d^{-1} is essentially independent of ω at audiofrequencies, in contrast with the behaviour predicted by equation (6.55). Although it is not therefore possible to obtain a reliable estimate of Q_d^{-1} from this result, it may be expected that Q_d^{-1} will be significant in very pure, well annealed metals which should have typical dislocation segment lengths of 3×10^{-5} cm or longer (Nowick and Berry, 1972).

In the case of metal such as Nb in a highly purified and annealed state, a value of order 10^{-10} is obtained for Q_d^{-1} at a frequency of 1 kHz in the normal state. In the superconducting state, this predicted value of Q_d^{-1} decreases, as the dominant electronic contribution to the dislocation drag coefficient, B, is reduced below T_c (Granato, 1975). At the relatively high impurity levels of order 1000 at. p.p.m. of interest in this chapter, Q_d^{-1} should in any case be negligible at audio frequencies, since the 'free' dislocation segment lengths should be greatly reduced by pinning of the dislocations by these impurities (Nowick and Berry, 1972).

In the case of dislocation lines or segments whose direction is close to a close-packed crystallographic direction, further dislocation relaxation mechanisms may arise which are due to the existence of a significant potential energy barrier to the motion of such dislocations, known as the Peierls energy or Peierls potential (Hirth and Lothe, 1982; Read, 1953; Seeger, 1981). This energy barrier is essentially due to the discrete atomic nature of the crystal structure, which results in a periodic variation of the energy of the dislocation along the direction of motion across the widely separated crystal planes normal to the close-packed direction. Since the magnitude of the Peierls energy in bcc metals such as Nb is expected to be of order 0.1 eV 'per dislocated atom' (Hirth and Lothe, 1982), thermal activation would clearly be required for the motion under a small applied stress of dislocation segments which lie parallel to a close-packed direction. Since the characteristic wavelength of thermal fluctuations is much smaller than the typical length of such segments, however, it is highly improbable that thermally activated motion of an entire dislocation segment of this orientation will occur, even at high temperatures.

A more favoured kinetic mechanism for the motion of a dislocation across the Peierls potential barrier is represented schematically in figure 6.6(a). The thermally activated formation of two 'kinks' of atomic dimensions and of opposite

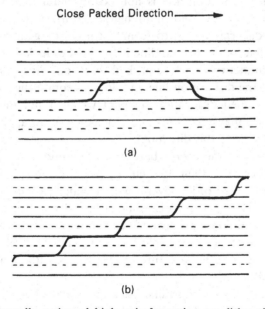

(a)

(b)

Figure 6.6. (a) Thermally activated kink pair formation on dislocation lying parallel to close-packed directions. Kinks separate along this direction under the influence of an applied stress. (b) Geometrical kinks on a dislocation at a small angle from the close-packed direction. The kinks minimise the total dislocation energy by allowing the dislocation to lie along 'valleys' of minimum energy parallel to this direction for most of its length (continuous lines), with relatively abrupt kinks crossing the 'hills' of maximum energy (dashed lines).

'polarity' allow the dislocation motion to proceed by the subsequent separation of the kinks under the influence of the external stress. The thermally activated nucleation of such kinks again involves an activation energy of order 0.1 ev per kink, but the process of kink formation by a local thermal fluctuation clearly has a much higher probability of occurrence than the thermally activated motion of the dislocation segment as a whole.

Several thermally activated relaxation processes observed in various metals between 30 K and 400 K have been tentatively identified as dislocation relaxation processes occurring by this kink pair formation mechanism (Fantozzi and Ritchie, 1981; Funk and Schultz, 1985). In the case of Nb, the process of kink pair formation on non-screw dislocations is believed to be responsible for the relaxation peak with an activation energy of about 0.1 eV which is observed at 50 K in the high purity hydrogen-free metal and which is shown in figure 6.5 (Funk and Schultz, 1985). This particular relaxation process will be almost completely 'frozen out' at liquid helium temperatures due to its high activation energy and since kink pair formation on a$\langle 111 \rangle$/2 screw dislocations in Nb has an even higher activation energy of order 0.67 eV (Seeger, 1981), and no further discussion of this mechanism will therefore be given here.

A dislocation relaxation mechanism with a much lower activation energy may however arise if the dislocation line is not exactly parallel to the close-packed direction but close to it. As shown in figure 6.6(b), it is then energetically favourable for the dislocation line to form 'intrinsic' or geometrical kinks of a single 'polarity' so that the dislocation line is for most of its length in a position of minimum energy parallel to the close-packed direction. Since such kinks are an intrinsic property of the dislocation line and do not require thermal activation for their formation, a pinned dislocation segment which is not quite parallel to a close-packed direction may relax under the influence of an applied stress by the motion of such kinks along this direction.

While the magnitude of the energy barrier to the motion of dislocation kinks is expected to be much less than the energy required for kink formation, a significant energy barrier of order several meV to kink motion is expected in the particular case of a$\langle 111 \rangle$/2 screw dislocations in the bcc structure (Seeger and Wüthrich, 1976). The thermally activated relaxation of these dislocations by this mechanism is believed to be responsible for the so-called α' relaxation observed in pure cold worked Ta at a temperature of about 15 K with an activation energy of 9 meV (Funk and Schultz, 1985).

Although no observations of a similar peak in Nb have been reported to date, it will be argued in section 6.4 that the relaxation peak observed at very much lower Q^{-1} levels in vacuum annealed and recrystallized Nb samples at temperatures of order 4 K may in fact be due to this mechanism. For this reason, a very brief summary of relevant features of the geometrical kink migration theory of Seeger and Wüthrich will now be given. Any further review of the microscopic theory of dislocations must however remain beyond the scope of this chapter.

In the case of screw dislocation segments with an average direction at a small angle ϕ from a close packed $\langle 111 \rangle$ direction and with distance l between pinning points, the theory predicts a contribution Q_s^{-1} to the internal friction which is of the usual 'Debye' form

$$Q_s^{-1} = \Delta \frac{\omega \tau}{1 + \omega^2 \tau^2} \tag{6.56}$$

if the kink migration is assumed to be a classical thermally activated process. The relaxation time, τ, then has a characteristic Arrhenius form, $\tau = \tau_0 \exp(H_a/kT)$, with pre-exponential factor, τ_0, given by

$$\tau_0 = \left(\frac{l \cos \phi}{b \pi} \right)^2 v^{-1}, \tag{6.57}$$

where b is the Burgers vector of the dislocation and n is a characteristic attempt frequency of the order of the Debye frequency (Seeger and Wüthrich, 1976). Physically, τ may be regarded as the time required for the diffusion of a kink across the segment length, l. If the density of such dislocation segments is denoted by ρ_{seg}, the theory also predicts a relocation strength Δ of the form

$$\Delta = \frac{8 l^3 a b^2}{\pi^4 k_B T} \sin \phi \cos^2 \phi G \rho_{seg} \tag{6.58}$$

where a is the kink height, and G is an elastic modulus.

It may reasonably be expected that the segment lengths in a real crystal will be randomly distributed about some mean length, $\bar{l}$, and that the corresponding expression for Q^{-1} will involve an integral over this distribution (Seeger and Wüthrich, 1976). The mean length $\bar{l}$, the pre-exponential factor τ_0, and the relaxation strength Δ all increase with sample purity, so that Q_s^{-1} and the peak height will be much greater in pure samples due to the absence of dislocation pinning. This conclusion is supported by the results of Kramer and Bauer (1967).

Experimental investigations of such intrinsic dislocation relaxation processes have to date been confined to cold worked crystals where the density ρ_{seg} may be expected to be large. However, the rapid increase of the relaxation strength, equation (6.58), with increasing l suggests that such a relaxation process may be observable also in highly annealed crystals, with the increase in l and the greatly reduced 'background' internal friction characteristic of the annealed state serving to counteract the effect of the greatly reduced dislocation segment density ρ_{seg}.

Finally, it may be noted that the interaction of dislocations with point defects also gives rise to a variety of anelastic relaxation processes (Nowick and Berry, 1972; Seeger, 1981). Recent careful work by Funk and Schultz has, for example, shown conclusively that the large internal friction peak observed in cold worked Nb at about $140 \, \text{K}$ is due to the interaction of dislocations with interstitial hydrogen (Funk and Schultz, 1985). Discussion of these complex relaxation effects is, however, beyond the scope of this chapter.

This brief discussion of 'intrinsic' dislocation relaxation mechanisms therefore concludes this review section. Discussion of some of the many important topics not reviewed in this chapter, such as surface and suspension losses, are given by Braginsky et al. (1985).

6.4   Measured internal friction in niobium and other high Q materials

In this section, the measured Q^{-1} values of niobium and other high Q materials will be interpreted in terms of the anelastic model where this is possible within the limitations of the model outlined earlier. In the next section the experimental results to be discussed here will help to determine the suitability of these materials for the construction of high Q resonators such as transducers (Paik, 1976) as well as cryogenic resonant bar antennae and high Q, room temperature pendulum suspensions for mirrors in large-scale laser interferometer gravitational wave detectors.

6.4.1   Internal friction in polycrystalline niobium

(i) Experimental methods
As mentioned previously in section 6.1, an extensive investigation of the effects of various metallurgical pretreatments on the internal friction of Nb at low temperatures has been carried out in conjunction with the gravitational radiation detection project at the University of Western Australia. At liquid helium temperatures, the internal friction mechanisms which have been identified include the thermoelastic effect, conduction electron relaxation and dislocation kink motion. Q^{-1} at liquid helium temperatures depends sensitively on the annealing treatment after machining and cold working and also on the residual hydrogen content. At higher temperatures, internal friction peaks due to O(N)–H defect complex relaxation (Cannelli and Cantelli, 1982) and the interaction of hydrogen with dislocations introduced by machining (hydrogen cold work peak, Maul and Schultz, 1981; Verdini and Baci, 1980) have also been observed.

The measurements were made using small disc samples (typically 90 mm to 125 mm diameter, 3 mm to 6 mm thick) vibrating in flexure with free edge and fixed centre. The flexural modes for which measurements were made have the form

$$w_n(r, \theta) = w_0 \cos n\theta [J_n(k_n r) + \beta_n I_n(k_n r)], \tag{6.59}$$

for $n \geq 2$. Here w_n is the transverse displacement, J_n and I_n are, respectively, the Bessel function and modified Bessel function of the first kind of order n, and the constant β_n is related to the eigenvalue k_n (Rayleigh, 1894). Modes of the form (6.59) have n nodal diameters and no nodal circles. It was found that such modes had significantly lower acoustic losses than modes with one or more nodal circles. Measurements of Q^{-1} for modes of the above form are therefore a more accurate measure of the internal friction in Nb. It was also found that modes with $n \geq 3$ showed consistently lower Q^{-1} than the fundamental $n = 2$ mode. The samples were suspended by means of a fine wire passing through a central suspension shaft, as shown in figure 6.7.

The purity of the samples was typically 99.9 at.%. Measurements of Q^{-1} were

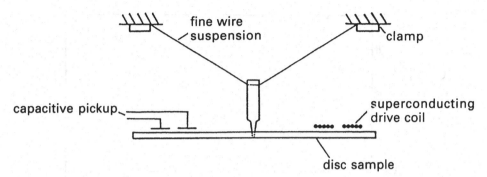

Figure 6.7. Sample suspension with press-fitted central suspension rod used for those samples which were machined from rolled Nb plate.

made for samples in the machined, etched and vacuum annealed conditions. The annealing treatments were carried out at a residual gas pressure of about 2×10^{-5} Torr. These annealing treatments also reduced the residual hydrogen concentration in the samples to less than 10 at. p.p.m., which has a negligible effect on the observed helium temperature Q^{-1} levels of interest in this section. As mentioned in previous sections, however, the trapping of hydrogen by oxygen and nitrogen impurities does lead to mechanisms which contribute significantly to Q^{-1} at higher residual H concentrations of order 50 at. p.p.m. or more (Cannelli and Cantelli, 1982).

(ii) Results and discussion

Vacuum annealing of machined samples above the 900°C to 950°C recrystallisation temperature of Nb results in an internal friction peak at liquid helium temperatures. This peak is absent for lower annealing temperatures and its height increases with increasing annealing temperature and degree of recrystallisation in the range from 900°C to about 1200°C (Ferreirinho, 1986). This and other evidence to be discussed later in this section shows that the peak is a dislocation relaxation peak. This peak is also identifiable with larger peaks observed in higher purity samples in both the as grown and deformed states by Kramer and Bauer (1967).

Typical Q^{-1} values obtained for various flexural modes of annealed disc samples are shown in figures 6.8 and 6.9. A major part of the residual acoustic loss levels Q_r^{-1} shown in these figures is believed to be due to surface and suspension losses in the small disc samples, so that the intrinsic Q^{-1} level in annealed Nb is much lower. This is confirmed by the low value of $Q^{-1} = 4.3 \times 10^{-9}$ measured in the 1.5 tonne Nb antenna at liquid helium temperatures (Veitch et al., 1987), as shown in figure 6.15.

The internal friction peak observed in the recrystallised samples annealed above 900°C shows a marked deviation from 'classical' thermal activation at temperatures below 5 K, and this deviation is greatest at low frequencies. This deviation is evident in figure 6.10, where a normalised dislocation relaxation

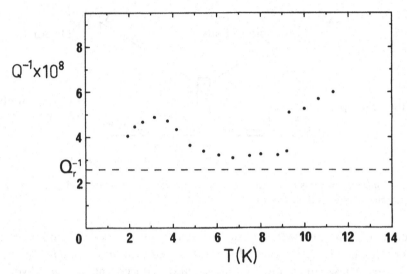

Figure 6.8. Temperature dependence of Q^{-1} of 1.6 kHz mode of a disc sample annealed for 2 h at 750°C and a further 2 h at 980°C after machining from rolled Nb plate. The dashed line shows the estimated residual or background acoustic loss level Q_r^{-1} due mainly to surface and suspension losses. The 'discontinuity' in Q^{-1} at the superconducting transition temperature $T_c = 9.25$ K is due to an abrupt change to the thermoelastic contribution to Q^{-1} at T_c (Blair and Ferreirinho, 1982).

contribution $Q_N^{-1} = (Q^{-1} - Q_r^{-1})/(Q_{peak}^{-1} - Q_r^{-1})$ is plotted as a function of temperature for the same results as shown in figure 6.9. The widths of the relaxation peaks and the deviation from a thermally activated relaxation process symmetrical in the variable T^{-1} (see equation 6.33) are too great to be modelled by a thermally activated relaxation process involving a spectrum of activation energies or relaxation times. The deviation can therefore only be interpreted as a transition from thermal activation to quantum mechanical tunnelling as the dominant mode of dislocation relaxation at the lower temperatures.

The very large pre-exponential factor $\tau_0 = 6 \times 10^{-10}$ s obtained by fitting the high temperature side of the peaks by a Debye curve with thermally activated relaxation time $\tau = \tau_0 \exp(H_a/kT)$ is about four orders of magnitude greater than the characteristic time scale for atomic vibrations in a crystal. This is very strong evidence for the identification of geometrical kink migration on a $\langle 111 \rangle /2$ screw dislocations (Seeger and Wüthrich, 1976) as the relaxation mechanism responsible for the internal friction peaks, since such long relaxation characteristic times are expected for dislocation kink migration, due to the diffusion of the kinks over dislocation segment lengths of order 100 atomic spacings or more (Seeger, 1981).

The measurements of Q^{-1} of annealed Nb disc samples were extended down to about 50 mK to 100 mK by the use of the ^{3}He–^{4}He dilution refrigerator facility at the Australian Defence Force Academy, Canberra. Some typical results are shown in figure 6.11. The approach of Q^{-1} to a temperature independent value

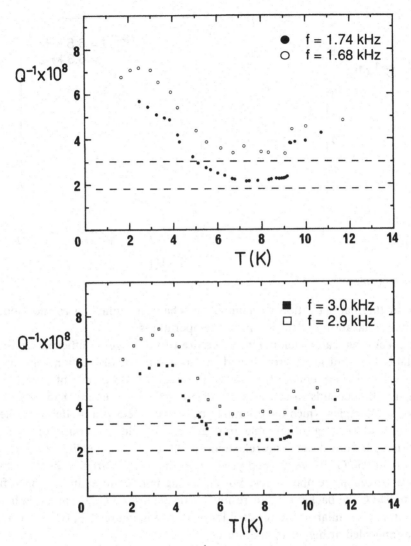

Figure 6.9. Temperature dependence of Q^{-1} of various flexural modes of a recrystallised disc sample machined from rolled Nb plate. Dashed lines show the estimated residual or background acoustic loss levels Q_r^{-1} for two annealing treatments. Note the appearance of a significant conduction electron relaxation contribution immediately below T_c for the highest frequency (6.3 kHz) mode. Full symbols: annealed for 1 h at 1080°C. Open symbols: after further anneal for 1 h at 1350°C.

below 1 K which is well above the background value Q_r^{-1} is the result of the attainment of full coherence of the tunnelling of the dislocation kinks at these low temperatures due to the rapid decrease in the short wavelength phonon fluctuations which limit the coherence at higher temperatures. The ultralow temperature measurements have therefore confirmed the 'phonon-driven' transi-

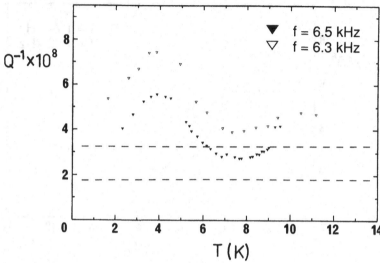

Figure 6.9. (*cont.*)

tion from thermal activation to quantum mechanical tunnelling as the mode of dislocation relaxation at sufficiently low temperatures.

The precise annealing temperature required to achieve a sufficient degree of recrystallisation and grain growth and the formation of dislocation segments of sufficient length (see equation 6.58) to produce the observed internal friction peak naturally depends to some extent on the previous treatment history of the sample. For samples which have initially been subjected to a moderate to heavy degree of cold working, observable grain growth and the appearance of the liquid helium temperature dislocation relaxation peak occur at annealing temperatures above about 950°C. As mentioned previously, the internal friction peak is absent for lower annealing temperatures. For annealing temperatures in the range from 900°C to 950°C the background or residual acoustic loss level Q_r^{-1} at liquid helium temperatures is similar to the residual levels shown in figures 6.8, 6.9 and 6.11 for samples annealed at higher temperatures.

An optimum pretreatment which minimises the internal friction of polycrystalline Nb at liquid helium temperatures has thus been determined. This pretreatment consists of a moderate to heavy degree of cold work (equivalent to a reduction in area of order 80% or more), followed by vacuum annealing at a temperature between 900°C and 950°C. A sufficient degree of cold work may already be present initially due to the original rolling or extrusion processes involved in the manufacture of the stock material. Such material is typically characterised by a mean crystal grain size of about 100 μm (Ferreirinho, 1986). This conclusion regarding the optimal pretreatment for minimising liquid helium temperature internal friction in Nb has recently been supported by the work of Solomonson (1990).

It is expected that temperature independent Q^{-1} levels of order 10^{-8} or less could be achieved at liquid helium temperatures by this treatment for large

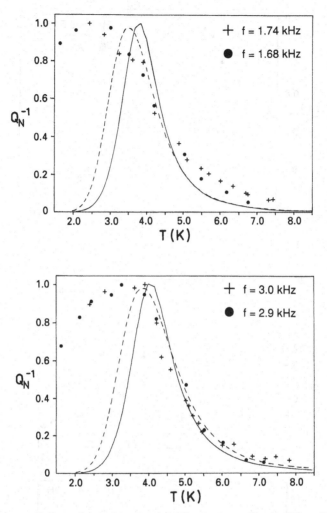

Figure 6.10. Approximate thermal activation fit to normalised dislocation relaxation contribution Q_N^{-1} obtained from data shown in figure 6.9 after subtraction of background. The full curve shows a 'simple' Debye relaxation curve with $H_a = 2.8$ meV and characteristic time (pre-exponential factor) $\tau_0 = 1.6 \times 10^{-8}$ s. The dashed line shows the fit obtained with an exponential or Poisson distribution of τ_0 with the same activation energy. Note that the above values of H_a and τ_0 differ significantly from the 'best fit' values of $H_a = 3.4 \pm 0.5$ meV and $\tau_0 = 3.6 \times 10^{-9}$ s. The accuracy of the fits increases with increasing frequency. $+$: annealed 1 h at 1080°C; $\bullet$: further annealed for 1 h at 1350°C.

samples where surface and suspension losses may be minimised. This conclusion is supported by the behaviour of Q^{-1} in a sample cut from a cold worked Nb bar and vacuum annealed at a temperature of about 1000°C, as shown in figure 6.12, where an estimated level $Q_{int}^{-1} = Q^{-1} - Q_r^{-1}$ of this order has been obtained.

It should be noted that although the above pretreatment minimises the liquid helium temperature internal friction in Nb by minimising the mean screw dislocation segment length and thus also the low-temperature dislocation relaxa-

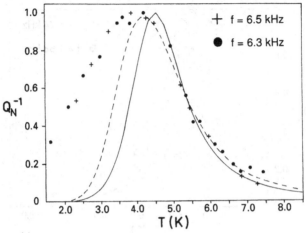

Figure 6.10. (*cont.*)

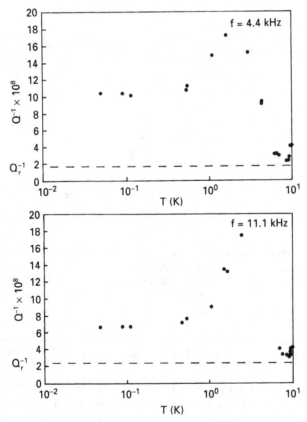

Figure 6.11. Temperature dependence of Q^{-1} of flexural modes below 1 K for a disc sample machined from a cylindrical Nb bar and annealed for 1 h at 1350°C. The transition from thermal activation to quantum mechanical tunnelling causes Q^{-1} to approach a limiting value well in excess of the background value Q_r^{-1} in the limit $T \to 0$.

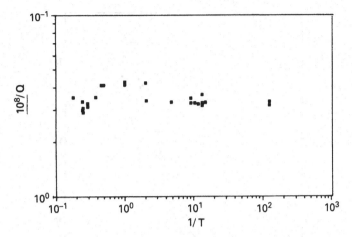

Figure 6.12. Temperature dependence of Q^{-1} of flexural modes below 1 K for a disc sample machined from a cylindrical Nb bar and annealed at 1000°C for 1 h.

tion contribution to Q^{-1} due to geometrical kink migration on these segments, it does not simultaneously minimise Q^{-1} at higher temperatures up to room temperature. The internal friction at temperatures between about 30 K and room temperature is by contrast minimised by vacuum annealing at the highest possible temperature, or by careful machining of the sample from a cast ingot produced from molten material. These treatments all result in mean crystal grain sizes of the order of several mm (Ferreirinho, 1986).

Since the 1.5 tonne GR antenna at UWA was machined from an ingot cast by electron beam melting, the detailed prior investigation of Q^{-1} in a disc sample machined and cut from a similar ingot was of particular importance to the GR project. The behaviour of Q^{-1} in this test sample at various temperatures and after varying degrees of chemical etching to remove superficial machining damage is shown in figures 6.13 and 6.14. This sample had a typical crystal grain size which was large compared to the 6 mm thickness of the disc, so that the sample was effectively a thin section through about 12 large crystals of a mean diameter of order 60 mm (Ferreirinho, 1986).

It is interesting to note that heavy etching significantly reduces Q^{-1} in this sample at room temperature, while the corresponding reduction at liquid helium temperatures is only marginal. This disparate effect of etching on Q^{-1} at low and high temperatures is believed to be due to the pinning at liquid helium temperatures of the high density surface dislocations by hydrogen, which is also introduced by the mechanical machining process (Ferreirinho, 1986). These surface dislocations do not therefore contribute significantly to the observed Q^{-1} at low temperatures, while the high mobility of both the hydrogen and the surface dislocations due to their complete thermal activation at room temperature prevents pinning and results in a significant contribution to the room temperature background Q^{-1} by the surface dislocations.

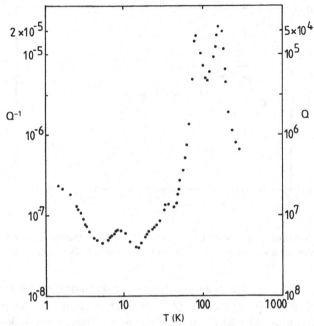

Figure 6.13. Temperature dependence of $n = 2$ mode ($f = 1.3$ kHz) of test sample offcut from electron beam melted bar ingot and successive etching of $100 \, \mu$m and $20 \, \mu$m layers from the original machined surface after cutting. The O(N)–H compelx peak at 90 K indicates hydrogen contamination of order 300 at. p.p.m. introduced by the machining process.

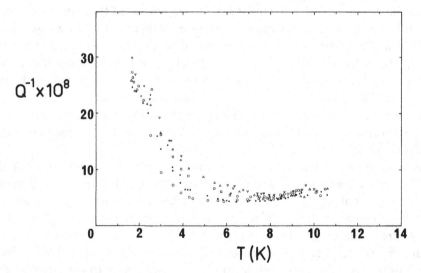

Figure 6.14. Temperature dependence of various modes of test sample offcut from electron beam melted bar ingot and successive etching of $100 \, \mu$m, $20 \, \mu$m and $200 \, \mu$m layers from the original machined surface after cutting. $\bigcirc$ $n = 2$, $f_{4.2} = 1242.94$ Hz; $\bullet$ $n = 3$, $f_{4.2} = 3099.99$ Hz; $\square$ $n = 4$, $f_{4.2} = 5358.88$ Hz; $\triangle$ $n = 5$, $f_{4.2} = 8035.96$ Hz; ∇ $n = 6$, $f_{4.2} = 11282.03$ Hz.

The presence of a significant amount of interstitial hydrogen in this machined and etched sample is readily deduced from figure 6.13, since the presence of a large hydrogen cold work peak at 150 K and a large O(N)–H relaxation peak at 90 K is clearly evident. It may be estimated on the basis of the results of Cannelli and Cantelli (1982) that the height of the O(N)–H relaxation peak in this sample at 90 K is indicative of an interstitial H concentration in this sample of order 300 at. p.p.m. As mentioned above, this hydrogen contamination is introduced by the machining process which involves the use of hydrogenated lubricants (Shibata and Koiwa, 1980).

The rapid increase in Q^{-1} below 4 K in this machined test sample apparent in figure 6.14 is indicative of the high temperature side of a relaxation peak which occurs at some temperature significantly lower than the relaxation peak temperature of the process occurring in the vacuum annealed and recrystallised samples. Since the hypothesised Q^{-1} peak temperature of the relaxation process in the machined sample was too low to be attained in the cryostat used, a detailed theoretical analysis of the mechanism involved is difficult. Analysis of the data shown in figure 6.13 again shows some evidence of a deviation from thermal activation (Ferreirinho, 1986), and it is considered likely that some superposition or interaction of processes involving dislocation kink motion and atomic hydrogen is involved.

Qualitatively similar behaviour to that shown in figure 6.13 for the small test sample is also evident in figure 6.15, which shows the temperature dependence of Q^{-1} of the 1.5 tonne Nb GR antenna at UWA (Veitch *et al.*, 1987). A

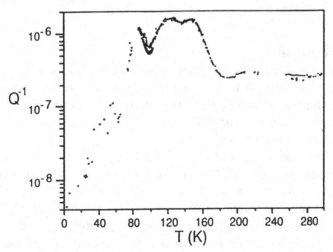

Figure 6.15. Results obtained for temperature dependence of Q^{-1} of fundamental longitudinal mode of the 1.5 tonne gravitational radiational antenna at the University of Western Australia (Veitch *et al.*, 1987). The presence of a significant O(N)–H peak at 90 K indicates an interstitial hydrogen concentration of order 20 at. p.p.m. introduced by machining, and the hydrogen cold work peak indicates a high dislocation density near the surface also due to the machining treatment.

Table 6.7. *Some typical Q^{-1} levels measured by Suzuki et al. (1978) for 20 kHz flexural modes of aluminium alloy discs at different temperatures.*

Al alloy design-ation	Impurities (at. %)	Q^{-1} (4.2 K)	Q^{-1} (77 K)	Q^{-1} (300 K)
1050	<0.5 total	7.7×10^{-7}	5.3×10^{-6}	5.0×10^{-6}
2017	4.0 Cu, 0.7 Mn, 0.12 Cr	1.9×10^{-7}	5.0×10^{-7}	4.0×10^{-6}
5056	5.1 Mg, 0.12 Mn, 0.12 Cr	2.5×10^{-8}	3.5×10^{-7}	4.0×10^{-6}
6061	1.0 Mg, 0.6 Si, 0.27 Cu, 0.20 Cr	3.5×10^{-7}	8.5×10^{-7}	4.0×10^{-6}

quantitative comparison of the heights of the 90 K O(N)–H peak and 150 K hydrogen cold work relaxation peak and the magnitude of the background or residual acoustic loss at liquid helium temperatures in the two specimens shows that these are all reduced in the large antenna in approximate proportion to the corresponding reduction in the surface to mass ratio of the antenna relative to the small test sample. Since the interstitial hydrogen contamination and cold work in these two samples were introduced by machining, which is essentially a surface treatment, this scaling behaviour of these high temperature peaks is not surprising. The height of the 90 K O(N)–H peak in the 1.5 tonne antenna is indicative of a hydrogen contamination level of order 20 at. p.p.m. It also suggests that some, if not most, of the contribution to the residual or background acoustic loss in the small test sample at liquid hleium temperatures is also due to a 'surface' mechanism.

6.4.2   Aluminium alloys
The mobility of dislocations in fcc metals such as aluminium is high even at low temperatures. The internal friction in these metals is usually dominated by the dislocation contribution rather than the electronic contribution, although the electronic contribution may dominate in ultrapure, well recrystallised samples with low dislocation density (Lax, 1959).

As might be expected, it was found by Suzuki *et al.* that aluminium alloys with an impurity content of about 5% have much lower Q^{-1} than higher purity alloys at room temperature (Suzuki *et al.*, 1978) due to pinning of the dislocation by impurities. The aluminium alloy 5056 ($\approx$5.1% Mg, 0.12% Cr, 0.12% Mn) has a significantly lower Q^{-1} level at liquid helium temperatures than the other alloys investigated. Some typical results obtained by Suzuki *et al.* are summarised in table 6.7.

It should be noted here that these alloys are sufficiently impure for the electronic contribution to be negligible, with an electronic relaxation time of 10^{-14} to 10^{-15} s.

Table 6.8. *Typical Q^{-1} levels measured by Braginsky et al. for longitudinal modes of polished sapphire monocrystal cylinders at various temperatures (Braginsky et al., 1985) at a frequency of 38 kHz.*

Temperature (K)	Q^{-1}
4.2	2×10^{-10}
77	1.8×10^{-9}
300	3×10^{-9}

6.4.3 Sapphire

Sapphire is the crystalline form of Al_2O_3 and has the lowest internal friction yet measured in any material at audiofrequencies. This is primarily due to the strongly directional covalent bonds in this material. Since dislocation motion requires the 'breaking' and remaking of the relatively rigid and directional covalent bonds, there is a large energy barrier to dislocation motion or 'Peierls energy' in sapphire and in other covalently bonded insulators and semiconductors such as quartz and silicon.

The large energy barrier of order 1 eV renders the dislocation immobile in the presence of an acoustic wave even at room temperature so that there is a negligible dislocation contribution to the internal friction. The small thermal expansion coefficient of sapphire results in relatively small thermoelastic and phonon relaxation losses as well (Braginsky *et al.*, 1985). These properties, together with the absence of an electronic contribution to Q^{-1}, are responsible for the very low internal friction of sapphire over a broad temperature range from liquid helium temperatures up to room temperature.

The Q^{-1} values achieved with polished sapphire monocrystals at various temperatures (Braginsky *et al.*, 1985) are shown in table 6.8.

Although the very low internal friction of sapphire makes this material suitable for the construction of Weber bar antennae and other high Q mechanical resonators sapphire does have some disadvantages in comparison with metals such as Nb and 5056 Al.

These disadvantages include the difficulty in fabricating very large single crystals and polycrystals with large grain size, and the relative difficulty of chemical etching at room temperature (Braginsky *et al.*, 1985).

6.4.4 Quartz

The physical properties of crystalline quartz (SiO_2) are similar to those of sapphire, although quartz differs from sapphire in that it exhibits a piezoelectric effect. As shown in table 6.9, crystalline quartz has a very low Q^{-1} level at liquid helium temperature, although the 77 K and room temperature Q^{-1} values are much higher then the corresponding results for sapphire shown in table 6.8 (Braginsky *et al.*, 1985).

Table 6.9. *Typical Q^{-1} levels measured by Braginsky et al. for 1 MHz modes of a quartz resonator.*

Temperature (K)	Q^{-1}
4.2	2×10^{-10}
77	3×10^{-7}
300	1×10^{-7}

Table 6.10. *Typical Q^{-1} levels measured by McGuigan et al. (1978) for a 20 kHz longitudinal mode of a 4.9 kg monocrystal Si cylinder at various temperatures.*

Temperature (K)	Q^{-1}
4.2	6×10^{-10}
77	5×10^{-9}
300	2.8×10^{-8}

6.4.5   Silicon

Although silicon is a semiconductor rather than an insulator, the strongly directional covalent bonds ensure that, as for sapphire and quartz, large dislocation densities are energetically unfavourable in a crystal grown from the melt. As in the case of sapphire and quartz, the small number of residual dislocations are rendered immobile by a large Peierls energy. The thermal expansion coefficient of silicon is also small at low temperatures, again minimising the thermoelastic and phonon relaxation contributions to Q^{-1}.

Internal friction measurements in silicon bars of typical mass 5 kg have been made by McGuigan *et al.* (1978) confirming a very low Q^{-1} value for this material at liquid helium temperatures. Typical Q^{-1} values obtained at various temperatures by these authors are summarised in table 6.10. It is believed that a significant contribution to the helium temperature internal friction in Si is due to electronic relaxations associated with impurities such as oxygen, boron and carbon, which modify the electron band structure. A very pure silicon bar ($\ll 1$ at. p.p.m. total impurity level) could therefore have even lower internal friction at liquid helium temperatures.

6.5   Summary and comparison of relevant properties of high Q materials

6.5.1   Covalently bonded materials – sapphire, quartz and silicon

As seen in the previous section, crystalline sapphire and quartz are covalently bonded dielectrics which have very high Q factors at liquid helium temperatures due to the very low dislocation density and dislocation mobility and the absence of electronic relaxation in high crystals grown from the melt. These remarks apply also to high purity silicon (McGuigan *et al.*, 1978). The low thermal expansion coefficients of sapphire and silicon also minimise the thermoelastic and phonon contribution to Q^{-1}, making these two materials particularly suitable for construction of high Q mechanical resonators at room temperatures.

The low dislocation mobility in these high Q dielectrics and semiconducting Si is however also responsible for their hardness and brittleness, which poses considerable technical difficulties in the fabrication by chemical machining and polishing of large resonators such as Weber bar antennae. Similar problems occur in the fabrication of more intricate resonators on a small scale. At liquid helium temperatures, the quality factor of these materials is sufficiently high that they may place the quantum limit on the measurement of the resonator amplitude and phase set by the Heisenberg uncertainty principle above the threshold sensitivity determined by the thermal or Brownian fluctuations.

The existence of the quantum limit and the relatively high helium temperature Q factors together with the ease of machining of these metals of Nb and to a lesser extent, 5056 Al, make these metals more suitable choices for the construction of mechanical resonators such as Weber bar antennae at low temperatures, while silicon and sapphire retain their attractiveness as choices for room temperature applications. The 5056 aluminium alloy also has the advantage of relatively low cost.

Since sapphire is more chemically inert than silicon, sapphire is the best choice for the construction of very high Q resonators at room temperature, in applications where ease of fabrication is not of primary importance. As noted in section 6.3.3(i), the attainment of high Q factors in room temperature resonators requires the use of materials with very low levels of interstitial impurities such as oxygen and nitrogen.

6.5.2 Suggestions for further work on other bcc transition metals

Finally, it should be noted that the bcc metals vanadium, tantalum and to a lesser extent molybdenum (Solomonson and Mann, 1987) are also promising high Q materials with possible applications in experimental GR research. Vanadium in particular is a transition metal which has similar physical and chemical properties to those of Nb. It is therefore reasonable to expect that, as in the case of Nb, it will be possible to optimise the annealing pretreatment to obtain a low internal friction level at liquid helium temperature or at room temperature.

A systematic study of the effects of low levels of interstitial hydrogen on the low temperature internal friction of Nb and other bcc transition metals is also desirable. It is not clear at present whether small amounts of hydrogen in well recrystallised samples such as the 1.5 tonne antenna at the University of Western Australia decrease the internal friction by pinning dislocations or whether the introduction of hydrogen-defect complex relaxations discussed in section 6.3.3(ii) results in an overall increase of Q^{-1} below 4 K.

References

Akhieser, A. (1939). *J. Phys. (U.S.S.R.)* **1**, 227.

Ashcroft, N. W. and Mermin, N. D. (1976). *Solid State Physics,* Holt, Reinhart and Winston, New York.

Baker, C. and Birnbaum, H. K. (1972). *Scr. Metall.* **6**, 851.

Baker, C. and Birnbaum, H. K. (1973). *Acta Metall.* **21**, 865.

Bardeen, J., Cooper, L. N. and Schrieffer, J. R. (1957). *Phys. Rev.* **108**, 1175.

deBatist, R. (1972). *Internal Friction of Structural Defects in Crystalline Solids,* North-Holland, Amsterdam.

Bellessa, G. (1983). *J. de Phys. Lett.* **44**, L387.

Benoit, W. and Gremaud, G. (eds.) (1981). *Proceedings of the 7th International Conference on Internal Friction and Ultrasonic Attenuation in Solids, J. de Phys.* **42**, Colloq. C-5.

Blair, D. G. and Ferreirinho, J. (1982). *Phys. Rev. Lett.* **49**, 375.

Bommel, H. E. and Dransfeld, K. (1960). *Phys. Rev.* **117**, 1245.

Bordoni, P. G. (1949). *Ricerca Sci.* **19**, 851.

Bordoni, P. G. (1954). *J. Acoust. Soc. Am.* **26**, 495.

Braginsky, V. B. and Manukin, A. B. (1977). *Measurements of Weak Forces in Physics Experiments,* University of Chicago Press.

Braginsky, V. B., Mitrofanov, V. P. and Panov, V. I. (1985). *Systems with Small Dissipation,* University of Chicago Press.

Bruner, L. J. (1960). *Phys. Rev.* **118**, 399.

Cannelli, G. and Cantelli, R. (1982). *Solid State Commun.* **43**, 567.

Cannelli, G., Cantelli, R. and Vertechi, G. (1982). *J. Less. Comm. Met.* **88**, 335.

Chambers, R. H. (1965). In *Physical Acoustics,* vol. III, ed. W. P. Mason, Academic Press, New York, p. 123.

Daniel, V. (1967). *Dielectric Relaxation,* Academic Press, London.

Debye, P. J. W. (1929). *Polar Molecules,* Chemical Catalog Co, reprinted by Dover Publications, New York.

Engelhard, J. (1979). *J. Phys. F* **9**, 2217.

Fantozzi, G. and Ritchie, I. G. (1981). In *Proceedings of the 7th International Conference on Internal Friction and Ultrasonic Attenuation in Solids,* eds. G. Fantozzi and A. Vincent, *J. de Phys.* **42**, Colloq. C-5, p. 3.

Fantozzi, G. and Vincent, A. (eds.) (1983). *Proceedings of the 4th European Conference on Internal Friction and Ultrasonic Attenuation in Solids, J. de Phys.* **44**, Colloq. C-9.

Fast, J. D. (1976). *Gases in Metals,* Macmillan, London.

Ferreirinho, J. (1986). Ph.D. Thesis, University of Western Australia.

Ferron, G. and Quintard, M. (1978). *Phys. Stat. Sol.* **46A**, K43.

Filson, D. H. (1959). *Phys. Rev.* **115**, 1516.

Flynn, C. P. and Stoneham, A. M. (1970). *Phys. Rev.* **1B**, 3966.

Funk, G., Maul, M. and Schultz, H. (1983). *Proceedings of the 4th European Conference on Internal Friction and Ultrasonic Attenuation in Solids, J. de Phys.* **44**, Colloq. C9–711.

Funk, G. and Schultz, H. (1985). *Z. Metallk.* **76**, 311.

Granato, A. V. (1975). In *Proceedings of the 5th International Conference on Internal Friction and Ultrasonic Attenuation in Crystalline Solids,* eds. D. Lenz and K. Lücke, Vol. 2, p. 33, Springer Verlag, Berlin.

Granato, A. V. and Lücke, K. (1966). In *Physical Acoustics,* ed. W. P. Mason, Vol. IVA, p. 225, Academic Press, New York.

Groh, P. and Schultz, H. (1981). In *Proceedings of the 7th International Conference on*

Internal Friction and Ultrasonic Attenuation in Solids, eds. G. Fantozzi and A. Vincent, *J. de Phys.* **42,** Colloq. C-5, p. 25.

Hayes, D. J. and Brotzen, F. R. (1974). *J. Appl. Phys.* **45,** 1271.

Hirth, J. P. and Lothe, J. (1982). *Theory of Dislocations,* John Wiley, New York.

Holstein, T. (1959). *Phys. Rev.* **32,** 97.

Huang, K. F., Granato, A. V. and Birnbaum, H. K. (1985). *Phys. Rev.* **32B,** 2178.

Kadanoff, L. P. and Pippard, A. B. (1966). *Proc. Roy. Soc.* **A292,** 299.

Kehr, K. W. (1978). In *Hydrogen in Metals,* eds. G. Alefeld and J. Volkl, Vol. 1, p. 197, Springer Verlag, Berlin.

Koehler, J. S. (1952). *Imperfections in Nearly Perfect Crystals,* ed. W. Shockley, John Wiley, New York.

Kramer, E. J. and Bauer, C. L. (1967). *Phys. Rev.* **163,** 407.

Landau, L. D. and Lifshitz, E. M. (1970a). *Theory of Elasticity,* Pergamon Press, Oxford.

Landau, L. D. and Lifshitz, E. M. (1970b). *Statistical Physics,* Pergamon Press, Oxford.

Landau, L. D. and Rumer, G. (1937). *Phys. D. Zelts, Sowjetunion* **11,** 18.

Lax, E. (1959). *Phys. Rev.* **115,** 1591.

Leibfried, G. and Ludwig, W. (1961). In *Solid State Physics,* eds. F. Seitz and D. Turnbull, vol. 12, p. 276, Academic Press, New York.

Lenz, D. and Lücke, K. (eds) (1975). *Proceedings of the 5th International Conference on Internal Friction and Ultrasonic Attenuation in Crystalline Solids,* Vol. 2, Springer Verlag, Berlin.

Love, A. E. H. (1944). *A Treatise on the Mathematical Theory of Elasticity,* Dover Publications, New York.

McGuigan, D. F., Lam, C. C., Gram, R. Q., Hoffman, A. W., Douglass, D. H. and Gutche, H. W. (1978). *J. Low Temp. Phys.* **30,** 621.

Maris, H. J. (1971). In *Physical Acoustics,* ed. W. P. Mason, Vol. VIII, p.280, Academic Press, New York.

Mason, W. P. (1955). *Phys. Rev.* **97,** 555.

Maul, M. and Schultz, H. (1981). In *Proceedings of the 7th International Conference on Internal Friction and Ultrasonic Attenuation in Solids,* eds. G. Fantozi and A. Vincent, *J. de Phys.* **42,** Colloq. C5–73.

Misner, C. W., Thorne, K. S. and Wheeler, J. A. (1973). *Gravitation,* W. H. Freeman and Co., San Francisco.

Morse, R. W. (1955). *Phys. Rev.* **97,** 1716.

Morse, R. W. (1959). *Progress in Cryogenics,* **1,** 221.

Morse, R. W. (1962). *IBM, J. Res. Develop.* **6,** 58.

Nowick, A. S. and Berry, B. S. (1972). *Anelastic Relaxation in Crystalline Solids,* Academic Press, New York.

Paik, H. J. (1974). Ph.D. thesis, HEPL report 743, Stanford University.

Paik, H. J. (1976). *J. Appl. Phys.* **47,** 1168.

Pippard, A. B. (1955). *Phil. Mag.* **46,** 1104.

Pippard, A. B. (1960). *Proc. Roy. Soc.* **A257,** 165.

Poker, D. B., Setser, G. G., Granato, A. V. and Birnbaum, H. K. (1979). *Z. Phys. Chem.* **116,** 439.

Powers, R. W. and Doyle, M. V. (1959). *J. Appl. Phys.* **30,** 514.

Qi, Z., Volkl, J. and Wipf, H. (1982). *Scr. Metall.* **16,** 859.

Rayleigh, J. W. S. (1894). *The Theory of Sound,* Macmillan Press, London. Reprinted by Dover Publications, New York, Vol. 1, 1945.

Read, W. T. (1953). *Dislocations in Crystals,* McGraw-Hill, New York.

Rees, M., Ruffini, R. and Wheeler, J. A. (1974). *Black Holes, Gravitational Waves and Cosmology,* Gordon and Breach, New York.

Richter, G. (1938). *Ann. Phys.* **32,** 683.

Routbort, J. L. and Sack, H. S. (1966). *J. Appl. Phys.* **37,** 4803.

Schiller, P. (1976). *Nuovo Cim.* **33B,** 226.

Schober, T. and Wenzl, H. (1978). In *Hydrogen in Metals,* eds. G. Alefeld and J. Volkl, vol. 2, p. 11, Springer Verlag, Berlin.

Seeger, A. (1981). *Proceedings of the 7th International Conference on Internal Friction and Ultrasonic Attenuation in Solids,* eds. G. Fantozzi and A. Vincent, *J. de Phys.* **42,** Colloq. C5-201.

Seeger, A. and Wüthrich, C. (1976). *Nuovo Cim.* **33B,** 38.

Shibata, K. and Koiwa, M. (1980). *Trans. Japan Inst. Metals* **21,** 639.

Snoek, J. L. (1939). *Physica* **6,** 591.

Solomonson, N. (1990). Ph.D. thesis, Louisianna State University.

Solomonson, N. and Mann, A. G. (1987). *Cryogenics* **27,** 587.

Suzuki, T., Tsubono, K. and Hirakawa, H. (1978). *Phys. Lett.* **67A,** 2.

Thompson, D. O. and Holmes, D. K. (1956). *J. Appl. Phys.* **27,** 213.

Thorne, K. S. (1980). *Rev. Mod. Phys.* **52,** 285.

Touloukian, Y. S., Powell, R. W., Ho, C. Y. and Klemens, P. G. (1970). *Thermophysical Properties of Matter,* IFI Plenum, New York.

Veitch, P. J., Ferreirinho, J., Blair, D. G. and Linthorne, N. (1987). *Cryogenics* **27,** 586.

Verdini, L. and Baci, D. (1980). *Nuovo Cim.* **59B,** 163.

Vineyard, G. H. (1957). *J. Phys. Chem. Solids* **3,** 121.

Weber, J. (1960). *Phys. Rev.* **117,** 306.

Woodruff, T. O. and Ehrenreich, H. (1961). *Phys. Rev.* **123,** 1553.

Yoshinari, O. and Koiwa, M. (1982a). *Acta Metall.* **30,** 1979.

Yoshinari, O. and Koiwa, M. (1982b). *Acta Metall.* **30,** 1987.

Yoshinari, O. and Koiwa, M. (1984). *Res. Mechanica* **11,** 27.

Zapp, P. E. and Birnbaum, H. K. (1980a). *Acta Metall.* **28,** 1275.

Zapp, P. E. and Birnbaum, H. K. (1980b). *Acta Metall.* **28,** 1523.

Zener, C. (1937). *Phys. Rev.* **52,** 230.

Zener, C. (1943). *Trans. AIME,* **152,** 122.

Zener, C. (1947). *Phys. Rev.* **71,** 34.

Zener, C. (1948). *Elasticity and Anelasticity of Metals,* University of Chicago Press.

Ziman, J. M. (1960). *Electrons and Phonons,* Clarendon Press, Oxford.

Zwikker, C. (1954). *Physical Properties of Solid Materials,* Pergamon Press, Oxford.

7

Motion amplifiers and passive transducers

J.-P. RICHARD AND W. M. FOLKNER

7.1 Introduction

Since the original use of piezoelectric crystals by Weber (1960, 1970) for the first resonant gravitational radiation antennas, more sensitive methods to measure the energy of the antenna have been sought. The necessity of cooling the antennas to cryogenic temperatures to reduce Brownian motion noise has made piezoelectric crystals less usable since some of their properties degrade at low temperatures. Rather than using strain sensors, most cryogenic antennas utilize transducers which respond to displacement of one end of the antenna. These transducers include capacitive sensors and inductance modulation sensors.

The minimum detectable change in the antenna energy is limited by the noise present in the detection system. The total noise includes the Brownian motion noise of the bar, characterized by its temperature T_1 and damping time τ_1, and the amplifier noise, characterized by the two noise sources $v_n(\omega)$ and $i_n(\omega)$. The transducer determines the efficiency of energy transfer between the antenna and the amplifier. If the transducer itself does not introduce noise, an approximate expression for the minimum detectable change ΔE_{min} in the energy of an antenna initially at rest is (Giffard, 1976; Richard, 1978):

$$\Delta E_{min} \cong 2k_B T_1 \frac{\tau_s}{\tau_1} + k_B T_n \left[\frac{2(\lambda + \lambda^{-1})}{\beta \tau \omega_1} + \frac{\beta \tau \omega_1}{2\lambda} \right], \qquad (7.1)$$

where ω_1 is the radial frequency of the lowest longitudinal mode of the antenna, τ_s is the time between samples of the antenna energy, τ is the electronic averaging time, λ is the ratio of the transducer output impedance to the amplifier noise impedance, β is the fraction of the antenna energy at the amplifier terminals, and k_B is Boltzman's constant. The noise impedance Z_n and noise temperature T_n of the amplifier are related to the amplifier voltage noise v_n and current noise i_n by

$$Z_n = v_n/i_n, \qquad (7.2)$$

$$T_n = v_n i_n/2k_B, \qquad (7.3)$$

The first term in equation (7.1) describes the fluctuations in the energy of the antenna due to the thermal noise in the antenna itself. This effect is reduced by operating at a low temperature and with antenna material with a very high quality factor $Q_1 = \tau_1 \omega_1 / 2$. The second term is the white noise from the amplifier and the third term is the back action from the amplifier resulting in additional fluctuations in the energy of the antenna.

In practice, $\tau_s \cong \tau$ is usually selected. Assuming that the output impedance of the transducer is equal to the noise impedance of the amplifier, $\lambda = 1$, and the coefficient of the amplifier noise term is minimized for

$$\beta \tau \omega_1 = 2\sqrt{2}. \tag{7.4}$$

As can be seen from equation (7.1), a small value for τ is desirable in order to reduce the antenna thermal noise. This also results in a large bandwidth ($\cong 1/\tau$). A small value for τ will minimize the amplifier noise only if the energy coupling can be made large enough. An important part of the experimental effort has been directed at increasing the effective value of β.

Exponentially tapered bars have been considered to achieve wide bandwidth and a large coupling between the antenna and the amplifier. The larger coupling is achieved through the magnified displacement of the tapered end (Bonazzola and Chevreton, 1973; Karim, 1984). The geometry of tapered antennas is not desirable in practice, however, since it does not efficiently fill the cylindrical volume of a large cryostat. An alternative arrangement which could be more suitable is a tapered transverse mode resonator, or 'whip' in which the longitudinal motion of the bar is coupled into a transverse flexure in a tapered beam (Blair, Giles and Zeng, 1987).

Mechanical resonators or motion amplifiers can increase the amount of energy coupling available and considerably improve the matching of the antenna to the amplifier. This is similar in principle to using an electrical resonant matching system such as the superconducting tank circuit used by Weber in conjunction with a room-temperature antenna (Weber, 1969). Detailed analyses using the optimum filtering method have shown that the observation bandwidth is extended by the multiple resonances of the motion amplifiers. This chapter discusses mechanical motion amplification, its use with various passive transducer mechanisms, and the signal-to-noise ratio calculations made to optimize the parameters of the overall system.

7.2 Multi-mode system analysis

7.2.1 Two-mode systems
The use of a resonant-small-mass transducer attached to the end of a cylindrical bar antenna was proposed by Lavrentev (1969) and Paik (1972). The resonance is used to transfer the energy from the antenna to the small mass and amplify the displacement of the face of the antenna. In addition, a small mass is more easily strongly coupled to an amplifier.

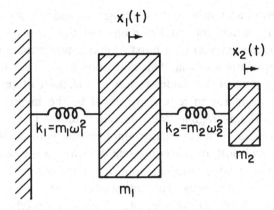

Figure 7.1. Model for an antenna with dynamic mass m_1 with a single resonator.

Figure 7.1 shows a model for a cylindrical antenna coupled to a second resonant mass. The antenna is modeled by the mass m_1 and the spring k_1. The mass m_1 represents the dynamic mass of the cylindrical antenna which is half of its total mass. The spring k_1 is chosen such that the resonant frequency $f_1 = (k_1/m_1)^{1/2}/2\pi$ equals the frequency of the fundamental longitudinal mode of the antenna. The mass m_2 of the resonator and its spring constant k_2 are chosen such that the frequency $f_2 = (k_2/m_2)^{1/2}/2\pi$ is equal to f_1. The Lagrangian for the coupled system in the absence of damping can be written in the form

$$\mathcal{L} = \frac{1}{2} T_{ij}\dot{x}_i\dot{x}_j - \frac{1}{2} V_{ij}x_ix_j, \tag{7.5}$$

where the matrices T and V are given in this case by

$$T = \begin{bmatrix} m_1 & 0 \\ 0 & m_2 \end{bmatrix} \tag{7.6}$$

$$V = \begin{bmatrix} k_1 + k_2 & -k_2 \\ -k_2 & k_2 \end{bmatrix}. \tag{7.7}$$

The frequencies of the normal modes of the system are found by solving the characteristic equation

$$\mathrm{Det}(V_{ij} - \omega^2 T_{ij}) = \begin{vmatrix} k_1 + k_2 - m_1\omega^2 & -k_2 \\ -k_2 & k_2 - m_2\omega^2 \end{vmatrix} = 0. \tag{7.8}$$

The approximate values are $f_+ = f_1(1 + \mu/2)$ and $f_- = f_1(1 - \mu/2)$, where the mass ratio $\mu^2 = m_2/m_1$ has been introduced. The mode unnormalized eigenvectors $\vec{a}_+$ and $\vec{a}_-$ are found to be:

$$\vec{a}_+ = 1 \cdot \hat{x}_1 - \mu^{-1} \cdot \hat{x}_2 \tag{7.9a}$$

$$\vec{a}_- = 1 \cdot \hat{x}_1 + \mu^{-1} \cdot \hat{x}_2, \tag{7.9b}$$

where $\hat{x}_1$ and $\hat{x}_2$ represent unit motion of the masses m_1 and m_2, respectively. The resonator motion is larger than the antenna motion by the factor μ^{-1}.

Energy deposited in the antenna by a burst of gravitational radiation results in an initial displacement of the antenna. This corresponds to an excitation of both normal modes. The beating of the modes results in a transfer of energy between the antenna and the resonator in a time τ_B equal to the inverse of the beat frequency $f_B = f_+ - f_-$. The averaging time τ used in equation (7.1) cannot be less than τ_B if the effectiveness of the resonator amplification is to be retained. This ultimately limits the bandwidth of the two-mode detector to the beat frequency. There is a tradeoff between large motion amplification and a large beat frequency in the selection of the resonator mass. The optimum choice for m_2 depends on the amplifier noise and on the electrical coupling which can be achieved with the selected transducer scheme.

In practical two-mode systems, the resonator must be rather precisely tuned to the antenna frequency, as shown by Paik (1975). If the resonator frequency f_2 differs from the antenna frequency f_1 by $\delta f = \mu f_1$, only half of the energy absorbed by the antenna appears later in the resonator. Since $\mu f_1 = f_B$ in the optimum case, the resonator tuning precision must be of the order of the ultimate detector bandwidth. Analyses have shown that systems with wider ultimate bandwidth have less stringent tuning requirements.

7.2.2   Three-mode systems

The idea of using more than one added resonant mass was suggested by Richard (1980, 1982). The motivation is to provide a faster response time and wider ultimate detection bandwidth for a given motion amplification. A smaller value of τ also reduces the Brownian noise contribution (first term in equation 7.1). The three-mode system modeled in figure 7.2 has been under development at the

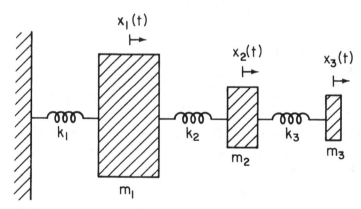

Figure 7.2. Model for an antenna with two attached resonators.

University of Maryland (Cosmelli and Richard, 1982; Folkner, 1987). The (symmetric) dynamic matrices T and V for this model are:

$$T = \begin{bmatrix} m_1 & 0 & 0 \\ 0 & m_2 & 0 \\ 0 & 0 & m_3 \end{bmatrix} \qquad (7.10)$$

$$V = \begin{bmatrix} k_1 + k_2 & -k_2 & 0 \\ -k_2 & k_2 + k_3 & -k_3 \\ 0 & -k_3 & k_3 \end{bmatrix}. \qquad (7.11)$$

The resonator masses m_2 and m_3 are chosen such that $m_2/m_1 = m_3/m_2 = \mu^2$, and the resonators are tuned such that the frequencies $f_2 = (k_2/m_2)^{1/2}/2\pi$ and $f_3 = (k_3/m_3)^{1/2}/2\pi$ are the same as f_1. To the next order in μ, the normal mode frequencies are found to be $f_- = f_1(1 - \mu/\sqrt{2})$, $f_m = f_1$, and $f_+ = f_1(1 + \mu/\sqrt{2})$. To the same accuracy the mode eigenvectors are

$$\vec{a}_m = 1 \cdot \hat{x}_1 + 0 \cdot \hat{x}_2 - \mu^{-2} \cdot \hat{x}_3 \qquad (7.12a)$$

$$\vec{a}_+ = 1 \cdot \hat{x}_1 - \mu^{-1}\sqrt{2} \cdot \hat{x}_2 + \mu^{-2} \cdot \hat{x}_3 \qquad (7.12b)$$

$$\vec{a}_- = 1 \cdot \hat{x}_1 + \mu^{-1}\sqrt{2} \cdot \hat{x}_2 + \mu^{-2} \cdot \hat{x}_3. \qquad (7.12c)$$

The motion for an initial displacement of the antenna is more complex than for the two-mode system, but it is found that the motion of the third mass is larger than the motion of the antenna by the mass ratio μ^{-2}. The energy transfer time τ_B is equal to the inverse of the beat frequency $f_B = f_+ - f_-$.

To compare the two-mode system to the three-mode system, consider the case where the final resonators have the same mass and the same electrical coupling to the output amplifier. Then same motion amplification is achieved and the three-mode system has a smaller mass ratio μ^2, a faster energy transfer time and a wider ultimate bandwidth. Alternatively, with the same μ ratio, transfer time and bandwidth, a three-mode system will use a smaller last resonator, making it easier to achieve a large coupling to the amplifier.

7.2.3 Generalization to n-mode systems

The analysis of the two-mode and three-mode systems can be extended to an arbitrary number n of coupled resonators (Richard, 1980, 1982, 1984), the largest of which is always the Weber cylindrical antenna which provides the coupling to the gravitational field. The ratio of the effective masses of two successive oscillators is a constant:

$$\mu^2 = m_{i+1}/m_i \qquad (7.13)$$

and is the analog to the taper of a truly exponential antenna. As is the case for an exponential antenna, the 'taper' μ imposes a practical upper limit to δv, the double-sided usable bandwidth of the multi-mode detector. That limit is estimated from the time spread at the nth resonator of a 'short' pulse deposited in the antenna (Davis and Richard, 1980; Richard, 1976), and, as is the case for the two- and three-mode systems, is given by:

$$\Delta f \approx 2\mu f_1. \tag{7.14}$$

As is the case in equation (7.1), the optimization of the multi-mode detector requires a condition on β. The qualitative discussion which follows gives a result very close to the one obtained from equation (7.1). Of the energy ΔE deposited in the detector, a fraction $\beta \Delta E$ is transferred to the amplifier in the time $1/(2f_1)$ corresponding to one half-cycle of the detector. The averaging time τ should be long enough to accumulate an energy ΔE at the amplifier. The resulting condition is $2\beta\tau f_1 \geq 1$. To use the bandwidth of equation (7.14), $\tau = 1/\Delta f$ and the condition on β is $\beta \geq \Delta f/(2f_1)$. In practical cases, one usually selects the smallest coupling possible. Then, $\Delta f \approx 2\beta f_1$, or $\beta\tau\omega_1 \approx \pi$. This also implies $\beta \approx \mu$. Again, for a bandwidth of the order of f_1, the coupling between the last resonator and the amplifier should be of the order of unity. In practice, when a specific transducer scheme has been selected, the condition $\beta \approx 1$ (for wide band systems) determines an upper value of the last resonator mass m_n. Then, n, m_1 and m_n determine the system mass ratio μ, the averaging time and the bandwidth. The optimum value of β which minimizes the noise of the amplifier is then evaluated.

7.3 Passive transducers and associated amplifiers

7.3.1 Capacitance transducer coupled to an FET

The modulated resonant dc biased capacitance transducer is one of the transducers which has been proposed and developed at Maryland (Davis and Richard, 1980; Richard, 1976). In that scheme a capacitance plate is mounted on a resonant suspension, attached to the bar antenna. The capacitance is biased at some dc voltage V_0 through a low noise resistor R_B whose value is large enough to make its own thermal noise current negligible. A low-leakage high electrical Q blocking capacitor C_B isolates the amplifier from the dc supply. The displacement of the bar is amplified at the resonant capacitance plate and generates an ac voltage which is detected with an FET amplifier. In this section, we study a system consisting of the last mass m_f of the multi-mode detector and its capacitance transducer.

The electrical circuit corresponding to the capacitance transducer is shown in figure 7.3, where C_1 is the transducer capacitance, C_2 is the sum of the input capacitance of the amplifier and parallel stray capacitances. R describes the losses of that circuit, v_n is the amplifier noise voltage, and i_n includes the amplifier current noise i_a and the current noise associated with the losses R:

$$i_n^2 = i_a^2 + 4k_B T_1/R. \tag{7.15}$$

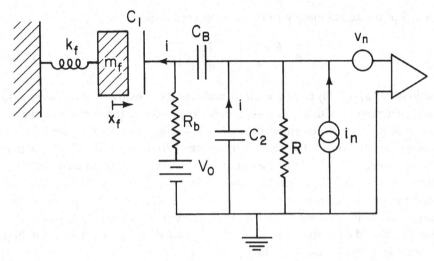

Figure 7.3. Schematic diagram for a capacitive sensor system.

The voltage v generated by the displacement x_f of m_f is given by:

$$v = -(V_0/s)x_r = -E_0 x_f, \tag{7.16}$$

where s is the equilibrium value of the capacitance plate spacing and E_0 is the corresponding value of the electrical field between the plates. The back action force f_e generated by the current i circulating through the capacitor satisfies:

$$df_e/dt = E_0 i. \tag{7.17}$$

Replacing dv/dt by its value $-E_0\,dx_r/dt$, and neglecting R and C_B since $R \gg 1/\omega C_2$ and $1/C_B \ll 1/C_2$, results in the current i being given by

$$i = C_s E_0 \dot{x}_f + (C_s/C_2)i_n, \tag{7.18}$$

where C_s is the series combination of C_1 and C_2. Inserting the value of i in equation (7.17) and integrating with respect to time:

$$f_e = E_0^2 C_s x_f + E_0(C_s/C_2)q_n, \tag{7.19}$$

where q_n is the time integral of i_n. The first term results from the additional stiffness provided by the capacitance transducer $k_e = -E_0^2 C_s$. The second term is the back action of the noise current.

The total voltage output v_0 of the system is obtained from the solution for x_f and from the following:

$$dv_0/dt = i_n/C_p - (C_1/C_p)E_0\dot{x}_f, \tag{7.20}$$

where C_p is the parallel combination of C_1 and C_2.

The signal energy $\mathscr{E}_e$ in the electrical circuit is given by

$$\mathscr{E}_e = \frac{1}{2} v_0^2 C_p = \frac{1}{2} E_0^2 x_f^2 \frac{C_1^2}{C_p}. \tag{7.21}$$

The ratio β of the signal energy to the total energy is given by

$$\beta = \frac{E_0^2 C_1^2}{m_f \omega_1^2 C_p} \left[1 - \frac{E_0^2 C_s}{m_f \omega_1^2}\right]^{-1}. \tag{7.22}$$

Equations (7.18)–(7.20) can be used to evaluate the sensitivity of a multi-mode detector instrumented with a resonant capacitor transducer through the optimum filter analysis procedure described in section 7.4. They can also be used for an approximate evaluation of the noise temperature from equation (7.1). In that case, τ_a is replaced by τ_{al}, the damping time of the system loaded with the capacitance transducer (in practical cases at Maryland, the additional damping produced by the capacitance transducer was negligible). Also, the noise current associated with R is now added to the noise current of the amplifier as done in equation (7.15) and corresponding values for T_n and Z_n are used. Again, the best noise temperature which can be achieved is $\approx 2 T_n$.

Since the best FET amplifiers have a $T_n \geq 0.050$ K, a lower noise amplifier such as a SQUID (superconducting quantum interference device) is desirable for the instrumentation of the resonant capacitor transducer. Such instrumentation is discussed briefly in section 7.3.3.

7.3.2 Inductance modulation transducer coupled to a SQUID
The mechanism presently used to convert the resonator motion into an electrical signal for the Maryland detector is the inductance modulation scheme. Paik obtained the coupling equations using flux quantization for a general case (Paik, 1976). The equations for the noise calculation are developed here.

Figure 7.4 shows schematically a last resonator m_f made of superconducting material located between two pancake coil inductors L_1 and L_2, each having the same inductance L_0. These inductors are connected to a SQUID current amplifier with input inductance L_s. A persistent current I_0 is stored in the loop containing the superconducting coils L_1 and L_2. A displacement x_f of the test mass produces

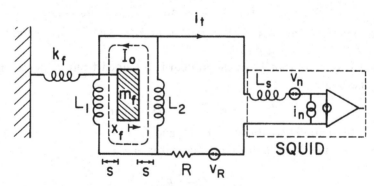

Figure 7.4. Schematic diagram for the inductance modulation sensor system.

the voltage v_t at the output of the transducer given by

$$v_t = \frac{L_0 I_0}{s} \dot{x}_f,$$

(7.23)

where s is the spacing between the coils and the resonator. This voltage drives a current i_t through the transducer and the SQUID input coil. The current through the transducer splits between the inductors L_1 and L_2 and produces a force on the test mass given by

$$f_e = \frac{L_0 I_0}{s} i_t.$$

(7.24)

An electrical spring term in the equations of motion results from the combination of equations (7.23) and (7.24) and is given by;

$$k_e = \left(\frac{L_0 I_0}{s}\right)^2 \frac{1}{L_t + L_s},$$

(7.25)

where $L_t = L_0/2$ is the parallel combination of L_1 and L_2. The value of k_e can be determined experimentally from the frequency shift of the resonator. The ratio β_e of electrical energy to the total energy of the system is given by the ratio of the electrical spring constant to the overall spring constant $k_f + k_e$. The amplifier energy fraction β is less than β_e by the factor $L_s/(L_s + L_t)$. Electrical coupling values ranging from 0.01 to 0.1 have been readily attained using the inductance modulation method with masses of the order of 100 g or less.

Losses associated with the inductance modulation scheme have been observed which result in additional damping and noise. The noise can be modeled by a resistor R with the associated voltage noise v_R in series with the amplifier input coil. The resistor R is related to the measured electrical quality factor Q_e by

$$R = \omega_f(L_t + L_s)/Q_e.$$

(7.26)

This voltage noise appears in the same manner as the SQUID voltage noise and is typically larger than the SQUID voltage noise. To estimate the sensitivity of the detector, equation (7.1) can be used with a modified amplifier noise temperature. A more precise sensitivity calculation uses the equations of motion of the system, which are:

$$m_f \ddot{x}_f + k_f x_f + c_f \dot{x}_f + \gamma i_t = f_n + f_{f-1}$$

(7.27)

$$(L_t + L_s)(di_t/dt) + i_t R - \gamma \dot{x}_f = v_R + v_n,$$

(7.28)

where k_f is the resonator mechanical spring constant, c_f is the mechanical damping coefficient, v_n is the voltage noise of the amplifier, f_n is the Brownian motion force noise due to the mechanical damping, and γ is the combination $L_0 I_0/s$ appearing in equations (7.23) and (7.24). f_{f-1} is the action from the previous resonator.

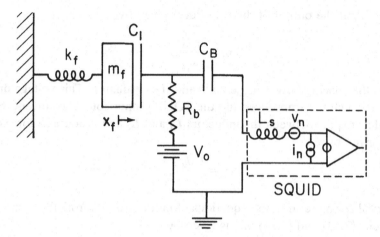

Figure 7.5. Schematic diagram for a capacitive sensor connected to a SQUID amplifier.

7.3.3 Capacitive transducer coupled to a SQUID amplifier

The transducer considered here has been proposed in Richard (1980, 1982) and is shown schematically in figure 7.5. As in section 7.3.1, the last resonator consists of a capacitance plate mounted on a resonant suspension attached to a cylindrical bar antenna. The capacitance is biased at some dc voltage and its output is coupled to an inductance, in turn coupled to a dc SQUID. Noise sources which need to be considered here are the noise current associated with the biasing resistor and the losses in the resonant capacitance, the shot noise associated with leakage current through the resonant blocking and stray capacitances, and the voltage noise and the current noise associated with the SQUID. The series LC circuit formed by C_1, C_B and L_s may or may not be resonant at the bar frequency, thus effectively forming a two- or three-mode system. The choice is determined from a noise analysis of the system and practical considerations. A system with a nonresonant LC circuit has been assembled for the instrumentation of a 2400 kg antenna at CERN (Carelli *et al.*, 1985).

7.4 Analysis of multi-mode systems

7.4.1 Signal-to-noise ratio with the optimum filter

The method of noise analysis described here closely follows that of Michelson and Taber (1984); see also Richard (1986). If a gravitational force $f_g(t)$ is applied to the bar antenna, and if the noise in the system reflected at the antenna is $n(t)$, the best signal-to-noise ratio ρ which can be obtained is (Wainstein and Zubakov, 1962, chapter 3):

$$\rho = \frac{2}{\pi} \int_0^\infty \frac{|F_g(\omega)|^2}{S_n(\omega)} \, d\omega, \tag{7.29}$$

where $F_g(\omega)$ is the Fourier transform of $f_g(t)$ and $S_n(\omega)$ is the power spectral density of the noise. The maximum signal-to-noise ratio requires filtering the detector output with the optimum filter. At time t_0, that filter is;

$$\mathcal{H}(\omega) = e^{-j\omega t_0} \frac{F_g^*(\omega)}{S_n(\omega)}. \tag{7.30}$$

The noise power S_n is due to the Brownian motion noise of the bar and resonators, the electrical damping noise, and the noise of the amplifier. Each noise source is assumed to be frequency independent and uncorrelated with any other source over the useful bandwidth of the detector. The evaluation of the noise at any point in the electro-mechanical circuit requires the introduction of transfer functions between the noise sources, so that the total noise power is

$$S_n(\omega) = \sum_i s_i(\omega) T_i(\omega). \tag{7.31}$$

The transfer functions $T_i(\omega)$ are determined from the solution of the equations of motion for the system.

The form of the signal must be known to achieve the maximum signal-to-noise ratio. If we consider broad band gravitational signals, $f_g(t)$ can be approximated as:

$$f_g(t) = (2\varepsilon m_1)^{1/2} \delta(t) \tag{7.32}$$

so that

$$F_g(\omega) = (2\varepsilon m_1)^{1/2}. \tag{7.33}$$

Here ε is the amount of energy the gravitational wave deposits in a bar with dynamic mass m_1, and $\delta(t)$ is the Dirac delta function. For a signal-to-noise ratio of $\rho = 1$, the energy ε equals $k_B T_d$, where T_d is defined as the noise temperature of the detector for the detection of short pulses.

7.4.2 Analysis of a three-mode system

The electro-mechanical diagram for the Maryland three-mode system is given in figure 7.6. The models for the bar, resonators and transducer coupling are given in the previous sections. The SQUID amplifier is represented by an ideal current amplifier of input inductance L_s with independent current and voltage noise sources. The voltage noise for contemporary SQUIDs is too small to be measured. In practice, the voltage noise V_R associated with the electrical damping is much larger than the expected SQUID voltage noise*. Note that the electrical coupling is between the pancake coils mounted on the mass m_1 and the final mass m_3.

---

* An electrical Q of 3.8×10^6 has been obtained recently with a 10 g resonator inductively coupled to a dc SQUID. Such a value is adequate for operation with a quantum limited SQUID if the resonator temperature is 50 mK or less. This result is reported in Folkner, Moody and Richard (1989).

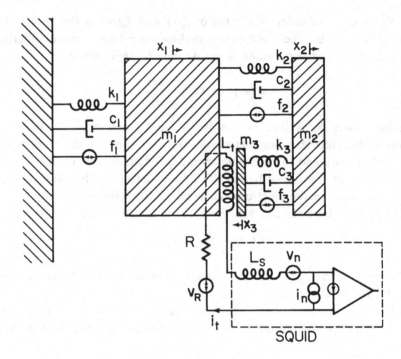

Figure 7.6. Electro-mechanical diagram for a three-mode system with an inductance modulation sensor.

The equations of motion for the system are;

$$m_1\ddot{x}_1 + k_1 x_1 + c_1\dot{x}_1 + k_2(x_1 - x_2) + c_2(\dot{x}_1 - \dot{x}_2) - \gamma i_t = f_1 - f_2, \qquad (7.34a)$$

$$m_2\ddot{x}_2 + k_2(x_2 - x_1) + c_2(\dot{x}_2 - \dot{x}_1) + k_3(x_2 - x_3) + c_3(\dot{x}_2 - \dot{x}_3) = f_2 - f_3, \qquad (7.34b)$$

$$m_3\ddot{x}_3 + k_3(x_3 - x_2) + c_3(\dot{x}_3 - \dot{x}_2) + \gamma i_t = f_3, \qquad (7.34c)$$

$$(L_t + L_s)\frac{di_t}{dt} + i_t R - \gamma(\dot{x}_3 - \dot{x}_1) = v_R + v_n, \qquad (7.34d)$$

where γ is the transducer parameter $L_0 I_0 / s_0$, and f_1, f_2, and f_3 are the Brownian motion noise forces. The current noise i_n is additive at the SQUID input. The differential equations are transformed into algebraic equations by assuming each quantity varies as $e^{j\omega t}$. Putting in the time dependence and rearranging the terms gives the following:

$$(j\omega L + R)i_t + (j\omega\gamma)x_1 + (O)x_2 + (-j\omega\gamma)x_3 = v_R + v_n, \qquad (7.35a)$$

$$\gamma i_t + (O)x_1 + (-j\omega c_3 - k_3)x_2 + (-m_3\omega^2 + k_3 + j\omega c_3)x_3 = f_3, \qquad (7.35b)$$

$$\gamma i_t + (-j\omega c_2 - k_2)x_1 + (-m_2\omega^2 + k_2 + j\omega c_2)x_2 + (-m_3\omega^2)x_3 = f_2, \qquad (7.35c)$$

$$(O)i_t + (-m_1\omega^2 + k_1 + j\omega c_1)x_1 + (-m_2\omega^2)x_2 + (O)x_3 = f_1, \qquad (7.35d)$$

with $L = L_t + L_s$. The algebraic equations can be solved numerically for i_t at each

Table 7.1. *Transducer parameters for Maryland three-mode transducer.*

Mechanical parameters at $T = 4.6\,K$

Masses	$m_1, m_2, m_3 = 600, 1.47, 0.0036$ g
Frequencies	$f_1, f_2, f_3 = 1770, 1770, 1770$ Hz
Quality factors	$Q_1, Q_2, Q_3 = 3.5 \times 10^6, 3.5 \times 10^6, 3.5 \times 10^6$

Transducer parameters

Inductors	$L_1, L_2 = 11 \times 10^{-6}$ H
Spacings	$s_1, s_2 = 2.0 \times 10^{-4}$ m
Stored current	$I_0 = 5.5$ A
Coupling	$\beta_e = 0.017$
Electrical Q	$Q_e = 5 \times 10^4$

SQUID parameters

Input inductance	$L_s = 2 \times 10^{-6}$ H
Energy resolution	$E_r = 2.6 \times 10^{-30}$ J/Hz
Current noise	$i_n = 1.6 \times 10^{-12}$ A/$\sqrt{\text{Hz}}$
Noise temperature	$T_n = 1 \times 10^{-4}$ K (estimated)

angular frequency ω using Cramer's rule to give:

$$i_t(\omega) = K_1(\omega)f_1 + K_2(\omega)f_2 + K_3(\omega)f_3 + K_v(\omega)(v_n + v_R) + i_n. \qquad (7.36)$$

The total noise power S_t at the SQUID input is then

$$S_t = |K_1|^2 S_1 + |K_2|^2 S_2 + |K_3|^2 S_3 + |K_v|^2 S_v + S_i \qquad (7.37)$$

where $S_i = i_n^2$, $S_v = v_R^2 + v_n^2$, $S_1 = f_1^2$, etc., since the noise sources are uncorrelated. To refer the total noise to the input of the bar, the inverse of the coefficient of S_1 is used:

$$S_n = S_1 + |K_1|^{-2}(|K_2|^2 S_2 + |K_3|^2 S_3 + |K_v|^2 S_v + S_i). \qquad (7.38)$$

For a wide band signal, as in equation (7.33), the inverse of the total noise is proportional to the frequency dependent signal-to-noise ratio. The integral of equation (7.29) is performed numerically to give the system noise temperature T_d.

The amount of calculation required to find the noise temperature makes it difficult to examine different configurations analytically. To examine the effects of varying parameters, equation (7.1) can be used as a guide while a series of computer calculations is needed to see the exact effects. In an actual detector, many of the parameters may not be well known prior to the final cooling. In particular, the frequency of each resonator depends on the temperature and the mounting method. Also, SQUIDs are very sensitive instruments whose performance is easily degraded by stray radio-frequency interference or temperature variations.

The design parameters for the Maryland three-mode system are given in table 7.1. The quality factors are derived from independent measurements, and the inductance modulation constant is set by the area of the third mass and the size of the wire used for the pancake coils. The SQUID current noise in table 7.1 is the specification value for a commercial dc SQUID (Biomagnetic Technologies, Inc., San Diego, CA). The electrical Q value is the one exhibited by our current three-mode system and is less than recently measured on a prototype resonator for a 2400 kg–800 Hz bar antenna (see footnote on p. 179).

The maximum amount of electrical coupling is set by the critical magnetic field of the material used for the third mass, if there is no other limitation on the amount of persistent current capable of being stored.

Figure 7.7 shows the noise sources, referred to the bar input, for the three-mode system at the maximum practical coupling. The Brownian motion noise of the bar is constant in frequency, while the Brownian noise of the first and second resonators has one and two minima, respectively. The system resonant frequencies appear as minima of the SQUID current noise, which is the solid line of figure 7.7. The voltage noise minima occur at the resonant frequencies of the mechanical system, i.e. with no electrical spring term. This is because the cofactor for the voltage noise in equation (7.37) does not contain the electrical spring term.

The signal-to-noise ratio as a function of frequency is given in figure 7.8. It can be seen that the region of significant signal-to-noise ratio is centered about the

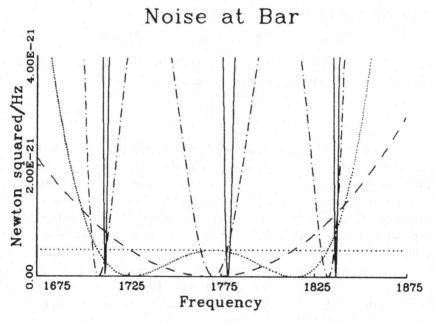

Figure 7.7. Plot of the noise components versus frequency for the three-mode system under development at Maryland: bar Brownian motion noise (dotted line), intermediate resonator noise (dashed line), final resonator noise (fine dotted line), SQUID current noise (solid line), combined voltage noise (dot–dashed line).

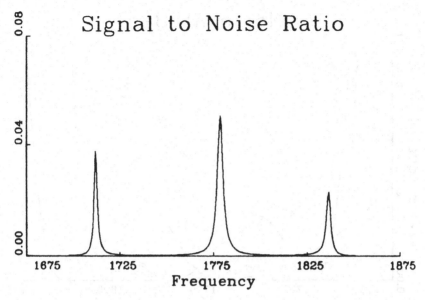

Figure 7.8. Plot of the signal-to-noise ratio for the three-mode system under development at Maryland.

three resonant frequencies. The calculated noise temperature for the entire bandwidth is 2.2 mK. If only the central peak is used for observations, the noise temperature is 5.0 mK. For a properly tuned system, any of the three peaks can be used which gives three possible detection frequencies of slightly different sensitivity.

Figure 7.9 shows the noise at the bar for the three-mode system of table 7.1 but using a SQUID with $200\,\hbar$ noise and a slightly higher electrical coupling β_e of 0.05. The detection temperature is 0.24 mK, which is near the Brownian noise limit. The usable bandwidth in figure 7.9 is 120 Hz, and is limited by the beat frequency of the system.

7.4.3 Analysis of a five-mode system

Five-mode systems have been studied in detail. First, an analysis of practical systems was made (Richard, 1984) using the approximate expression for T_d given in equation (7.1). Then, a detailed optimum-filter analysis was made which confirmed the validity of the approximate analysis (Richard, 1987). These studies show that a wide bandwidth with approximately uniform sensitivity (fluctuations of the order of 50% or less) can be achieved with multi-mode systems. More precisely, a bandwidth of the order of 40% of the antenna frequency is a realistic goal. In addition, the small value of the averaging time used in such systems makes for significantly lower Q/T_i requirements. Quality factors of $Q \cong 10^7$ at $T_i \cong 0.05\,\text{K}$ bring the fluctuations resulting from the thermal noise in the mechanical system to the level of one phonon. The impact of deviations from

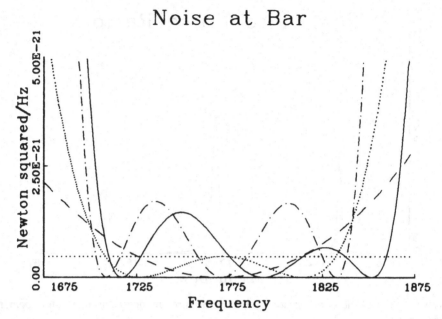

Figure 7.9. Plot of the noise sources versus frequency for the three-mode system with a more sensitive SQUID: bar Brownian motion noise (dotted line), intermediate resonator noise (dashed line), final resonator noise (fine dotted line), SQUID current noise (solid line), combined voltage noise (dot–dashed line).

design parameters has also been studied. Deviations from the design parameters of the order of 10% in the frequencies and as high as 50% in the dynamic masses have little impact on the bandwidth and on the uniformity of the sensitivity across it for a massive five-mode system with $\mu^2 = 0.1$.

References

Blair, D. G., Giles, A. and Zeng, M. (1987). *J. Phys. D* **20**, 162–8.
Bonazzola, S. and Chevreton, M. (1973). *Phys. D* **8**, 359.
Carelli, P., Castellano, M. G., Cosmelli, C., Foglietti, V. and Modena, I. (1985). *Phys. Rev. A* **32**, 3258.
Cosmelli, C. and Richard, J.-P. (1982). *Rev. Sci. Instrum.* **53**, 674.
Davis, W. S. and Richard, J.-P. (1980). *Phys. Rev. D* **22**, 2297.
Folkner, W. M. (1987). Thesis, University of Maryland (unpublished).
Folkner, W. M., Moody, M. V. and Richard, J.-P. (1989). *J. Appl. Phys.* **65**, 887.
Giffard, R. P. (1976). *Phys. Rev. D* **14**, 2578.
Karim, M. (1984). *Phys. Rev. D* **30**, 2031.
Lavrentev, G. Y. (1969). *JETP Lett.* **10**, 318.
Michelson, P. F. and Taber, R. C. (1984). *Phys. Rev.* **29**, 2149.
Paik, H. J. (1972). In *Proceedings of the International School of Physics 'Enrico Fermi', Varenna, Italy*, Academic Press, New York.

Paik, H. J. (1975). Ph.D. Thesis, Stanford University (unpublished).

Paik, H. J. (1976). *J. Appl. Phys.* **47**, 1168.

Richard, J.-P. (1976). *Rev. Sci. Instrum.* **47**, 423.

Richard, J.-P. (1978). *Acta Astronautica* **5**, 63.

Richard, J.-P. (1980). In *Gravitational Radiation, Collapsed Objects and Exact Solutions,* ed. C. Edwards, Lecture Notes in Physics 124, p. 370, Springer, Berlin.

Richard, J.-P. (1982). In *Proceedings of the 2nd Marcel Grossman Meeting on General Relativity,* ed. R. Ruffini, pp. 1239–44, North Holland, Amsterdam.

Richard, J.-P. (1984). *Phys. Rev. Lett.* **52**, 165.

Richard, J.-P. (1986). *J. Appl. Phys.* **60**, 3807.

Richard, J.-P. (1987). University of Maryland Technical Report No. PP-88-246, July 27, 1987, unpublished.

Wainstein, L. A. and Zubakov, V. D. (1962). *Extraction of Signal from Noise,* Prentice-Hall, Englewood Cliffs, NJ.

Weber, J. (1960). *Phys. Rev.* **117**, 306.

Weber, J. (1969). *Phys. Rev. Lett.* **22**, 1320.

Weber, J. (1970). *Phys. Rev. Lett.* **24**, 276.

8

Parametric transducers

P. J. VEITCH

8.1 Introduction

To be able to detect gravitational radiation, resonant mass antennae must achieve a dimensionless strain sensitivity of $\sim 10^{-19}$–10^{-20} (Thorne, 1987). Such a high sensitivity can only be obtained by the use of well isolated, massive, high acoustic Q antennae which are cooled to liquid-helium temperatures, and use 'quantum limited' transducers to read out the antenna's vibrations. Modern resonant mass antennae generally consist of a high Q cylindrical bar to which is attached one or more smaller masses which are resonant at the antenna frequency, to form a two-mode or multi-mode antenna (see Richard and Folkner's chapter 7). The coupled resonators mechanically amplify the bar's vibrations thereby reducing the effect of transducer wideband noise. It is important that the acoustic Q of the entire antenna be high so as to minimise noise due to Brownian motion of the masses.

Initially, resonant mass antennae used passive PZT crystal transducers which were mounted near or around the girth of the bar. These were subsequently superseded by passive, modulated inductance and capacitance transducers which have proven to be much more sensitive, and are still being developed (see chapter 7). More recently, several groups have started to investigate another class of transducers: the parametric or active transducer (Bordoni *et al.*, 1986; Braginsky, Panov and Popel'nyuk, 1981; Oelfke and Hamilton, 1983; Tsubono, Ohashi and Hirakawa, 1986; Veitch *et al.*, 1987). This type of transducer differs from passive transducers in that it requires an external power source (a pump oscillator), and it has intrinsic power gain. The power gain is obtained by upconverting the antenna signal to a much higher frequency, generally microwave frequencies.

The basic operation of a parametric transducer is indicated schematically in figure 8.1. Antenna vibrations modulate the resonant frequency of a high Q resonant circuit which, in turn, phase (and amplitude) modulates the pump signal, producing sidebands which are displaced from the pump by the antenna frequency. The modulated pump is amplified using a low-noise amplifier and then demodulated to baseband using part of the original pump signal as the reference. All parametric transducers being developed at present operate by modulating the

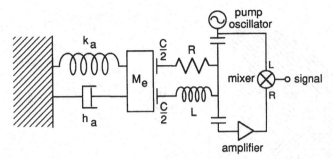

Figure 8.1. A simplified model of a resonant mass antenna with a parametric transducer. The parameters M_e, k_a and h_a can be related to the antenna mass, frequency and acoustic Q. The transducer parameters C, L and R are determined by the transducer geometry, resonant frequency and electrical Q.

capacitance of the high Q circuit. This is generally accomplished by using a reentrant, or capacitively loaded coaxial or toroidal, cavity.

Parametric transducers have several advantages over passive transducers. Firstly, their operating frequency may be chosen to give the optimum performance using the best pump oscillators and amplifiers. Passive transducers, on the other hand, are restricted to using amplifiers which operate at the antenna frequency. As can be seen from figure 8.2, there were considerable benefits to be gained by operating the transducer at microwave frequencies. This advantage has only recently been diminished by the development of a more sensitive dc SQUID audio amplifier.

The second advantage is that parametric transducers should not substantially increase the antenna noise due to Brownian motion. Whereas passive transducers resistively damp the antenna, decreasing its Q and increasing the Brownian motion noise contribution, parametric transducers reactively cold-damp the antenna and only slightly degrade the ratio of the mode temperature to acoustic Q, provided that the noise temperature of the transducer is sufficiently small. Two further benefits accrue as a result of the cold-damping: the reduced Brownian motion amplitude alleviates the stringent requirements on the dynamic range of the amplifiers and analogue-to-digital converters, and the dead-time which would occur after the excitation of the antenna by a high energy event is decreased because of the reduced antenna decay time. Finally, an advantage which is perhaps more of a convenience: parametric transducers are self calibrating. That is, the sensitivity of the antenna can be determined without the need for an auxiliary calibration transducer.

An obvious disadvantage of parametric transducers is the need for a power source. Clearly, the pump signal must have ultra low–low levels of phase and amplitude noise. The solution to this problem requires the development of new ultra-clean frequency sources (Mann, 1984; Veitch et al., 1987). The stringent requirements on the purity of the pump can be relaxed slightly by the

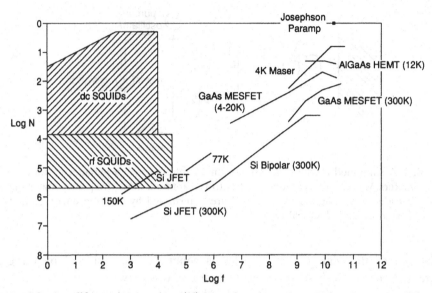

Figure 8.2. Amplifier noise number (N) as a function of operating frequency (f). The noise level of the high frequency amplifiers is the white-noise level or, equivalently, the noise obtained for signal frequencies sufficiently far from the carrier or operating frequency. Low frequency noise in the amplifiers upconverts to produce excess noise at signal frequencies close to the carrier. The cross-hatched regions representing the performance of SQUIDs each include several distinct devices. Note that the low frequency '1/f' noise associated with all SQUIDs is largely hidden by this representation. All SQUIDs have '1/f' noise; and it is generally worse than that shown for the dc SQUID, delimiting the upper boundary of that region. The devices delimiting the lower and upper boundaries of the RF SQUIDs are the BTi 330X RF SQUID and a 9 GHz RF SQUID (Hollenhorst and Giffard, 1979), respectively. There is some overlap of RF and dc SQUID performance with the BTi 440 dc SQUID having $\log N \approx 4.5$. The performance of dc SQUIDs has recently been significantly improved (Awschalom et al., 1988). The inductance of the input coil was very low (~100 nH), however, which could present problems when trying to couple a gravity wave transducer to this device, and may result in a degraded energy sensitivity. The semiconductor amplifier performance curves are generally also compilations of several authors' work: Si JFET 300 K (Bordoni and Pallotino, 1977 and L. D. Mann, 1988, private communication); Si JFET 150 K (Bordoni et al., 1981); Si JFET 77 K (L. D. Mann, 1988, private communication); Si Bipolar (Cooke, 1971; L. D. Mann, 1988, private communication; Weinreb, 1982); GaAs MESFET 300 K (Goronkin, 1985, and Wang and Wang, 1987); GaAs MESFET 4–20 K (Bocko, 1984; Long, Clark and Prance, 1980; Weinreb, 1982); 4 K Maser (Weinreb, 1982); AlGaAs HEMT 12 K (Pospieszalski et al., 1986; van der Ziel, 1986); Josephson Paramp (Yurke et al., 1988).

use of a pair of balanced transducers (Oelfke and Hamilton, 1983). Another problem is that the low-noise amplifiers which are available at microwave frequencies saturate easily. Thus, the non-signal part of the modulated pump, or carrier, must be suppressed without adding any noise. A scheme for accomplishing this will be described later.

The remainder of this chapter is divided into several sections. The first, section 8.2, discusses the Manley–Rowe equations which provide a general description of parametric devices. While these equations describe some of the salient features of parametric transducers, they do not consider the operation of the transducer in conjunction with a resonant mass gravitational radiation antenna. This is discussed in section 8.3, where the impedance matrix description of the antenna and transducer is considered. Using results from that section, the modification of the antenna's frequency and acoustic Q by the transducer is investigated in section 8.4. The impedance matrix description yields equations expressing the antenna sensitivity in terms of various generalised noise sources. To be useful, these sources must be related to actual noise sources in the transducer. Sections 8.5 and 8.6 are devoted to a discussion of the transducer sensitivity and noise characteristics, and to general comments about a typical signal processing circuit for a parametric transducer. Having discussed the concepts behind the operation and optimisation of a parametric transducer, we then summarise the practical implementation of this class of transducers by various groups in section 8.7.

8.2   The Manley–Rowe equations

The basic operation of a parametric transducer is described by the Manley–Rowe equations (Manley and Rowe, 1956). These equations relate the power at different frequences in a non-linear lossless reactance.

If two signals having frequencies ω_1 and ω_2 are applied to a non-linear reactance then frequency conversion by the reactance will produce signals at all integral harmonics $\pm|m\omega_1 + n\omega_2|$. Using $P_{m,n}$ to denote the average power flowing *into* the reactance at the frequencies $\pm|m\omega_1 + n\omega_2|$, the Manley–Rowe equations can be written:

$$\sum_{m=0}^{\infty} \sum_{n=-\infty}^{\infty} \frac{m P_{m,n}}{m\omega_1 + n\omega_2} = 0 \tag{8.1a}$$

$$\sum_{m=-\infty}^{\infty} \sum_{n=0}^{\infty} \frac{n P_{m,n}}{m\omega_1 + n\omega_2} = 0. \tag{8.1b}$$

Consider a double sideband upconverter where power is restricted to flow at $\omega_1 = \omega_a$ (the antenna frequency), $\omega_2 = \omega_p$ (the pump frequency), $\omega_2 + \omega_1 = \omega_+$ and $\omega_2 - \omega_1 = \omega_-$ by the use of filters (the high Q resonant circuit for example). Denoting $P_{1,0} = P_a$, $P_{0,1} = P_p$, $P_{1,1} = P_+$ and $P_{-1,1} = P_-$, Equations (8.1) give

$$\frac{P_a}{\omega_a} + \frac{P_+}{\omega_+} - \frac{P_-}{\omega_-} = 0 \tag{8.2a}$$

$$\frac{P_p}{\omega_p} + \frac{P_+}{\omega_+} + \frac{P_-}{\omega_-} = 0. \tag{8.2b}$$

From equation (8.2a) we see that

(a)   If $P_+/\omega_+ > P_-/\omega_-$ then $P_a/\omega_a < 0$. That is, if more power is extracted from the lower sideband than the upper then net power is injected into the antenna. This can lead to parametric instability where the transducer causes the antenna to oscillate if the power injected into the antenna exceeds the acoustic losses. This case corresponds to the pump frequency being above the centre of resonance.

(b)   If $P_+/\omega_+ < P_-/\omega_-$ then $P_a/\omega_a > 0$. That is, power is absorbed from the antenna if the pump frequency is below the centre of resonance. In this configuration the antenna is cold-damped.

The transducer power gain, G_t, is given by (Oelfke, Hamilton and Darling, 1981) $G_t = -(P_+ + P_-)/P_a$, which can be expressed in terms of the power in the sidebands:

$$G_t \approx \left(\frac{\omega_p}{\omega_a}\right)\left(\frac{P_+\omega_- + P_-\omega_+}{P_+\omega_- - P_-\omega_+}\right) \quad \text{if} \quad \omega_p \gg \omega_a. \tag{8.3}$$

There are several cases of interest. If the pump frequency is centred on the resonant circuit response then $P_+ = P_-$ and $G_t \approx -(\omega_p/\omega_a)^2$. If the pump frequency is not sufficiently close to the centre of resonance, however, then

$$G_t \approx \left(\frac{\omega_p}{\omega_a}\right)\left(\frac{P_+ + P_-}{P_+ - P_-}\right),$$

which reduces to the gain observed for single sideband upconverters, $\pm(\omega_p/\omega_a)$, as the pump frequency moves towards the edge of resonance and one sideband is suppressed. In the case $P_+/\omega_+ = P_-/\omega_-$ there is no net power flow from the antenna (see equation (8.2a)) and the gain apppears to be infinite. However, this is an artefact of the model which can be remedied by including the second-order sidebands in equations (8.2).

8.3   Impedance matrix description

In this section we will review Giffard's impedance matrix description (Giffard, 1976) for estimating the sensitivity of a resonant bar gravitational radiation antenna. We will start by summarising the application of this formalism to a single-mode antenna using a generalised transducer, and then apply the results to an antenna using a parametric transducer. Finally, we discuss the extension of the results to allow them to be used to estimate the sensitivity of a two-mode antenna.

The single-mode antenna can be modelled as shown in figure 8.3, where the bar is represented by a damped spring–mass system and the transducer is modelled as

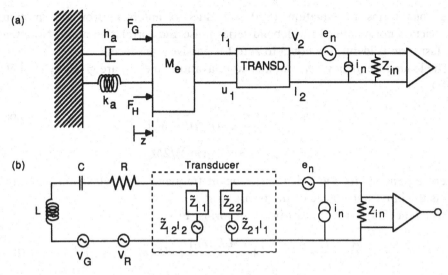

Figure 8.3. (a) A model of the resonant mass antenna. The effective mass M_e, spring constant k_a and damping h_a are related to the antenna properties through $M_e \approx \frac{1}{2}M_a$, $k_a = M_e\omega_a^2$ and $h_a = \omega_a M_e/Q_a$, where M_a, ω_a and Q_a are the mass, resonant frequency and Q of the antenna. The forces F_G and F_H represent the forces acting on the antenna due to the gravity wave and Nyquist noise forces. e_n and i_n represent the equivalent voltage and current noise generators of the transducer and amplifier. Z_{in} is the input impedance of the amplifier. (b) The equivalent electrical circuit of (a).

an electromechanical twoport. Using conventional twoport theory, the transducer can be characterised by an impedance matrix $[Z_{ij}]$:

$$\begin{pmatrix} f(t) \\ V(t) \end{pmatrix} = \begin{pmatrix} Z_{11} & Z_{12} \\ Z_{21} & Z_{22} \end{pmatrix} \begin{pmatrix} u(t) \\ I(t) \end{pmatrix}, \tag{8.4}$$

which relates the force $f(t)$ and velocity $u(t)$ at the input of the transducer to the voltage $V(t)$ and current $I(t)$ at the output.

Using this model the equivalent energy sensitivity of an antenna can be shown (Blair, 1979; Giffard, 1976; Veitch, 1986), to be given by

$$E_n(\omega = \omega_1) = \frac{2k_B T_a \tau_s}{\tau_1} + \frac{|Z_{12}|^2 S_i \tau_s}{2M_e} + \frac{2M_e[|Z_{22}|^2 S_i + S_v]}{|Z_{21}|^2 \tau_s} + \frac{2|Z_{21}| |Z_{12}| \operatorname{Re}(Z_{22})S_i}{|Z_{21}|^2},$$

$$\tag{8.5}$$

where

T_a is antenna physical temperature,
τ_s is the electronics integration and sampling time,
ω_1 is the loaded antenna frequency,
τ_1 is the loaded antenna amplitude decay time,
S_i is the spectral density of amplifier current noise, and
S_v is the spectral density of amplifier voltage noise.

The four terms in equation (8.5) are due to antenna Brownian motion, back-action noise, additive wideband series noise and correlation noise. Note that the last term is identically equal to zero for passive transducers.

The loaded frequency, ω_1, and loaded relaxation time, τ_1, are given by (Veitch, 1986)

$$\omega_1^2 = \frac{\omega_a^2}{1 + [\mathrm{Im}(Z_{11}(\omega_a))/\omega_a M_e]} \tag{8.6a}$$

$$\tau_1^{-1} = \tau_a^{-1} + [\mathrm{Re}(Z_{11}(\omega_a))/2M_e], \tag{8.6b}$$

where ω_a and τ_a are the unloaded resonant frequency and relaxation time, and Z_{11} is the input impedance of the transducer.

Equation (8.5) can be recast into two forms. Firstly,

$$E_n = \frac{2k_B T_a \tau_s}{\tau_1} + \frac{S_f(\omega_1)\tau_s}{2M_e} + \frac{2M_e S_u(\omega_1)}{\tau_s} + 2[S_f(\omega_1)S_{u,f}(\omega_1)]^{1/2}, \tag{8.7}$$

where

$S_f(\omega_1)$ is the spectral density of back-action force noise acting on the antenna,
$S_u(\omega_1)$ is the spectral density of equivalent velocity noise,
$S_{u,f}(\omega_1)$ is that part of S_u that correlates with the force noise.

This is more convenient than equation (8.5) because most noise sources can be more easily related to force and velocity noise than current and voltage noise.

Alternatively,

$$E_n = \frac{2k_B T_a \tau_s}{\tau_1} + \frac{k_B T_N}{2} \frac{|Z_{12}|}{|Z_{21}|} \left[\frac{\beta\omega_1\tau_s |\xi|}{2} + \frac{2(|\xi|^2 + 1)}{\beta\omega_1\tau_s |\xi|} + 2\,\mathrm{Re}(\xi) \right], \tag{8.8}$$

where

$T_N = 2k_B^{-1}(S_i S_v)^{1/2}$ is the amplifier noise temperature,
$\beta = |Z_{12}|\,|Z_{21}|/M_e\omega_1\,|Z_{22}|$ is the electromechanical energy coupling coefficient,
$\xi = Z_{22}/R_N$ is the normalised output impedance, and
$R_N = (S_v/S_i)^{1/2}$ is the optimum source resistance of the amplifier.

Note that for passive transducers ξ is imaginary while for active transducers ξ is predominantly real (see Paik, 1982, for example).

Since a parametric transducer produces two output sidebands which can have different amplitudes, the impedance matrix for this class of transducer is slightly more complicated than that shown in equation (8.4):

$$\begin{pmatrix} V_-(t) \\ f(t) \\ V_+(t) \end{pmatrix} = \begin{pmatrix} Z_{--} & Z_{-1} & 0 \\ Z_{1-} & Z_{11} & Z_{1+} \\ 0 & Z_{+1} & Z_{++} \end{pmatrix} \begin{pmatrix} I_-(t) \\ u(t) \\ I_+(t) \end{pmatrix}. \tag{8.9}$$

The matrix elements for a capacitance modulated parametric transducer are given by (Blair, 1979; Giffard and Paik, 1977):

$$Z_{11} = \frac{C_0 V_p^2}{4\Omega_0 x_0^2} \left[\frac{(Q_0\Omega_+/\omega_a)(1 - j2Q_0\Delta_+)}{(1 + 4Q_0^2\Delta_+^2)} - \frac{(Q_0\Omega_-/\omega_a)(1 + j2Q_0\Delta_-)}{(1 + 4Q_0^2\Delta_-^2)} \right]$$

$$Z_{1\pm} = -\frac{V_p Q_0}{2\Omega_0 x_0} \left[\frac{1 \pm j2Q_0\Delta_\pm}{1 + 4Q_0^2\Delta_\pm^2} \right]$$

(8.10)

$$Z_{\pm 1} = \pm \frac{\Omega_\pm}{\omega_a} \frac{V_p Q_0}{2\Omega_0 x_0} \left[\frac{1 - j2Q_0\Delta_\pm}{1 + 4Q_0^2\Delta_\pm^2} \right]$$

$$Z_{\pm\pm} = \frac{Q_0}{\Omega_0 C_0} \left[\frac{1 - j2Q_0\Delta_\pm}{1 + 4Q_0^2\Delta_\pm^2} \right],$$

where

Ω_0 and Q_0 are the electrical resonant frequency and Q of the transducer,
C_0 is the unperturbed capacitance of the capacitor used to sense the motion,
$x_0 = C_0^{-1}(dC/dx)$ is the unperturbed dynamic plate spacing of the capacitor,
V_p is the peak voltage across the capacitor,

$$\Delta_\pm = \frac{1}{2}\left(\frac{\Omega_\pm}{\Omega_0} - \frac{\Omega_0}{\Omega_\pm}\right) \approx \left(\frac{\delta\Omega \pm \omega_a}{\Omega_0}\right) \quad \text{if } |\delta\Omega \pm \omega_a| \ll \Omega_0,$$

$\Omega_\pm = \Omega_p \pm \omega_a$, $\delta\Omega = \Omega_p - \Omega_0$, and Ω_p is the pump frequency.

The electromechanical coupling coefficient can be generalised to

$$\beta_\pm = |Z_{1\pm}| |Z_{\pm 1}| / (M_e\omega_a |Z_{\pm\pm}|)$$
$$= \frac{C_0 V_p^2 Q_0}{4M_e\omega_a^2 x_0^2} \frac{\Omega_\pm/\Omega_0}{[1 + 4Q_0^2\Delta_\pm^2]^{1/2}}.$$

(8.11)

By letting $\beta = \beta_+ + \beta_-$, we obtain the electromechanical coupling coefficient appropriate to equation (8.8).

The usual operating regime for a parametric transducer is $|Q_0\Delta_\pm| \ll 1$. That is, the Q of the resonant circuit is low and the sidebands are well within the resonator bandwidth. In this limit the impedances can be simplified to:

$$Z_{11} = \frac{C_0 V_p^2 Q_0}{2\Omega_0 x_0^2}$$

$$Z_{1\pm} = -\frac{V_p Q_0}{2\Omega_0 x_0}$$

$$Z_{\pm 1} = \pm \left(\frac{\Omega_\pm}{\omega_a}\right)(-Z_{1\pm}) \approx \pm \frac{\Omega_p}{\omega_a}(-Z_{1\pm})$$

(8.12)

$$Z_{\pm\pm} = \frac{Q_0}{\Omega_0 C_0}$$

$$\beta = \frac{C_0 V_p^2 Q_0}{2M_e\omega_a^2 x_0^2}.$$

Note that the forward transductances $(Z_{\pm 1})$ are a factor Ω_p/ω_a ($\gg 1$) greater than the reverse transductances $(Z_{1\pm})$. When combined with the possibility of using non-reciprocal power transmission devices (see section 8.6) at RF and microwave frequencies, this significantly reduces the importance of back-action forces. Also, even though the capacitance of the transducer must be decreased to enable it to resonate at high frequencies, the coupling coefficient of a parametric transducer is generally larger than for a passive transducer because of the additional factor $Q_0/2$.

Since the antenna was modelled as a single-mode spring–mass system, the equations obtained thus far would appear to only apply to a single-mode antenna. They can, however, after simple changes, also be used to estimate the sensitivity of a two-mode antenna. In a two-mode antenna a second, smaller mass mechanical oscillator which is resonant at the antenna frequency is attached to the bar so as to mechanically amplify the bar vibrations, or equivalently, increase the electromechanical coupling (see Richard and Folkner's chapter 7). The transducer is positioned so as to monitor either the motion of the smaller mass or the relative motion of the two masses. To use the equations given here to estimate the sensitivity of a two-mode antenna, simply replace the effective mass M_e by the quantity (m_e/G_e^2) (Veitch, 1986), where m_e is the effective mass of the second mode and G_e is the effective amplitude gain: the ratio of the actual amplitude of the second oscillator to the effective amplitude (Paik, 1974). Note that this reduces the series noise when referred to the bar, at the expense of increased back-action noise. The antenna decay time to be used is just the decay time of the normal mode being monitored.

8.4 Modification of the antenna's frequency and acoustic quality factor by the transducer

Having determined the input impedance of the transducer we can now use equations (8.6) to investigate the modification of the antenna frequency and Q by the transducer. The results will be extended to yield the modification of the normal mode frequencies and Qs of a two-mode antenna by assuming that only one of the mechanical oscillators is affected by the transducer (as is usually the case since they have such different effective masses) and substituting the results for the single-mode antenna into the equations which determine the normal mode frequencies and Qs. Before doing that, however, we will qualitatively describe the mechanism responsible for these effects (Braginsky, 1970; Veitch, 1986).

The capacitor plates are attracted with a force F:

$$F = \frac{\varepsilon_0 A V^2}{2x^2},$$
(8.13)

where

A is the area of the capacitor plates,
V is the peak potential difference between the plates, and
x is the separation of the plates.

Vibrations of the antenna change the plate spacing, thereby retuning the cavity and changing V. The force between the plates is thus strongly dependent on the instantaneous plate spacing. This produces an additional spring constant which adds to that of the antenna and changes the antenna frequency.

The spring constant is introduced with a time lag $\sim (Q_0/\Omega_0)$, the storage time of the resonator, leading to the regeneration or degeneration of the antenna vibrations, which manifests itself as a change in the acoustic Q. An effect of this type was alluded to earlier in the discussion of the Manley–Rowe equations.

Substituting Z_{11} into equations (8.6) gives to second order in $\delta\Omega^2$ and ω_a^2 (Veitch, 1986)

$$\omega_1 \approx \omega_a \left[1 + \frac{\left(\frac{\beta}{4Q_0}\right)\left(\frac{2Q_0}{\Omega_0}\right)^2 \left(\delta\Omega \cdot \Omega_0 + \frac{1}{2}(\delta\Omega^2 + \omega_a^2)\right)}{1 + 8Q_0^2\left(\frac{\delta\Omega^2 + \omega_a^2}{\Omega_0^2}\right)} \right] \tag{8.14a}$$

$$\tau_1^{-1} \approx \tau_a^{-1} + \frac{\left(\frac{\beta\omega_a^2}{2\Omega_0}\right)\left[1 - 4Q_0\left(\frac{2Q_0\delta\Omega}{\Omega_0}\right) + \left(\frac{2Q_0}{\Omega_0}\right)^2(\delta\Omega^2 + 2\omega_a^2)\right]}{1 + 8Q_0^2\left(\frac{\delta\Omega^2 + \omega_a^2}{\Omega_0^2}\right)}, \tag{8.14b}$$

where ω_1 and τ_1 are the loaded frequency and amplitude decay time and ω_a and τ_a are the unloaded values. (Recall that $\tau_i = 2Q_i/\omega_i$.)

These equations can be simplified to

$$\omega_1 \approx \omega_a[1 + (\beta/2)(2Q_0\delta\Omega/\Omega_0)] \tag{8.15a}$$

$$\tau_1^{-1} \approx \tau_a^{-1} + \left(\frac{\beta\omega_a^2}{2\Omega_0}\right)\left[1 - 4Q_0\left(\frac{2Q_0\delta\Omega}{\Omega_0}\right)\right] \tag{8.15b}$$

in the limit of small offset frequencies, $|\delta\Omega \pm \omega_a| \ll \Omega_0/2Q_0$. Equation (8.15b) can be further simplified to

$$\tau_1^{-1} \approx \tau_a^{-1} - \beta\omega_a\left(\frac{2Q_0\omega_a}{\Omega_0}\right)\left(\frac{2Q_0\delta\Omega}{\Omega_0}\right) \tag{8.15c}$$

if $Q_a \ll \Omega_0/(\beta\omega_a)$ – a condition which is usually satisfied. Note that the system is unconditionally stable only if the pump frequency is below resonance ($\delta\Omega < 0$). Equation (8.15c) describes the cold-damping of the antenna by the transducer. The damping originates in the *reactive* properties of the transducer, and the magnitude of the damping increases as the resistive losses in the transducer reduce ($Q = \omega L/R$).

All transducers can be used to cold-damp mechanical oscillators. Parametric transducers simultaneously readout and reactively feedback to damp the antenna, while passive transducers, on the other hand, must use electromechanical feedback to a second transducer to damp the antenna. In the process of reading-out the antenna motion the passive transducer resistively damps the antenna, increasing the Brownian motion noise term. The resistive damping of the antenna by parametric transducers (the $\beta\omega_a^2/2\Omega_0$ term in equation (8.15b)) is negligible.

Cold-damping has been extensively analysed by the Tokyo group (see Tsubono's chapter 9). The reduction in Q and the change of the mode temperature to Q ratio are described by

$$Q_c^{-1} = Q_a^{-1} + \beta Q_e^{-1} \tag{8.16a}$$

$$\frac{T_c}{Q_c} = \frac{T_a}{Q_a} + \frac{\beta T_e}{Q_e}, \tag{8.16b}$$

where Q_c and T_c are the cold-damped Q and mode temperature, Q_a and T_a are the undamped (but still including any resistive damping by the transducer) quantities, and Q_e and T_e are the effective Q and noise temperature of the damping circuit.

Using equations (8.16) we can determine the maximum allowable noise temperature of the feedback circuit used to cold-damp the antenna. For a high antenna Q, the cold-damped Q will be determined mostly by the transducer (i.e. $Q_c^{-1} \approx \beta Q_e^{-1}$). Equation (8.16b) then gives

$$\frac{T_c}{Q_c} = \frac{T_a}{Q_a} + \frac{T_e}{Q_c}. \tag{8.17}$$

Substituting $T_a = 4\,\mathrm{K}$, $Q_a = 10^7$, $Q_c = 10^5$ we find $T_c/Q_c = 4 \times 10^{-7} + T_e \times 10^{-5}$. Thus, if $T_e \ll 40\,\mathrm{mK}$ (as would almost certainly be the case) then the Brownian motion noise term would be unaffected by the cold-damping circuit. It is interesting to note that, by comparing equations (8.15c) and (8.16a), we find (in the limit of small offset frequencies):

$$Q_c^{-1} \approx -2(2Q_0\omega_a/\Omega_0)(2Q_0\delta\Omega/\Omega_0), \tag{8.18}$$

indicating that there is a lower limit on the cold-damped Q (typically $\sim 10^3$).

To determine the modification of the normal-mode frequencies and amplitude decay times of a two-mode antenna by the transducer, we must first consider the relationship between the respective normal-mode and uncoupled-mode quantities.

It is convenient to express the uncoupled resonant frequencies by

$$\omega_{1,2} = \omega_0(1 \pm \delta/2), \tag{8.19}$$

where δ is a measure of the imperfect tuning of the second oscillator to the bar frequency. Here we will assign $\omega_1(+)$ and $\omega_2(-)$ to be the bar frequency (ω_a)

and second oscillator frequency, respectively. The two normal-mode frequencies of the coupled oscillators are given, to leading order in δ^2 and m_e/M_e, by (Paik, 1974)

$$\omega_\pm^2 = \omega_0^2\left[1 \pm \left(\delta^2 + \frac{m_e}{M_e}\right)^{1/2}\right], \tag{8.20}$$

where M_e and m_e are the effective masses of the bar and second oscillator, respectively. From equations (8.19) and (8.20), a change in either ω_1 or ω_2 will cause ω_0 and δ to change, leading to changes in ω_+ and ω_- which are generally unequal.

The parametric transducer will mostly act to change the resonant frequency of the second oscillator since this will have the smaller effective mass and spring constant. To determine the dependence of the normal-mode frequencies on β and $(2Q_0\,\delta\Omega/\Omega_0)$, first substitute equation (8.15a) into the equations for ω_0 and δ derived from equation (8.19), and then substitute the resulting equations into equation (8.20). This gives

$$\omega_{\pm1} \approx \omega_{\pm0} + K_\pm\beta\left(\frac{2Q_0\delta\Omega}{\Omega_0}\right) + O\left(\beta^2\left(\frac{2Q_0\delta\Omega}{\Omega_0}\right)^2\right) \tag{8.21a}$$

$$K_\pm = \frac{\omega_{20}(2\omega_{\pm0}^2(\delta_0^2 + m_e/M_e)^{1/2} \mp \delta_0\omega_{00}^2(2 + \delta_0))}{8\omega_{00}\omega_{\pm0}(\delta_0^2 + m_e/M_e)^{1/2}}, \tag{8.21b}$$

where equation (8.21b) has been simplified using

$$\omega_{00} = (\omega_{10} + \omega_{20})/2$$

$$\delta_0 = (\omega_{10} - \omega_{20})/\omega_{00}$$

$$\omega_{\pm0}^2 = \omega_{00}^2[1 \pm (\delta_0^2 + m_e/M_e)^{1/2}].$$

Here, ω_{10}, ω_{20}, ω_{00}, $\omega_{\pm0}$ and δ_0 are unmodified ($\beta = 0$ or $\delta\Omega = 0$) frequencies and detuning factor.

The dependence of the normal-mode decay times on β and $(2Q_0\delta\Omega/\Omega_0)$ is similarly affected by their relationship to the uncoupled-mode decay times (Paik, 1974):

$$2\tau_\pm^{-1} = \left[1 \pm \frac{\delta}{(\delta^2 + m_e/M_e)^{1/2}}\right]\tau_1^{-1} + \left[1 \mp \frac{\delta}{(\delta^2 + m_e/M_e)^{1/2}}\right]\tau_2^{-1}. \tag{8.22}$$

Substituting equation (8.15c) into equation (8.22), and assuming $\delta = \delta_0 + \delta_1\beta(2Q_0\delta\Omega/\Omega_0)$ gives to leading order

$$\tau_{\pm1}^{-1} \approx \tau_{\pm0}^{-1} - \left(1 \mp \frac{\delta_0}{(\delta_0^2 + m_e/M_e)^{1/2}}\right)\left(\frac{\beta\omega_{20}}{2}\right)\left(\frac{2Q_0\delta\Omega}{\Omega_0}\right)\left(\frac{2Q_0\omega_{20}}{\Omega_0}\right), \tag{8.23}$$

where $\tau_{\pm0}$ are the unmodified normal-mode decay times.

Measurements of $\omega_{\pm 1}$ and $\tau_{\pm 1}^{-1}$ as a function of $\delta\Omega$ (see figure 8.4) can be analysed using equations (8.21) and (8.23) to provide a unique determination of m_e, δ_0, β and Q_0. This is extremely elegant since the determination is independent of estimates of parameters such as transducer capacitance or inductance, plate spacing, pump drive amplitude, transmission line losses, electrical couplings, etc.

The ratio of the slopes of the '$\omega_{\pm 1}$ vs. $\delta\Omega$ graphs' (see figures 8.4a and b) yields K_+/K_-. Since the unmodified normal-mode frequencies ω_{-0} and ω_{+0} are known, equation (8.20) can be used to calculate ω_{00} and $(\delta_0^2 + m_e/M_e)$. Assuming a value for m_e (M_e can be calculated (Paik, 1974)), we can calculate a value for δ_0^2. The sign of δ_0 is obtained by comparing the relative slopes of the two graphs: if $K_+ > K_-$ then, from equation (8.21b) and since $|\delta_0| < 1$, δ_0 must be negative ($\omega_{20} > \omega_{10}$); similarly, if $K_+ < K_-$ then δ_0 must be positive ($\omega_{20} < \omega_{10}$). Having fully determined δ_0, equation (8.19) can be used to calculate ω_{20}. These quantities can then be substituted into equation (8.21b) and m_e varied until the calculated value of K_+/K_- agrees with the measured value. As an example, consider the graphs in figures 8.4a and b, which show the dependence of the normal-mode frequencies of the UWA antenna on $\delta\Omega$ (Linthorne, Veitch and Blair, 1990). Using the above procedure, we find $m_e \approx 0.52 \pm 0.04$ kg – a value which compares favourably with the estimated value of ~ 0.6 kg (the effective mass of the UWA second mode cannot be calculated accurately because of its irregular shape). Having determined m_e, the actual value of δ_0 can be calculated.

To determine the other quantities we compare the slope of the ω_{+1} (or ω_{-1}) graph (figures 8.4(a) and (b)), which is proportional to βQ_0, and the slope of the $\tau_{+1}^{-1}(\tau_{-1}^{-1})$ graph (figures 8.4(c) and (d)), which is proportional to βQ_0^2. Since all other parameters are now known, the ratio yields Q_0 which in turn enables us to calculate β. Using the graphs shown in figures 8.4(b) and (d) we calculate (Linthorne, Veitch and Blair, 1990) $Q_0 \approx (1.2 \pm 0.1) \times 10^5$ and $\beta \approx (1.4 \pm 0.2) \times 10^{-3}$, which agree roughly with the expected values.

8.5 Calculation of the transducer sensitivity and noise characteristics using the equivalent electrical circuit

The transducer resonant circuit can be represented by either a parallel or series RLC circuit provided that the resonant frequencies are not too close together. The choice of either circuit is arbitrary and affects the values ascribed to the lumped components. At low frequencies (below RF) the choice is generally dictated by the actual components used.

As an aside, at high frequencies the equivalent circuit can be related to the position on the transmission line from which the cavity is viewed. If viewed from a 'detuned short position' where there exists a standing wave minimum when the cavity is detuned (from resonance at the pump frequency) then a parallel RLC

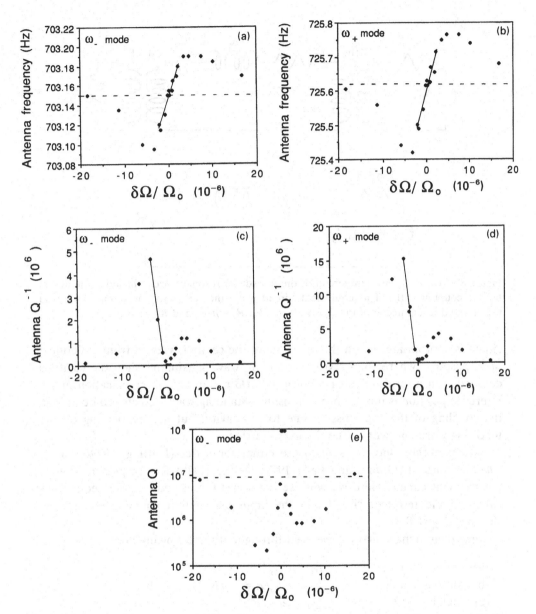

Figure 8.4. Graphs of the variation of the normal-mode frequencies of the antenna ((a) and (b)), the inverse of the normal-mode Qs ((c) and (d)) and a normal-mode Q(e) as a function of the normalised frequency offset of the pump from the cavity resonant frequency. (a), (c) and (e) correspond to the lower normal mode having an unmodified frequency and Q of 703.15 Hz and 9.0×10^6. (b) and (d) correspond to the upper normal mode having a frequency and Q of 725.62 Hz and 2.5×10^6. Note that, as expected, the frequencies and Q return to the unmodified values – indicated by the horizontal lines, as the offset becomes large. The sloping straight lines indicate the regions over which equations (8.21a) and (8.23) are fitted. In (e), the points at 10^8 correspond to oscillatory instability where the 'cold-damped' antenna acoustic loss, τ_c^{-1}, ≤ 0.

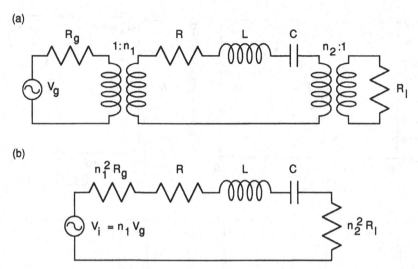

Figure 8.5. (a) An equivalent circuit of the transducer resonant circuit including input and output couplings. (b) The equivalent circuit with the generator or pump, and load transformed to the inside of the cavity. $V_i = n_1 V_g$, $R_1 = n_1^2 R_g$ and $R_2 = n_2^2 R_1$.

circuit is appropriate. If, on the other hand, the cavity is viewed from a 'detuned open position' where there exists a standing wave maximum when the cavity is detuned then a series circuit is appropriate (Ginzton, 1957). Specification of the reference position is important when using a standing-wave detector to determine the coupling of the transmission line to the cavity, but has no bearing on the following work and will not be discussed further.

Various authors have chosen to use either the series (Darling, 1979; Veitch, 1986) or parallel (Giffard and Paik, 1977; Oelfke, 1978) RLC representation of the resonant circuit. Here we will use the series representation. The equivalent circuit of the resonant circuit and the input and output coupling elements is shown in figure 8.5.

Some definitions which will be used throughout the following are:

(a)   input coupling $\beta_1 = n_1^2 R_g / R = R_1 / R,$
(b)   output coupling $\beta_2 = n_2^2 R_1 / R = R_2 / R,$
(c)   unloaded Q $Q_0 = \Omega_0 L / R,$
(d)   loaded Q $Q_1 = Q_0 / (1 + \beta_1 + \beta_2),$
(e)   resonant frequency $\Omega_0 = (LC_0)^{-1/2}.$

In the following subsections we will show, using the equivalent electrical circuit of the transducer, how to determine the transducer sensitivity, calibration of the transducer, modification of the pump noise by the transducer and the Nyquist noise generated by the transducer. Since the resonant circuit of a parametric transducer can be used either in transmission or reflection, we will determine the formulae for both configurations.

8.5.1 Transducer sensitivity

Relative motion of the antenna and transducer modulates the capacitance (or inductance) of the resonant circuit which, in turn, phase and amplitude modulates the pump power. The antenna signal is recovered by demodulating the power from the transducer, using some of the unmodulated pump as the reference (this will be discussed in section 8.6).

The current flowing in the equivalent circuit shown in figure 8.5(b) obeys the differential equation

$$V_i = iR_1 + iR + L\frac{di}{dt} + \frac{1}{C}\int i\,dt + iR_2. \tag{8.24}$$

Expressing equation (8.24) in terms of the charge stored on the capacitor plates, q, gives

$$V_i = \dot{q}R_1 + \dot{q}R + L\ddot{q} + \frac{q}{C} + \dot{q}R_2. \tag{8.25}$$

If we assume that the relative motion of the antenna and transducer changes the plate spacing according to $d = x_0(1 + \alpha \sin \omega_a t)$, where $|\alpha| \ll 1$, then the capacitance variation is given by $C = C_0(1 + \alpha \sin \omega_a t)^{-1}$. Since α is very small $(\sim 10^{-9})$ essentially all the output signal power will be contained in the first-order sidebands. That is, the steady state solution will be of the form

$$q = q_0[\cos \Omega_p t + a_+ \sin \Omega_+ t + a_- \sin \Omega_- t + b_+ \cos \Omega_+ t + b_- \cos \Omega_- t], \tag{8.26}$$

where $\Omega_\pm = \Omega_p \pm \omega_a$.

Equation (8.26) can also be written in terms of amplitude and phase modulation indices, m_{AM} and m_ϕ:

$$i = i_0\{[1 + m_{AM} \cos(\omega_a t + \theta_c)] \sin \Omega_p t + m_\phi \cos(\omega_a t + \theta_d) \cos \Omega_p t\},$$

where

$$i_0 = -q_0\Omega_p,$$

$$m_{AM}^2 = \left(\frac{q_0}{i_0}\right)^2 [(b_+\Omega_+ + b_-\Omega_-)^2 + (a_+\Omega_+ - a_-\Omega_-)^2], \tag{8.27a}$$

$$m_\phi^2 = \left(\frac{q_0}{i_0}\right)^2 [(b_+\Omega_+ - b_-\Omega_-)^2 + (a_+\Omega_+ + a_-\Omega_-)^2], \tag{8.27b}$$

$$\tan \theta_c = -\frac{(a_+\Omega_+ - a_-\Omega_-)}{(b_+\Omega_+ + b_-\Omega_-)},$$

$$\tan \theta_d = \frac{(b_+\Omega_+ - b_-\Omega_-)}{(a_+\Omega_+ + a_-\Omega_-)}.$$

Since the phase of the charge (current) is specified, the pump signal driving the circuit must be of the form

$$V_i = v_i \sin(\Omega_p t + \phi) = v_{ia} \sin \Omega_p t + v_{ib} \cos \Omega_p t, \tag{8.28}$$

where $v_i^2 = v_{ia}^2 + v_{ib}^2$ and $\tan \phi = v_{ib}/v_{ia}$.

Substituting equations (8.26) and (8.28) into equation (8.25), and then equating terms in Ω_p, Ω_+ and Ω_- gives

$$a_\pm = \frac{\pm \alpha \Omega_0 Q_1^2 \Delta_\pm}{\Omega_\pm (1 + 4Q_1^2 \Delta_\pm^2)}, \tag{8.29a}$$

$$b_\pm = \frac{\pm \alpha \Omega_0 Q_1}{2\Omega_\pm (1 + 4Q_1^2 \Delta_\pm^2)}, \tag{8.29b}$$

$$v_{ia} = -q_0 \left[\Omega_p (R_1 + R + R_2) + \frac{\alpha}{2C_0} (b_+ - b_-) \right], \tag{8.29c}$$

$$v_{ib} = -q_0 \left[L\Omega_0 \Omega_p 2\Delta_p + \frac{\alpha}{2C_0} (a_- - a_+) \right], \tag{8.29d}$$

where

$$\Delta_\pm = \frac{1}{2} \left[\frac{\Omega_\pm}{\Omega_0} - \frac{\Omega_0}{\Omega_\pm} \right] \approx \frac{\delta \Omega \pm \omega_a}{\Omega_0}.$$

Equations (8.29c) and (8.29d) can be used to calculate the magnitude of the input voltage:

$$v_i = -\Omega_p q_0 (R_1 + R + R_2)(1 + 4Q_1^2 \Delta_p^2)^{1/2}, \tag{8.30}$$

where

$$\Delta_p = \frac{1}{2} \left[\frac{\Omega_p}{\Omega_0} - \frac{\Omega_0}{\Omega_p} \right] \approx \frac{\delta \Omega}{\Omega_0}$$

and terms of order α have been ignored.

Since $i_0 = -q_0 \Omega_p$, we can use equation (8.30) to determine the voltage across the output load, $v_0 = i_0 R_2$:

$$v_0 = v_i \frac{R_2}{(R_1 + R + R_2)} (1 + 4Q_1^2 \Delta_p^2)^{-1/2}, \tag{8.31a}$$

which can be made more applicable by replacing the equivalent resistances by couplings:

$$\frac{v_0}{v_i} = \frac{\beta_2}{(1 + \beta_1 + \beta_2)} (1 + 4Q_1^2 \Delta_p^2)^{-1/2}. \tag{8.31b}$$

The power dissipated in the output load is, therefore, given by

$$P_0 = \frac{4\beta_1 \beta_2}{(1 + \beta_1 + \beta_2)^2} \frac{P_i}{(1 + 4Q_1^2 \Delta_p^2)}. \tag{8.32}$$

where $P_i = \frac{1}{4}(\frac{1}{2} v_i^2 / R_i)$ is the maximum power which can be obtained from the pump. This is, as would be expected, just the standard formula for the power transmitted through a resonant cavity.

The amplitude and phase modulation indices can be calculated using equations

(8.27), (8.29a) and (8.29b). To leading order in $\delta\Omega$ and ω_a, they are given by

$$m_{AM}^2 = \alpha^2 Q_1^2 \left(\frac{2Q_1\delta\Omega}{\Omega_0}\right)^2 \left[1 - 2\left(\frac{2Q_1\omega_a}{\Omega_0}\right)^2\right],$$

giving

$$m_{AM} \approx \pm\alpha Q_1 \left(\frac{2Q_1\delta\Omega}{\Omega_0}\right), \tag{8.33}$$

and

$$m_\phi^2 = \alpha^2 Q_1^2 \left[1 - \left(\frac{2Q_1\omega_a}{\Omega_0}\right)^2 - 2\left(\frac{2Q_1\delta\Omega}{\Omega_0}\right)^2\right],$$

giving

$$m_\phi \approx \pm\alpha Q_1 \left[1 - \left(\frac{2Q_1\delta\Omega}{\Omega_0}\right)^2\right]. \tag{8.34}$$

A graph of the modulation indices as a function of the fractional frequency offset $(2Q_1\delta\Omega/\Omega_0)$ is shown in figure 8.6.

Another, more simple, method of estimating the modulation indices is to make a quasi-static approximation and employ transfer function formalism (Veitch, 1986). The quasi-static approximation assumes that the component values vary sufficiently slowly that the system is essentially always in equilibrium and hence one may analyse the circuit using a transfer function. In practice this requires that $\omega_a \ll (\Omega_0/2Q_1)$.

The transfer function, $Y(j\Omega)$, of the circuit shown in figure 8.5(b) is given by

$$Y(j\Omega) = \frac{\beta_2}{(1 + \beta_1 + \beta_2)} \frac{e^{j\theta}}{[1 + (2Q_1\delta\Omega/\Omega_0)^2]^{1/2}}, \tag{8.35}$$

where $\tan\theta = -2Q_1\delta\Omega/\Omega_0$.

If, as before, we assume that $C = C_0(1 + \alpha\sin\omega_a t)^{-1} \approx C_0(1 - \alpha\sin\omega_a t)$ then the transfer function may be modified to

$$Y(j\Omega) = \frac{\beta_2}{(1 + \beta_1 + \beta_2)} \frac{e^{j\theta}}{[1 + Q_1^2((2\delta\Omega/\Omega_0) - \alpha\sin\omega_a t)^2]^{1/2}}, \tag{8.36}$$

where $\tan\theta = -Q_1((2\delta\Omega/\Omega_0) - \alpha\sin\omega_a t)$.

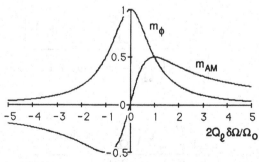

Figure 8.6. Graph of the normalised PM (m_ϕ) and AM (m_{AM}) modulation indices for a resonant cavity operated in transmission as a function of the fractional frequency offset.

If the pump voltage is described by $V_i = v_i \cos \Omega_p t$ then the voltage across the load resistor is given by $V_{\text{out}} = v_i |Y| \cos (\Omega_p t + \theta)$ (Middleton, 1960). The modulation indices are then given, to first order in α, by

$$m_{\text{AM}} \approx \frac{-\alpha Q_1 (2Q_1 \delta\Omega/\Omega_0)}{1 + (2Q_1 \delta\Omega/\Omega_0)^2} \approx -\alpha Q_1 (2Q_1 \delta\Omega/\Omega_0), \qquad (8.37a)$$

$$m_\phi \approx \frac{-\alpha Q_1}{1 + (2Q_1 \delta\Omega/\Omega_0)^2} \approx -\alpha Q_1 [1 - (2Q_1 \delta\Omega/\Omega_0)^2], \qquad (8.37b)$$

which agree with equations (8.33) and (8.34). Note, however, that this method determines the signs of the modulation coefficients. Thus, the quasi-static approximation can be employed to obtain an accurate description of the transducer output.

To determine the sensitivity when the transducer resonant circuit (cavity) is operated in reflection we model the system as a transmission line which is terminated by the resonant cavity. We then apply the quasi-static approximation to the reflection coefficient of the terminated transmission line to obtain the modulation indices.

The equivalent circuit of the cavity operated in reflection is just that shown in figure 8.5 but with the output transformer and the load impedance removed. The characteristic impedance of the transmission line (transformed to the inside of the cavity) is given by $Z_0 = n_1^2 R_g$. If, as before, we assume that the capacitor plate spacing varies according to $d = x_0(1 + \alpha \sin \omega_a t)$, then the impedance (referred to the input of the cavity) which terminates the transmission line is given by $Z_{\text{term}} = R[1 + jQ_0((2\delta\Omega/\Omega_0) - \alpha \sin \omega_a t)]$.

Substituting the characteristic impedance of the transmission line and the terminating impedance into the formula for the reflection coefficient (Liao, 1987):

$$\Gamma(j\Omega) = \frac{Z_{\text{term}} - Z_0}{Z_{\text{term}} + Z_0} \equiv |\Gamma| e^{j\theta}, \qquad (8.38)$$

we obtain

$$|\Gamma| = \frac{\left[\left(\dfrac{1 - \beta_1}{1 + \beta_1}\right)^2 + Q_1^2\left(\dfrac{2\delta\Omega}{\Omega_0} - \alpha \sin \omega_a t\right)^2\right]^{1/2}}{\left[1 + Q_1^2\left(\dfrac{2\delta\Omega}{\Omega_0} - \alpha \sin \omega_a t\right)^2\right]^{1/2}} \qquad (8.39a)$$

and

$$\tan \theta = \frac{\dfrac{2\beta_1}{(1 + \beta_1)} Q_1 \left(\dfrac{2\delta\Omega}{\Omega_0} - \alpha \sin \omega_a t\right)}{\left(\dfrac{1 - \beta_1}{1 + \beta_1}\right) + Q_1^2\left(\dfrac{2\delta\Omega}{\Omega_0} - \alpha \sin \omega_a t\right)^2}. \qquad (8.39b)$$

To first order in α

$$|\Gamma| = \frac{\left[\left(\frac{1-\beta_1}{1+\beta_1}\right)^2 + \left(\frac{2Q_1\delta\Omega}{\Omega_0}\right)^2\right]^{1/2}}{\left[1 + \left(\frac{2Q_1\delta\Omega}{\Omega_0}\right)^2\right]^{1/2}} (1 - m_{AM} \sin \omega_a t), \qquad (8.40)$$

where the amplitude modulation index, m_{AM}, is given by

$$m_{AM} = \frac{\frac{4\beta_1}{(1+\beta_1)^2}\left(\frac{2Q_1\delta\Omega}{\Omega_0}\right)\alpha Q_1}{\left[\left(\frac{1-\beta_1}{1+\beta_1}\right)^2 + \left(\frac{2Q_1\delta\Omega}{\Omega_0}\right)^2\right]\left[1 + \left(\frac{2Q_1\delta\Omega}{\Omega_0}\right)^2\right]} \qquad (8.41)$$

$$\approx \frac{4\beta_1}{(1-\beta_1)^2}\left(\frac{2Q_1\delta\Omega}{\Omega_0}\right)\alpha Q_1 \quad \text{in the limit } |2Q_1\delta\Omega/\Omega_0| \ll 1.$$

The phase modulation index, m_ϕ, is given by

$$m_\phi = \frac{\frac{2\beta_1}{(1+\beta_1)}\left[\left(\frac{1-\beta_1}{1+\beta_1}\right) - \left(\frac{2Q_1\delta\Omega}{\Omega_0}\right)^2\right]\alpha Q_1}{\left[\left(\frac{1-\beta_1}{1+\beta_1}\right)^2 + \left(\frac{2Q_1\delta\Omega}{\Omega_0}\right)^2\right]\left[1 + \left(\frac{2Q_1\delta\Omega}{\Omega_0}\right)^2\right]}$$

$$\approx \frac{2\beta_1}{(1-\beta_1)}\alpha Q_1 \quad \text{in the limit } |2Q_1\delta\Omega/\Omega_0| \ll 1. \qquad (8.42)$$

The apparent singularity of these indices as $\beta_1 \to 1$ when $\delta\Omega = 0$ is due to the fact that the reflected carrier power $\to 0$ in this limit. It is not significant as the power in the AM and PM sidebands is finite. The dependence of the AM and PM sideband power on fractional frequency offset ($2Q_1\delta\Omega/\Omega_0$) is plotted in figure 8.7. Figure 8.7 shows that the maximum value of the power in the PM sideband is

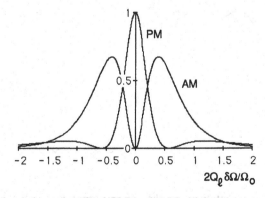

Figure 8.7. Graph of the normalised PM and AM sideband power for a resonant cavity operated in reflection as a function of the fractional frequency offset. The microwave coupling, β_1, has been assumed to be 0.5.

greater than that in the AM sideband. Clearly, therefore, we would like to operate the resonant circuit near resonance ($\delta\Omega \approx 0$), thereby obtaining the largest possible signal from the transducer. This also applies when the resonant circuit is operated in transmission.

8.5.2   Calibration of the transducer

As mentioned in section 8.1, one of the advantages of parametric transducers is that they are self-calibrating. That is, the sensitivity of the antenna can be determined without having to use an additional calibration transducer to inject a known amount of energy into the antenna. The calibration method makes use of the pump-frequency response of the transducer (see figure 8.8(a)); and assumes that the unperturbed spacing of the capacitor plates, x_0, is known.

If the pump can be easily tuned and the cavity resonant frequency and Q are independent of power, as is usually the case, then the transducer can be calibrated by, firstly, measuring ΔV and $\Delta\Omega$ as shown in figure 8.8(a). Next, calculate $d\Omega/dx$ from the design of the resonant circuit. Generally $\Omega = \Omega(x) \approx$

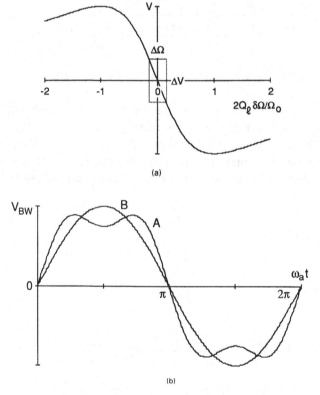

Figure 8.8. Schematic representations of the mixer output as (a) the pump frequency sweeps across resonance, and (b) the antenna vibrates with an amplitude greater than the cavity bandwidth (A), and equal to the cavity bandwidth (B).

kx^n whence $d\Omega/dx = n(\Omega_0/x_0)$. For an ideal capacitance-modulated parametric transducer $n = \frac{1}{2}$. The transducer calibration factor $\Delta V/\Delta x$ can then be calculated using $\Delta V/\Delta\Omega$ and $d\Omega/dx$.

If the pump cannot easily be tuned then the calibration is more complicated. In this case one must first measure the loaded quality factor, Q_1, of the resonant circuit. This can be used to calculate the position bandwidth of the cavity $\Delta x_{\mathrm{BW}} \approx (n^{-1})x_0/Q_1$, which can in turn be used to calibrate the output of the mixer when the antenna is excited so as to vibrate with an amplitude $> \Delta x_{\mathrm{BW}}$.

The transducer Q can be determined by amplitude or phase modulating the microwave pump, and increasing the modulation frequency until the modulation index of the transmitted/reflected sidebands has changed by the appropriate amount. This frequency can then be related to the cavity bandwidth, $(\Omega_0/2Q_1)$, which will give Q_1.

The calculation of the relationship between the incident and transmitted/reflected sidebands is straightforward, and uses the transfer function formalism discussed earlier. For definiteness we will assume that the incident signal is amplitude modulated and can be described by $V_{\mathrm{in}} = V_0(1 + a\cos\omega t)\cos\Omega_0 t$. We will assume that the pump frequency is sufficiently close to Ω_0 ($\delta\Omega \ll \omega$, where ω will be $\sim(\Omega_0/2Q_1)$) that the difference may be ignored.

Cavity in transmission
The power transmitted through the cavity is given by

$$V_{\mathrm{out}} = V_0\,\mathrm{Re}\left[e^{j\Omega_0 t}Y(j\Omega_0) + \frac{a}{2}e^{j(\Omega_0+\omega)t}Y(j(\Omega_0+\omega)) + \frac{a}{2}e^{j(\Omega_0-\omega)t}Y(j(\Omega_0-\omega))\right],$$

where $Y(j\Omega)$ is given by equation (8.36). This can be simplified to

$$V_{\mathrm{out}} = V_0\frac{\beta_2}{(1+\beta_1+\beta_2)}\left[1 + \frac{a}{[1+(2Q_1\omega/\Omega_0)^2]^{1/2}}\cos(\omega t + \theta)\right]\cos\Omega_0 t, \quad (8.43)$$

where $\tan\theta = -2Q_1\omega/\Omega_0$.

Cavity in reflection
The power reflected from the cavity is given by

$$V_{\mathrm{refl}} = V_0\,\mathrm{Re}\left[e^{j\Omega_0 t}\Gamma(j\Omega_0) + \frac{a}{2}e^{j(\Omega_0+\omega)t}\Gamma(j(\Omega_0+\omega)) + \frac{a}{2}e^{j(\Omega_0-\omega)t}\Gamma(j(\Omega_0-\omega))\right],$$

where $\Gamma(j\Omega)$ is given by equation (8.39). This can be simplified to

$$V_{\mathrm{refl}} = V_0\left(\frac{1-\beta_1}{1+\beta_1}\right)[1 + a_{\mathrm{refl}}\cos(\omega t + \theta)]\cos\Omega_0 t, \qquad (8.44)$$

where

$$a_{\mathrm{refl}} = \frac{\left[1 + \left(\frac{1+\beta_1}{1-\beta_1}\right)^2\left(\frac{2Q_1\omega}{\Omega_0}\right)^2\right]^{1/2}}{\left[1 + \left(\frac{2Q_1\omega}{\Omega_0}\right)^2\right]^{1/2}}\,a$$

and

$$\tan \theta = \frac{\left(\dfrac{2\beta_1}{1+\beta_1}\right)\left(\dfrac{2Q_1\omega}{\Omega_0}\right)}{\left(\dfrac{1-\beta_1}{1+\beta_1}\right)+\left(\dfrac{2Q_1\omega}{\Omega_0}\right)^2}.$$

Thus, if the cavity was operated in transmission then the modulation index would decrease by $\sqrt{2}$ when $\omega = (\Omega_0/2Q_1)$. The change in the index for a cavity operated in reflection depends on the input coupling.

To complete the transducer calibration, the antenna must be excited so as to vibrate with an amplitude greater than $(\Delta x_{BW}/2)$. The output of the microwave mixer will then look like curve A in figure 8.8(b). The mixer output can approach a distorted square-wave shape if the vibration amplitude is sufficiently large. As the amplitude of the antenna decays the coefficients of the higher odd-order sidebands decrease (the coefficients of the even-order sidebands are zero) and the waveform becomes more rounded as shown by curve B in figure 8.8(b).

When the amplitude of the sinusoidal shaped curve is maximum the output of the transducer is due only to the first-order sidebands and corresponds to the antenna oscillating between the cavity 3 dB points. The shape of the curve can be calculated using the quasi-static approximation and transfer function method described in section 8.5.1 but letting $\delta\Omega$ vary slowly (at ω_a) and putting $\alpha = 0$. Since the mixer is being used as a phase detector, the output, which is $\propto |Y| \sin \theta$, is given by either

$$V_{out} \propto \frac{(4\beta_1\beta_2)^{1/2}}{(1+\beta_1+\beta_2)} \frac{(-2Q_1\delta\Omega/\Omega_0)}{[1+(2Q_1\delta\Omega/\Omega_0)^2]} \quad \text{(transmission)}$$

or
$$\text{(8.45)}$$

$$V_{out} \propto \left(\frac{2\beta_1}{1+\beta_1}\right) \frac{2Q_1\delta\Omega/\Omega_0}{[1+(2Q_1\delta\Omega/\Omega_0)^2]} \quad \text{(reflection)}.$$

The calibration factor (sensitivity) of the transducer for a given pump frequency offset, K, can be related to the average sensitivity $K_{av} = V_{BW}/\Delta x_{BW}$ by

$$|K| = 2K_{av} \frac{[1 - (2Q_1\delta\Omega/\Omega_0)^2]}{[1+(2Q_1\delta\Omega/\Omega_0)^2]^2} \text{ (volts/metre)}. \quad \text{(8.46)}$$

This calibration procedure will only work if the cavity Q is independent of the power in the cavity since the Q must remain constant as the antenna vibrates across the cavity bandwidth. If the Q was power dependent then the modulation of the power in the cavity produced by the vibration of the antenna (since the frequency offset of the pump from the centre of resonance is being modulated) would cause the Q to vary.

8.5.3 Modification of pump noise by the transducer

The antenna signal will only be discernible if the amplitude and phase noise of the pump is sufficiently small. To determine the acceptable level of pump noise we must first consider the modification of the noise by the transducer.

The resonant circuit of the transducer modifies the pump noise in several ways. Firstly, on transmission (reflection) of the pump by the resonant circuit, the noise sidebands are attenuated (amplified) relative to the carrier due to the filter characteristics of the circuit. The noise sidebands are also phase shifted relative to the carrier. Since the phase shift of an upper sideband is equal and opposite to that of the corresponding lower sideband, the character (AM or PM) of the noise is unchanged. However, the phase shift added to the sidebands will degrade the antenna sensitivity by reducing the degree of cancellation of correlated noise in the subsequent signal processing (see section 8.6). Since we will only be concerned with noise sidebands which are offset from the carrier by ω_a, and since these are usually well within the resonant circuit bandwidth, the phase shift will be small and the output noise can be expressed as a combination of in-phase (with the input) and quadrature noise. The decorrelation of the pump noise occurs for any pump frequency offset but is largest for $\delta\Omega = 0$.

If $\delta\Omega \neq 0$ then there is also AM/PM interconversion since the phase shift of the two sidebands is unequal and the character of the noise is therefore changed.

We will begin by considering what happens to PM noise sidebands at the antenna frequency when they interact with a high Q circuit. The results for AM sidebands will, by symmetry, be the same. We will use the transfer function formalism discussed in section 8.5.1. The input will be assumed to be described by

$$V_{in} = V_0 \cos(\Omega t + m \sin \omega_a t) \tag{8.47a}$$

$$\approx V_0[\cos \Omega t - m \sin \Omega t \sin \omega_a t] \quad \text{since} \quad |m| \ll 1 \tag{8.47b}$$

$$= V_0[\cos \Omega t - (m/2)\cos(\Omega - \omega_a)t + (m/2)\cos(\Omega + \omega_a)t]. $$

Cavity in transmission

The output is given by

$$V_{out} = V_0 \, \mathrm{Re}[e^{j\Omega t}Y(j\Omega) - (m/2)e^{j(\Omega_0 - \omega_a)t}Y(j(\Omega - \omega_a))$$

$$+ (m/2)e^{j(\Omega_0 + \omega_a)t}Y(j(\Omega + \omega_a))]$$

$$\approx V_0 \frac{\beta_2}{(1 + \beta_1 + \beta_2)} \frac{1}{[1 + (2Q_1\delta\Omega/\Omega_0)^2]^{1/2}} \Bigg[\cos(\Omega t + \theta_1)$$

$$\times \left[1 - m_{AM}\cos\left(\omega_a t + \frac{\theta_3 - \theta_2}{2}\right)\right] - \sin(\Omega t + \theta_1)m_\phi \sin\left(\omega_a t + \frac{\theta_3 - \theta_2}{2}\right)\Bigg]$$

$$\tag{8.48}$$

where

$$m_{AM} \approx m\left(\frac{2Q_1\delta\Omega}{\Omega_0}\right)\left(\frac{2Q_1\omega_a}{\Omega_0}\right), \tag{8.48a}$$

$$m_\phi \approx m\left[1 + \left(\frac{2Q_1\omega_a}{\Omega_0}\right)^2\right]^{-1/2}, \tag{8.48b}$$

$$\tan\theta_1 = -2Q_1\delta\Omega/\Omega_0,$$

$$\tan\theta_2 = -2Q_1(\delta\Omega - \omega_a)/\Omega_0,$$

$$\tan\theta_3 = -2Q_1(\delta\Omega + \omega_a)/\Omega_0.$$

Comparison of equations (8.47b) and (8.48) yield a PM→AM conversion factor of $(2Q_1\delta\Omega/\Omega_0)(2Q_1\omega_a/\Omega_0)$.

If $\delta\Omega = 0$ then

$$V_{out} \approx V_0 \frac{\beta_2}{(1+\beta_1+\beta_2)}\left[\cos\Omega t - \sin\Omega t\, m\left(\sin\omega_a t - \left(\frac{2Q_1\omega_a}{\Omega_0}\right)\cos\omega_a t\right)\right], \tag{8.49}$$

and comparison of equations (8.47b) and (8.49) gives a (maximum) quadrature conversion factor of $(2Q_1\omega_a/\Omega_0)$.

Cavity in reflection

The reflected signal is given by

$$V_{refl} = V_0\,\mathrm{Re}[e^{j\Omega t}\mathrm{T}(j\Omega) - (m/2)e^{j(\Omega-\omega_a)t}\mathrm{T}(j(\Omega-\omega_a))$$
$$+ (m/2)e^{j(\Omega+\omega_a)t}\mathrm{T}(j(\Omega+\omega_a))],$$

which can be expanded and simplified in a similar manner to give a PM→AM conversion factor of

$$\frac{4\beta_1}{(1-\beta_1)^2}\left(\frac{2Q_1\delta\Omega}{\Omega_0}\right)\left(\frac{2Q_1\omega_a}{\Omega_0}\right),$$

and a quadrature conversion factor of

$$\frac{2\beta_1}{(1-\beta_1)}\left(\frac{2Q_1\omega_a}{\Omega_0}\right).$$

The application of these conversion factors will be discussed in section 8.6.

8.5.4 Power dissipated in the transducer

Dissipation of pump power in the resonant circuit could cause the temperature of the transducer and antenna to increase, thereby producing a change in the cavity resonant frequency and Q. This process would occur slowly and could result in (for example) (a) a reduction in the size of the signal, (b) a change in the phase and amplitude of the pump carrier transmitted-through/reflected-from the cavity which could, in turn, degrade the subsequent signal processing by disrupting the carrier suppression (see section 8.6).

Using the equivalent circuit shown in figure 8.5(a), the power dissipated in the cavity, P_{dis}, is given by

$$P_{dis} = \frac{4\beta_1}{(1+\beta_1+\beta_2)^2} \frac{P_i}{[1+(2Q_1\delta\Omega/\Omega_0)^2]}. \tag{8.50}$$

8.5.5 Nyquist noise produced by the transducer resonant circuit

Losses in the transducer produce Nyquist noise which, together with the Nyquist noise produced by lossy elements in the signal processing, competes directly with the antenna signal. The noise which appears at the output of the cavity, P_N, can be determined by considering the equivalent circuit of figure 8.5(b) with the pump source replaced by a voltage noise generator with amplitude $V_{RMS} = 4k_B TRB$, where B is the bandwidth; whence

$$P_N = \frac{4\beta_2}{(1+\beta_1+\beta_2)^2} \frac{k_B TB}{[1+(2Q_1\delta\Omega/\Omega_0)^2]} \quad \text{(transmission)}$$

$$\tag{8.51}$$

$$P_N = \frac{4\beta_1}{(1+\beta_1)^2} \frac{k_B TB}{[1+(2Q_1\delta\Omega/\Omega_0)^2]} \quad \text{(reflection)}.$$

8.6 Noise analysis: general comments

In this section we will discuss the noise performance and optimisation of the generalised signal processing circuit for a parametric transducer, shown in figure 8.9. This circuit contains elements which are common to most parametric transducer systems. Briefly, it consists of a pump oscillator which supplies power to the transducer, and a 'phase bridge' which demodulates the power transmitted-through/reflected-from the transducer, to yield the antenna signal. The phase bridge will be described further below. A more detailed set of equations from a noise analysis of the UWA transducer can be found elsewhere (Veitch, 1986).

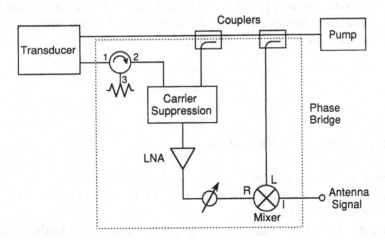

Figure 8.9. Generalised signal processing circuit for a parametric transducer.

Generally, the pump frequency is chosen to be just less than the electrical resonant frequency of the transducer so as to prevent parametric oscillations (see section 8.4) but ensure the largest possible phase modulation signal sidebands (see section 8.5.1). The pump oscillator must have ultra-low levels of amplitude and phase noise, and have sufficient long term stability that its frequency does not drift significantly relative to the resonant circuit bandwidth.

The pump and transmission lines connecting the pump to the transducer should, ideally, all be in the same dewar and be as short as possible so as to minimise Nyquist noise which adds amplitude and phase noise to the carrier. In general, a transmission line or device with loss L and noise temperature T adds noise power $k_B T(L-1)B$, where B is the bandwidth, when referred to the input of the line or device. This power can be split evenly between amplitude and phase noise since these two modulations are in quadrature.

The (effective) pump phase noise is of prime importance since the antenna signal must compete with this directly. There will be significant cancellation of the pump phase noise in both the carrier suppression and across the double balanced mixer (this will be discussed further below). The quadrature component of this phase noise (see section 8.5.3), produced by the dispersion of the resonant circuit, does not cancel, however. The phase noise produced by AM/PM conversion in the resonant circuit (see section 8.5.3) and carrier suppression is also uncorrelated (with the noise on the 'local oscillator (LO) line') and therefore does not cancel in the mixer. This effect imposes limits on the allowable pump–cavity offset $\delta\Omega$, carrier suppression phase error $\langle \delta\phi^2 \rangle$ and AM noise.

Amplitude noise can, in addition to the effect just discussed, degrade the antenna sensitivity in several more ways. Errors in setting the relative phase of the LO and RF inputs to the mixer, and uncorrelated low frequency phase modulation of those inputs result in AM sensitivity of the mixer. Also, due to imbalance in the mixer diodes there is some feed through of LO AM noise.

The amplitude noise also produces back-action noise. Using the formula for the force between the capacitor plates (see section 8.4), it can be seen that the spectral density of force noise is given by

$$S_{f,AM} = \left(\frac{C_0 V^2}{2x_0} \right)^2 S_{AM}. \tag{8.52}$$

8.6.1   Description of the phase bridge

The transducer output first passes through a circulator or isolator, which isolates the transducer from noise in the RF line and prevents additional series and back-action noise.

A circulator uses the gyromagnetic properties of magnetised ferrite to produce non-reciprocal transmission of electrical power (Baden Fuller, 1987). Power incident on port 1 is coupled to port 2 only, while port 3 is isolated (ideally). Similarly, power incident on port 2 is coupled to port 3 only. In figure 8.9 the direction of circulation is indicated by the arrow. Clearly, if port 3 is terminated (with the transmission line characteristic impedance) then any power reflected from a device on line 2 or generated by losses in that line or device will not couple into line 1. That is, the transducer is isolated from impedance mismatches and noise in line 2. Only Nyquist noise generated by the termination $k_B TB$ is incident on the transducer.

The transducer output is then carrier-suppressed by combining it with some unmodulated pump power having nearly the same amplitude and exactly (ideally) opposite phase. The remaining power can then be amplified using a low noise amplifier, and demodulated using a double balanced mixer as a phase detector. Carrier suppression is necessary to enable the use of the lowest possible noise amplifiers (masers, GaAs MESFET amplifiers, paramps) while being able to use sufficiently large power (compared to the input saturation power of the amplifier) to maintain high electromechanical coupling and minimise the effect of Nyquist and mixer noise.

As before, it is important that all transmission lines are kept as short as possible to minimise Nyquist noise. Likewise, the first stage (at least) of amplification should occur at low temperatures.

Carrier suppression should be performed passively, if possible, to prevent the addition of extra electronics noise. At microwave frequencies it can be achieved using a 'hybrid T', a four-port device which sums and differences two inputs. The principle of carrier suppression is illustrated in figure 8.10. The amplitudes of the input carriers are represented by V_t (from the transducer) and $V_{cs} = (1 - \varepsilon)V_t$ $|\varepsilon| \ll 1$, and the phase difference differs from π by $\delta\phi$ – due to either the finite resolution of the phase shifter used to adjust the phase difference or a low

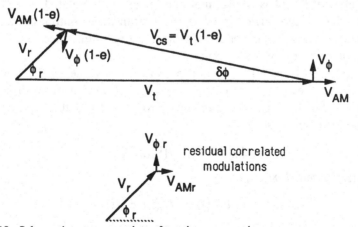

Figure 8.10. Schematic representation of carrier suppression.

frequency phase modulation produced by the transducer. The resultant output carrier amplitude, V_r, and its phase, ϕ_r, relative to the incident signal from the transducer, are given, for $|\delta\phi| \ll 1$, by

$$V_r = [\varepsilon^2 + (1 - \varepsilon)\,\delta\phi^2]^{1/2}V_t \qquad (8.53a)$$

and

$$\sin \phi_r = \frac{(1 - \varepsilon)}{[\varepsilon^2 + (1 - \varepsilon)\,\delta\phi^2]^{1/2}}\,\delta\phi \qquad (8.53b)$$

to order $\delta\phi$. The residual correlated amplitude and phase modulations, relative to the transducer output carrier (since the mixer will be operated in quadrature relative to that carrier), are given to order $\delta\phi$ by

$$V_{AMr} = \varepsilon V_{AM} - (1 - \varepsilon)\,\delta\phi\,V_\phi$$

giving

$$S_{AMr} = \varepsilon^2 S_{AM} + (1 - \varepsilon)^2 \langle \delta\phi^2 \rangle_B S_\phi, \qquad (8.54a)$$

and

$$V_{\phi r} = \varepsilon V_\phi + (1 - \varepsilon)\,\delta\phi\,V_{AM}$$

giving

$$S_{\phi r} = \varepsilon^2 S_\phi + (1 - \varepsilon)^2 \langle \delta\phi^2 \rangle_B S_{AM}, \qquad (8.54b)$$

where S_{AMr} and $S_{\phi r}$ are the spectral densities of residual correlated AM and PM noise relative to the transducer output carrier, and $\langle \delta\phi^2 \rangle_B$ denotes the average of $\delta\phi^2$ over a bandwidth B.

These equations show that the phase error limits the maximum possible carrier suppression to $\delta\phi$. Further, it causes interconversion of correlated AM and PM noise. Depending on the relative sizes of ε^2, $\langle \delta\phi^2 \rangle_B$, S_{AM} and S_ϕ, it could substantially nullify the otherwise advantageous cancellation of correlated noise during carrier suppression. A possibly more serious effect is that this interconversion could degrade the cancellation of correlated phase noise in the mixer (see later) since most of the phase noise might no longer be correlated.

The phase shift of the carrier (relative to the transducer output carrier) could also complicate setting the mixer for quadrature operation since this would no longer coincide with zero mixer output.

To investigate the transformation and production of noise by the mixer, it is useful to consider the operation of the mixer. The IF output of an ideal double balanced mixer (DBM) is given, assuming appropriate RF and LO power levels, by (Mouw and Fukuchi, 1969)

$$V_{IF} \propto V_{RF}V_{LO}/|V_{LO}|. \qquad (8.55)$$

We will write (Veitch, 1986)

$$V_{RF} = V_R(1 + a_R(t)) \sin(\Omega_p t + \phi_R(t)) \qquad (8.56a)$$

$$V_{LO} = V_L(1 + a_L(t)) \sin(\Omega_p t + \phi_L(t) + \phi_0), \qquad (8.56b)$$

where $a_{R(L)}(t)$ and $\phi_{R(L)}(t)$ are amplitude and phase modulations (including the signal). Expanding and simplifying by assuming that all modulations are small ($\ll 1$), and letting $\phi_0 = \pi/2 + \delta\phi_0$ where $\delta\phi_0$ is a slowly varying phase error, gives

$$V_{IF} \propto \frac{1}{2} V_R[(\phi_R - \phi_L) + (1 + a_R + a_L)(\delta\phi_r + \delta\phi_0)]. \tag{8.57}$$

Thus, the antenna signal must compete with residual uncorrelated microwave pump and Nyquist phase noise and, due to non-zero slowly varying phase errors, residual pump and Nyquist amplitude noise.

In a real DBM we must also consider other effects. Firstly, a real DBM is only sensitive to LO AM noise by virtue of circuit imperfections which unbalance the mixer. The output due to LO amplitude noise is given by

$$V_{IF}(\text{LO AM}) = K'P_L^{1/2}a_L \tag{8.58}$$

where P_L is the LO carrier power and K' is the open circuit voltage LO AM sensitivity which is related to the mixer LO–IF isolation, I, by $I = 400/(K')^2$ for a 50Ω device. Another difference is that there is an additive output voltage noise, V_m, generated by the mixer.

Thus, the output of the DBM can be rewritten

$$V_{IF} = KP_R^{1/2}[(\phi_R - \phi_L) + (1 + a_R)(\delta\phi_r + \delta\phi_0)] + K'P_L^{1/2}a_L + V_m \tag{8.59}$$

where P_R is the incident RF carrier power and K is the open circuit voltage sensitivity, which is dependent upon P_L and related to the (single-sideband) conversion loss L_{SSB} by $L_{SSB} = 400/K^2$.

The mixer voltage noise can be related to the effective input noise temperature of the mixer, T_m, by $K^2 k_B T_m = S_{Vm}$ where S_{Vm} is the mixer output voltage noise spectral density. Due to diode flicker noise, T_m and S_{Vm} are frequency dependent below about 1 MHz and cannot be easily related to the specified single-sideband noise figure of the mixer. They can, however, be determined by terminating the RF input ($P_R = 0$) and driving the LO with sufficiently clean ($a_L \approx 0$) power at a specified level (P_L) and measuring the spectrum of output voltage noise.

Thus, the antenna signal must compete with noise terms due to

(a) residual pump phase noise,
(b) quadrature pump phase noise,
(c) Nyquist phase noise,
(d) pump and Nyquist amplitude noise,
(e) pump and Nyquist LO amplitude noise,
(f) mixer voltage noise.

The total of these noise terms determine the wideband series noise of the antenna.

While pump noise has been shown to be considerably reduced by the carrier suppression system and mixer, it can be further reduced by the use of a balanced transducer configuration. Here two cavities are pumped by a single source and positioned such that their antenna signals are out of phase. The mixer IF signals can then be subtracted to yield twice the antenna signal and a reduced noise level due to the pump. The level of suppression of pump noise which can be obtained by this method depends on the accuracy with which the cavities can be balanced.

Usually, one is concerned with reducing the quadrature pump phase noise contribution (the residual pump phase noise can be reduced by decreasing $\langle \delta\phi^2 \rangle_B$), and so the quantities β, $(2Q_1\omega_a/\Omega_0)$, Ω_0 and $\langle \delta\phi_0^2 \rangle_B$ must be matched. Matching these quantities would also reduce the AM$\rightarrow$PM reentrant cavity noise and AM pump noise contributions. If AM$\rightarrow$PM conversion in the carrier suppression is significant then one must also match (or reduce) $\langle \delta\phi^2 \rangle_B$.

The back-action noise produced by the pump AM noise can also be reduced by matching the parameters of the cavities. More specifically, the quantity (C_0V^2/x_0) where V, the potential difference across the capacitor plates, is determined by β, Q_1, Ω_0, C_0 and P_i, must be matched.

8.7 Practical implementation of parametric transducers

In the following subsections we will discuss the practical implementation of parametric transducers by various groups. Here, we will describe the figures of merit which will be used in the discussion. Firstly, the transducer displacement sensitivity, $S_x^{1/2}$, expressed in m/Hz$^{1/2}$ at a given frequency, is a measure of the performance of the transducer in the absence of the antenna. S_x, the spectral density of displacement noise is measured independently of the antenna by using a frequency at which the antenna is nonresonant. For a given transducer, the displacement sensitivity generally improves as the measurement frequency increases because electrical '$1/f$' noise decreases. The displacement sensitivity can be combined with the spectral density of force noise produced by the transducer, S_f, which is either estimated or, if it is large, calculated from the increase in antenna mode temperature, to give the transducer noise temperature $T_t = 2k_B^{-1}(S_xS_f)^{1/2}$. Finally, the antenna noise temperature, which is determined by analysing the antenna output as discussed by Pallottino and Pizzella in chapter 10, indicates the antenna energy sensitivity. Intimately associated with the antenna noise temperature is a sampling or integration time which is a measure of the temporal resolution of the antenna.

8.7.1 The UWA transducer
A schematic of the UWA antenna is shown in figure 8.11. The antenna consists of a 1.5 tonne superconducting niobium bar to which is bonded a bending flap, resonant at the bar's fundamental longitudinal mode, and a niobium microwave X-band reentrant cavity transducer (Veitch et al., 1987, 1988). The transducer is shown in more detail in figure 8.12.

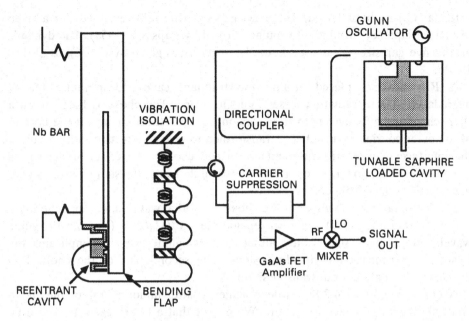

Figure 8.11. Schematic of the UWA antenna.

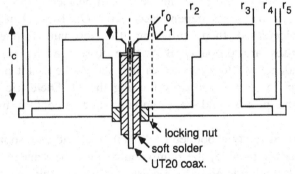

Figure 8.12. Cross section of the UWA reentrant cavity parametric transducer. The cavity is cylindrically symmetric. The cavity dimensions are $r_0 = 0.4$ mm, $r_1 = 0.8$ mm, $r_2 = 4.0$ mm, $r_3 = 11.5$ mm, $r_4 = 14.0$ mm, $r_5 = 15.0$ mm, $l = 1.5$ mm and $l_c = 8.25$ mm. The electric field coupling probe enters the cavity through a 0.9 mm diameter hole. The hole is countersunk to facilitate accurate setting of the coupling.

The reentrant cavity consists of two regions. The central region ($<r_2$) forms a resonant cavity, the resonant frequency of which is essentially determined by the inductance of the central post in the coaxial cavity and the capacitance between the end of the post and the bending flap. The capacitance of the radial waveguide part of the choke (see later) may be ignored since it is much larger than that of the post. The central post, which has slightly rounded edges to prevent premature electric field breakdown and is tapered for mechanical strength, protrudes 0.05 mm past the top of the cavity. The gap between the end of the post and the

antenna (x_{os}) is about 12 μm. The exact gap size can be determined, for a given resonant frequency, using the formulae derived by Fujisawa (1958). The dynamic or effective gap size x_0, as used in earlier sections, is given by $C_0(dC_0/dx_{os})^{-1} \approx$ 15 μm.

A RF choke is placed around the resonant cavity to minimise the Q degradation due to radiation losses (Veitch, 1986). The choke is based upon a short-circuited half wavelength transmission line; the short circuit (at the bottom of the outer coaxial section) being transformed to give an effective short circuit at the interface between the resonant cavity and choke. The effect of the gap is minimised by placing it one-quarter wavelength from the short circuit where imperfections have little effect.

The first quarter wavelength of the choke is a radial waveguide. Since the inner and outer radii (r_2 and r_3) of the waveguide are $\ll c/(\Omega_0/2\pi)$, (i.e. λ_0), the guide wavelength changes significantly along the waveguide, which complicates the accurate determination of its dimensions. The second part of the choke is a quarter wavelength coaxial transmission line.

A Q of 3.9×10^5 at 4.2 K has been achieved after minor tuning of the choke and only minimal surface treatment. We expect that a Q of nearly 10^6 could be obtained by further tuning (Veitch, 1986).

The mechanism limiting the Q of the cavity can be identified by investigating the temperature and frequency dependence of the Q. If the Q increases as the temperature is decreased then, providing the resonant frequency remains constant, the losses are probably BCS type losses in the superconductor. Frequency dependence of the Q indicates that the choke is not properly tuned and that radiation losses are probably limiting the Q. Dielectric losses in the oxide layer on the post should be essentially independent of frequency and temperature.

Since the reentrant cavity is bonded to the antenna, and its resonant frequency is therefore not easily tunable, the pump source must be tunable and yet have ultra-low levels of amplitude and phase noise. To achieve this we are developing a *t*unable *s*apphire *lo*aded *s*uperconducting *c*avity (T-SLOSC) microwave source. The T-SLOSC consists of a niobium cavity which is loaded with a sapphire mushroom and a thin sapphire disc which can be moved so as to tune the resonant frequencies of the cavity. The high dielectric constant and low loss of the sapphire enable extremely high Q's to be obtained in this type of resonator with only modest cleaning and surface preparation (Jones and Blair, 1987).

The ultra-pure microwave pump is produced by servo controlling the frequency of a free-running low noise Gunn oscillator to a high Q mode in the T-SLOSC using a Pound stabiliser servo system (Stein, 1974). Alternatively, a loop oscillator may be formed by connecting the T-SLOSC, an amplifier and a phase shifter in a loop; or a Pound stabilised loop oscillator, which combines both methods, may be used. The microwave power can be further filtered by using the T-SLOSC as a high Q filter, as indicated in figure 8.11. We are currently testing the T-SLOSC to determine the phase noise limit due to the vibration of the tuning mechanism; using the loop oscillator configuration we have measured $S_\phi(1\,\text{kHz}) < -126\,\text{dBc/Hz}$ (D. K. Ramm, private communication). The amplitude noise of a loop oscillator and Pound stabilised Gunn oscillator are $-133\,\text{dBc/Hz}$ and $-140\,\text{dBc/Hz}$, respectively.

The pump power is transmitted along semi-rigid coaxial cable to the reentrant cavity. The cable between the circulator and reentrant cavity has an outer diameter of 0.020″, and is bonded to a low-loss three-stage mechanical filter to prevent excitation of the antenna by ambient vibrations.

The reentrant cavity is operated in reflection to provide carrier suppression of the reflected power and to simplify the vibration isolation. Since it is impossible

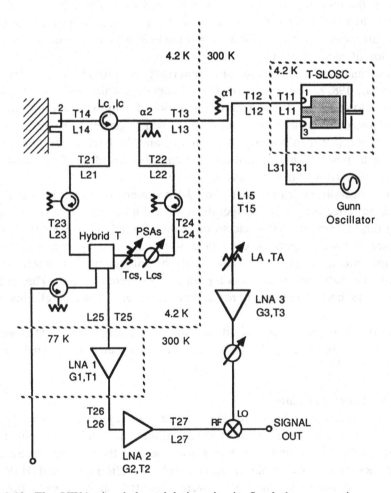

Figure 8.13. The UWA signal demodulation circuit. Symbols representing some of the parameters which determine the noise contributed by each element are shown alongside. The generic 'T' and 'L' symbols represent effective noise temperatures and losses. The symbols $\alpha1$ and $\alpha2$ represent directional couplers.

to set the cavity coupling (carrier suppression) with sufficient accuracy, additional carrier suppression must be used. This additional carrier suppression uses a 'hybrid T' (see figure 8.13) to add some unmodulated carrier to the modulated carrier. The phase and amplitude of the unmodulated carrier can be adjusted using cryogenic phase shifter/attenuators (PSAs) (Mann and Blair, 1983) to give at least 40 dB of stable and precise carrier nulling.

After carrier suppression the signal is amplified using a low noise cryogenic GaAs MESFET preamplifier (LNA1) and a conventional room-temperature low noise amplifier (LNA2). It is then demodulated using a double balanced mixer (DBM).

The microwave circuit currently used to provide pump power for the transducer and process the resulting signal is shown in figure 8.13. Ultimately, the T-SLOSC will be installed in the same dewar as the antenna. Alongside each of the circuit elements are some of the associated parameters which determine the noise contribution of each element.

We have analysed this system and numerically investigated its optimisation (Veitch et al., 1987). There are two 'free' parameters which may be adjusted to produce the lowest noise for a given configuration. Firstly, the pump power level can be adjusted to balance the series noise, which generally decreases as the power increases (for a sufficiently clean pump), and the back-action noise, which increases with power. The integration time can also be adjusted to balance the wideband and narrowband noise terms.

The investigation indicated that the UWA antenna should be capable of obtaining an effective noise temperature of $\sim 1\,\text{mK}$ with a relatively short sampling time of $\sim 0.1\,\text{s}$ and an electromechanical coupling of $\sim 3 \times 10^{-3}$, in the near future. Thus far spurious excitation of the antenna by noise transmitted through the antenna suspension has increased the antenna mode temperature and prevented the antenna from attaining this noise temperature. The antenna suspension has been recently improved (Veitch et al., 1988) to eliminate this problem.

A transducer displacement sensitivity of $10^{-17}\,\text{m/Hz}^{1/2}$ at $1\,\text{kHz}$, and an antenna noise temperature of $40\,\text{mK}$ have been measured in a much simpler system (Blair and Mann, 1981).

8.7.2 The Tokyo transducer

The Tokyo group have been involved in the development of two types of parametric transducers. We will start by describing the 210.7 MHz split doubly reentrant cavity which was developed for use on their room temperature low-frequency mass quadrupole antenna (Tsubono, Hiramatsu and Hirakawa, 1977). The two halves of the cavity were mounted opposite each other on adjacent quadrants of the antenna. A loop oscillator was formed by putting the cavity into the feedback loop of an amplifier. Relative motion of the quadrants changed the resonant frequency of the cavity which, in turn, changed the oscillation frequency. Some of the oscillator power was demodulated in a mixer using a crystal local oscillator at 200 MHz. The IF signal was then amplified and converted into AM using a discriminator. The resulting signal was phase sensitively detected at the antenna mechanical frequency.

This system achieved a transducer sensitivity of 7×10^{-16} m/Hz$^{1/2}$ at 145 Hz and could detect the Brownian motion of the antenna with a signal to noise ratio of 24 dB using a 3 s integration time. The sensitivity was limited by the phase noise generated by the loop oscillator.

A balanced microwave X-band reentrant cavity transducer for use on a cryogenic disk antenna operating at 15 kHz (Tsubono, Ohashi and Hirakawa, 1986) is also being developed by the Tokyo group. The antenna and one of the cavities is shown schematically in figure 8.14. The Al disk has four flat faces machined on its cylindrical surface to remove the two-fold degeneracy and, after niobium foil has been glued to two of the adjacent surfaces, provide sense surfaces for the transducers.

The reentrant cavity design is different to that of the UWA group. Specifically, the resonant region has a large outer diameter and is deeper resulting in a larger cavity inductance. The actual and effective capacitance gap sizes are 23 μm and 52 μm, respectively. Also, the choke region consists of two quarter-wavelength coaxial chokes. The cavities have Qs of about 5.7×10^4 at 4.2 K with an untreated surface.

The resonant frequencies of the cavities are adjusted using a SCSO (Pound stabiliser) scheme (Stein, 1974) but with the cavity frequency servo-controlled to the pump oscillator frequency, as shown in figure 8.14. A pump modulation frequency of 20 kHz (well inside the cavity bandwidth) is used.

A Pound stabilised Gunn oscillator was used as the pump source. The reference cavity was a niobium TM$_{010}$ cavity with a Q of 6.7×10^5 at 4.2 K. The

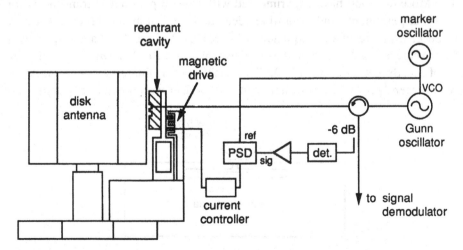

Figure 8.14. Schematic of the Tokyo antenna and transducer.

phase noise of the Gunn oscillator was suppressed by 36 dB at 15 kHz giving $S_\phi(15\,\text{kHz}) = -123\,\text{dBc/Hz}$.

Further reduction of the noise due to the pump phase noise was obtained by using the pump modulation, which was used for the reentrant cavity servo-control, as a marker frequency in a balanced cavity system. The IF signals from the two mixers were subtracted and then phase sensitively detected at the modulation frequency to provide an error signal which was used to adjust the phase and amplitude of one of the channels. This reduced the noise level by more than 16 dB.

This system had a transducer sensitivity of $2.7 \times 10^{-17}\,\text{m/Hz}^{1/2}$ at 15 kHz and a transducer noise temperature of 23 mK. The sensitivity was limited by pump phase noise.

More recently, the Tokyo group have investigated the stabilisation of a Gunn oscillator by a mechanically modulated superconducting TE_{011} cavity (Ohashi, Tsubono and Hirakawa, 1986). A driven high Q mechanical oscillator which forms part of the cavity provides the frequency modulation necessary for the operation of the stabilisation scheme and for servo-controlling the positions of the reentrant cavities. The stabilisation cavity had a Q of 2.5×10^6 at 4.2 K. This scheme produced 40 dB of active stabilisation and 2 dB (only – since the measurement frequency is inside the cavity bandwidth) of passive stabilisation, by using the cavity as a high Q filter, to give $S_\phi(1.5\,\text{kHz}) = -99\,\text{dBc/Hz}$.

8.7.3   The Moscow transducer

The sapphire gravitational radiation antenna being developed by the group led by Braginsky at Moscow State University is shown schematically in figure 8.15. The ends of the 'horns' are coated with niobium and a niobium sleeve is placed concentrically around this region to form a resonant reentrant cavity.

The Moscow group have experimented with several parametric transducers for use as displacement and vibration detectors. A niobium reentrant cavity transducer with a resonant frequency of 750 MHz, Q of $\sim 10^5$ and a gap spacing of 6.5 µm achieved a displacement sensitivity of $1.5 \times 10^{-17}\,\text{m}$ with an averaging time of 3 s (Braginsky et al., 1977). The sensitivity was limited by pump noise.

More recently a 3 GHz reentrant cavity transducer with a gap size of 3 µm and

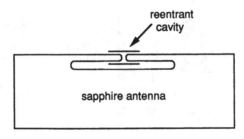

Figure 8.15. Schematic of the Moscow antenna and transducer.

a Q of 5×10^4 has been developed. The pump power was provided by injection locking a reflex klystron to a tunable superconducting sapphire resonator. The pump noise was $S_{AM}(8\,\text{kHz}) = -177\,\text{dBc/Hz}$ and $S_{\phi}(8\,\text{kHz}) = -153\,\text{dBc/Hz}$. The transducer, which was operated in its AM modulation mode (i.e. $|\delta\Omega| = \Omega_0/2Q$), had a displacement sensitivity of $6 \times 10^{-19}\,\text{m}$ for an integration time of 1 s, and $2 \times 10^{-19}\,\text{m}$ for an integration time of 10 s (Braginsky, Panov and Popel'nyuk, 1981).

8.7.4 The LSU transducer

The parametric transducer which has been developed at Louisiana State University (LSU) (Oelfke and Hamilton, 1983; Oelfke, Hamilton and Darling, 1981) is shown schematically in figure 8.16. It consists of two balanced niobium cavities separated by a thin niobium diaphragm. The cavities can be tuned to resonate at the pump frequency (nominally 600 MHz) by a superconducting magnetic loudspeaker type tuner which adjusts the capacitance gap. The Q and gap size are $\sim 10^5$ and $10^{-5}\,\text{m}$, respectively. The mechanical resonant frequency of the niobium diaphragm is set by room temperature adjustment of the tension in the diaphragm. If tuned to the antenna frequency the diaphragm mechanically amplifies the antenna motion.

Mechanical motions with amplitudes $<10^{-16}\,\text{m}$ have been observed with this type of transducer when using a single cavity and driving a diaphragm, which was resonant at 7 kHz, at 2 kHz (Darling, Hamilton and Oelfke, 1982).

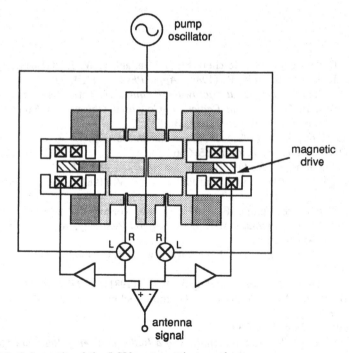

Figure 8.16. Schematic of the LSU parametric transducer.

8.8   Conclusion

We have shown that, in spite of the only relatively recent development and use of parametric transducers on resonant mass gravitational radiation antennae, they are achieving impressive sensitivities. Parametric transducers are clearly complex, and are susceptible to a large number of noise sources. All of these can be modelled, however, as shown in sections 8.5 and 8.6. They have the important advantage of not degrading the acoustic Q of the antenna, along with the convenience of cold-damping which removes the severe dynamic range requirement on the audio amplifiers and A/D converters in the signal recording circuit. Thus, parametric transducers appear to be a realistic alternative to the more conventional passive transducers hitherto used on antennae.

Acknowledgements

This work was supported by the Australian Research Council. I would also like to thank members of the UWA gravity research group for useful discussions; in particular, David Blair for many helpful suggestions, and Nick Linthorne for proof reading the manuscript and providing figure 8.4.

References

Awschalom, D. D., Rozen, J. R., Ketchen, M. B., Gallagher, W. J., Kleinsasser, A. W., Sandstrom, R. L. and Bumble, B. (1988). *Appl. Phys. Lett.* **53,** 2108.
Baden Fuller, A. J. (1987). *Ferrites at Microwave Frequencies*, Peter Peregrinus, London.
Blair, D. G. (1979). In *Gravitational Radiation, Collapsed Objects and Exact Solutions*, ed. C. Edwards, p. 299, Springer-Verlag, Berlin.
Blair, D. G. and Mann, A. G. (1981). *Il Nuovo Cimento* **61,** 73.
Bocko, M. F. (1984). *Rev. Sci. Instrum.* **55,** 256.
Bordoni, F. and Pallotino, G. V. (1977). *Rev. Sci. Instrum.* **48,** 757.
Bordoni, F., Maggi, G., Ottaviano, A. and Pallotino, G. V. (1981). *Rev. Sci. Instrum.* **52,** 1079.
Bordoni, F., De Panfilis, S., Fuligni, F., Iafolla, V. and Nozzoli, S. (1986). In *Proceedings of the Fourth Marcel Grossmann Meeting on General Relativity*, ed. R. Ruffini, p. 523, North Holland, Amsterdam.
Braginsky, V. B. (1970). *Physical Experiments with Test Bodies*, NASA Technical Translation F-672.
Braginsky, V. B., Panov, V. I. and Popel'nyuk, V. D. (1981). *JETP Lett.* **33,** 405.
Braginsky, V. B., Panov, V. I., Petnikov, V. G. and Popel'nyuk, V. D. (1977). *Instr. Exp. Tech.* **20,** 269.
Cooke, H. F. (1971). *Proc. IEEE* **59,** 1163.
Darling, D. H. (1979). *A Perturbation Analysis of a Radio Frequency Resonant Modulated Capacitance Accelerometer*. LSU Gravitational Radiation Group Technical Memorandum #12 (unpublished).

Darling, D. H., Hamilton, W. O. and Oelfke, W. C. (1982). In *Proceedings of the Second Marcel Grossmann Meeting on General Relativity*, ed. R. Ruffini, p. 1179, North Holland, Amsterdam.

Fujisawa, K. (1958). *IRE Trans. MTT* **6,** 344.

Giffard, R. P. (1976). *Phys. Rev. D* **14,** 2478.

Giffard, R. P. and Paik, H. J. (1977). 'Derivation of relations between mechanical and electrical amplitudes for modulated resonator transducers', Stanford University Report (unpublished), June, 1977.

Ginzton, E. (1957). *Microwave Measurements,* chap. 9, McGraw-Hill, New York.

Goronkin, H. (1985). *IEEE Elect. Dev. Lett.* **EDL-6,** 47.

Hollenhorst, J. N. and Giffard, R. P. (1979). *IEEE Trans. Mag.* **MAG-15,** 474.

Jones, S. K. and Blair, D. G. (1987). *Electron. Lett.* **23,** 817.

Liao, S. Y. (1987). *Microwave Circuit Analysis and Amplifier Design,* p. 16, Prentice-Hall, New Jersey.

Linthorne, N. P., Veitch, P. J. and Blair, D. G. (1990). *J. Phys. D* **23,** 1–6.

Long, A. P., Clark, T. D. and Prance, R. J. (1980). *Rev. Sci. Instrum.* **51,** 8.

Manley, J. M. and Rowe, H. E. (1956). *Proc. IRE* **44,** 904.

Mann, A. G. (1984). *IEEE Trans. Microwave Theory Tech.* **MTT-33,** 51.

Mann, L. D. and Blair, D. G. (1983). *J. Phys. E* **16,** 119.

Middleton, D. (1960). *Introduction to Statistical Communication Theory,* p. 93, McGraw-Hill, New York.

Mouw, R. B. and Fukuchi, S. M. (1969). *The Microwave Journal,* March 1969, p. 131.

Oelfke, W. C. (1978). *Analysis of Re-entrant Cavity Modulated Capacitance Displacement Transducer,* LSU Gravitational Radiation Group Technical Memorandum #1 (unpublished).

Oelfke, W. C. and Hamilton, W. O. (1983). *Rev. Sci. Instrum.* **54,** 410.

Oelfke, W. C., Hamilton, W. O. and Darling, D. (1981). *IEEE Trans. Mag.* **MAG-17,** 853.

Ohashi, M., Tsubono, K. and Hirakawa, H. (1986). *Jap. J. Appl. Phys.* **25,** L687.

Paik, H. J. (1974). PhD Thesis, Stanford University, HEPL Report #743, 37.

Paik, H. J. (1982). In *Proceedings of the Second Marcel Grossmann Meeting on General Relativity,* ed. R. Ruffini, p. 1193, North Holland, Amsterdam.

Pospieszalski, M. W., Weinreb, S., Chao, P. C., Mishra, U. K., Palmateer, S. C., Smith, P. M. and Hurang, J. C. M. (1986). *IEEE Trans. Electr. Dev.* **ED-33,** 218.

Stein, S. R. (1974). PhD Thesis, Stanford University, HEPL Report 741.

Thorne, K. S. (1987). In *Three Hundred Years of Gravitation,* eds. S. W. Hawking and W. Israel, pp. 330–458, Cambridge University Press.

Tsubono, K., Hiramatsu, S. and Hirakawa, H. (1977). *Jap. J. Appl. Phys.* **16,** 1641.

Tsubono, K., Ohashi, M. and Hirakawa, H. (1986). *Jap. J. Appl. Phys.* **25,** 622.

Veitch, P. J. (1986). PhD Thesis, University of Western Australia (unpublished).

Veitch, P. J., Blair, D. G., Linthorne, N. P., Mann, L. D. and Ramm, D. K. (1987). *Rev. Sci. Instrum.* **58,** 1910.

Veitch, P. J., Blair, D. G., Linthorne, N. P., Buckingham, M. J., Edwards, C., Prestage, N. P. and Ramm, D. K. (1988). *Proceedings of the Fifth Marcel Grossmann Meeting on General Relativity,* eds. D. G. Blair and M. J. Buckingham, p. 157, World Scientific, Singapore.

Wang, K. G. and Wang, S. K. (1987). *IEEE Trans. Elect. Dev.* **ED-34,** 2610.

Weinreb, S. (1982). *IEEE MTT-S Digest,* p. 10.

Yurke, B., Kaminsky, P. G., Miller, R. E., Whittaker, E. A., Smith, A. D., Silver, A. H. and Simon, R. W. (1988). *Phys. Rev. Lett.* **60,** 764.

van der Ziel, A. (1986). *Noise in Solid State Devices and Circuits,* p. 281, Wiley, New York.

9

Detection of continuous waves

KIMIO TSUBONO

Generally, gravitational radiation (GR) is divided into three classes according to its nature: burst, continuous and stochastic waves. Continuous waves can be described as a sinusoidal stationary train of metric perturbation for a sufficiently long time, in contrast with burst waves, which are characterized by their short duration. The third type of GR, stochastic waves, is characterized by its random nature of statistics of arrival, regardless of wave form.

In searching for continuous waves with resonant antennas, various kinds of detecting methods and signal analyses are employed which are different from those used in the detection of burst waves. For example, a resonant antenna should be tuned precisely to the frequency of the source in order to obtain the best sensitivity. Also, long-time integration of the signal output from the detector is a necessary technique for distinguishing a coherent signal buried in a noise. Under these circumstances, usually, the sensitivity of a detector for continuous waves is determined by the level of Brownian motion of the antenna. These features are not common with the case of burst events.

Continuous sources, such as pulsars and binaries, have rather low frequencies, except for rapid pulsars (Backer *et al.*, 1982) or new-born pulsars. Since the pioneering work of J. Weber (1969) bars have been widely used as resonant antennas in detecting burst waves. The resonant frequency of a bar is given by the sound velocity in the material divided by twice the length of the cylinder; usually, the available frequency range is above $\sim$700 Hz (Amaldi *et al.*, 1986; Boughm *et al.*, 1982). Hence, other types of resonant antennas are required for detecting low-frequency GR. The design of low-frequency antennas with a high quality factor is essential in this field. Moreover, an anti-vibration system with good performance is crucial in order to overcome low-frequency seismic noise.

In this chapter, the properties of low-frequency antennas are described. Then, the techniques necessary for tuning an antenna to a continuous source are discussed. Also, cold damping, which is a technique for obtaining a wider bandwidth for resonant antennas, is reviewed. Finally, the sensitivity of a resonant antenna to continuous waves is discussed. Most of the content of this chapter is related to papers written by the late Hiromasa Hirakawa and his colleagues.

9.1 Antenna properties

Though bar antennas are widely used as resonant detectors, there are several other types of antennas, especially in the low-frequency region. The torsion antenna (Kimura, Suzuki and Hirakawa, 1981) is one example of a low-frequency detector which has been developed by the Tokyo group. When we discuss the properties of antennas, it is desirable that all types of resonant antennas be treated in the same way. Hirakawa, Narihara and Fujimoto (1976) have presented a general treatment of antennas for GR by introducing a quantity, effective area A_G, which expresses the effective cross section for GR. By using the effective area, all the various properties of the antennas, such as sensitivity and directivity, can be discussed in general. An outline of the theory is as follows.

Under the assumption that the eigenfunctions of a solid body antenna are known, one can write an equation of motion of the antenna for a certain eigenmode as follows, using a displacement field $\mathbf{u}(\mathbf{r}, t)$,

$$\frac{d^2\mathbf{u}}{dt^2} + \frac{\omega_M}{Q_M}\frac{d\mathbf{u}}{dt} + \omega_M^2\mathbf{u} = \frac{\mathbf{u}\int\rho\mathbf{fu}\,dV}{\int\rho\mathbf{u}^2\,dV}, \tag{9.1}$$

where $\omega_M/2\pi$ is the eigenfrequency, ρ and Q_M are the density of the antenna material and the quality factor of the mode, and $\mathbf{f}(\mathbf{r}, t)$ is the external force (per unit mass) acting on the antenna. This equation shows the motion of a damped harmonic oscillator. By introducing a generalized coordinate $\xi(t)$ which satisfies

$$\mathbf{u}(\mathbf{r}, t) = \xi(t)\mathbf{v}(\mathbf{r}), \tag{9.2}$$

the antenna system can be treated as one-dimensional. It should be noticed that by specifying ξ (for example, ξ is defined as a displacement at an end-wall of a bar antenna or the torsion angle of a torsion antenna), $\mathbf{v}(\mathbf{r})$ is uniquely given over the entire antenna. Then, by rewriting equation (9.1) we obtain

$$\mu[\ddot{\xi} + (\omega_M/Q_M)\dot{\xi} + \omega_M^2\xi] = \int\rho\mathbf{fv}\,dV, \tag{9.3}$$

where

$$\mu = \int\rho\mathbf{v}^2\,dV \tag{9.4}$$

is the reduced mass related to coordinate ξ. Let ϕ and U be potentials such that

$$\mathbf{f} = -\text{grad }\phi, \tag{9.5}$$

and

$$U = \int\rho\phi\,dV. \tag{9.6}$$

Then, equation (9.3) can be rewritten using a generalized force $-\partial_v U$ as

$$\mu[\ddot{\xi} + (\omega_M/Q_M)\dot{\xi} + \omega_M^2\xi] = -\partial_v U. \tag{9.7}$$

The right-hand side of the above equation has the following meaning:

$$\partial_v U = \lim_{\delta \to 0} \frac{U(\mathbf{r} + \delta\mathbf{v}) - U(\mathbf{r})}{\delta}. \tag{9.8}$$

The mass quadrupole Q_{ij} of the antenna and the dynamic part q_{ij} of Q_{ij} are defined as

$$Q_{ij} = \int \rho\left(x_i x_j - \frac{1}{3}\delta_{ij}r^2\right)dV \tag{9.9}$$

and

$$q_{ij} = \partial_v Q_{ij} = \int \rho\left(v_i x_j + x_i v_j - \frac{2}{3}\delta_{ij}\mathbf{vr}\right)dV. \tag{9.10}$$

The potential ϕ can be expanded around the center of mass of the antenna; then, the potential U is expressed as

$$U = \frac{1}{2}\sum \frac{\partial^2\phi}{\partial x_i \partial x_j} Q_{ij} \tag{9.11}$$

by neglecting higher-order terms. Finally, the substitution of equation (9.11) into equation (9.7) yields the following equation of the motion of the antenna:

$$\mu[\ddot{\xi} + (\omega_M/Q_M)\dot{\xi} + \omega_M^2\xi] = -\frac{1}{2}\sum \frac{\partial^2\phi}{\partial x_i \partial x_j} q_{ij}. \tag{9.12}$$

This equation expresses a quadrupole–quadrupole interaction between the antenna and the field.

In a weak gravitational field of metric perturbation h_{ij}, we can regard that there exists a gravitational field potential ϕ which satisfies the following relation (Misner, Thorne and Wheeler, 1973):

$$\frac{\partial^2\phi}{\partial x_i \partial x_j} = -\frac{1}{2}\frac{d^2 h_{ij}}{dt^2}. \tag{9.13}$$

We can then obtain the following equation of the motion for an antenna under a weak gravitational field:

$$\mu[\ddot{\xi} + (\omega_M/Q_M)\dot{\xi} + \omega_M^2\xi] = \frac{1}{4}\sum \frac{d^2 h_{ij}}{dt^2} q_{ij}. \tag{9.14}$$

The effective area A_G of an antenna is given by the dynamic part q_{ij} of the quadrupole moment of the vibration mode as

$$A_G = \frac{2\sum q_{ij}^2}{\mu M}, \tag{9.15}$$

where M is the antenna mass. Notice that A_G is the proper value for each antenna; the value is independent of the choice of the generalized coordinate ξ.

Next, polarization matrices $\mathbf{e}^+$ and $\mathbf{e}^\times$ are constructed to represent the two independent polarizations of GR which propagate along the z-direction:

$$\mathbf{e}^+(\hat{z}) = \begin{pmatrix} 1 & 0 & 0 \\ 0 & -1 & 0 \\ 0 & 0 & 0 \end{pmatrix}, \quad \mathbf{e}^\times(\hat{z}) = \begin{pmatrix} 0 & 1 & 0 \\ 1 & 0 & 0 \\ 0 & 0 & 0 \end{pmatrix}. \tag{9.16}$$

In general, when the incident direction of waves is $\mathbf{k}(\theta\phi)$, by using a rotation matrix the above matrices become

$$\mathbf{e}^+(\mathbf{k}) = \begin{pmatrix} \cos^2\theta\cos^2\theta - \sin^2\phi & \cos^2\theta\cos\phi\sin\phi + \cos\phi\sin\phi & -\cos\theta\sin\theta\cos\phi \\ \cos^2\theta\cos\phi\sin\phi + \cos\phi\sin\phi & \cos^2\theta\sin^2\phi - \cos^2\phi & -\cos\theta\sin\theta\sin\phi \\ -\cos\theta\sin\theta\cos\phi & -\cos\theta\sin\theta\sin\phi & \sin^2\theta \end{pmatrix}, \tag{9.17}$$

$$\mathbf{e}^\times(\mathbf{k}) = \begin{pmatrix} -\cos\theta\sin 2\phi & \cos\theta\cos 2\phi & \sin\theta\sin\phi \\ \cos\theta\cos 2\phi & \cos\theta\sin 2\phi & -\sin\theta\cos\phi \\ \sin\theta\sin\phi & -\sin\theta\cos\phi & 0 \end{pmatrix}. \tag{9.18}$$

By using these matrices, waves with definite polarization and propagation direction can be represented as $h_{ij}^A(t) = h^A(t)e_{ij}^A$ $(A = +, \times)$. Considering the polarization and the incident angles of GR, the directivity function $d^A(\theta\phi)(A = +, \times)$ is defined as

$$d^A(\theta\phi) = \frac{5(\sum q_{ij}e_{ij}^A(\mathbf{k}))^2}{4\sum q_{ij}^2}. \tag{9.19}$$

For unpolarized waves,

$$d(\theta\phi) = d^+(\theta\phi) + d^\times(\theta\phi) \tag{9.20}$$

is used as a directivity pattern. When this function $d(\theta\phi)$ is averaged over an entire solid angle

$$\overline{d(\theta\phi)} = \frac{1}{4\pi}\int d(\theta\phi)\, d\Omega = 1. \tag{9.21}$$

After a proper coordinate transformation, the tensor of the dynamic part of quadrupole moment, q_{ij}, becomes

$$q_{ij} = \text{const} \times \begin{pmatrix} 1 & & \\ & -(1-\eta)/2 & \\ & & -(1+\eta)/2 \end{pmatrix}, \tag{9.22}$$

$$0 \leq \eta \leq 1,$$

where η is the asymmetry parameter of q_{ij}. Most antennas have one or more symmetry axes. An antenna operating in the A-mode ($\eta = 0$) has the directivity pattern

$$d(\theta\phi) = \frac{15}{8} \sin^4 \theta, \tag{9.23}$$

as shown in figure 9.1(a). For an antenna in the B-mode($\eta = 1$),

$$d(\theta\phi) = \frac{5}{2} \cos^2 \theta + \frac{5}{8} 2 \sin^4 \theta \cos^2 \phi, \tag{9.24}$$

as shown in figure 9.1(b). For an A-mode antenna, d and d^A are less than or equal to 15/8, while those of a B-mode antenna are less than or equal to 5/2. Thus, B-mode antennas have a sharper directivity than A-mode type. In table 9.1 four types of resonant antennas (Kimura, Suzuki and Hirakawa, 1981; Paik, 1977; Tsubono, Ohashi and Hirakawa, 1986) with their calculated presentation of q_{ij}, $d(\theta\phi)$ and A_G are given.

One can determine the energy absorbed by an antenna when GR is incident onto the antenna which has a resonant frequency ω_M, an effective area A_G and a directivity $d^A(\theta\phi)$. First, considering the burst wave, when GR with a polarization A ($A = +, \times$) and an energy flux spectrum $F^A(\omega)$ (energy per unit area per hertz) hits the antenna from the ($\theta\phi$) direction, the antenna obtains the following energy:

$$E = \frac{\pi G}{10c^3} \omega_M^2 M A_G d^A(\theta\phi) F^A(\omega_M). \tag{9.25}$$

If the above energy is averaged over the polarization of incident waves which have $F^+(\omega) = F^\times(\omega) = F(\omega)$, we obtain

$$E = \frac{\pi G}{20c^3} \omega_M^2 M A_G d(\theta\phi) F(\omega_M), \tag{9.26}$$

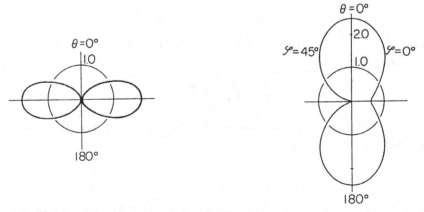

Figure 9.1. (a) Directivity pattern, equation (9.23), of an A-mode antenna. (b) Directivity pattern, equation (9.24), of a B-mode antenna.

Table 9.1. *For four types of typical resonant antennas, the evaluated values of q_{ij}, $d(\theta\phi)$ and A_G are given.*

Type of antenna	Dynamic part of quadrupole, q_{ij}	Directivity pattern $d(\theta\phi)$	Effective area A_G
Dumb-bell $\ell = 2a$	constant $\times$ $\begin{pmatrix} -\frac{1}{2} & & \\ & -\frac{1}{2} & \\ & & 1 \end{pmatrix}$	$\frac{15}{8}\sin^4\theta$	$\frac{4}{3}l^2 = 1.70\pi a^2$
Bar $\ell = 2a$	$\begin{pmatrix} -\frac{1}{2} & & \\ & -\frac{1}{2} & \\ & & 1 \end{pmatrix}$	$\frac{15}{8}\sin^4\theta$	$\frac{128}{3}\left(\frac{l}{\pi^2}\right)^2 = 0.558\pi a^2$
Disk + mode × mode $2a$	$+$ *mode* $\begin{pmatrix} 1 & & \\ & -1 & \\ & & 0 \end{pmatrix}$ $\times$ *mode* $\begin{pmatrix} & 1 & \\ 1 & & \end{pmatrix}$	$\frac{5}{2}\cos^2\theta + \frac{5}{8}\sin^4\theta\cos^2\phi$ $\frac{5}{2}\cos^2\theta + \frac{5}{8}\sin^4\theta\sin^2 2\phi$	$A_G^+ = A_G^\times = 0.563\pi a^2$ The two-fold degeneracy of quadrupole vibration mode is supposed to be removed here. Poisson ratio $\sigma = 0.33$ is assumed.
Torsion $2a$	$\begin{pmatrix} 1 & & \\ & -1 & \\ & & 0 \end{pmatrix}$	$\frac{5}{2}\cos^2\theta + \frac{5}{8}\sin^4\theta\cos^2 2\phi$	$0.270\pi a^2$ The shape is optimized to have a best sensitivity.

Moreover, averaging over all possible incident directions the following result from equation (9.21) is obtained:

$$E = \frac{\pi G}{20c^3}\,\omega_M^2 M A_G F(\omega_M). \tag{9.27}$$

Also, the following equation

$$\sigma_T = \frac{\pi G \omega_M^2 M}{20c^3} A_G \tag{9.28}$$

relates the effective area A_G with the total cross section σ_T appearing in *Gravitation* by Misner, Thorne and Wheeler (1973).

The energy flux s^A (energy per unit area per second) carried by continuous GR with polarization A and amplitude $h^A(t)$ can be expressed as

$$s^A = \frac{c^3}{16\pi G} \overline{h^A(t)^2}. \tag{9.29}$$

When the frequency of a resonant antenna and that of continuous GR are the same, the energy obtained by the antenna at the equilibrium state is

$$E = \frac{2\pi G}{5c^3} M Q_M^2 A_G d^A(\theta\phi) s^A. \tag{9.30}$$

9.2   Frequency tuning

Frequency tuning is an important technique for detecting continuous GR. Since resonant antennas with very high quality factors have quite narrow bandwidths, the initial setting and tracking of the resonant frequency are indispensable.

Usually the frequency of GR from a continuous source is modulated by the Doppler effect caused by both the orbital and diurnal (spin) motion of the earth. In addition, the source itself has deceleration or acceleration components due to various causes: for example, pulsars usually have a small spin down/up rate. Sometimes they show sudden frequency jumps with an unpredictable nature. As an illustrative example, the arrival frequency of the expected Crab pulsar GR (twice the frequency of the optical pulse) is plotted in figure 9.2 for a time period around 1986 (Owa, 1987). The arrival frequency of $\sim$30 Hz optical pulse shows a spin down of $\sim$0.01 Hz/year and a maximum $\pm$0.03 Hz variation owing to the orbital motion of the earth. Moreover, the diurnal spin motion of the earth causes a frequency shift of about $\pm 2 \times 10^{-5}$ Hz. Regarding the bandwidth of the detector, one should distinguish two kinds of bandwidths: response bandwidth and the SNR (signal-to-noise ratio) bandwidth. The response bandwidth is the bandwidth in which the output response for the input signal is flat and the SNR bandwidth is the bandwidth in which the SNR is flat. Usually resonant mass detectors have a much wider SNR bandwidth than that of response. However, provided that both the resonant frequency and the quality factor of the antenna are precisely known, we can deduce the signal force exerted on the antenna from what is observed at the output by considering the transfer function of the antenna. Thus, the response bandwidth is not so serious as long as the parameters of the antenna are fully known. A bandwidth restriction only arises when the coupling of the antenna to an electronic device is considered (Michelson and Taber, 1984). The SNR bandwidth of a resonant antenna is roughly expressed as

$$\Delta\omega \sim \beta\omega, \tag{9.31}$$

where β is the coupling constant between the antenna and the transducer. The

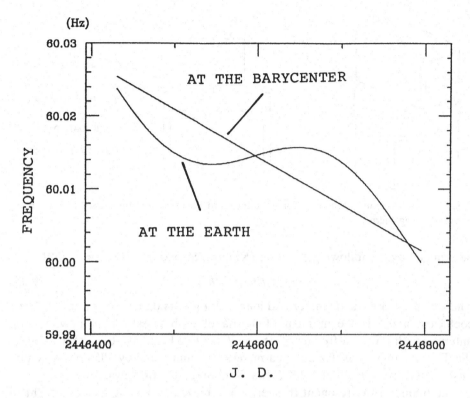

Figure 9.2. The expected frequency of the Crab pulsar GR arriving at the earth is plotted for the time period around 1986.

above relation remains valid only when the antenna and the transducer are properly optimized.

Resonant antennas can be divided into two classes by their nature (Press and Thorne, 1972). Class I is distributed antennas; the restoring forces and inertial forces are distributed more or less uniformly throughout the antenna body and the resonant frequency is determined (approximately) by the sound travel time across the mass. Examples are bars, disks and spheres. Class II is lumped resonant antennas; the main restoring force and main inertial force are contributed by different parts of the system. The resonant frequency of the fundamental mode can be made much lower than that of a class-I antenna of the same size. Examples are dumb-bells, square antennas and torsion antennas.

Class-II antennas are suitable as detectors of low-frequency continuous GR, because they can be tuned to an almost arbitrary frequency by adjusting their parameters. For example, a torsion antenna is used to detect low-frequency GR; it consists of an upper and a lower mass and a torsion bar connecting them (figure 9.3). In the fundamental torsion mode, both masses rotate in opposite directions around the axis of symmetry. The resonant frequency ω_M of a torsion antenna is

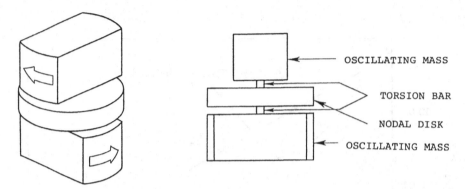

Figure 9.3. A torsion antenna comprises upper and lower oscillating masses and a torsion bar connecting them.

determined by the following equation (Kimura, Suzuki and Hirakawa, 1981)

$$\omega_M = d^2(\pi\mu/32hI)^{1/2}, \tag{9.32}$$

where d and h are the diameter and length of the torsion bar, μ is Lame's elastic coefficient and I is the moment of inertia of each mass. Up to now, torsion antennas with resonant frequencies from 40 Hz to 3 kHz have been fabricated by the Tokyo group. A 60 Hz antenna made of aluminium alloy 5056 has a quality factor of $Q_M = 3.6 \times 10^7$ at 4.2 K (Owa, Tsubono and Hirakawa, 1983).

Fine tuning of the resonant frequency is achieved by adding a small weight to some part of the antenna. As a result, the resonant frequency always decreases. Also, the resonant frequency can be slightly shifted by changing the elastic constant of the antenna body through the temperature. For example, it has been reported that a 60 Hz aluminium antenna has a thermal sensitivity -1×10^{-2} Hz/K at 300 K (Hirakawa, Tsubono and Fujimoto, 1978; Oide, Hirakawa and Fujimoto, 1979).

The above-mentioned methods for changing the resonant frequency are not suitable for cryogenic antennas. Another method is to apply an electromechanical force to the antenna, thus modifying the effective force constant of the structure associated with its deformation. This electromechanical tuning works under various circumstances. There are two methods to couple the antenna to the electrical circuit: one is by using an electrical field and the other is by using a magnetic field.

Electrostatic tuning is considered here. An electrostatic condenser comprising two electrodes with a capacitance C at an equilibrium distance d is mounted on the antenna as shown in figure 9.4. The condenser is charged at a voltage V, which makes an electric field with strength $E = V/d$ inside the condenser. The stored energy in the condenser changes with the distance of the electrodes; thus, a restoring force is produced on the plates. The resonant frequency of the antenna changes with the voltage V on the electrodes as

$$\omega_M = \omega_{M0}(1 - \beta)^{1/2}, \tag{9.33}$$

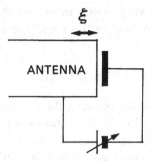

Figure 9.4. Schematic diagram of an electrostatic tuning circuit for a resonant antenna.

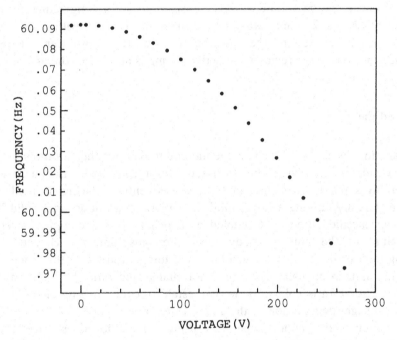

Figure 9.5. Resonant frequency of a torsion antenna obtained by the electrostatic tuning method.

where $\beta = CE^2/\mu\omega_{M0}^2$ is the coupling coefficient between the condenser and the antenna and ω_{M0} and μ are the resonant frequency without electromechanical coupling and the reduced mass of the antenna, respectively.

In the Crab pulsar experiment (Owa *et al.*, 1986) electrostatic tuning of the antenna was used to track the arrival frequency of the expected GR from the Crab pulsar. The antenna is a torsion type with a mass of 74 kg and a resonant frequency of $\omega_M/2\pi \sim 60$ Hz. The obtained frequency change by this method is shown in figure 9.5. This frequency range covers the expected two-year variation of the Crab pulsar. It was possible to tune the frequency with an accuracy of 3×10^{-6} Hz.

Magnetic coupling (Paik, 1976) can be also used to tune the resonant frequency of the antenna. The effect is very similar to that of electrostatic coupling, except that the restoring force is repulsive in this case. The resonant frequency changes as

$$\omega_M = \omega_{M0}(1 + \beta)^{1/2}, \tag{9.34}$$

where β is the coupling coefficient between the magnetic circuit and the antenna.

Furthermore, frequency tuning of the antenna can be achieved by the active feedback device (Ogawa, Oide and Hirakawa, 1979). The antenna deformation is sensed by a transducer, amplified, and then fed back as a driving force on the antenna. In principle, frequency tuning by a passive device is free from additional noise; however, active tuning is very sensitive to noise originating from the transducer. The noise temperature of the antenna increases with a shift of the resonant frequency due to the disturbance from the transducer. Therefore, transducer noise is very crucial when active tuning is applied to the antenna.

9.3   Cold damping

Cold damping, using a cold load, is one method to dampen the vibrational motion of the oscillator. Applying this technique to a solid-body antenna we can considerably suppress Brownian motion, consequently reducing its mode temperature. Usually, among a large number of vibrational modes in a solid body, only some specified modes are cooled as a result of cold damping; thus, the temperature of the antenna body, itself, remains almost unchanged. Cold damping reduces the quality factor Q and the temperature T of the mode at the same time and, in an ideal case, does not change the ratio T/Q. Actually, by preparing a sufficiently cold load, we can establish cold damping of the antenna with only a slight degradation of the T/Q of the mode.

A reduction of the mode temperature using this method does not result in a direct improvement of the signal-to-noise ratio (SNR) in the measurement of the external force exerted on the antenna. This can be understood by the fact that cold damping does not improve the T/Q of the antenna. However, this method is quite useful by the very fact that it does not cause much degradation in the SNR. There are several merits in the cold damped state of an antenna, especially in the application of GR experiments. The reduced Brownian motion or coherent excitation removes severe requirements on the dynamic range of amplifiers and A/D (analog-to-digital) converters. The inconvenience associated with a long relaxation time of a high-Q antenna (for example (Kimura, Suzuki and Hirakawa, 1981), 44 Hz antenna with a Q-value 4×10^7 has an amplitude decay time $2Q/\omega_M = 2.9 \times 10^5$ s $= 3.4$ days) is eliminated. Furthermore, the widened response bandwidth makes it easy for a high-Q antenna to tune to an external coherent signal.

Cold damping of an oscillator is realized by loading the system with a cooled dissipative element (Hirakawa, Hiramatsu and Ogawa, 1977; Hirakawa, Oide and Ogawa, 1978; Oide, Ogawa and Hirakawa, 1978). An electromechanical coupling is used to incorporate the load to the antenna. With a cold load a coupled system has an effective quality factor Q_e which satisfies the following relation:

$$\frac{1}{Q_e} = \frac{1}{Q_M} + \frac{\beta}{Q_E},$$ (9.35)

where Q_M and Q_E are the quality factors of the antenna itself and the electrical load, respectively; β is the coefficient of the electromechanical coupling. On the other hand, the effective temperature T_e is determined by the following rate equation of the energy transfer:

$$-\frac{d}{dt} T_e = \frac{\omega}{Q_M} (T_E - T_M) + \frac{\beta\omega}{Q_E} (T_e - T_E),$$ (9.36)

where T_M and T_E are the physical temperature of the antenna and the electric load, respectively. In a steady state condition, the above equation reduces to

$$\frac{T_e}{Q_e} = \frac{T_M}{Q_M} + \frac{\beta T_E}{Q_E}.$$ (9.37)

Equations (9.35) and (9.37) are the fundamental equations of cold damping.

With a resistor dipped into liquid helium at 4.2 K and an electrostatic coupling circuit, the Brownian motion of a 1400 kg square antenna was damped at $\omega_M/2\pi = 145$ Hz (Hirakawa, Hiramatsu and Ogawa, 1977). A reduced effective temperature of $T_e = 20$ K was obtained by this system. The schematic diagram of the whole system is shown in figure 9.6. The coupling coefficient β and the quality factor Q_E of the electric circuit are obtained as

$$\beta = \frac{|Z_{12}Z_{21}|}{\mu\omega |Z_{22}|} = \frac{CE^2}{\mu\omega^2} \frac{\omega CR}{[1 + (\omega CR)^2]^{1/2}}$$ (9.38)

and

$$\frac{1}{Q_E} = \frac{\mathrm{Re}Z_{22}}{|Z_{22}|} = [1 + (\omega CR)^2]^{-1/2},$$ (9.39)

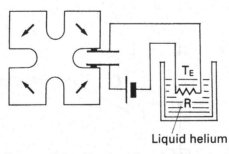

Liquid helium

Figure 9.6. Schematic diagram of the damping of an oscillator by loading a cooled resistor dipped into liquid helium.

where Z_{ij} are components of the 2×2 impedance matrix of the transducer (Giffard, 1976; chapter 8 of this book), μ and ω_M are the reduced mass and the resonant frequency of the antenna, and C and E are capacitance and the strength of the electric field in the capacitor plates. The lowest effective temperature $T_e = 20\,\text{K}$ is obtained at $T_E = 4.2\,\text{K}$, $T_M = 280\,\text{K}$, $Q_M = 1.4 \times 10^5$, $C = 2700\,\text{pF}$ and $\beta = 2.6 \times 10^{-4}$.

A suppression of the thermal fluctuation (Johnson noise) in resistors by means of a suitable feedback method could result in reduced noise, thereby lowering the noise temperature of the resistors (Hirakawa, Oide and Ogawa, 1978; Oide, Ogawa and Hirakawa, 1978). This electronic-type artificial cold load could replace a cryogenic cold load in all applications, except those at very low temperatures, where the amplifier internal noise supersedes the Nyquist noise. In practice, an electronic cold resistor has an obvious advantage over a cryogenic cold resistor regarding the maintenance procedure. The use of electronic cooling circuits based on this principle has been limited in practice only by the presence of amplifier internal noise and the limited bandwidth.

The Johnson noise across an impedance Z in thermal equilibrium at temperature T_E is given by the Nyquist formula,

$$\langle v^2 \rangle = 4kT\,\text{Re}(Z), \tag{9.40}$$

where $\langle v^2 \rangle$ is the mean square noise voltage in a unit bandwidth. Consider the circuit shown in figure 9.7, which comprises a resistor R and an amplifier of voltage gain G. The apparent impedance R' across AA',

$$R' = \frac{R}{1+G}, \tag{9.41}$$

has an effective temperature T'_E given by

$$T'_E = \frac{T_E}{1+G}. \tag{9.42}$$

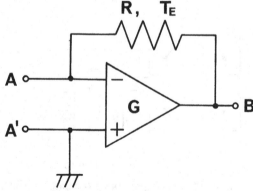

Figure 9.7. Electrical circuit representing an artificial cold resistor across AA'.

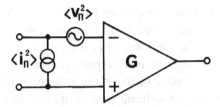

Figure 9.8. Amplifier internal noise is represented by input voltage and current noise sources.

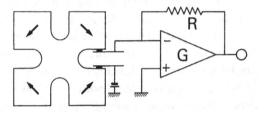

Figure 9.9. An artificial cold resistor is coupled to an antenna with an electrostatic transducer.

The impedance R' behaves as if it is an actual resistor cooled in a cryostat at temperature T'_E. In a practical circuit one has to take into account the amplifier internal noise, which is represented by the two sources shown in figure 9.8. Using the mean square noise voltage $\langle v_n^2 \rangle$ and the mean square noise current $\langle i_n^2 \rangle$ of the sources, both in a unit bandwidth, we obtain the effective noise temperature of the impedance R',

$$T'_E = \frac{T_E}{1+G} + \frac{1}{4kR} + \frac{G^2\langle v_n^2 \rangle + R^2\langle i_n^2 \rangle}{1+G}. \tag{9.43}$$

Now, cold damping with an artificial cold resistor is often used for a low-frequency mechanical resonator. The first demonstration of this (Oide, Ogawa and Hirakawa, 1978) was performed for a gravitational radiation antenna with $\mu = 300$ kg, $\omega_M/2\pi = 145$ Hz, $T_M = 290$ K and $Q_M = 1.5 \times 10^5$. The antenna is coupled through an electrostatic transducer (see figure 9.9) of $\beta = 2.6 \times 10^{-4}$ to an artificial cold resistor, $R' = 2.1 \times 10^5$ Ω, $T'_E = 2.2$ K, obtained with a resistor $R = 10^8$ Ω and an FET amplifier of $G = 410$. The observed effective quality factor $Q_e = 1.2 \times 10^4$ and the effective temperature $T_e = 25$ K determined from the Brownian motion agree with the calculation.

9.4 Detector sensitivity

The sensitivity of a resonant antenna for continuous GR emitted from a coherent source is limited by the Brownian motion of the antenna and the noise originating from the transducer. While searching for a continuous signal in the output of the detector, a different algorithm is employed from that which is used for burst GR. Since a rather long observation time is usually available, owing to the nature of the signal, the noise from the transducer has little effect on the final sensitivity.

A Fourier integration of the antenna output over an extended period results in a SNR proportional to the quality factor of the antenna (Fujimoto and Hirakawa, 1979). For an antenna with an extremely large quality factor, however, it is no longer an easy task to exploit the full merit of its characteristics when the practically available observation time is not sufficiently long compared with the antenna decay time. There are difficulties in attaining equilibrium between the field and the antenna and keeping the antenna at precise resonance with the source. The previously described method of cold damping is useful in this connection, increasing the bandwidth of the antenna without losing its sensitivity.

Here, the sensitivity for a continuous signal is discussed while assuming that the observation time is sufficiently long compared with the decay time of the antenna; thus, the antenna always maintains an equilibrium state during observations. In considering the noise, the SNR is well understood in terms of the equivalent noise forces acting on the antenna. Taking into account the noise forces, equation (9.14) can be rewritten as

$$\mu[\ddot{\xi} + (\omega_M/Q_M)\dot{\xi} + \omega_M^2\xi] = f_G + f_M + f_{BA} + f_{WB}, \tag{9.44}$$

where f_G, f_M, f_{BA} and f_{WB} are the driving forces representing the gravitational field, Brownian motion, back-action from the circuit of the transducer and wide-band noise arising from the transducer, respectively. The signal force $f_G(t)$ caused by the continuous GR h_{ij}^A with a polarization A (A = +, ×) is written as

$$f_G = \frac{1}{4}\sum \frac{d^2 h_{ij}^A}{dt^2} q_{ij}. \tag{9.45}$$

Using equations (9.45) and (9.29), the mean square signal force $\overline{f_G^2}$ can be obtained as

$$\overline{f_G^2(t)} = \frac{2\pi G}{5c^3} M\mu\omega_G^2 A_G d^A s^A, \tag{9.46}$$

where ω_G is the frequency of the incident continuous GR. The noise force from the thermal fluctuation of the antenna has the following power spectrum density:

$$\langle f_M^2(\omega)\rangle = \frac{4\mu\omega_M k T_M}{Q_M}. \tag{9.47}$$

Regarding the back-action force, one can define the noise temperature T_{BA}, in terms of which the power spectrum density is expressed as

$$\langle f_{BA}^2(\omega)\rangle = \frac{4\mu\omega_M k T_{BA}}{Q_M}. \tag{9.48}$$

Notice that the two noise forces mentioned above are assumed to have a white

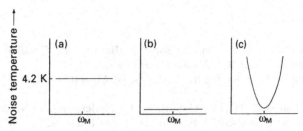

Figure 9.10. Power spectra of noise sources in terms of equivalent driving forces. (a) Brownian motion; (b) back-action; (c) wide-band noise.

spectra in the vicinity of the resonant frequency of the antenna. Wide-band noise from the electric circuit is given in terms of the equivalent displacement noise $\langle \xi_N^2 \rangle$ as

$$\langle f_{WB}^2(\omega) \rangle = \mu^2 \omega_M^4 [(1 - (\omega/\omega_M)^2)^2 + (\omega/Q_M\omega_M)^2]\langle \xi_N^2 \rangle. \qquad (9.49)$$

The noise contribution from each origin is shown in figure 9.10. From this figure it can be easily understood that tuning to the source frequency is necessary only because of the existence of the transducer wide-band noise. The characteristic bandwidth of the Fourier integration over the observation period τ_{obs} is $1/\tau_{obs}$; therefore, the SNR can be written as

$$\text{SNR} = \frac{\tau_{obs}\overline{f_G^2(t)}}{\langle f_M^2(\omega) \rangle + \langle f_{BA}^2(\omega) \rangle}, \qquad (9.50)$$

assuming that wide-band noise can be neglected compared with other noise. The above equation can be rewritten in terms of equations (9.46)–(9.48) in the form

$$\text{SNR} = \frac{\pi G}{10c^3} \frac{\omega_G \tau_{obs} Q_M MA_G d^A s^A}{k(T_M + T_{BA})}$$

$$= \frac{\omega_G^3 \tau_{obs} Q_M MA_G d^A \overline{h^A(t)^2}}{160k(T_M + T_{BA})}. \qquad (9.51)$$

Finally, it is noted that the sensitivity limitation imposed by the quantum mechanical uncertainty principle is less serious in the detection of continuous GR than in the case of burst events. The observed quantity related to burst events is the change in the Brownian motion, which must be compared with the energy of a phonon. While in the experiment on continuous GR, the antenna motion, containing many phonons, is integrated over a sufficiently long time in order to search for specified components. Therefore, even if a signal has an energy that is less than that of one phonon, the integrated output can contain information of the signal as a statistical property.

References

Amaldi, E., Pizzella, G., Rapagnani, P., Ricci, F., Bonifazi, P., Cavallari, G., Coccia, E. and Pallottino, G. V. (1986). *Nuovo Cimento* **9C,** 829.

Backer, D. C., Kulkarni, S. R., Heiles, C., Goss, W. M. and Davis, M. M. (1982). *Nature* **300,** 615.

Boughm, S. P., Fairbank, W. M., Giffard, R. P., Hollenhorst, J. N., Mapoles, E. R., McAshan, M. S., Michelson, P. F., Paik, H. J. and Taber, R. C. (1982). *Astrophys. J.* **261,** L19.

Fujimoto, M.-K. and Hirakawa, H. (1979). *Jpn. J. Appl. Phys.* **46,** 703.

Giffard, R. P. (1976). *Phys. Rev.* **14D,** 2478.

Hirakawa, H., Hiramatsu, S. and Ogawa, Y. (1977). *Phys. Lett.* **63A,** 199.

Hirakawa, H., Narihara, K. and Fujimoto, M.-K. (1976). *J. Phys. Soc. Jpn.* **41,** 1093.

Hirakawa, H., Oide, K. and Ogawa, Y. (1978). *J. Phys. Soc. Jpn.* **44,** 337.

Hirakawa, H., Tsubono, K. and Fujimoto, M.-K. (1978). *Phys. Rev. D* **17,** 1919.

Kimura, S., Suzuki, T. and Hirakawa, H. (1981). *Phys. Lett.* **81A,** 302.

Michelson, P. F. and Taber, R. C. (1984). *Phys. Rev. D* **29,** 2149.

Misner, C. W., Thorne, K. S. and Wheeler, J. A. (1973). *Gravitation,* chaps. 35–7, Freeman, San Francisco.

Ogawa, Y., Oide, K. and Hirakawa, H. (1979). *Jpn. J. Appl. Phys.* **18,** 1565.

Oide, K., Hirakawa, H. and Fujimoto, M.-K. (1979). *Phys. Rev. D* **20,** 2480.

Oide, K., Ogawa, Y. and Hirakawa, H. (1978). *Jpn. J. Appl. Phys.* **17,** 429.

Owa, S. (1987). PhD thesis, Tokyo University (unpublished).

Owa, S., Tsubono, K. and Hirakawa, H. (1983). *Jpn. J. Appl. Phys.* **22,** L452.

Owa, S., Fujimoto, M.-K., Hirakawa, H., Morimoto, K., Suzuki, T. and Tsubono, K. (1986). *Proceedings of 4th Marcel Grossman Meeting on General Relativity,* ed. R. Ruffini, p. 571, North-Holland, Amsterdam.

Paik, H. J. (1976). *J. Appl. Phys.* **47,** 1168.

Paik, H. J. (1977). *Phys. Rev. D* **15,** 409.

Press, W. H. & Thorne, K. S. (1972). *Ann. Rev. Astron. Astrophys.* **10,** 335.

Tsubono, K., Ohashi, M. and Hirakawa, H. (1986). *Jpn. J. Appl. Phys.* **25,** 622.

Weber, J. (1969). *Phys. Rev. Lett.* **22,** 1320.

10

Data analysis and algorithms for gravitational wave antennas

G. V. PALLOTTINO AND G. PIZZELLA

10.1 Introduction

The analysis of the sensitivity of a gravitational wave detector shows that the role of the algorithms used for the analysis of the data is comparable to that of the experimental apparatus.

In this chapter we shall discuss the response of an antenna to an incoming gravitational wave, deriving a mathematical model of the system, which is used for examining various algorithms of data analysis, including the optimum filter and the matched filter. This will be followed by a brief discussion of coincidence techniques, using two detectors.

10.2 The antenna response to a gravitational wave

A cylindrical resonant gravitational wave antenna (g.w.a.) is equivalent to a system of elementary oscillators with angular resonance frequencies

$$\omega_k = (2k+1)\frac{\pi v}{L} \quad (k = 0, 1, 2 \ldots), \tag{10.1}$$

where v is the sound velocity in the bar ($v = 5400$ m/s in Al at a temperature of 4.2 K) and L is the bar length. As customary we limit ourselves here to the fundamental mode oscillator with resonant angular frequency

$$\omega_0 = \frac{\pi v}{L}. \tag{10.2}$$

We now consider a polarized g.w. with metric perturbation $h(t)$ impinging on the bar perpendicularly. We indicate with $\eta(t)$ the displacement of the bar end face, equal to the displacement of our equivalent elementary oscillator. We have

$$\ddot{\eta} + 2\beta_1\dot{\eta} + \omega_0^2\eta = \frac{2}{\pi^2}L\ddot{h}, \tag{10.3}$$

where

$$2\beta_1 = \omega_0/Q, \tag{10.4}$$

243

Q being the merit factor (expressing the losses); the numerical factor on the right of equation (10.3) is obtained when solving the problem for the continuous bar (Pizzella, 1975).

Introducing the Fourier transforms $H(\omega)$ and $\eta(\omega)$ we obtain from equation (10.3)

$$\eta(\omega) = \frac{2}{\pi^2} L \frac{\omega^2 H(\omega)}{(\omega^2 - \omega_0^2) - 2i\omega\beta_1}. \tag{10.5}$$

The quantity

$$T(\omega) = \frac{2L}{\pi^2} \frac{\omega^2}{(\omega^2 - \omega_0^2) - 2i\omega\beta_1} \tag{10.6}$$

is the transfer function of our oscillator.

We get $\eta(t)$ by calculating the inverse Fourier transform

$$\eta(t) = \frac{1}{2\pi} \int_{-\infty}^{\infty} T(\omega)H(\omega)e^{i\omega t}\, d\omega. \tag{10.7}$$

In the cases that $H(\omega)$ have no poles, the poles of the integrand of equation (10.7) are

$$\omega_{1,2} \simeq i\beta_1 \pm \omega_0. \tag{10.8}$$

In most cases $H(\omega_1) = H(\omega_2)$. Then we get

$$\eta(t) = -\frac{2L}{\pi^2} e^{-\beta_1 t} H(\omega_1)\omega_0 \sin \omega_0 t. \tag{10.9}$$

It is interesting to consider the case of g.w. short bursts that can be expressed with a Dirac function $h(t) = H_0 \delta(t)$. In this case $H(\omega) = H_0$ independent of frequency.

Formula (10.9) expresses the two fundamental points of a resonant g.w.a.: (a) the antenna detects the Fourier component $H(\omega)$ of the metric perturbation at its resonance frequency; (b) the antenna has a 'memory' for a time of the order of β_1^{-1}. The last point allows the use of convenient algorithms for the data analysis. As far as point (a), if we want to determine the value of $h(t)$ from $H(\omega_0)$ we must make assumptions on the wave spectrum. The simplest one, very rough, is to consider a short burst with duration τ_g and obtain $h(t)$ from the measured $H(\omega_0)$ by means of

$$h(t) \simeq H(\omega_0) \Delta\omega \simeq \frac{H(\omega_0)}{\tau_g} \tag{10.10}$$

considering a bandwidth $\Delta\omega \sim 1/\tau_g$.

Examining equation (10.9) it is evident that the interesting part of the signal is the amplitude of the response at the antenna resonance frequency. There is no need to record $\eta(t)$ at all t values for determining its sinusoidal behavior, which would require, according to the Nyquist condition, a sampling time

$$\Delta t \leq \frac{1}{2\omega_0}, \tag{10.11}$$

that is $\Delta t \sim 10^{-4}$ s for the frequencies of interest.

To reduce the amount of recorded data one can eliminate the sinusoidal part at the ω_0 frequency, either by means of an appropriate computer program operating in real time, or by means of lock-ins. Here we shall consider the last case.

10.3   The basic block diagram and the wide band electronic noise

The antenna mechanical vibrations are transformed into an electrical signal by means of an electromechanical transducer (i.e. a piezoelectric ceramic, an inductive or capacitive diaphragm, etc.). See Pallottino and Pizzella (1981) for more details. We consider

$$V(t) = \alpha\eta(t), \tag{10.12}$$

where α is the transducer constant and $V(t)$ is the output voltage (this is for a voltage-wise signal; the treatment for a current-wise signal is similar).

This voltage is very small and needs to be amplified with a low-noise electronic amplifier. The amplifier noise, although small, is one of the most important limitations to the antenna sensitivity. The noise can be characterized by the voltage and current power noise spectra, V_n^2 and I_n^2, or by the following combinations of them

$$T_n = V_n I_n / k \tag{10.13}$$

$$R_n = V_n / I_n, \tag{10.14}$$

where k is the Boltzmann constant, T_n the amplifier noise temperature (V_n^2 and I_n^2 in bilateral form) and R_n the amplifier noise resistance.

Indicating by Z the output impedance of the transducer, we have the matching parameter between transducer and amplifier

$$\lambda = \frac{R_n}{|Z|}, \tag{10.15}$$

whose optimum value will be shown to be of the order of unity.

Another important quantity which will be used later is the ratio of the energy stored in the transducer to the energy stored in the bar of mass M. For a PZT or capacitive transducer, indicating the output capacity with C, we have

$$\beta = \frac{\frac{1}{2}C(\alpha\eta)^2}{\frac{1}{4}M\omega^2\eta^2} = \frac{2C\alpha^2}{M\omega^2} = \frac{2\alpha^2}{M\omega^3|Z|}, \tag{10.16}$$

where the last equality holds in general.

The voltage noise at the amplifier output is of wide-band type

$$S_0 = V_n^2 + I_n^2 |Z|^2 = V_n^2(1 + 1/\lambda^2) = kT_n |Z| (1/\lambda + \lambda). \qquad (10.17)$$

The voltage noise at the output is just V_n^2 only if the amplifier input is shorted ($|Z| = 0$).

The amplifier (with gain G) is followed by the lock-in driven by a synthesizer at ω_0. The lock-in is an electronic device that produces, at its output, the two following quantities:

$$x(t) = \frac{1}{t_0} \int_{-\infty}^{t} GV(t')e^{-(t-t')/t_0} \, \text{sign}[\cos \omega_0 t'] \, dt' \qquad (10.18)$$

$$y(t) = \frac{1}{t_0} \int_{-\infty}^{t} GV(t')e^{-(t-t')/t_0} \, \text{sign}[\sin \omega_0 t'] \, dt'. \qquad (10.19)$$

This is done in two steps. The first step (for $x(t)$) consists in multiplying $V(t)$ by sign[$\cos \omega_0 t$]; with the second step the signal is filtered by means of a RC filter with time constant t_0, which produces the exponential weighting, i.e. the integration.

To describe the operation of the lock-in amplifier mathematically, particularly in regard to the noise, we introduce the autocorrelation function of $V(t)$:

$$R_V(\tau) = \lim_{T \to \infty} \frac{1}{2T} \int_{-T}^{T} V(t) \cdot V(t + \tau) \, dt \equiv E\{V(t) \cdot V(t + \tau)\}$$

(where $E\{\cdots\}$ stands for expectation$\{\cdots\}$). The power spectrum $S_V(\omega)$ is the Fourier transform of $R_V(\tau)$:

$$S_V(\omega) = \int_{-\infty}^{\infty} R_V(\tau)e^{-i\omega\tau} \, d\tau. \qquad (10.20)$$

Expanding sign($\cos \omega_0 t$) in series:

$$\text{sign}(\cos \omega_0 t) = \frac{4}{\pi} \left[\cos \omega_0 t + \frac{1}{3} \cos 3\omega_0 t + \cdots \right], \qquad (10.21)$$

we calculate the autocorrelation of $V(t)(4/\pi) \cos \omega_0 t$

$$R'(\tau) = \frac{16}{\pi^2} \frac{1}{2} R_V(\tau) \cos \omega_0 \tau \qquad (10.22)$$

and the corresponding power spectrum

$$S'(\omega) = \int_{-\infty}^{\infty} R'(\tau)e^{-i\omega\tau} \, d\tau = \frac{4}{\pi^2} [S_V(\omega + \omega_0) + S_V(\omega - \omega_0)]. \qquad (10.23)$$

The additive property gives the power spectrum of $V(t)$ sign($\cos \omega_0 t$)

$$S''(\omega) = \frac{4}{\pi^2} \left[(S_V(\omega + \omega_0) + S_V(\omega - \omega_0) + \frac{1}{9} (S_V(\omega + 3\omega_0) + S_V(\omega - 3\omega_0)) + \cdots \right].$$

The contribution due to the higher order harmonics ($\pm 3\omega_0$ and higher) can be neglected because the amplifier preceding the lock-in has a bandwidth which strongly filters $V(t)$ at frequencies different from ω_0.

Thus, with a very good approximation, we consider $S'(\omega)$ given by equation (10.23) as the power spectrum of $V(t) \cdot \text{sign}[\cos \omega_0 t]$.

The RC filter of the second step has a transfer function

$$W_2(\omega) = \frac{\beta_2}{\beta_2 + \omega}, \tag{10.24}$$

where $\beta_2 = t_0^{-1}$. According to general theorems, the power spectrum of the signal at the lock-in output is

$$S_x(\omega) = S'(\omega) |W_2(\omega)|^2 = S'(\omega) \frac{\beta_2^2}{\beta_2^2 + \omega^2}. \tag{10.25}$$

Its inverse Fourier transform gives the autocorrelation of $x(t)$

$$R_{xx}(\tau) = \frac{1}{2\pi} \int_{-\infty}^{\infty} S_x(\omega) e^{i\omega\tau} \, d\omega. \tag{10.26}$$

Similar expressions hold also for the quantity $y(t)$. As already mentioned these expressions will be used for evaluating the noise.

As far as the effect of the lock-in on a signal of the type given by equation (10.9), considering the signal in phase with the synthesizer, $\beta_1 \ll \beta_2$ (which is easily verified), $t_0 \gg 2\pi/\omega_0$ and putting $V_s = (2L/\pi^2)(e^{-\beta_1 t})H(\omega_1)\omega_0$, we obtain

$$\left\{
\begin{aligned}
y(t) &= \frac{1}{t_0} \int_{-\infty}^{t} V_s \sin \omega_0 t' e^{-(t-t'/t_0)} \frac{4}{\pi} \sin \omega_0 t' \, dt' \simeq \frac{2V_s}{\pi} \\
x(t) &= \frac{1}{t_0} \int_{-\infty}^{t} V_s \sin \omega_0 t' e^{-(t-t'/t_0)} \frac{4}{\pi} \cos \omega_0 t' \, dt' \simeq 0
\end{aligned}
\right\}.$$

Thus, apart from the factor $2/\pi$ that is usually included in the lock-in amplification gain, the lock-in has extracted the amplitude of the signal at ω_0. If the signal is not in phase with the synthesizer both $y(t)$ and $x(t)$ are, in general, different from zero, such that $x^2 + y^2 = (2V_s/\pi)^2$.

The lock-in operation can be combined with the resonant bar transfer function to obtain a much simpler basic scheme as follows. We apply the result expressed by equation (10.23) to the bar transfer function (10.6), substituting in equation (10.6) ω with $\omega \pm \omega_0$

$$T(\omega \pm \omega_0) = \frac{2L}{\pi^2} \frac{(\omega \pm \omega_0)^2}{\omega(\omega \pm 2\omega_0) - 2i\beta_1(\omega \pm \omega_0)}.$$

For the cases of interest we have $\beta_1 \ll \omega_0$; thus the important range of ω is when $|\omega| \ll \omega_0$, where $T(\omega \pm \omega_0)$ becomes very large. In this region we have

$$T(\omega \pm \omega_0) \simeq \frac{\omega_0^2}{\pm 2\omega\omega_0 \mp 2i\beta_1\omega_0} = \frac{i\omega_0}{2\beta_1} \frac{\beta_1}{\beta_1 + i\omega}. \tag{10.27}$$

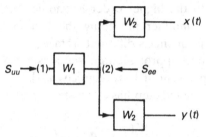

Figure 10.1. Basic scheme representing the mathematical model of an antenna and the associated instrumentation.

Therefore the bar becomes a low pass filter with time constant β_1^{-1}, which we represent by the transfer function:

$$W_1(\omega) = \frac{\beta_1}{\beta_1 + i\omega}. \qquad (10.27a)$$

The basic scheme, then, is shown in figure 10.1. At port 1 the g.w. excitation is applied together, as we shall see in the next section, with the Brownian noise. The signal is then integrated by the bar. The amplifier electronic noise is applied at port 2. The total signal is then integrated by the lock-in RC filter described by equation (10.24) providing the two quantities in quadrature $x(t)$ and $y(t)$.

In what follows, according to the block diagram of figure 10.1, we shall consider all signals as referred to the transducer. This means that all the amplifiers (including the lock-in) are assumed to have unity gain.

10.4   The narrow band noise and the total noise

There are two types of narrow band noise; one is of Brownian origin and is related to the bar thermodynamical temperature T, the other one is due to the electronic amplifier that heats up the mode of the bar at the resonance ω_0 (backaction) see Pallottino and Pizzella (1981) for more details. In both cases the noise can be represented by a white noise at the bar input, port 1 of figure 10.1. The effect of the low pass filter that represents the bar consists in narrowing the band of the noise at the bar output, producing a power spectrum of Lorentzian type peaked at zero frequency with bandwidth depending on the merit factor Q.

We indicate the white spectrum at port 1 by S_{uu}. At the bar output this noise becomes

$$S_{nb}(\omega) = S_{uu} \frac{\beta_1^2}{\beta_1^2 + \omega^2} \qquad (10.28)$$

(S_{uu} includes the factor $(i\omega_0/2\beta_1)^2$ of equation (10.27)). Its autocorrelation is

$$R_{nn}(\tau) = \frac{1}{2\pi} \int_{-\infty}^{\infty} S_{uu} \frac{\beta_1^2}{\beta_1^2 + \omega^2} e^{+i\omega t} \, d\omega = \frac{S_{uu}\beta_1}{2} e^{-\beta_1|\tau|}. \qquad (10.29)$$

The mean square value of the voltage noise is

$$V_{nb}^2 = R_{uu}(0) = \frac{S_{uu}\beta_1}{2}.$$ (10.30)

A direct way to obtain V_{nb}^2 for an elementary oscillator with mass m and transducer constant α is by means of the energy equipartition principle at the temperature T

$$\frac{1}{2}m\omega_0^2\overline{\eta^2} = \frac{1}{2}kT,$$

from which we get

$$\overline{V^2} = \alpha^2\overline{\eta^2} = \frac{\alpha^2 kT}{m\omega_0^2} = \beta\omega\,|Z|\,kT.$$

In our case the temperature T is increased because of the backaction of the electronic amplifier. For a voltage-wise signal the heating is generated by the current noise I_n^2 of the voltage amplifier (it would be due to the voltage noise V_n^2 of the current amplifier for a current-wise signal). It can be shown that this temperature increase is $4\alpha^2 I_n^2 Q/(m\omega_0^3)$. Thus we define an 'equivalent temperature'

$$T_e = T\left(1 + \frac{\alpha^2 I_n^2 Q}{2m\omega_0^3 kT}\right) = T\left(1 + \frac{\beta Q T_n}{2\lambda T}\right).$$ (10.31)

The overall mean square voltage due to narrow band noise is finally

$$V_{nb}^2 = \frac{S_{uu}\beta_1}{2} = \frac{\alpha^2 kT_e}{m\omega_0^2} = \beta\omega\,|Z|\,kT_e.$$ (10.32)

In this equation we have used the mass m of the equivalent elementary oscillator; this is also called the reduced mass of the antenna and it is one half of the real mass of the bar,

$$m = \frac{M}{2},$$ (10.33)

since the energy of the bar has the form $\frac{1}{4}M\omega_0^2\overline{\eta^2}$.

From equation (10.32) we obtain the white spectrum S_{uu} needed for representing the observed mean square voltage due to the Brownian and backaction noises

$$S_{uu} = \frac{2\alpha^2 kT_e}{m\omega_0^2\beta_1}.$$ (10.34)

The power spectrum at the lock-in output due to the noise S_{uu} will be

$$S'_{nb}(\omega) = S_{uu}\frac{\beta_1^2}{\beta_1^2 + \omega^2}\frac{\beta_2^2}{\beta_2^2 + \omega^2}.$$

The corresponding autocorrelation is found to be

$$R'_{zz}(\tau) = \frac{S_{uu}\beta_1\beta_2}{2} \frac{\beta_2 e^{-\beta_1|\tau|} - \beta_1 e^{-\beta_2|\tau|}}{\beta_2^2 - \beta_1^2} \qquad (10.35)$$

To obtain the total noise we must add the contribution of the wide band noise at port 2 due to the amplifier. This noise is given by equation (10.17). The lock-in, with its first step operation expressed by equation (10.23), produces a white noise with power spectrum

$$S_{ee} = 2S_0. \qquad (10.36)$$

At the lock-in output the noise spectrum is

$$S'_{ee}(\omega) = S_{ee} \frac{\beta_2^2}{\beta_2^2 + \omega^2},$$

and the corresponding autocorrelation is

$$R'_{ee}(\tau) = S_{ee} \frac{\beta_2}{2} e^{-\beta_2|\tau|}. \qquad (10.37)$$

The total noise, in conclusion, has autocorrelation

$$R_{nn}(\tau) = \frac{S_{uu}\beta_1}{2} \left[\frac{\beta_2}{\beta_2^2 - \beta_1^2} (\beta_2 e^{-\beta_1|\tau|} - \beta_1 e^{-\beta_2|\tau|}) + \Gamma e^{-\beta_2|\tau|} \right], \qquad (10.38)$$

where we have introduced the quantity

$$\Gamma = \frac{S_{ee}}{S_{uu}}, \qquad (10.39)$$

This quantity, as we shall see later, is of fundamental importance for the data analysis algorithms and for characterizing the antenna basic properties (sensitivity and bandwidth).

Other convenient expressions for Γ are obtained from equations (10.32), (10.16) and (10.17):

$$\Gamma = \frac{S_0\beta_1}{V_{nb}^2} = \frac{T_n}{2\beta Q T_c} (\lambda + 1/\lambda). \qquad (10.40)$$

10.5   Data filtering

The output data of a g.w.a., as they are provided by the instrumentation, namely the quantities $x(t)$ and $y(t)$ discussed in section 10.4, do not lend themselves to physical analysis, since the possible signals are deeply embedded in the background noise.

A preliminary filtering is therefore required, which is based on linear techniques, in order to improve as much as possible the signal to noise ratio

(SNR). This is done by exploiting the statistical properties of both the signal and the noise, as recognized by Hawking and Gibbons in their pioneering paper (Gibbons and Hawking, 1971).

Different filtering techniques, of course, are used according to the nature, mainly the duration, of the signals which stems from their physical origin. We can distinguish between short pulses or bursts, whose duration is much smaller than the main time constants of the detection apparatus, and long trains of periodical waves, which maintain coherence over appreciable time intervals.

Both of the above situations will be dealt with in this section, where we shall also consider the intermediate case of transient signals of relatively long duration.

10.5.1 Detection of short bursts

Here we only consider signals whose duration is smaller than both the decay time β_1^{-1} of the mechanical mode used for capturing the incoming gravitational radiation and the dominant characteristic time of the electronics, that is the time constant t_0 of the integration filter used in the lock-in amplifier. See Bonifazi et al. (1978) for more details. The latter is also important since a sampling time Δt of the same value is usually adopted, with values ranging from some tens of milliseconds to about a second.

Such signals are expected to impart a sudden variation to the energy status of the mechanical mode, which is to be compared with the spontaneous variations due to the various noise effects.

It can be shown that a short burst, independently of its actual shape, can be modeled as a delta function. Being of short duration, its spectral content will cover a frequency range more extended than the spectral observation window of the detection apparatus, where its contribution has the typical white spectrum of a delta function.

The main statistical properties of the background noise, as observed at the output of the antenna instrumentation, have been discussed in section 10.4. The overall spectrum of the noise can be viewed as the superimposition of two spectra: one representing the white noise of the electronics, which is band limited by the lock-in amplifier, the other one representing the narrow band thermal noise of the mechanical oscillator plus the backaction effect.

The power spectrum of the two components $x(t)$ and $y(t)$, that represents the output of the lock-in amplifier, can be written as:

$$S(\omega) = S_{uu} \frac{\beta_1^2}{\omega^2 + \beta_1^2} \frac{\beta_2^2}{\omega^2 + \beta_2^2} + S_{ee} \frac{\beta_2^2}{\omega^2 + \beta_2^2}, \tag{10.41}$$

where the spectra S_{uu} and S_{ee} have been defined in equations (10.34) and (10.36), and the characteristic angular frequencies, β_1 and β_2, have been defined in equations (10.4) and (10.24). Their ratio

$$\gamma = \beta_1/\beta_2 \tag{10.42}$$

is much smaller than unity.

A representation of the noise equivalent to the spectrum (10.41) is provided by its inverse Fourier transform, that is the autocorrelation function, as given by equation (10.38), which we repeat here for convenience:

$$R(\tau) = V_{nb}^2 \left[\frac{e^{-\beta_1|\tau|} - \gamma e^{-\beta_2|\tau|}}{1 - \gamma^2} + \frac{\Gamma}{\gamma} e^{-\beta_2|\tau|} \right]. \tag{10.43}$$

V_{nb}^2 is the narrow band noise variance at the transducer and Γ is the spectral ratio defined by equation (10.39). We notice that the main contribution is strongly correlated, with characteristic time $1/\beta_1$.

From equation (10.43) we obtain the variance of both $x(t)$ and $y(t)$:

$$R(0) = V_{nb}^2 \left(\frac{1}{1 + \gamma} + \frac{\Gamma}{\gamma} \right), \tag{10.44}$$

the latter term in the parentheses representing the effect of the wide band noise.

The probability density function of both $x(t)$ and $y(t)$ follows the normal law with zero mean and standard deviation $\sqrt{R(0)}$.

The energy status of the antenna mode in the observation bandwidth is described by the variable

$$r^2(t) = x^2(t) + y^2(t), \tag{10.45}$$

whose expected value can be expressed as

$$\overline{r^2} = E\{x^2(t) + y^2(t)\} = 2R(0) \tag{10.46}$$

in the case of the absence of any signal as well as of any disturbance of origin different from the noise considered above.

This quantity is distributed according to the Boltzmann function

$$F(r^2) = \frac{1}{\overline{r^2}} e^{-r^2/\overline{r^2}} \tag{10.47}$$

If we denote by V_s the amplitude (referred to the transducer) of the response to a possible gravitational signal that occurs at $t = 0$, we have (neglecting the noise):

$$r_s^2(t) = \frac{V_s^2 [e^{-\beta_1 t} - e^{-\beta_2 t}]^2}{(1 - \gamma)^2} \tag{10.48}$$

with maximum value

$$r_s^2 = K_d V_s^2 \tag{10.49}$$

where the coefficient K_d, usually slightly smaller than unity is given by

$$K_d = \left[\frac{\gamma^{\gamma/(1-\gamma)} - \gamma^{1/(1-\gamma)}}{1 - \gamma} \right]^2. \tag{10.50}$$

Here the signal-to-noise ratio is

$$SNR = K_d V_s^2 / 2R(0). \tag{10.51}$$

This can be considerably improved by detecting the variations of the energy status of the mechanical mode, since most of the noise, being strongly correlated, is not likely to contribute to fast variations of $r^2(t)$ as it does to $r^2(t)$ itself.

An algorithm that is able to detect such variations of energy, independently of the energy stored in the mode (as well as of the noise level), is the so-called zero-order prediction (ZOP) algorithm:

$$\rho_z^2(t) = [x(t) - x(t - \Delta t)]^2 + [y(t) - y(t - \Delta t)]^2, \qquad (10.52)$$

where Δt is the sampling time.

This is sensitive to variations both of the energy level of the mode and of its phase in the phase plane x, y (an incoming pulse may increase or decrease the initial energy, as well as only modify its status in the phase plane).

The noise variance σ_z^2 of the filtered variables $x_z(t) = x(t) - x(t - \Delta t)$; $y_z(t) = y(t) - y(t - \Delta t)$ can be expressed in terms of the autocorrelation, equation (10.43), as

$$\sigma_z^2 = 2R(0) - 2R(\Delta t), \qquad (10.53)$$

which shows the role of the correlation of the noise for reducing the variance.

The above expression can be written, using equation (10.43), as follows:

$$\sigma_z^2 = \frac{2V_{nb}^2}{e} \left[\frac{\gamma}{1 - \gamma^2} + \frac{\Gamma}{\gamma}(e - 1) \right]. \qquad (10.54)$$

We notice that the narrow band noise is reduced by the factor $\gamma = \beta_1/\beta_2$ (which can be rather small) while the wide band noise is increased by the inverse of the same factor.

The variance σ_z^2 can be minimized with a convenient choice of γ. With

$$\beta_2^{opt} = \frac{\beta_1}{\sqrt{(\Gamma(e - 1))}} \qquad (10.55)$$

the two contributions (narrow band and wide band noise, respectively) appearing in the right side of equation (10.54) are made equal and we have

$$(\sigma_z^2)^{min} = \frac{4V_{nb}^2}{e} \sqrt{(\Gamma(e - 1))}. \qquad (10.56)$$

The quantity ρ_z^2, due to the noise only, follows an exponential distribution, whose parameter $\overline{\rho_z^2}$ represents the expected value

$$\overline{\rho_z^2} = 2\sigma_z^2. \qquad (10.57)$$

Using this algorithm, the signal undergoes a reduction of amplitude, but to a much smaller extent than the noise. We have

$$\rho_{zs}^2 = K_z V_s^2, \qquad (10.58)$$

where the coefficient K_z, of the order of $1/e$, is given by

$$K_z = \left[\frac{e^{-\gamma} - e^{-1}}{1 - \gamma} \right]^2.$$

(10.59)

The signal to noise ratio, in the optimum condition ($\beta_2 = \beta_2^{\text{opt}}$), is:

$$(\text{SNR})_z = \rho_{zs}^2 / \overline{\rho_z^2} = \frac{K_z e V_s^2}{8 V_{\text{nb}}^2 \sqrt{(\Gamma(e-1))}}.$$

(10.60)

The above filtering method is both very powerful and very simple. One can, however, think of a more general method that uses the information provided by a number (larger than two) of data samples. The framework for the development of an optimum algorithm is provided by the Wiener–Kolmogoroff (WK) filtering theory (Papoulis, 1977), which will be applied, in what follows, to the estimation of the input acting on a g.w. detector, in terms of short impulses.

More specifically, the best linear estimate of the two orthogonal components of the input force is

$$\hat{u}_x = \int_{-\infty}^{+\infty} x(t - \alpha) w(\alpha) \, d\alpha$$

(10.61)

$$\hat{u}_y = \int_{-\infty}^{+\infty} y(t - \alpha) w(\alpha) \, d\alpha.$$

(10.62)

Here $w(t)$ is the weighting function, that is the impulse response, of the optimum WK filter, which is determined by minimizing the mean square deviation

$$\sigma_w^2 = E[(u_i(t) - \hat{u}_i(t))^2] \qquad i = x, y.$$

(10.63)

By applying the orthogonality principle of linear mean square estimation between the deviation and the observation (Papoulis, 1977), we have for the x component

$$E[(u_x(t) - \hat{u}_x(t)) x(t')] = 0, \qquad \forall t',$$

(10.64)

that is

$$R_{ux}(\tau) = \int_{-\infty}^{+\infty} R_{xx}(\tau - \alpha) w(\alpha) \, d\alpha, \qquad \forall \tau$$

(10.65)

where $\tau = t - t'$, and there is a similar expression for the y component.

The Fourier transform of the above provides the transfer function of the optimum filter to be applied to the data $x(t)$ and $y(t)$

$$W(i\omega) = \frac{S_{ux}(\omega)}{S_{xx}(\omega)}.$$

(10.66)

Here S_{xx} is the spectrum, equation (10.41), and S_{ux} is the cross spectrum of the signals $u_x(t)$ and $x(t)$, which can be expressed as follows in terms of the transfer functions W_1 and W_2 of the antenna and of the lock-in (see section 10.3), respectively:

$$S_{ux} = S_{uu}W_1^*W_2^* \tag{10.67}$$

Using equations (10.41) and (10.67) we obtain the following expression:

$$W(i\omega) = \frac{W_1^*W_2^*}{|W_1|^2|W_2|^2 + \Gamma|W_2|^2} = \frac{1}{W_1W_2}\frac{1}{1 + \Gamma/|W_1|^2}, \tag{10.68}$$

which shows that the optimum filter is composed of two parts: the first one is an inverse filter that cancels the dynamics of the apparatus (including any delay); the second provides the smoothing (band limiting) required for minimizing the effect of the wide band electronic noise.

By introducing the characteristic angular frequency of the smoothing section

$$\beta_3 = \beta_1\sqrt{(1 + 1/\Gamma)} \tag{10.69}$$

we have from equation (10.68):

$$W(i\omega) = \frac{\gamma}{\Gamma}\frac{(\beta_1 + i\omega)(\beta_2 + i\omega)}{\omega^2 + \beta_3^2}, \tag{10.70}$$

which clearly shows the non-causal nature of the filter (see equations (10.61) and (10.62)) which operates on the past as well as on the future data samples. This is not, of course, a problem, since the analysis is usually performed on data stored in a magnetic tape. A quasi real time filter can be also applied to the output data of a detector, which provides a good approximation of the desired estimate with a delay of the order of a few units of $1/\beta_3$, as required to operate on the 'future' data over a time range of this order, beyond which the weighting function $w(t)$ of the filter decays to negligible values.

The function $w(t)$ is obtained by performing the inverse Fourier transform of equation (10.70):

$$w(t) = \frac{\gamma}{\Gamma}\left[\delta(t) + \frac{(\beta_2 \pm \beta_3)(\beta_1 \pm \beta_3)e^{\pm\beta_3 t}}{2\beta_3}\right] \tag{10.71}$$

where the $-$ sign is for $t > 0$ and the $+$ sign for $t < 0$.

The output noise variance of the optimum filter, for each one of the two components, is:

$$\sigma_w^2 = S_{uu}\frac{\beta_1^2}{2\beta_3\Gamma} = V_{nb}^2\frac{\beta_1}{\beta_3\Gamma}. \tag{10.72}$$

Here the quantity considered for the analysis is

$$\rho_w^2(t) = \hat{u}_x^2(t) + \hat{u}_y^2(t), \tag{10.73}$$

which, in the absence of signal, has expected value

$$\overline{\rho_w^2} = 2\sigma_w^2 \tag{10.74}$$

and exponential distribution with parameter $\overline{\rho_w^2}$.

The response to an input delta function signal

$$S(t) = \frac{V_s \beta_1}{2\beta_3 \Gamma} e^{-\beta_3 |t|} \tag{10.75}$$

is only determined by the smoothing section of the optimum filter (due to the action of the inverse filter): an input delta function, without any delay, is converted into a two-sided exponential. The characteristic time $1/\beta_3$ expresses the available time resolution for optimum SNR. For signals well above the noise, however, one can perform the smoothing with a characteristic time smaller than $1/\beta_3$, thereby improving the time resolution while reducing the SNR.

We remark here that this analysis considers continuous time signals, while the data submitted to filtering are sequences of samples. Therefore β_3 cannot exceed $1/\Delta t$, that is β_2.

With the WK filter the signal to noise ratio is, from equations (10.72) and (10.75):

$$(\text{SNR})_w = \frac{S^2(0)}{2\sigma_w^2} = \frac{V_s^2}{8V_{nb}^2} \frac{1}{\sqrt{(\Gamma(1+\Gamma))}} . \tag{10.76}$$

The SNR improvement obtained with this filter can be expressed in terms of reduction of the temperature of the noise. If the noise temperature of the unfiltered data is T_e (as given by equation (10.31)) the noise temperature after the optimum filtering is

$$T_{\text{eff } w} = 4T_e \sqrt{(\Gamma(1+\Gamma))}. \tag{10.77}$$

The above expression clearly shows the essential role of the electronic noise (expressed by the spectral ratio Γ) for defining the temperature of the noise, that is the sensitivity of the detector for short bursts.

By substituting equations (10.31) and (10.40) into equation (10.77) we obtain the following expression:

$$T_{\text{eff } w} = 2T_n \sqrt{\left[\left(1 + \frac{1}{\lambda^2}\right)\left(1 + \frac{2\lambda T}{\beta Q T_n}\right)\right]}, \tag{10.77a}$$

which is of basic importance as regards the design and the optimization of a detector (Pallottino and Pizzella, 1981). From the above, two 'matching conditions' can be derived in order to approach the Giffard limit $2T_n$ (Giffard, 1976), dictated by the noise temperature of the electronic amplifier. The effects of the different filters as discussed above are illustrated in figure 10.2.

10.5.2   Detection of longer bursts

We come now to a brief discussion of the filtering techniques aimed at detecting bursts of longer duration, that is inputs extended over time intervals larger than one sampling period. These cannot, of course, be modeled as a delta function.

The optimum solution, in this case, is represented by the so called 'matched filter' (Whalen, 1971) that is a linear filter which performs on the observed data an operation similar to that of equations (10.61) and (10.62), aimed at detecting the occurrence of signals of known shape.

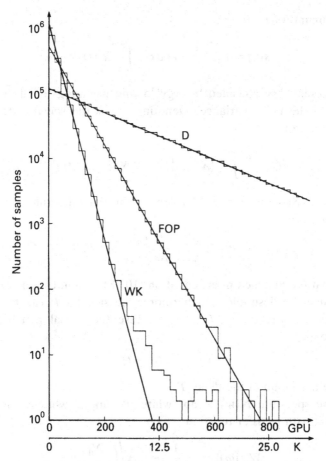

Figure 10.2. Frequency distributions of the experimental data of a small cryogenic antenna (operating at 7523 Hz). D represents the unfiltered quality r^2, FOP the data filtered with an algorithm similar to the ZOP, WK the data filtered with the Wiener–Kolmogoroff filter.

If a signal $x(t)$ with limited temporal support is embedded in white noise $n(t)$, it has been shown that the corresponding matched filter has impulse response equal to the time-reversed signal

$$w(t) = x(-t) \tag{10.78}$$

and transfer function

$$W(i\omega) = X^*(i\omega) \tag{10.79}$$

where $X(i\omega)$ is the Fourier transform of $x(t)$.

The output signal $s(t)$ of this filter is given by the following convolution integral:

$$s(t) = \int_{-\infty}^{+\infty} w(\tau)x(t-\tau)\,d\tau = \int_{-\infty}^{+\infty} x(-\tau)x(t-\tau)\,d\tau, \tag{10.80}$$

which is maximum for $t = 0$

$$s(0) = \int_{-\infty}^{-\infty} x^2(-\tau)\, d\tau = \int_{-\infty}^{+\infty} x^2(t)\, dt, \tag{10.81}$$

where it represents the so-called 'energy' of the incoming signal. This is to be compared with the noise variance. Denoting by N_0 the spectral density of the noise, the variance is

$$\sigma_n^2 = \frac{N_0}{2\pi} \int_{-\infty}^{+\infty} |X(i\omega)|^2\, dw = N_0 \int_{-\infty}^{+\infty} x^2(t)\, dt \tag{10.82}$$

where we have used the Parseval theorem. The corresponding signal to noise ratio (at $t = 0$) is:

$$\mathrm{SNR} = \frac{S^2(0)}{\sigma_n^2} = \frac{1}{N_0} \int_{-\infty}^{+\infty} x^2(t)\, dt. \tag{10.83}$$

We note that the matched filter, equation (10.78), is non-causal, while a causal realization might be desirable. Furthermore, the signal $x(t)$ can be assumed to vanish outside an interval $(0, T)$. In this case the causal matched filter has impulse response

$$w(\tau) = x(T - t) \tag{10.84}$$

providing the maximum at the delay T.

If the noise spectrum $N(w)$ is not white, it can be whitened by passing it through a filter with transfer function

$$|W_w(i\omega)| = \frac{1}{|W_N(i\omega)|} = \sqrt{\left(\frac{N_0}{N(\omega)}\right)} \tag{10.85}$$

This is the inverse filter corresponding to the filter $W_N(i\omega)$ that has shaped a white noise source with flat spectrum N_0 providing the actual noise with spectrum $N(\omega)$

$$N(\omega) = |W_N(i\omega)|^2 N_0. \tag{10.86}$$

At the output of the whitening filter, however, the shape of signal $x_w(t)$ is distorted: its Fourier transform is

$$X_w(i\omega) = X(i\omega) W_w(i\omega). \tag{10.87}$$

The corresponding matched filter, therefore, has transfer function

$$W_m(i\omega) = X^*(i\omega) W_w^*(i\omega) \tag{10.88}$$

and the overall filter is

$$W(i\omega) = W_w(i\omega) W_m(i\omega) = X^*(i\omega) \frac{N_0}{N(\omega)}. \tag{10.89}$$

The maximum signal to noise ratio, at $t = 0$, is given by the celebrated Dwork's formula (Dwork, 1950):

$$\text{SNR} = \frac{S^2(0)}{\sigma_n^2} = \frac{1}{2\pi} \int_{-\infty}^{+\infty} \frac{|X(i\omega)|^2}{N(\omega)} \, d\omega. \tag{10.90}$$

We remark that if the signal is a delta function (or the response to a delta function) the matched filter is formally equivalent to the Wiener–Kolmogoroff filter. This can be shown by letting $X(i\omega) = W_1(i\omega)W_2(i\omega)$, $N(\omega) = S_{uu}|W_1 W_2|^2 + S_{ee}|W_2|^2$ and substituting into equation (10.89); the result is equation (10.68).

The WK filtered data, as given by equations (10.61) and (10.62), can be directly applied to the detection of signals of duration longer than a sampling period Δt. One only has to perform on them a further filtering, matched to the specific shape of the signal considered.

In fact, if the input signal acting on the g.w.a. has a given shape $u(t)$, with Fourier transform $U(i\omega)$, and the observed signal is $x(t)$, with Fourier transform

$$X(i\omega) = U(i\omega)W_1(i\omega)W_2(i\omega) \tag{10.91}$$

the corresponding overall filter has transfer function:

$$W(i\omega) = \frac{U^*(i\omega)W_1^*(i\omega)W_2^*(i\omega)}{S_{uu}|W_1 W_2|^2 + S_{ee}|W_2|^2}; \tag{10.92}$$

If a Wiener filter with transfer function (10.68) has already been applied to the observation $x(t)$, what remains to be done is to apply to the data a filter with transfer function $U^*(i\omega)$, that is precisely a filter matched to the specific shape of the input.

The implementation of the above procedure, however, requires detailed knowledge of the input signal shapes for the two components $u_x(t)$ and $u_y(t)$ of the Rice representation ω_R of the actual input $u(t)$, where

$$u(t) = u_x(t) \cos \omega_R t + u_y(t) \sin \omega_R t, \tag{10.93}$$

as well as of the phase of the reference signal used in the lock-in.

This problem can be approached in two ways. First, one can apply the final matched filtering to the $\rho_z^2(t)$ quantity, instead of $x_w(t)$ and $y_w(t)$, at the price of a small decrease of the SNR. Here, of course, the template is matched to $u^2(t)$, rather than to $u(t)$. Second, one can use the 'direct sampling method', developed by S. Frasca (Frasca, Pallottino and Pizzella, 1986), that consists in processing the data sampled at the output of the amplifiers (without passing through the lock-in). The sampling theorem, in fact, which is usually applied for low pass signals, holds as well for bandpass signals of the same bandwidth.

10.5.3 Detection of periodic signals

We conclude this section by mentioning the detection problem for periodic signals (Pallottino and Pizzella, 1984). In this case, both the WK filter and the matched filter frameworks lead to the same intuitive result: one should filter the data over a vanishing bandwidth centered at the frequency of interest. Here the attainable SNR is maximum, for a given observation bandwidth Δv, if the frequency of the signal coincides with the resonance of the mechanical mode of the antenna, where the contribution of the narrow band noise is dominant (being $\Gamma \ll 1$).

The observation bandwidth depends on the duration of the measuring $t_m : \Delta v = 1/t_m$. The SNR, for a sinusoidal signal of amplitude V_s and angular frequency ω, is:

$$\text{SNR}(\omega) = \frac{V_s^2/2}{S_{uu}\left\{1 + \Gamma\left[\frac{\omega^2}{\omega_R^2} + Q^2\left(1 - \frac{\omega^2}{\omega_R^2}\right)\right]\right\}}. \tag{10.94}$$

This is maximum at ω_R and it is larger than $\text{SNR}(\omega_R)/2$ in the frequency range

$$\beta_3 \simeq \frac{\omega_R}{2\pi Q\sqrt{\Gamma}} \tag{10.95}$$

that is much larger than the mechanical bandwidth of the antenna.

This means that the WK filtered data can be used as well as for the detection of periodic signals by performing on them the required spectral analysis. The advantage over using the $x(t)$ and $y(t)$ data is twofold: (a) the noise variance of the data is smaller, (b) the amplitude of the possible spectral lines thus obtained is not to be corrected according to the Lorentzian shape of the mechanical frequency response curve of the antenna.

10.6   The cross-section and the antenna sensitivity

In order to detect g.w. it is necessary that the detector absorbs some energy from the wave. As a first step to estimate the cross-section let us consider the energy carried by the g.w. It can be shown that the energy carried per unit time across the unit area is given by

$$I(t) = \frac{c^3}{16\pi G}[\dot{h}_+(t)^2 + \dot{h}_\times(t)^2] \quad \text{J s}^{-1}\,\text{m}^{-2}, \tag{10.96}$$

where $h_+(t)$ and $h_\times(t)$ indicate the two polarization states of the g.w. The total energy per unit area is

$$I_0 = \int_{-\infty}^{\infty} I(t)\,dt \quad \text{J m}^{-2}. \tag{10.97}$$

For simplicity we consider one polarization state only, $h_\times(t)$ or $h_+(t)$, and indicate it by $h(t)$. We indicate by $H(\omega)$ the Fourier transform of $h(t)$,

$$H(\omega) = \int_{-\infty}^{\infty} h(t)e^{-i\omega t}\,dt. \tag{10.98}$$

From equation (10.96) we get

$$I_0 = \frac{c^3}{16\pi G} \int_{-\infty}^{\infty} |\omega H(\omega)|^2\,dv, \tag{10.99}$$

where $v = \omega/2\pi$ is the frequency. The quantity

$$f(\omega) = \frac{c^3}{16\pi G} |\omega H(\omega)|^2 \quad \text{J m}^{-2}\,\text{Hz}^{-1} \tag{10.100}$$

is called the 'spectral energy density', written here in bilateral form (frequencies from $-\infty$ to $+\infty$).

As a special case, we consider a g.w. burst of duration τ_g that can be described by a sinusoidal wave with angular frequency ω_0 and amplitude h_0 for $|t| < \tau_g/2$ and zero value for $|t| > \tau_g/2$. Since we have $H(\omega) \simeq h_0\tau_g/2$, from equation (10.100) we get

$$f(\omega) = \frac{c^3}{64\pi G} \omega_0^2 h_0^2 \tau_g^2 \quad \text{J m}^{-2}\,\text{Hz}^{-1}. \tag{10.101}$$

Another interesting case is a g.w. burst of the type $h(t) = h_0 e^{-\beta_w |t|} \cos \omega_0 t$. This wave has duration of the order of $\tau_g = 2/\beta_w$. If $\tau_g \ll 2\pi/\omega_0$ then the Fourier transform

$$H(\omega) \simeq \frac{h_0}{\omega_w} = \frac{h_0\tau_g}{2}$$

and we obtain again the result equation (10.101).

In order to obtain the total amount of energy per unit area we consider that the frequency bandwidth for a duration τ_g is $1/\tau_g$. Multiplying equation (10.101) for it we get

$$I_0 = f(\omega_0)/\tau_g = \frac{c^3}{64\pi G} \omega_0^2 h_0^2 \tau_g \quad \text{J m}^{-2}. \tag{10.102}$$

Finally an interesting case is also a g.w. burst of a δ-type, $h(t) = H_0\delta(t)$, $\delta(t)$ being the Dirac function. Since $H(\omega) = H_0$ we obtain from equation (10.100)

$$F(\omega) = \frac{c^3}{16\pi G} \omega^2 H_0^2 \quad \text{J m}^{-1}\,\text{Hz}^{-1}. \tag{10.103}$$

In order to compute the value of h_0 on the Earth due to a g.w. burst of duration τ_g that occurs at a distance R, indicating with $M_{GW}c^2$ the total g.w.

energy, we multiply equation (10.102) by $4\pi R^2$, we obtain

$$h_0 = \sqrt{\left(\frac{16GM_{GW}c^2}{c^3R^2\omega_0^2\tau_g}\right)} = 1.38 \times 10^{-17} \frac{1000\,\text{Hz}}{v} \frac{1000\,\text{pc}}{R} \sqrt{\left(\frac{M_{GW}}{10^{-3}M_\odot} \frac{10^{-3}\,\text{s}}{\tau_g}\right)}.$$

(10.104)

If we consider a sinusoidal g.w. of angular frequency ω with amplitude h_0 we obtain the average power per unit area from equation (10.96)

$$W_0 = \frac{c^3}{32\pi G}\,\omega^2 h_0^2 \quad \text{W m}^{-2}.$$

(10.105)

Indicating with W the total power irradiated by the source, at distance R we obtain

$$h_0 = \sqrt{\left(\frac{8G}{c^3}\frac{W}{R^2\omega^2}\right)} = 2.29 \times 10^{-41} \frac{1000\,\text{pc}}{R} \frac{1000\,\text{Hz}}{v}\sqrt{W}.$$

(10.106)

The cross-section Σ is defined such that, multiplied by the incident spectral energy density $f(\omega_0)$, it gives the energy deposited in the bar

$$\varepsilon = \Sigma f(\omega_0).$$

(10.107)

The energy ε is calculated from equation (10.9)

$$\varepsilon = \frac{1}{4}M\omega_0^2\left(\frac{2L}{\pi^2}\,\omega_0 H(\omega_0)\right)^2 = \frac{M\omega_0^2 H(\omega_0)v^2}{\pi^2},$$

(10.108)

where $v = \omega L/\pi$ is the sound velocity in the bar. Making use of equation (10.100) for the spectral energy density we obtain the cross-section

$$\Sigma = \frac{16}{\pi}\left(\frac{v}{c}\right)^2 \frac{G}{c}M \quad \text{m}^2\,\text{Hz}$$

(10.109)

in bilateral form.

We derive now another expression which relates $H(\omega_0)$, the quantity measured with a resonant g.w. antenna, to T_{eff} which expresses the sensitivity of the apparatus. From equations (10.108) and (10.77) we obtain

$$[H(\omega_0)]_{min} = \frac{L}{v^2}\sqrt{\left(\frac{kT_{eff}}{M}\right)} \quad \text{Hz}^{-1}.$$

(10.110)

In general if we detect an energy innovation $\Delta\varepsilon$, the corresponding value of $H(\omega_0)$ is

$$H(\omega_0) = \frac{L}{v^2}\sqrt{\left(\frac{\Delta\varepsilon}{M}\right)}.$$

(10.111)

It must be stressed that with a resonant antenna we measure $H(\omega_0)$ and *not* $h(t)$. If we want to have a feeling about possible values for $h(t)$ we must make assumptions about the $h(t)$ spectrum. For instance, for a flat spectrum from 0 to v_g, which we can think of as being due to a burst of duration $\tau_g \sim 1/v_g$, we can put

$$h(t) \simeq H(\omega_0)v_g = \frac{H(\omega_0)}{\tau_g}. \qquad (10.112)$$

10.7 Coincidence techniques

When examining the data of a g.w.a. we consider as candidate events the signals whose amplitude exceeds the background distribution.

We do not discuss here the statistical techniques that can be applied to establish the detection of an event, in terms of false alarm probability, false dismissal probability, and so on. We just note that any large amplitude signal, observed by an antenna, might be due to local disturbances, of various physical origin, which do not follow the statistical distribution of well behaved noise. It is, therefore, common practice to perform coincidence experiments with two or more g.w.a.s in order both to improve the detection statistics (by reducing the probability density of the background, thereby increasing the SNR) and to reduce drastically the effect of local disturbance.

The standard technique consists in using the two time series provided by two detectors (located, if possible, at great distance) to construct a new time series, where the value of each sample is the smallest of the corresponding samples of the two original time series. The basic idea is that a large g.w. excitation should give rise to large signals in both detectors at the same time. It is, of course, necessary to normalize the values of the two sequences, according to the sensitivity of the detectors, prior to the creation of the new sequence.

It can be easily shown that the probability distribution of the background of the new data is the product of the two distributions of the original data. This means that if both follow the Boltzmann distribution with parameters T_1 and T_2, the resulting parameter is:

$$T_c = \frac{T_1 T_2}{T_1 + T_2}, \qquad (10.113)$$

which means a reduction of the noise temperature to one half in the case of $T_1 = T_2$.

This analysis is usually performed by first thresholding at suitable levels the original data (thus reducing the amount of data to be processed) and then searching for possible events well above the background.

In order to take into account possible effects of non-stationarity of the background, the probability of the candidate events being due to chance is usually evaluated using 'local' distributions (obtained by performing the coincidence operation, only near the times of the events, on the two original sequences, one displaced with respect to the other by a small delay).

This procedure, however, improves the SNR considerably only if the responses of the two detectors to a given input signal have the same, or nearly the same, amplitudes. This is not the case, even for detectors operating at the same frequency and of similar sensitivity, if they have not the same orientation in space.

In order to take into account the effect of the geometrical response pattern of the detectors (this follows the law $\sin^4 \theta$, θ being the angle between the axis of the antenna and the direction of the source), as well as other effects, several methods of analysis have been introduced. One of them consists in normalizing the antenna responses with respect to the direction for a given source (as, for instance, the galactic center) before performing the coincidence operation. Each sample of the original time series is divided by the factor $\sin^4 \theta_i(t)$, where $\theta_1(t)$ and $\theta_2(t)$ are the angles between the axes of the two detectors and the direction of the source at time t.

Other methods, which do not require any assumption about the direction of the source, are based on different ways of combining the original data.

One of them consists in summing the samples of the two original time series (Pizzella, 1988). Here the idea is to use at best the signals available, which may differ not only because the detectors have different orientations but also because they operate at different frequencies, where the spectral content of a signal might be considerably different.

The meaning of the resulting sequence, as regards possible signals, is in terms of total energy recorded by the detectors involved. The obvious vulnerability to local disturbances of this method can be partially circumvented by accepting as candidate signals only the data whose ratio, in the original sequences, is contained in a preassigned range.

Another solution consists in creating a new data sequence by performing the product, rather than the sum, of the original time series. This has the additional advantage of being insensitive to possible calibration errors of the detectors. The noise rejection performance of this method is better than the sum method, while the capability of dealing with signals of different amplitude on the two detectors is inferior.

References

Bonifazi, P., Ferrari, V., Frasca, S., Pallottino, G. V. and Pizzella, G. (1978). *Nuovo Cimento C* **1C,** 465–87.

Dwork, B. M. (1950). *Proc. IRE* **38,** 771.

Frasca, S., Pallottino, G. V. and Pizzella, G. (1986). 'Spectral domain data analysis techniques for a gravitational wave antenna', in *Signal Processing III*, pp. 597–600, North Holland, Amsterdam.

Gibbons, G. W. and Hawking, S. W. (1971). *Phys. Rev. D* **4,** 2191–7.

Giffard, R. (1976). *Phys. Rev. D* **14,** 2478–86.

Pallottino, G. V. and Pizzella, G. (1981). *Nuovo Cimento* **4C,** 237–83.

Pallottino, G. V. and Pizzella, G. (1984). *Nuovo Cimento* **7C,** 155–68.

Papoulis, A. (1977). *Signal Analysis,* McGraw Hill, New York.

Pizzella, G. (1975). *Rivista Nuovo Cimento* **5,** 369–97.

Pizzella, G. (1988). *Nuovo Cimento* **102B,** 471.

Whalen, A. D. (1971). *Detection of Signals in Noise,* Academic Press, New York.

PART III

Laser interferometer antennas

11
A Michelson interferometer using delay lines

WALTER WINKLER

11.1 Principle of measurement

Gravitational waves manifest themselves as a variation of the metric of space-time. From an experimental point of view this can be considered as a time-dependent strain in space, which can be observed optically by registering the travel time of light between free test masses. Such experiments were first proposed by Gertsenshtein and Pustovoit (1963) and investigated in more detail by Weiss (1972) and Forward (1978). The corresponding arrangements are broadband in nature, as the effect of a gravitational wave onto the propagation of light *between* essentially free test masses is to be observed. No frequency is preferred, unless the storage time of the light inside the interferometer becomes comparable to the periods of the signals to be observed. Resonances of the test masses, for instance, are unwanted side-effects in this context. Since the strain in space introduced by gravitational waves has opposite signs in two directions perpendicular to each other, an ideal instrument is a Michelson interferometer (figure 11.1a). The signal at its output is a function of the path difference between the two arms. The beamsplitter and the mirrors serve as test masses. A gravitational wave with optimal polarization and direction of propagation would be incident perpendicularly on to the plane of the interferometer, making one arm shrink and the other one grow during half of a period; for the next half cycle the signs change. The response of the antenna to other polarizations and orientations can be described by a characteristic directivity pattern, see e.g. Forward (1978) and Schutz and Tinto (1987, 1989).

For optimum alignment of the interferometer the wavefronts of the two interfering beams are arranged to be parallel. In this case no spatial fringe patterns are to be seen at the two output ports: the intensity profile is that of the illuminating laser beam; but the power leaving through each of the two output ports is a function of the path difference (figure 11.1b). Variations in path difference can therefore be registered by measuring the corresponding variations of the output power with photodiodes.

The practical realization of the signal readout is done by means of a nulling method in order to be insensitive to power fluctuations of the illuminating laser

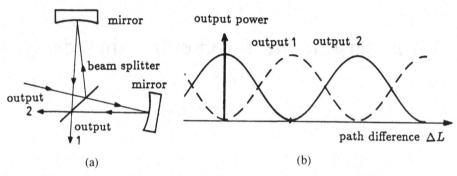

Figure 11.1. (a) Schematic diagram of a Michelson interferometer. (b) Light power in the two output ports 1 and 2 as a function of path difference ΔL in case of perfect interference.

beam. One possible technique uses a modulation scheme together with an optical servo loop that keeps the interferometer at the desired operating point, a dark fringe. For this purpose the beams in the two arms are sent through Pockels cells (Pockels cells use crystalline materials that change their index of refraction for light when an electric field is applied). These Pockels cells serve a twofold purpose: they are used to modulate the phase difference between the two interfering beams and to keep their path difference constant. At the modulation frequency there is a signal proportional to any (small) deviation from the operating point on the null fringe and this is used for measurement. This signal is demodulated and, after appropriate amplification, applied to the Pockels cells to compensate for deviations from the operating point. The voltage produced at the Pockels cells by the servo loop serves as the output signal of the interferometer. Since the voltage necessary to produce a path difference of one wavelength can be determined very accurately, the calibration of the set-up is easy and reliable. The output signal eventually is analyzed in the frequency window of interest between, say, a hundred hertz and several thousand hertz.

For the following text a few definitions should be given. The mirror separation of an optical delay line or of a Fabry–Perot resonator will be denoted by 'l'. Neglecting the small distance between the beamsplitter and the near mirror in one arm, this quantity is equal to the armlength of the interferometer. For N beams in each arm the total optical path L is related to l by $L = Nl$. Small and possibly fluctuating deviations of a quantity from its reference value are indicated by a 'δ', for instance 'δl', whereas a static and possibly very large deviation is denoted by 'Δ', for instance 'ΔL'.

11.2   Sensitivity limits

Since a gravitational wave introduces a strain $\delta L/L$ in space, the path difference δL to be measured increases with optical path L. Therefore one wants to make the light path as long as possible. An optimum is reached when the storage time of the light inside the interferometer is equal to half a period of the gravitational wave. For longer storage times one does not gain any more, since the gravitational wave changes sign during the travel time of the light, and part or all of the effect cancels. Taking 1 kHz as a typical signal frequency one ends up with optical path lengths in the order of 100 km.

Such long light paths – and even longer ones – can certainly be realized in outer space. Studies relevant to this have been carried out at Joint Institute for Laboratory Astrophysics, University of Colorado, Boulder (JILA) (Decher *et al.*, 1980).

In a ground-based experiment it is necessary for cost reasons to implement a multi-reflection scheme: each arm of the interferometer is traversed many times successively by the same light beam; for not too short pulses the effect on the phase of the beam accumulates in the same way as in an unfolded arrangement.

One possibility for a multi-reflection scheme is an optical delay line (Herriot, Kogelnik and Kompfner, 1964) proposed for use in gravitational wave detectors by Rainer Weiss (1972), or a Fabry–Perot cavity, first introduced in this context by R. W. P. Drever (Drever *et al.*, 1983a). The Fabry–Perot system will be described in one of the following chapters. Here we will deal mainly with optical delay lines. In this case the total light path L is well defined as Nl, where l denotes the mirror separation and N the number of beams. With an armlength of several kilometers less than a hundred beams are required to realize the path length of 100 km. The concept of the optical delay line will be discussed below in more detail.

The effective path length in an interferometer can be further increased by inserting a partially transmitting mirror at the output (Meers, 1988). This mirror and the interferometer together form a cavity, which can be tuned to a given frequency and a given bandwidth. In this way the interferometer can be optimized also for low frequency signals where otherwise the large number of reflections would require too big mirrors (see section 11.3.3).

The optical path length is one quantity which determines the sensitivity of a Michelson interferometer for strain measurements. The second quantity is the capability to resolve variations δL in path difference. If all relevant noise sources are suppressed sufficiently, the fundamental limit is set by Heisenberg's uncertainty relation. It is reached when the back action of the measuring process (in our case the variation of the test-mass positions due to the fluctuating light pressure) is of the same magnitude as the signal simulated by the photon statistics at the output of the interferometer. The photon statistics of the impinging light shows up in the photo current of the photodiode that monitors the output of the interferometer, if the quantum efficiency is close to 100%. In the present experiments the back action is negligible, and the photon counting error dominates. As the technique of squeezed states of light (see sections 15.5–15.8 of chapter 15) is not yet advanced enough, the present situation is that an interferometer can operate at best at the so-called shot-noise limit, corresponding to the Poisson statistics of the light leaving the interferometer. Poisson statistics leads to the well known $\sqrt{n}$ argument: within a given resolution time, on average n photons are counted with an uncertainty of $\sqrt{n}$. On the other hand a signal due to mirror displacements is proportional to n. Therefore the signal to noise ratio increases with the square root of the light power inside the interferometer, if the sensitivity is shot-noise limited.

As the measurement is performed within a given frequency window, for instance between one hundred and several thousand hertz, the resolution time is not the proper quantity to refer to. In this case fluctuations of a variable are better described in the frequency domain by a spectral density $S(f)$, defined as the squared deviations per unit bandwidth. To get a linear measure, it has become common use to take the square root of such a quantity; in the following this 'linear' spectral density shall be denoted by a tilde above the symbol of the variable. For example: the spectral density of fluctuations δL in path length is denoted by $S_{\delta L}(f)$; it has the units m²/Hz. The corresponding linear quantity is denoted by $\widetilde{\delta L}$ with the units m/$\sqrt{(\text{Hz})}$.

If the photo current shows only shot noise, the spectral density of the corresponding current fluctuations is given by the Schottky equation:

$$S_i(f) = 2ei, \tag{11.1}$$

where e is the charge of the electron and i the photo current. Assuming perfect interference and negligible noise contributions from other sources, these fluctuations simulate variations in phase difference φ between the two interfering beams with a spectral density of

$$S_\varphi(f) = \frac{2e}{i}, \tag{11.2}$$

or, expressed as fluctuations δL in optical path difference,

$$S_{\delta L}(f) = \frac{\hbar c}{\pi} \frac{\lambda}{\eta P}. \tag{11.3}$$

Here the relation

$$i = \frac{e\eta P}{\hbar \omega} = \frac{e\eta P\lambda}{2\pi\hbar c}$$

has been used, with $\hbar$ = Planck's constant/2π, c = speed of light, P = light power leaving the interferometer and η = quantum efficiency of the photodiode – usually about 50%. Note, however, that the effective quantum efficiency can be enhanced considerably if the light beam hits the photodiode at the Brewster angle and if the reflected light remaining is sent back onto the diode with mirrors.

To minimize $S_{\delta L}(f)$ one has to choose $\lambda/\eta P$ as small as possible. So far, mainly argon-ion lasers have been used, with $\lambda = 514.5$ nm and up to 5 W single mode output power. The efficiency of these lasers in transforming electrical power into light power is only on the order of 10^{-4}. Nd-YAG lasers may provide more light power at much better efficiency; but their fundamental wavelength is $\lambda = 1.064\ \mu$m. Using frequency doubling 10 to 50 W of single mode light power in the visible seems to be possible within a few years.

To increase the light power further, the beams of several lasers may be added coherently. A prerequisite for this scheme to work is that the lasers oscillate phase-locked at the same frequency. Preliminary work for this purpose has already been done with the technique of 'injection locking' (Kerr and Hough, 1989; Man and Brillet, 1984). This keyword describes the fact that a laser tends to oscillate at the frequency of some injected light, if that frequency is close enough to a possible eigenfrequency. The beam of a well stabilized master laser may be split up and sent into the resonators of several slave lasers, whose beams are eventually added in phase.

Finally, the light power traveling inside the interferometer can be increased by reusing the light that leaves the interferometer. This technique was first described by Drever (Drever *et al.*, 1983b; Hough *et al.*, 1984) and found independently by the Garching group as a variant of their setup of a 'second loop' for frequency stabilization (Billing *et al.*, 1983). The idea behind this so-called recycling technique is the following: the interferometer is operated at minimum power in the measurement output port. Almost all of the light therefore leaves through the other output port and may be used again by adding it coherently to the illuminating laser beam. From another point of view the beamsplitter for adding the output beam to the input beam can be considered as the coupling mirror of a Fabry–Perot cavity containing the Michelson interferometer. If the losses can be kept small, considerable light power may build up inside this cavity. The power enhancement relative to the illuminating laser beam is given by the inverse relative losses. Losses in this context are e.g. scattering and absorption, the finite interference contrast due to some mismatch between the wavefronts of interfering beams, and the fraction of light sent to the photodiode for the readout of the signal.

The feasibility of the recycling technique has already been shown by two groups, see, e.g., Maischberger *et al.* (1988) and Man *et al.*, (1989). In a small scale set-up built up in Garching the optical components have been suspended as pendulums. Because of their moderate optical quality, the power enhancement was only around a factor of ten, but encouragingly accompanied by almost the expected reduction in shot noise. To a great extent the enhancement factor was limited by losses and wavefront deformations caused by the Pockels cells. A much larger enhancement factor, about 50, was reached by the group in Orsay using a rigid arrangement with Pockels cells outside the recycling path.

From all these facts one can conclude that, with a straightforward extension of present technology, an effective light power of 500 W in the visible seems to be feasible. Together with a light path of 100 km this would allow a strain sensitivity at the 10^{-21} level:

$$\frac{\delta L}{L} = 1 \times 10^{-21} \sqrt{\left(\frac{500 \text{ W}}{\eta P} \frac{\Delta f}{1 \text{ kHz}}\right) \frac{100 \text{ km}}{L}}, \tag{11.4}$$

thus providing a realistic chance to see gravitational waves.

The possibility of reducing the photon counting error by the use of squeezed states was proposed by C. Caves (1981); see also chapter 15.

11.3   The optical delay line

Already in the early sixties Herriot, Kogelnik and Kompfner (1964) described the essential features of an optical delay line. Rainer Weiss from MIT proposed to implement such delay lines in a Michelson interferometer for the detection of gravitational radiation (Weiss, 1972).

11.3.1   A laser beam in an optical delay line
In the simplest case an optical delay line consists of two spherical mirrors with equal radii of curvature R (corresponding to a focal length $f = R/2$), facing each other at a separation l. The optical axis of the delay line is defined as the connecting line between the two centers of curvature. The light path of a beam in such an arrangement is equivalent to the light path of a beam through an infinite series of lenses, spaced at a distance l and each having a focal length of $f = R/2$.

For the description of the behavior of a light beam inside an optical delay line it is advantageous to consider separately the path of the beam as defined by the beam axis, and the beam shape given e.g. by the width and by the radius of curvature of the wavefront. Position and orientation of the beam axis after the subsequent reflections are described by geometrical optics, whereas the beam shape has to be treated by wave optics.

Let us first consider the position of the subsequent reflections on the mirrors. The optical axis is assumed to be the z-axis and the x- and y-directions are chosen such as to provide an orthogonal reference frame.

In a symmetric delay line the position of the reflection spots on the mirrors is given by

$$x_n = x_0 \sin(n\Theta + \alpha),$$
$$y_n = y_0 \sin(n\Theta + \beta). \tag{11.5}$$

x_0, y_0, α and β are given by position and orientation of the input beam. Θ is defined by the radius of curvature R of the mirrors and the mirror separation l:

$$\cos \Theta = 1 - \frac{l}{R}. \tag{11.6}$$

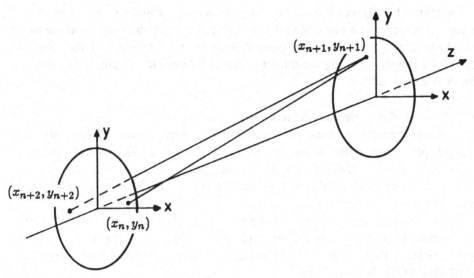

Figure 11.2. Light path in an optical delay line.

Equations (11.5) and (11.6) were first derived by Pierce in describing the path of an electron beam when guided by magnetic fields in an accelerator (Pierce, 1954).

For a proper choice of the input conditions $(y_0 = x_0, \alpha = \pi/2, \beta = 0)$ the reflections can be arranged on a circle:

$$x_n = x_0 \cos n\Theta,$$
$$y_n = x_0 \sin n\Theta. \tag{11.7}$$

Even n's describe the position of the beam at the 'near' mirror (including $n = 0$ for the coupling hole), and odd ones the position at the 'far' mirror. In this circular arrangement the meaning of Θ is illustrated very clearly: the successive reflections on the opposite mirrors are rotated against each other by Θ, the subsequent reflection on the same mirror by 2Θ.

For particular values of l/R an even multiple of Θ is equal to a multiple of 2π:

$$N\Theta = m \times 2\pi. \tag{11.8}$$

In this case the Nth beam falls into the coupling hole:

$$(x_N, y_N) = (x_0, y_0)$$

and the so-called 're-entrance condition' is fulfilled. This expression was introduced for the historical delay lines without a coupling hole. Since $(x_N, y_N) = (x_0, y_0)$ and $(x_{N+1}, y_{N+1}) = (x_1, y_1)$, the axis of the reflected beam coincides with that of the input beam. As one can see from equation (11.14), the two beams also match in shape.

Consequently the beam leaving the delay line in any respect behaves as if it were reflected from the back of the entrance mirror coating.

Equation (11.6) gives the simple relation between l and Θ. A variation in l leads in first order to a tangential displacement of the reflection spots on the circle in which they are arranged. A desired number N of beams can therefore be chosen by simply adjusting the mirror separation l until the re-entrance condition for N beams is fulfilled.

11.3.2   Imperfect spherical mirrors

For the considerations made above, the mirror surface was assumed to be a perfect sphere. In reality, however, there will be some deviations from the ideal shape. The effect of these deviations on the light propagation depends on their spatial wavelength across the surface and, of course, on their amplitude. There are three regions to be considered.

Surface irregularities with lateral dimensions smaller than the beam diameter are named 'micro-roughness'. They lead to scattering of the light. The relative loss at each reflection due to this effect is sometimes called 'total integrated scattering' (TIS). It is given by

$$\text{TIS} = (2h_{\text{rms}}k)^2 = \left(\frac{4\pi h_{\text{rms}}}{\lambda}\right)^2.$$

Present technology is able to produce impressively small micro-roughness, rms values in the order of 10^{-10} m, allowing reflection losses of less than 10^{-5} as far as this loss mechanism is concerned.

Surface irregularities with spatial wavelengths comparable to the beam diameter modify the beam shape and consequently deteriorate the interference quality. Relative deviations in the order of $\lambda/100$ between the interfering wavefronts, or differences between the beam diameters caused by an irregularity of the same magnitude at any of the reflection spots diminish an otherwise perfect visibility to 99.8%. This value reduces already the enhancement in light power that would be possible by recycling, if the low reflection losses provided by present technology were limiting. The rms-value of the surface irregularities with wavelengths comparable to the beam diameter has therefore to be kept below $\lambda/100$.

Finally there are the deviations of an ideal sphere with spatial wavelengths bigger than the beam diameter. Their main effect lies in a small tilt of the surface with respect to the ideal shape, leading to a slight misorientation of the reflected beam. One can show that in a 3 km set-up the tolerable misorientation of the surface at the different reflections is less than 10^{-7} radian, if statistically oriented (Winkler, 1983).

The simplest, almost unavoidable deviations from a symmetric delay line are:

(1)   the mirrors, though still spherical, have slightly different radii of curvature, or

(2)   there is some astigmatism, that is different curvature in the x- and y-directions.

(i)   Differently curved mirrors

A delay line consisting of mirrors with radii of curvature R_1 and R_2, respectively, can be described in the same way as a symmetric delay line, if only Θ is chosen according to

$$\cos^2 \Theta = \left(1 - \frac{l}{R_1}\right)\left(1 - \frac{l}{R_2}\right). \tag{11.9}$$

If R_1 and R_2 are very close to each other such that the geometric and the arithmetic mean between the two can be considered to be the same, say R, then the delay line can be treated as a symmetric one with mirrors having this radius of curvature R.

Despite the present advanced state of the art for producing high quality mirrors, one would have to tolerate a fairly large path difference in an interferometer with two mirror delay lines for the following reason: the mirrors have radii of curvature of the order of the mirror separation – in the planned large interferometer several kilometers – so they are almost flat. It is very difficult to manufacture such mirror surfaces with a small tolerance in curvature. With present technology a radius of curvature of 3 km can be produced with a tolerance of a few meters (Carl Zeiss, Oberkochen, private communication). Even for matched pairs of mirrors the average radii of curvature in the two arms are expected to differ by about one meter. The mirror separation has to be chosen to be different by that amount in order to fulfil the re-entrance condition separately in each arm (see equations (11.6) to (11.8)); otherwise the beams coming back to the beamsplitter will not superimpose properly or will even miss the coupling hole. Consequently, a path difference between 10 and 100 m is likely to occur.

There have been proposals to bend the mirrors properly for fine adjustment of the curvature by applying forces to the substrate. But this seems to be a difficult task as one has to avoid any mechanical resonance inside the frequency window of measurement, and, in addition, one has to maintain the high mechanical quality factor of the substrate.

Some of the problems arising from the deviations of the mirror surfaces with respect to the ideal sphere can be reduced by making position and orientation of the output beam adjustable. The simplest way is to use two different holes for the input and for the output beams – an arrangement which has been investigated theoretically at MIT and at the MPQ. Another possibility is a multi-mirror delay line. A short description can be found at the end of this chapter.

(ii)   Astigmatism

Astigmatic mirrors have different curvatures in different directions, and therefore the x- and y-coordinates of the successive reflections are described by different Θ-values. In general it is not possible to fulfil the re-entrance condition simultaneously for both coordinates. If, for instance, the re-entrance condition is fulfilled for the y-direction, that is $N\Theta_y = m2\pi$, then in general $N\Theta_x$ will not be a multiple of 2π. Thus a displacement in the x-direction remains which cannot be compensated for by changing the mirror separation or by some other kind of adjustment. The acceptable amount of astigmatism can be calculated once the demand on the quality of the interference is given quantitatively. A lateral displacement Δx of the interfering beams with respect to each other gives a relative minimum of

$$\frac{P_{\min}}{P_0} = \frac{1}{2}\left[1 - \exp\left(-\frac{(\Delta x)^2}{2w^2}\right)\right] \approx \left(\frac{\Delta x}{2w}\right)^2, \tag{11.10}$$

with P_0 the total light power. The beam radius w denotes that distance from the beam axis where the intensity is down to $1/e^2$.

(iii)   Local heating by absorption of light
From equation (11.4) it is clear that a high laser power is mandatory for the sensitivity required to observe gravitational waves. Even when squeezed states of light can be implemented to improve the sensitivity beyond the classical shot-noise limit, it is desirable to have as much light hitting the mirrors as possible, eventually many kilowatts of continuous light power.

In a real set-up this high light power causes severe problems because of absorption at the mirror surfaces, for instance at not perfectly oxidized metallic constituents inside the surface layers, or at adsorbed dust particles. The absorbed power locally heats the mirror, leading to a spatially non-uniform temperature profile. This temperature profile around the reflection spot is responsible for a non-uniform expansion of the mirror substrate. This expansion in first order changes the curvature of the mirror surface at the reflection spots. The magnitude of this effect can be estimated in the following way. Let us assume an almost confocal arrangement and a symmetric mode of the laser light (that is with equal beam diameter at the mirrors). The curvature depth h of the mirror measured across the beam diameter is in this case independent of the radius of curvature. It is given by

$$h = \frac{\lambda}{2\pi},$$

where λ is the wavelength of the light. In the usual case of a positive expansion coefficient the heating deforms the mirror surface locally by δh in the direction of making the mirror flat:

$$\delta h \approx \frac{\alpha}{2\pi\lambda_h} P_a, \tag{11}$$

where α is the linear expansion coefficient, λ_h is the thermal conductivity of the substrate material and P_a is the absorbed light power. An absorbed light power of $P_a = \lambda\lambda_h/\alpha$ will expand the mirror locally such that its original curvature is compensated; the mirror can be considered as being flat at this reflection spot. In the experimental set-up only much smaller deformations are tolerable. This means that the absorption loss at each reflection spot has to be kept below a relatively small fraction of a watt. This statement holds independently of the fact that the mode diameter increases with increasing mirror separation.

One can not expect the wavefront deformations in the two interferometer arms to be very much the same in order to get compensation; the absorption will at least partly be due to some unwanted irregularities such as dust deposits appearing at not well defined conditions. The mirror heating will therefore tend to degrade the interference quality and particularly the power enhancement achievable with recycling techniques.

To minimize the effect of mirror deformation by local heating the absorption has certainly to be kept as low as possible (in the past, in most cases the reflection losses were dominated by scattering and the greater part of the development went into the reduction of scattering; absorption was of minor importance). Further, the ratio α/λ_h should also be chosen as small as possible. Most of the materials that are possible candidates for mirror substrates because of their low internal damping have values for α/λ_h close to 10^{-7} m/W. Fused silica for instance has 3.5×10^{-7} m/W and sapphire 1.5×10^{-7} m/W. For delay lines the opaque material silicon with the best value of 5×10^{-8} can also be used.

In the context of local heating by the laser beam, delay lines have two other advantages over the Fabry–Perot systems. Since in resonators $N/2$ reflections are put on top of each other, the intensity there is higher; in an otherwise comparable set-up the mirror deformation is then also larger. Thus a given laser power may be prohibitive for resonators, whereas in delay lines it may only slightly change the local curvature of the mirrors at each reflection spot, as if mirrors with somewhat different radii of curvature were used, at least as far as the curvature of the wavefront is concerned.

In the case of a Fabry–Perot where the beam is transmitted through the locally heated substrate, there is in addition the so-called 'thermal lensing'. The temperature profile gives rise to a corresponding gradient in the index of refraction, leading to a wavefront deformation of the transmitted beam.

11.3.3 Mirror size

In contrast to a Fabry–Perot resonator, in the original two mirror delay line the different reflection spots are more or less separated from each other. For very long light paths one ends up with fairly large mirror diameters – certainly a disadvantage. For a calculation of the mirror size needed let us assume the arrangement to be not too far from confocal spacing. One starts out with the so-called matched case, where the wavefront of the beam at each reflection and the mirror surface have the same curvature. The beam shape is reproduced at each reflection; the reflection spots all have the same diameter

$$2w_{\mathrm{m}} = \sqrt{\left(\frac{4\lambda l}{\pi}\right)}. \tag{11.12}$$

For the assumptions made above the beam waist with a diameter $2w_m/\sqrt{2}$ is always in the middle between the mirrors.

The circular area assigned to each reflection has to be chosen with a diameter larger than $2w_m$, say by a factor S, in order to avoid truncation of the beam by the coupling hole or by the edge of the mirror. S is usually taken to be somewhere between two and three.

For a circular arrangement of the reflection spots with a matched beam the mirror diameter is given by

$$D_M = S\sqrt{\left(\frac{N\lambda L}{\pi^3}\right)} = 1.22\,\mathrm{m}\,\frac{S}{3}\,\sqrt{\left(\frac{N}{102}\frac{\lambda}{5\times 10^{-7}\,\mathrm{m}}\frac{L}{10^5\,\mathrm{m}}\right)},\qquad (11.13)$$

where, as usual, N denotes the number of beams and L is the total optical path. This mirror diameter is certainly too large, for at least two reasons: the vacuum tubes must have a diameter D_T which is larger than D_M, with the price for the tubes going up rather steeply for diameters of more than one meter; and secondly the eigenfrequencies of the mirror substrates come down into the frequency window of interest around 1 kHz, and this has to be avoided because of the thermal excitation of the eigenmodes (see section 11.5).

For a large number of reflections it helps to give up the condition of equal spot sizes. The beam is sent with a diameter $2w_e$ smaller than in the matched case through the coupling hole, which can then be made smaller as well:

$$2w_H = S2w_e \qquad (11.14)$$

In the case where the curvature of the wavefront of the input beam at the input mirror matches that of the mirror, and with the re-entrance condition fulfilled, the two spots neighboring the coupling hole have the same size. This can be deduced from equation (97) in Kogelnik (1965), describing the radius w_n of the nth reflection. For the conditions just mentioned, this equation reduces to

$$\left(\frac{w_n}{w_e}\right)^2 = \frac{1}{2}\left[1 + \left(\frac{w_m}{w_e}\right)^4\right] + \frac{1}{2}\left[1 - \left(\frac{w_m}{w_e}\right)^4\right]\cos 2n\Theta. \qquad (11.15)$$

In figure 11.3 the sizes of the reflection spots are drawn for $N = 46$, that is 23 spots per mirror. A reduction of the beam width inside the coupling hole leads to a corresponding widening of the spot size elsewhere:

$$w_e w_{max} = w_m^2. \qquad (11.16)$$

The distribution of the beam sizes as shown in figure 11.3 is the same at the near and at the far mirror. The minimum diameter of the mirrors can now be calculated if the following three conditions are taken into account:

(1)   no truncation of the input beam by the coupling hole,
(2)   separation of the neighboring spots from the coupling hole,
(3)   no truncation of the large reflection spots by the mirror edge.

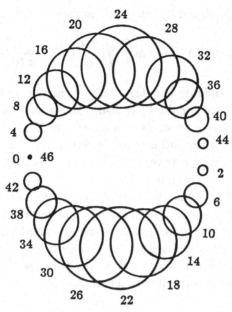

Figure 11.3. Distribution and size of the reflection spots at the delay line mirrors for a minimization of the mirror diameter.

As a result we get for the minimum mirror diameter

$$D_M = S \frac{2 + \sqrt{2}}{\pi} \sqrt{(\pi L)} = 0.74 \frac{S}{3} \sqrt{\left(\frac{\lambda}{5 \times 10^{-7} \text{ m}} \frac{L}{10^5 \text{ m}} \right)}. \qquad (11.17)$$

As equation (11.17) shows, in this case the minimum mirror diameter depends only on the total light path L, and not on the mirror separation or the number of beams. Thus, with a mirror diameter significantly less than one meter an optical path of 100 km can be realized, particularly if the safety factor S is chosen somewhat smaller than three.

To get the minimum diameter given in equation (11.17), the diameter of the input beam has to be chosen as

$$2w_e = \sqrt{\left(\frac{4\pi}{N} \right)} 2w_m = \sqrt{\left(\frac{16\lambda l}{N} \right)} = \frac{4}{N} \sqrt{(\lambda L)}. \qquad (11.18)$$

For the reduction in mirror size one has to pay with a certain overlap of reflection spots, leading to increased contributions of scattered light interfering with the main beam (see section 11.7 and figure 11.3). It is worthwhile to note that for $N \leq 40$ the diameter of the mirrors cannot be reduced by focussing the beam at the coupling hole; see equations (11.13) and (11.17).

The detailed calculation as well as a description of several other arrangements of the reflection spots can be found in Winkler (1983). There, for instance, it is also shown that the mirror diameter cannot be reduced drastically using a

Lissajous pattern, as produced by astigmatic mirrors:

$$D_M = S\frac{2+\sqrt{2}}{\pi}\sqrt[4]{\left(\frac{2}{N}\right)}\sqrt{(\lambda L)} = 0.53\,\mathrm{m}\,\frac{S}{3}\sqrt[4]{\left(\frac{126}{N}\right)}\sqrt{\left(\frac{\lambda}{5\times 10^{-7}\,\mathrm{m}}\frac{L}{10^{5}\,\mathrm{m}}\right)}. \quad (11.19)$$

A possible reduction would be significant only for numbers of beams much larger than the 30 to 40 planned for the first set-up. But a large number of reflections has to be avoided because of various noise effects, for instance the thermally driven motions of the mirror surface (see section 11.5), which appear multiplied by that number. In addition it is very difficult to produce astigmatic mirrors with the accuracy necessary for this purpose.

The simplest solution is therefore an arrangement as shown in figure 11.3; possibly with a second hole opposite to the first one, in this case at the position of reflex number 44. Here the beam may again be separated from its neighbors and could be sent back to the beamsplitter for recombination or reflected back into itself – an arrangement which doubles the light path ($N = 88$), and thus allows smaller mirrors for a given optical path length.

11.3.4   Misalignment and path length variations

In the case where the re-entrance condition is fulfilled, the optical path length inside the delay line is to a very good approximation given by Nl, where N is the number of beams and l the mirror separation, measured along the optical axis. The optical axis is defined as the connecting line through the centers of curvature of the two mirrors.

In 1972 Rainer Weiss described, for a symmetrical delay line, the variation in optical path length occurring as a consequence of a misalignment of one of the mirrors (Weiss, 1972). The effect of a small tilt by an angle ε is given by

$$\Delta L = -\frac{N}{2}\frac{R(l-R)}{2R-l}\varepsilon^2 \approx -\frac{N}{2}(l-R)\varepsilon^2, \quad (11.20)$$

and a lateral displacement by Δy leads to

$$\Delta L = -\frac{N}{2}\frac{(\Delta y)^2}{2R-l} \approx -\frac{N}{2}\frac{(\Delta y)^2}{R}. \quad (11.21)$$

As the signal is quadratic in the particular displacement, its magnitude should be very small. In reality, however, there will be certain deviations from ideal conditions, e.g. a slight violation of the re-entrance condition. As a consequence there is a linear but still very weak dependence of the path length L on motions in other degrees of freedom, see for instance Fattaccioli et al. (1986). Let us consider a particular example. A displacement of the far mirror by Δl in longitudinal direction leads to a lateral displacement of the output beam by Δy, tangentially to the circle of radius w_c formed by the reflection spots. In a nearly confocal arrangement Δy is related to Δl by

$$\Delta y = N w_c \frac{\Delta l}{R}. \tag{11.22}$$

Under these conditions a lateral displacement of the input beam by δy or δx leads to a change in path length of

$$\delta L = \frac{\Delta y}{l} \, \delta y,$$

or

$$\delta L = -\frac{\Delta y}{l} \, \delta x, \tag{11.23}$$

respectively. As the static displacement Δy of the output beam is allowed to be only a small fraction of the beam diameter – otherwise the beams of the two arms would not interfere properly – fluctuations in position of the input beam lead to very small fluctuations in optical path. In the planned large scale experiments the factor of reduction $\Delta y/l$ will be about 10^{-6} or even less. Similar relations hold for other kinds of misalignment of the delay line. Therefore cross coupling of motions in degrees of freedom other than longitudinal ones into the optical path length seems to be a minor problem – provided proper filters as developed by all experimental groups are implemented: mechanical filters to reduce the motion of the optical components and mode cleaners to reduce motions of the laser beam (see below).

11.4   Mechanical noise

The effect of a gravitational wave on a Michelson interferometer can be considered as a variation in mirror separation, under favorable conditions with different signs in the two arms. The difference in optical path length is exactly the quantity measured by a Michelson interferometer.

The most obvious noise source to spoil the measurement of optical path lengths is mechanical vibration that finds its way to the optical components. At frequencies f greater than a few hertz the 'natural' motion of the floor in a place with medium noise level, for instance the laboratory in Garching, can be described by the linear spectral density of the displacement

$$\widetilde{\delta z} \approx 10^{-7} \left(\frac{1 \, \text{Hz}}{f} \right)^2 \frac{\text{m}}{\sqrt{(\text{Hz})}}. \tag{11.24}$$

Most critical for the experiment is a displacement δl of the mirrors in longitudinal direction – the degree of freedom finally to be observed – as that displacement is multiplied by the number N of beams to give the variation in path length.

Horizontal vibrations therefore have to be reduced very effectively. This is done by suspending all optical components of the interferometer as pendulums. Above the pendulum frequency f_0 the transfer function is proportional to the inverse square of the frequency $\widetilde{\delta l}/\widetilde{\delta z} \propto (f_0/f)^2$ because of the inertia of the suspended mass. This relation is valid in the frequency range between the pendulum frequency and the eigenfrequency of the lowest transverse eigenmode of the suspension wire, the so-called violin string mode. Above that frequency a $1/f$ dependence takes over for the depth of the minima between the successive transverse eigenmodes. At the resonance frequencies the transfer function assumes large values, depending on the internal friction of the wire (Shoemaker et al., 1988). In most prototypes the length of the pendulum is chosen to be between 0.2 and 1 m, corresponding to an eigenfrequency f_0 around 1 Hz. Such a choice of the pendulum length does not allow the transverse eigenfrequencies to be above the frequency window of observation, because of the finite tension any material can withstand without breaking. For that reason in future set-ups the suspension wire may be designed to be as short as feasible.

As the sensitivity of the prototype interferometers improved with time, a single pendulum was no longer sufficient to reduce the mechanical noise, particularly at the lower end of the frequency window. In most prototype detectors an isolation of the suspension point is provided by alternating layers of metal and soft plastics – the well known acoustic filters as used already by Joseph Weber in his pioneering bar experiments.

In the set-up at Garching, however, all optical components are suspended from a heavy platform, which in turn is suspended with springs as a pendulum. It is important to choose the mass of this platform comparable to or bigger than the mass of the suspended components in order to get good filtering.

It is not sufficient to filter out only the horizontal vibrations. Due to finite cross coupling, motions in other degrees of freedom lead to interferometer signals as well. Vertical motions for instance produce spurious signals if the beamsplitter is slightly misaligned (see preceding section), if the beamsplitter or the Pockels cells inside the interferometer show a finite wedge, or the optical axis is not perfectly perpendicular to the vertical.

The latter case cannot be avoided in principle, since the vertical direction is defined as pointing towards the center of the earth, and that direction is different for mirrors separated by a large distance l.

Because of this type of cross coupling into the mirror separation the vertical motions have to be filtered out as well. In Garching this is done by suspending the platform not by wires but rather by coil springs. This provides simultaneously a good horizontal and vertical decoupling. The lower pendulums act also as vertical pendulums; the restoring force being provided by the elastically elongated wire. The corresponding vertical resonance frequency is close to 10 Hz. Therefore the vertical motion is also filtered with two stages, but less effectively because of the higher resonance frequencies.

Up to now this double pendulum design was sufficient to reduce the direct coupling of seismic or acoustic vibrations to the optical components, at least at frequencies above several 100 Hz. Further reduction of the shot noise or extension of the frequency window to lower frequencies may raise the necessity of further filtering – either by adding further pendulum stages (Del Fabbro *et al.*, 1987) or implementing active systems to suppress motions of the suspension point (see, for instance, chapters 13 and 14).

Even if all the optical components are well isolated, mechanical noise still may find its way into the interferometer signal via scattered light, if it is reflected at some moving object, e.g. the inner surface of the vacuum tubes, and eventually interferes with the main beam. Mechanisms of this type will be described in section 11.7.

11.5   Thermal mechanical noise

After the mechanical noise 'from outside' has been reduced sufficiently by filters, one is left with the thermal noise in the eigenmodes of the optical components. To describe these eigenmodes in closed form is very complicated, if not impossible. Fortunately there exist already numerical solutions for the eigenmodes of simple structures, for example for cylinders (Hutchinson, 1979).

An exact description of the light path has to take into account the distribution of the thermal amplitudes of the different modes across the surface of the mirrors, folded with the distribution of the reflection spots. Only the thickness modes show a synchronous motion of the entire reflecting surface, whereas in other modes parts of the surface move forward and others backward, thus elongating the path of some beams whereas others become shorter. Therefore for these modes part of the effect cancels out. In figure 11.8 (at the end of this chapter) the peak at 6.3 kHz is due to a thermally excited bending mode of the mirrors – their lowest resonance. A calculation shows that this peak is already reduced considerably due to the effect just mentioned.

For the planned long light paths the storage time of the light becomes comparable to the period of the lowest eigenmodes. Therefore the position of the differently moving parts of the surface cannot be considered to be sensed simultaneously by the laser beam. The real situation depends strongly on the shape of the mirror substrate, the arrangement of the reflection spots and the frequency range of observation.

For an estimate of the thermal amplitudes one can take the model of a damped harmonic oscillator driven by a stochastic force with a spectral density of (Weiss, 1972)

$$S_{\mathrm{F}}(f) = \frac{4kTM\omega_0}{Q}. \tag{11.25}$$

The resulting motion of the mass M can be described by the spectral density of the displacement δl

$$S_{\delta l}(f) = \frac{4kT}{MQ\omega_0^3} \frac{1}{\left[1 - \left(\frac{\omega}{\omega_0}\right)^2\right]^2 + \frac{1}{Q^2}\left(\frac{\omega}{\omega_0}\right)^2}. \tag{11.26}$$

Integration over all frequencies gives a mean value of the surface displacement of

$$\sqrt{(\langle \delta l^2 \rangle)} = \sqrt{\left(\frac{kT}{M\omega_0^2}\right)}, \tag{11.27}$$

as expected.

As an example let us take a resonance frequency $\omega_0/2\pi$ of, say, 1 kHz and a mass in the order of 100 kg. Then one gets from equation (11.27): $\sqrt{(\langle \delta l^2 \rangle)} = 10^{-15}$ m. This quantity has to be compared with the variation in mirror separation introduced by a gravitational wave: $\delta l_{GW} = \frac{1}{2}hl$. Assuming $h = 10^{-21}$ (in the more distant future even lower values are hoped to be detectable) and $l = 3 \times 10^3$ m, we get

$$\delta l_{GW} = \frac{1}{2} \times 10^{-21} \times 3 \times 10^3 \text{ m} = 1.5 \times 10^{-18} \text{ m}.$$

Thus the thermally induced amplitudes are more than three orders of magnitude too large, if they fall into the frequency range of interest.

Cooling of the components does not help so much, as the thermally induced amplitudes scale only with the square-root of the temperature. Besides, cold areas act as traps for residual gas molecules inside the vacuum system and the high quality optical surfaces would become contaminated. A more elegant solution is to shift all resonance frequencies out of the frequency window of observation. For this purpose one has, for instance, to avoid the usual mirror mounts, as the use of screws or clamps invariably introduces resonances in the kilohertz region. Certainly motions of the mirror surfaces are most critical, as their amplitudes appear in the signal output multiplied with the number of beams. But as the thermal amplitudes are so far above the tolerable level, also thermally driven eigenmodes of other components (beamsplitter, Pockels cells, etc.) produce spurious signals.

The best approach is to keep all components as simple and rigid as possible. In the Garching set-up the bare delay line mirrors and the beamsplitter have been suspended separately in a wire sling, similar to the suspension of the original Weber bars. The thickness of the mirror substrate was chosen big enough to keep the lowest resonance – a bending mode – slightly above 6 kHz (see figure 11.8). Even for the present sensitivity it turned out to be necessary to suspend also the other optical components independently.

In Glasgow the best results have been obtained by optically contacting the small Fabry–Perot mirrors to a rigid mass, thus forming a simple structure with resonance frequencies above 20 kHz.

Even if all resonances are shifted out of the frequency window of interest, the 'tails' of the resonances may lead to spurious signals as well. For a quantitative analysis equation (11.26) has to be considered in three frequency ranges: below, around and above the resonance frequency.

Well below the resonance, equation (11.26) reduces to

$$S_{\delta l}(f) = \frac{4kT}{MQ\omega_0^3}, \qquad \omega \ll \omega_0. \tag{11.28}$$

The mirror displacement δl has a white (frequency independent) noise spectrum.

Close to the resonance frequency the spectral density can be approximated by

$$S_{\delta l}(f) = \frac{4kTQ}{M\omega_0^3} \frac{1}{1 + 4Q^2\left(1 - \frac{\omega}{\omega_0}\right)^2}, \qquad \omega \approx \omega_0. \tag{11.29}$$

The higher the mechanical quality factor Q is, the better are the thermal motions concentrated in a narrow band around the resonance frequency.

Above resonance the spectral density is proportional to $1/Q$ just as in the subresonant case:

$$S_{\delta l}(f) = \frac{4kT\omega_0}{MQ\omega^4}, \qquad \omega \gg \omega_0. \tag{11.30}$$

The relations (11.28)–(11.30) tell us how to proceed for the construction of the system: any relevant resonance has to be shifted out of the frequency range of interest, and the mechanical quality factor has to be made high. This implies that all components must have low internal damping, and additional losses introduced by the suspension have to be carefully avoided.

To give some figures let us first consider the subresonant case, relevant for instance for the eigenmodes of the mirror substrate. Assuming a cubic shape, for the lowest thickness mode equation (11.28) can be rewritten as

$$S_{\delta l}(f) = \frac{4kT}{\pi^3 \rho v_s^3 Q}, \tag{11.31}$$

where ρ denotes the density and v_s the velocity of sound. For many applications a preferred material for the mirror substrates is Zerodur, a glass ceramic mixture with extremely low thermal expansion. Because of a slight hysteretic behavior in expansion after thermal cycling (Jacobs, Johnston and Hansen, 1984) and because of the low Q value of only about 1000 it cannot be used in a future installation. Fused silica is more stable in shape (Jacobs, Shough and Connors, 1984) and has a Q-value of 10^5. As four mirrors are involved, whose amplitudes have to be added quadratically, and as several eigenmodes contribute, equation (11.31) has to be multiplied roughly by a factor of 10 if it is compared with an apparent mirror displacement introduced by a gravitational wave. Including this correction factor and assuming fused silica as the substrate material, equation (11.31) gives

$\widetilde{\delta l}_{th} = 1 \times 10^{-20}$ m/$\sqrt{}$(Hz). A gravitational wave can only be seen if it dominates this value. The hope is finally to see – in some more distant future – strains of 10^{-22} in a bandwidth of 1 kHz, or, with a mirror separation of 3 km, a linear spectral density in the apparent mirror displacement of $\widetilde{\delta l}_{GW} = 1 \times 10^{-20}$ m/$\sqrt{}$(Hz). Thus, thermal motion is able to become a limiting factor at the 10^{-22} level for strain sensitivity, even with a mirror separation of 3 km.

There are monocrystals such as sapphire or silicon with Q values up to 10^8 (at room temperature), that are also very stable in shape. These materials could be used as mirror substrates – in delay lines even opaque ones are allowed, as the beam is transmitted through a coupling hole. So far only monocrystals of much smaller size than required for delay lines do exist, but several industrial companies are actively pursuing the possibility of producing mirror substrates with appropriate dimensions.

There is also at least one case in which the high frequency wing of a resonance is relevant, namely the pendulum mode of the suspension of the optical components. The corresponding thermal noise is closely related to the loss mechanism for the pendulum motion. Losses at the suspension point or even at stages further up are less critical, as the noise appearing there is filtered by the suspension wire until it reaches the optical component, for instance a mirror. But if the damping occurs at the suspended mass itself, for instance because of the friction between the moving mass and the residual gas molecules surrounding it, then this particular thermal noise is limiting at the 10^{-22} level up to 300 Hz for the following parameters: $\omega_0 = 2\pi \times 0.5$ Hz, $M = 200$ kg, $Q = 10^6$ and a mirror separation of 3 km. Certainly one has to proceed carefully in designing an interferometer for lower frequencies. As far as the pendulum mode of the suspended optical components is concerned, one has to provide a fairly good vacuum ($p \approx 10^{-8}$ mbar) and to build the suspensions with high $Q(\geq 10^8)$. Helpful may be the use of fibers of sapphire – the technology for their production exists already.

Finally one should mention the transverse modes of the suspension wire. Due to the conservation of momentum its motion δl_w will lead to a displacement δl of the suspended mass, which is reduced, however, at least by the mass ratio m/M of wire (m) and suspended mass (M).

The eigenfrequencies can be found approximately at

$$f_n = n\pi f_p \sqrt{\left(\frac{M}{m}\right)}, \qquad n = 1, 2, 3, \ldots, \tag{11.32}$$

where f_p is the pendulum frequency as defined by its effective length l_p and the gravitational acceleration g

$$f_p = \frac{1}{2\pi} \sqrt{\left(\frac{g}{l_p}\right)}. \tag{11.33}$$

In the Garching set-up the suspension is done with a wire sling; $l_p \approx 0.7\,\text{m}$, $f_p \approx 0.6\,\text{Hz}$. The transverse eigenfrequencies are at multiples of $220\,\text{Hz}$, and thus are just in the frequency range of interest.

But due to the high quality factor also for this degree of freedom the motion is again well concentrated around the resonance frequencies, and the corresponding signal could be cut out with notch filters. The wings of these resonances can be kept small enough. The low frequency wing of the lowest transverse mode for instance introduces a mirror motion of

$$S_{\delta l}(f) = \frac{m}{M^2} \frac{4kT}{Q\omega_0^3}.$$
(11.34)

Inserting typical values as $m = 10\,\text{g}$, $M = 200\,\text{kg}$, $Q \geq 10^5$ and $\omega_0/2\pi = 250\,\text{Hz}$ we get

$$\widetilde{\delta l} \geq 3 \times 10^{-21} \frac{\text{m}}{\sqrt{(\text{Hz})}},$$

a tolerable quantity.

As the transverse eigenfrequencies are inversely proportional to the pendulum length (see equations (11.32) and (11.33)), one could shift these resonances out of the relevant frequency window by use of a short suspension wire. But the corresponding increase in pendulum frequency aggravates the problems with the high frequency wing of the pendulum resonance at the low frequency end of the frequency window, see equation (11.30).

11.6   Laser noise and a Michelson interferometer with delay lines

The illuminating laser light shows three types of fluctuations: in power, in frequency and in geometry. A perfectly symmetrical interferometer would not be sensitive to any of these fluctuations; but because of the small asymmetries in a real set-up they give rise to spurious signals.

11.6.1   Power fluctuations

The power of large commercial lasers such as argon-ion lasers is fairly noisy up to frequencies of several megahertz. At still higher frequencies the laser power shows shot noise corresponding to the Poisson statistics of a coherent state. The modulation frequency mentioned in section 11.1 is therefore chosen to be within that frequency range. Low-frequency power fluctuations contribute to an interferometer signal only proportional to any deviation from the optimal point of operation, which is at minimum output power in the outport port that is used for measurement. A high open loop gain of the optical servo has therefore to be used to keep that deviation small. The residual deviation from the optimal operating point caused by a disturbance at a given frequency is determined by the deviation to be compensated by the servo, divided by the open loop gain at the frequency in question. The total deviation is obtained by integration over all frequencies.

Up to now the particular nulling method just mentioned was sufficient to reduce the influence of power fluctuations to below the present sensitivity level. If necessary the open loop gain could still be increased. At considerably higher sensitivity levels power fluctuations may lead to spurious signals via two further mechanisms: in combination with scattered light (see section 11.7), and if the fluctuating light pressure shakes the mirror or the beamsplitter differently in the two arms by a detectable amount due to some asymmetry. Certainly the laser power can be better stabilized than it is done by the manufacturers – up to now there was no need to do so.

Finally there are fluctuations in light pressure that are different in principle in the two arms. They are introduced by the vacuum fluctuations at the second input port. Probably they set the ultimate limits for the sensitivity (see Caves, 1981, and chapter 15, sections 15.5–15.8).

11.6.2   Frequency noise

Frequency noise gives rise to interferometer signals if there is a path difference between the two interfering beams or between the main beam and scattered light interfering with it (see section 11.7).

The effect of a static path difference ΔL between the two main beams is the following. The corresponding phase difference φ as measured by the interferometer is kept close to an integral multiple of 2π by the optical servo to maintain minimum output power:

$$\varphi = 2\pi \frac{\Delta L}{\lambda_L} = \omega_L \frac{\Delta L}{c}. \tag{11.35}$$

Variations in φ can be produced either by variations in ΔL – the quantity finally to be measured – or by a fluctuation of the laser frequency $\nu_L = \omega_L/2\pi$.

For a sufficient reduction of the latter noise contribution the frequency has to be stabilized to better than

$$\frac{\widetilde{\delta\nu}}{\nu} < \frac{\widetilde{\delta L}_{GW}}{\Delta L}, \tag{11.36}$$

where δL_{GW} is the variation in optical path produced by a gravitational wave.

In an interferometer with delay lines the static path difference ΔL is determined mainly by the difference between the average radii of curvature of the mirrors in the two delay lines and the number of beams (see section 11.3). Even for matched combinations of 'state of the art' mirrors one has to face a static path difference between ten and a hundred meters. At a sensitivity level for h of 10^{-21} the linear spectral density of the relative frequency fluctuations must then be kept below

$$\frac{\widetilde{\delta\nu}}{\nu} < \tilde{h}\frac{L}{\Delta L} = 3 \times 10^{-20} \frac{1}{\sqrt{(Hz)}}. \tag{11.37}$$

Around 1 kHz the frequency stability of the laser in the Garching prototype experiment is

$$\frac{\widetilde{\delta v}}{v} = 5 \times 10^{-18} \frac{1}{\sqrt{(\text{Hz})}} \, ,$$

sufficient for the present requirements there.

The higher stability level just mentioned has already been reached in Glasgow and at Caltech. The much longer optical path of the prototype experiments there served as a particularly sensitive reference for the laser frequency; in addition more light power has been used for the frequency stabilization loop.

Finally it is worthwhile to mention an effect that helps to stabilize the light frequency. In the case of recycling the light from the unused output port, a mirror of appropriate transmittance is inserted between the laser and the interferometer. In this way a Fabry–Perot cavity is formed containing the interferometer. Frequency fluctuations with periods shorter than the storage time of this cavity are reduced when entering the cavity.

In summarizing one can say that frequency fluctuations have to be taken very seriously. But it seems to be possible to stabilize the frequency to the required extent.

In this context one should note that the frequency stability quoted should not be taken as an absolute stability. Rather it is determined by the deviation of the laser frequency from the frequency defined by the length of the light path inside the interferometer. The error signal in the corresponding servo loop is to some extent a measure of this deviation.

11.6.3 Instabilities in beam geometry

A perfectly symmetric interferometer is insensitive to fluctuations in beam geometry, such as position, orientation, or shape of the beam. But a real experiment is never perfect. A simple example of asymmetry is a small angle α of misalignment of the beamsplitter with respect to the symmetry plane between the two near mirrors. Lateral displacements δy of the laser beam result in a variation in optical path difference of

$$\delta L = 4\alpha \, \delta y. \tag{11.38}$$

As an example the fluctuations in position of an Innova 90-5 laser beam are shown in figure 11.4. There are other types of argon lasers with much higher beam position noise. (Hough et al., 1984).

Within the frequency window of interest, the fluctuations δy shown in figure 11.4 are as large as $10^{-9} \text{m}/\sqrt{(\text{Hz})}$. To keep this noise source below a gravitational wave induced signal, α would have to be made smaller than 10^{-9} radian.

Such perfect alignment is very difficult to realize, especially for long periods of time. In addition, there are other kinds of geometrical instabilities of the laser beam producing spurious signals in combination with particular interferometer

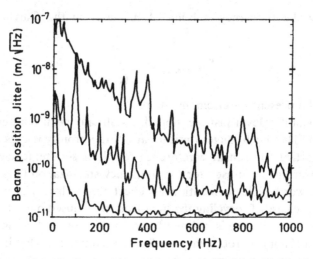

Figure 11.4. Fluctuations in the position of an Innova 90-5 laser beam, one of the two components measured with a quadrant diode. The diode was suspended in vacuum for isolation against mechanical vibrations. *Top curve*: beam directly from the laser. *Middle curve*: beam jitter suppressed by a particular Fabry–Perot mode cleaner. *Bottom curve*: beam jitter suppressed by a single mode glass fiber.

asymmetries, for instance a pulsation in beam width in connection with differently curved wave fronts of the two interfering beams.

As a remedy a mode cleaning device is inserted between the laser and the interferometer, either a mode selector (Rüdiger *et al.*, 1981) or a single mode glass fiber.

A mode selector is a special Fabry–Perot resonator with spherical mirrors in a non-confocal arrangement. The ratio between mirror separation and radius of curvature is chosen to allow resonance for the TEM_{00}-mode and the higher order modes simultaneously being out of resonance. The laser beam is matched to the TEM_{00}-mode of the mode selector; its power throughput is best if the mirrors have equal transmittance and low losses.

Variations in beam geometry can be described as admixtures of higher order modes of the mode cleaning device. In the case of a mode selector these modes are not in resonance – they are reflected. In the case of a single mode fiber the higher order modes are deflected laterally. The suppression of fluctuations in beam geometry is therefore determined by the capability of the mode cleaning device to reject higher modes.

The mode cleaner has to be isolated against mechanical vibrations – otherwise relative motions between the laser beam and the interferometer would again be introduced. As has been shown, the best solution is to suspend the whole fiber in vacuum. In Garching the output end of the glass fiber is rigidly connected to a suspended mass inside the vacuum chamber. For simpler handling, however, currently the input end is still situated at the laser table. In this case the shielding is not optimal, as that part of the fiber which is still under atmospheric pressure shows a slight, but up to now tolerable, microphonicity.

The admixture of higher modes describing the variations in beam geometry is very small. From figure 11.4 one can deduce amplitudes below $10^{-6}/\sqrt{(\text{Hz})}$ for the modes being responsible for lateral displacements. The amplitudes of higher modes have proven to be even smaller. The greater part of these time dependent contributions is removed by the mode cleaning device. As a consequence the transmitted beam should show power fluctuations. But these fluctuations are much smaller than the original technical noise of the laser power, whose influence on the interferometer signal is already minimized by the nulling method mentioned above.

11.7   Scattered light

In an interferometer capable of measuring variations in path differences smaller than 10^{-8} of a wavelength within one millisecond, very tiny effects usually buried in a variety of other noise contributions become important. Scattered light is responsible for such an effect.

11.7.1   Amplitudes of scattered light interfering with the main beam

If a laser beam hits a real surface or passes through an optical component, there are always some imperfections causing small fractions of the light beam to deviate from the light path as determined by undisturbed geometrical or wave optics. In this way part of the light originally contained in the gaussian beam is scattered into lateral tails whose intensity decreases with increasing angle of deviation. This happens to the beam before it enters the interferometer, as well as inside the interferometer, especially inside the delay lines. If such scattered light superimposes upon the main beam at some optical component, a small fraction of it is scattered back into the mode of the main beam. Interference then takes place between light components that possibly have a large path difference ΔL, or a correspondingly large phase difference Φ:

$$\Phi = \omega_L \frac{\Delta L}{c} = \omega_L \tau. \tag{11.39}$$

Let us denote the relative field amplitude interfering with the main beam by σ, then the phase shift φ of the resultant with respect to the original main beam is given by

$$\varphi = \sigma \sin \Phi. \tag{11.40}$$

See figure 11.5.

Variations in φ are sensed by superposition with the beam of the other arm. The resolution for measuring φ in the Garching experiment is close to the

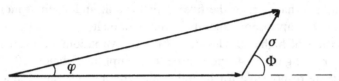

Figure 11.5. Phase shift φ of the main beam due to superposition with scattered light of relative amplitude σ.

shot-noise limit of 70 mA photo current, corresponding to

$$\widetilde{\delta\varphi} = \sqrt{\left(\frac{2e}{i_0}\right)} = 2 \times 10^{-9} \frac{\text{rad}}{\sqrt{(\text{Hz})}}.$$

Thus, for Φ fluctuating by radians, a relative power of scattered light of 10^{-17} – and even smaller fractions for improved sensitivity – could already be sufficient to limit the sensitivity!

For a complete treatment it is necessary first to determine the path difference ΔL and the relative field strength σ, then to describe the different mechanisms as to how scattered light produces spurious interferometer signals, and finally to point out possibilities for eliminating these noise sources.

For the measurement of σ and ΔL the laser frequency is increased linearly with time at a known rate $\dot{\nu}_L$. As a consequence Φ increases correspondingly, and $\tilde{\sigma}$ rotates with respect to the vector of the main beam at a frequency

$$f_r = \dot{\nu}_L \frac{\Delta L}{c}. \tag{11.41}$$

Thus a peak in the spectrum of the interferometer signal appearing at f_r immediately gives the path difference ΔL, and σ is equal to the amplitude of $\delta\varphi$. Figure 11.6 shows such a spectrum.

The spectrum of figure 11.6 was taken in the early Garching 30 m prototype with the reflection spots in the delay line well separated from each other. The strongest contributions of scattered light had a path difference of integer multiples of L, the total path length inside the delay line. They are produced in the following way: each reflection at the delay line mirrors adds some scattered light to the wings of the laser beam, thus forming a cone around the main beam. This cone is not focussed into the coupling hole, but rather hits the reflector surrounding it and starts a new round trip. It now forms a halo around the newly incoming beam and overlaps with it at subsequent reflections. Part of the scattered light is scattered back into the mode of the main beam and interference takes place. The corresponding path difference is L. The remaining cone is focussed again to the surrounding of the coupling hole and starts a further round trip giving rise to a contribution with path difference $2L$, and so on, until the power is sufficiently reduced by reflection losses. In this way scattered light components having a path difference of up to 10 km could be seen in the

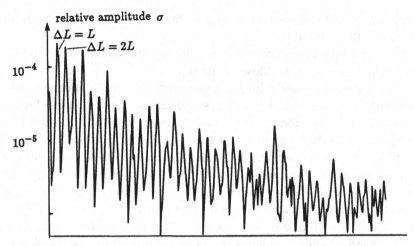

Figure 11.6. Spectrum of the interferometer output taken with linearly increasing laser frequency, showing peaks due to interference with scattered light.

Garching 3 m interferometer, despite the fact that the delay line mirrors had only a moderate reflectivity of 0.997.

The amplitudes of the different scattered light components can be calculated, if the scattering properties of the delay line mirrors are known. At small scattering angles ϑ the scattering behavior of the 3 m mirrors was measured to obey approximately the relation

$$\frac{\mathrm{d}P}{P} = \left(\frac{a_1}{\vartheta} + \frac{a_2}{\vartheta^2}\right) \frac{\mathrm{d}F}{R^2}, \tag{11.42}$$

where $\mathrm{d}P/P$ is the relative power picked up by a photodiode with an active area $\mathrm{d}F$ at a distance R from the scattering surface and ϑ is the deviation angle relative to the reflected beam; a_1 and a_2 were found to be 2×10^{-4} and 10^{-6}, respectively. Relation (11.42) is valid for ϑ between 5×10^{-5} rad and 10^{-1} rad (Winkler, 1983).

The amplitude σ_{mN}, that is the scattered light component with a path difference of $mNl = mL$ with respect to the main beam, is given by

$$\sigma_{mN} = \sqrt{(2)}\pi(N - 1)\rho^{mN}a_2\frac{w_m}{w_H}, \tag{11.43}$$

where ρ^2 denotes the reflectivity of the mirrors, $w_m = \sqrt{(\lambda l/\pi)}$ the mode radius and w_H the radius of the coupling hole. The contribution of a_1/ϑ in the first term on the right hand side of equation (11.42) is negligible; it describes the scattering at large scattering angles.

σ_{mN} is a function of the number N of beams, but is independent of the mirror separation, as w_H has to be chosen proportional to w_m. Thus, an extrapolation to the large interferometer planned is possible: assuming the same number of beams, the amplitudes of scattered light will be the same as in the existing prototypes, provided the mirror coating is the same. Typically, σ_N is found to be between 10^{-3} and 10^{-4}.

Mirrors with lower losses provide a smaller a_2, but the storage time of the scattered light is longer as ρ is closer to unity.

There is a second mechanism for the creation of scattered light components. In order to keep the mirror size in a long baseline set-up within tolerable limits (see section 11.3.3), the reflection spots will have to overlap partly. If a small fraction of the light in the overlap region is scattered into the mode of some neighboring beam, this leads to components having path differences equal to even multiples of the mirror separation. The crosstalk certainly is strongest between two nearby neighbors and decreases rapidly with distance for two reasons: the intensity of the beam profile diminishes steeply with distance from the beam axis and the angle necessary for backscattering becomes larger. Again the field strengths of these particular scattered light components can be calculated. They strongly depend on the degree of overlap between the reflection spots. For the present prototypes they are much weaker than σ_{mN}; but in an arrangement with more overlap they will become comparable.

Finally there are scattered light contributions coming from other surfaces besides the mirrors, for instance from the beamsplitter, the Pockels cells or the inner surfaces of the vacuum pipes. Up to now these contributions have been negligible, but especially scattering at the walls of the vacuum tubes may become a problem in the future, particularly at low frequencies of observation (see below).

11.7.2 Scattered light and spurious interferometer signals
There are several mechanisms via which scattered light may change the phase of the main beam in an interferometer, namely in connection with frequency fluctuations, nonlinearities, power fluctuations and mechanical noise.

(i) Frequency fluctuations and scattered light
The implications of frequency fluctuations in connection with scattered light became clear in 1978 when the Munich group started the operation of an interferometer with delay lines. The influence of scattered light possibly having very long path differences with respect to the main beam was described subsequently at the Texas Conference in Munich (Billing *et al.*, 1978). A treatment in more detail can be found in Billing *et al.* (1983) and in Winkler (1983). Quantitatively one can say: the phase shift (see figure 11.5) due to frequency fluctuations is given by

$$\widetilde{\delta\varphi} = \sigma \frac{\Delta L}{c} \cos \Phi \, \widetilde{\delta\omega}_{\text{L}}. \tag{11.44}$$

The cosine function may have any value between $+1$ and -1. For an estimate let us take $\cos \Phi = 1$. Then equation (11.44) can be rewritten as an equivalent strain in space as seen by the interferometer:

$$\frac{\overline{\delta L}}{L} \approx \sigma \frac{\overline{\delta v_L}}{v_L}, \tag{11.45}$$

where the path difference ΔL was assumed to have the typical value L. The simulated strain in space has to be smaller than the one induced by a gravitational wave, and therefore equation (11.45) tells what frequency stability is necessary to reduce this noise source sufficiently. For $\sigma = 10^{-3}$ it turns out to be of the same order of magnitude as that obtained in context with the static path difference.

It may be worth noting that the linearized relation (11.44) is valid only for differences in travel time $\tau = \Delta L/c$ that are smaller than the period of the frequency fluctuation in question. Otherwise the phases are obtained by integration. If, for instance, the laser frequency can be described by

$$\omega_L = \omega_0 + \Omega \sin \omega_M t \tag{11.46}$$

then the phase difference is given by

$$\Phi = \omega_0 \tau + 2 \frac{\Omega}{\omega_M} \sin \frac{\omega_M \tau}{2} \sin \omega_M \left(t - \frac{\tau}{2} \right). \tag{11.47}$$

(ii) Nonlinearities

If the phase difference $\Phi = \omega_L \Delta L/c$ between the scattered light vector $\vec{\sigma}$ and the main beam varies by more than 2π, $\vec{\sigma}$ rotates and thus modulates the phase φ of the resulting main beam at the frequency of rotation.

One reason for such large changes in Φ may be the usually rather large frequency excursions of the laser light at low frequencies. For the present effect they appear multiplied by the huge path difference of up to many times L, leading to interferometer signals in the whole frequency range up to many thousand hertz. Therefore at least a moderate degree of frequency stabilization is necessary also at low frequencies.

Another mechanism is the variation in path difference ΔL. If the individually suspended mirrors are damped only with respect to the surrounding, their relative motion at the pendulum frequency will be at least of the same magnitude as that of the floor, typically 10^{-7} m$_{rms}$. As scattered light components suffering many thousands of reflections inside the delay lines contribute to the interferometer signal (see above), the variation in path difference (at the pendulum frequency) is up to 1000 wavelengths – producing signals in the kilohertz region. Therefore a stabilization of the mirror separation has to be implemented.

In Garching the phase difference Φ is held constant by means of a second servo loop for the frequency stabilization (Shoemaker et al., 1986). Here, at low frequencies the laser frequency is taken as a reference and the mirrors are shifted correspondingly, whereas at higher frequencies the mirror separation serves as a reference for the laser frequency.

(iii)   Power fluctuations and scattered light

The direct influence of power fluctuations on the interferometer signal is eliminated sufficiently by a nulling method: for operation in the 'dark fringe' the field vectors of the two interfering beams are arranged to be antiparallel, and thus variations in their lengths do not change the condition for this interferometer output to stay in a minimum of output power. But the vector $\tilde{\sigma}$ of scattered light in general has arbitrary orientation with respect to the main beam. Because of the long path difference ΔL, there is interference of light components emitted by the laser at different times with $\Delta t = \tau = \Delta L/c$, thus their field vectors independently change in length. An upper limit for the corresponding phase shift $\delta\varphi$ of the main beam is given by

$$\delta\varphi \leq 2\sigma \frac{\delta E}{E} \sin(\pi f_e \tau), \qquad (11.48)$$

where $\delta E/E$ denotes the relative fluctuation of the field amplitude and f_e the frequency of the light fluctuation.

The currently used argon-ion lasers show relative power fluctuations of some $10^{-5}/\sqrt{(\text{Hz})}$ inside the frequency window in question. For the prototype experiments the effect described by relation (11.48) has been negligible, because the product $\sigma \cdot \sin(\pi f_e \tau)$ has been small enough. But in the planned large antennas a stabilization of the laser power by a moderate factor of about 100 may become necessary to reach the 10^{-21} level of sensitivity, as far as this noise source is concerned. This factor may even be smaller, if the recycling of light is realized. Again, just as it was the case for the frequency fluctuations, the cavity formed for recycling will also reduce the fluctuations of the light power inside the interferometer and thus further reduce the demands.

(iv)   Mechanical noise and scattered light

In an interferometer for the detection of gravitational waves the direct influence of mechanical noise on the optical path is eliminated by careful mechanical isolation of the optical components. But there is still another possibility for the production of spurious signals by mechanical vibrations.

The wings of the laser beam also hit the inner surface of the vacuum system, are reflected to a certain extent on the delay line mirrors and may be scattered back into the mode of the laser beam. In this way mechanical vibrations are able to produce interferometer signals even in the case of mechanically very well isolated optical components.

So far there has been no direct measurement of this effect, but several estimates show that it has to be taken seriously, see e.g. Leuchs *et al.* (1987). Let us assume the following conditions: typical motion of the floor of

$$\widetilde{\delta z} \approx 10^{-7} \left(\frac{1\,\text{Hz}}{f}\right)^2 \frac{\text{m}}{\sqrt{(\text{Hz})}}, \qquad (11.49)$$

scattered light produced only at the delay line mirrors (with scattering properties as of that implemented in Garching), scattering at tubes made of stainless steel or aluminium, and a mirror separation of the order of kilometers. The signal produced can then be estimated to be approximately

$$\frac{\widetilde{\delta L}}{L} \approx 10^{-20}\left(\frac{1\,\text{Hz}}{f}\right)^2 \frac{1}{\sqrt{(\text{Hz})}}.$$
(11.50)

This value seems to be tolerable, but one has to avoid strong scattered light components hitting the walls as well as big amplitudes of motion of the tubes, as caused for example by vacuum pumps. Certainly the effect can also be reduced by a sufficient number of properly shaped baffles, as considered in detail by Kip Thorne (1989). It is important to note that in a Fabry–Perot this particular noise source is expected to be of the same order of magnitude as in a delay line.

From the previous pages it may have become clear that scattered light has a considerable effect on measurements with interferometers at the high sensitivity levels envisaged. But certainly the problems can be overcome if

(a) high quality optical components are used to keep the power in the wings of the gaussian beam low,
(b) large vibrational amplitudes of the vacuum housing are avoided and properly shaped baffles to reduce scattered light are implemented,
(c) the laser is stabilized in frequency and power,
(d) the mirror separation is stabilized with respect to the wavelength of the laser light.

Further, another possibility to reduce scattered light effects has been proposed and demonstrated. The phase of the laser light can be modulated such that the coherence between scattered light and the main beam is destroyed. The idea behind this is closely related to that described above for measurement of scattered light components: if, for instance, the laser frequency is increased linearly with time, the scattered light vector rotates with respect to the main beam. If this rotation is fast enough, there is no contribution inside the frequency window of measurement (unless there is some noise at the frequency of rotation, which is then down-converted to lower frequencies).

But the dynamic range for shifting the laser frequency is limited. Therefore a modulation with limited swing has to be chosen. Several modulating functions for this purpose have been discussed: sinusoidal (Schilling et al., 1981), step functions (Schnupp et al., 1985), and several forms of white noise at MIT.

The implementation of such modulation schemes has several disadvantages: the reduction of scattered light effects is only possible to a certain degree (for instance if the path difference between main beam and scattered light is very small); the tolerable path difference between the two arms is fairly small; and it is very difficult to implement recycling with such heavily modulated light.

11.8   Multi-mirror delay line

One big advantage of delay lines of the Herriot–Kogelnik–Kompfner type is given by the fact that once the re-entrance condition is fulfilled, the position of the output beam is to first order independent of tilts or lateral displacements of the far mirror. The output beam behaves as if it were reflected from the back of the input mirror. Therefore the delay line is easy to align and to handle. However, a precondition for this behavior is a very good surface shape of the mirrors. Any deviation in the surface shape with respect to an ideal sphere in general leads to a deformation of the wavefront or to a displacement of the output beam, which cannot be compensated for by adjustment in order to get perfect interference. A typical example is a slight degree of astigmatism which real spherical mirrors are very likely to have. To fulfil the requirements for the full scale interferometers, the deviations from an ideal sphere have to be kept extremely small – well below $\lambda/100$.

It is very difficult to meet the demands in surface quality – either due to difficulties in producing the mirrors, or due to possible deformations of the surface with time, e.g. because of ageing processes inside the mirror substrate or because of non-uniform thermal expansion. The demands on the sphericity of the mirrors on scales bigger than the beam diameter can be reduced if more than two mirrors are used for each delay line – in the extreme case it might be useful to have one mirror for each reflection. Then the overlap of the two output beams and the relative orientation of the wavefronts at the beamsplitter for optimum interference are adjustable by the orientation of the last two mirrors in each arm. The requirements to the surface quality on scales comparable to or smaller than the beam diameter remain high, as they set limits to the interference quality and to the losses (see section 11.3.2).

There are several further advantages of multi-mirror delay lines. For instance, the thermal noise in the mirror substrates is less effective, as the surfaces of the separately suspended mirrors move independently and thus the thermal amplitudes add quadratically. If there are N mirrors, one mirror for each reflection, then the effective thermal amplitude increases proportionally to the square root of N, whereas in a two mirror delay line or in a Fabry–Perot the subsequent bounces on the same mirror see the same displacement, leading to an amplitude proportional to N. Further, as the mirrors in a multi-mirror delay line can be made smaller than in a two mirror delay line, the mechanical eigenfrequencies can easier be kept above the frequency range of observation.

For several reasons it is desirable to have equal path lengths in the two interferometer arms – for instance in order to reduce the coupling of frequency fluctuations to the interferometer signal via a static path difference. In a multi-mirror delay line this condition can be fulfilled, since the position of the output beam is no longer given by the ratio R/l as in the case of a two mirror delay line, but rather by adjustment of the mirror orientations. On the other hand

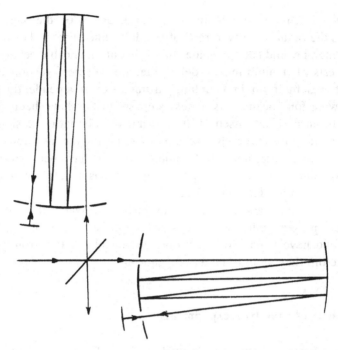

Figure 11.7. An interferometer with a modified version of the original two mirror delay line; the beam leaves through a second coupling hole and is reflected at a retro-mirror.

frequency fluctuations still produce spurious signals in connection with scattered light. The amplitudes of scattered light interfering with the main beam are strongly dependent on the overlap between the different reflection spots; they will be big in the situation defined by figure 11.7.

A simple and versatile version of a multi-mirror delay line that might be used for the full scale interferometer is shown in figure 11.7. The beam enters a two mirror delay line through a coupling hole in the near mirror and leaves after several bounces through a second hole in the same mirror. There it hits perpendicularly a retro-mirror and retraces its original path.

Particularly attractive in this case is the possibility of restoring the independence of position and orientation of the output beam from tilt and lateral motions of the far mirror. This feature of a Herriot two mirror delay line is usually destroyed by using more than two mirrors. In the present case it can be restored by chosing the radius of curvature R_r of the retro-mirror corresponding to

$$R_r = \frac{1 - \cos n\Theta}{1 - \cos n\Theta + \cos(n-1)\Theta - \cos\Theta}\, l. \qquad (11.51)$$

Hereby n is the number of the reflex that falls into the second coupling hole (input is number 0); l is the mirror separation and Θ was defined in equation (11.6).

For $n = N - 2$, that is $n = 44$ in figure 11.3, R_r is of the same order of magnitude as the radius of curvature R of the delay line mirrors. For other values of n, R_r in general would take on values that rule out practicable beam diameters.

For first tests of a multi-mirror delay line, the simplest version of such an arrangement was built up in Garching, using three independently suspended mirrors allowing four beams. As a next step six beams have been realized by sending the beam back into itself at the last mirror. The latter version simplifies the implementation of the recycling scheme by requiring the insertion of only one extra mirror between laser and interferometer. Results have been promising, and the shot-noise limit for 50 mA of photo current has almost been reached at frequencies between 400 Hz and 5 kHz.

Certainly, there are many different delay lines conceivable that might also be used for the present purpose. The ones described here are simple, well understood and have been proven to operate reliably in the prototype experiments. Therefore they seem to be a good choice.

11.9   Sensitivity of prototype experiments

As an example for the sensitivities reached so far, a plot taken from the output of the Garching prototype is shown in figure 11.8.

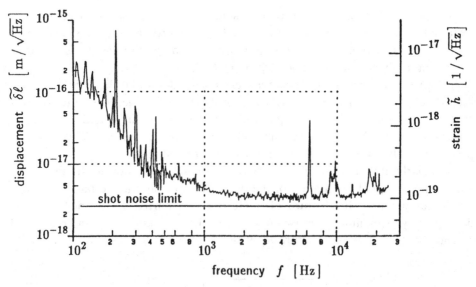

Figure 11.8. Spectrum of the noise at the output of the Garching interferometer. With a mirror separation of 30 m and with 90 beams an optical path of 2.7 km has been realized. The linear spectral density of a simulated mirror displacement $\tilde{\delta l}$ or a simulated strain in space $\tilde{h}$ is plotted versus frequency. For comparison the theoretical limit set by the shot noise of 70 mA of photo current is shown.

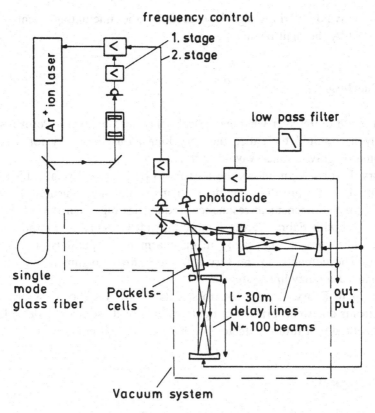

Figure 11.9. Block diagram of the Garching prototype interferometer.

Inside the frequency window of interest around 1 kHz the performance is close to the theoretical limit set by the shot noise of 250 mW light power available at the interferometer output.

A block diagram of the Garching prototype interferometer is shown in figure 11.9.

The light source is an argon-ion laser (Innova 90/5, Coherent Radiation). At low frequencies the laser frequency is stabilized with respect to a reference cavity (first stage) and at high frequencies, say above 10 Hz, with respect to the light path inside the interferometer (second stage, see also chapter 12). For suppression of fluctuations in beam geometry, the light is sent through a monomode glass fiber before it enters the interferometer. An optical servo-loop keeps the power in one of the two output ports at a minimum (see section 11.1). At frequencies above, say, 30 Hz the control elements are Pockels cells. At lower frequencies the positions of the two far mirrors are servoed with coil and magnet systems to keep the path difference constant. With a mirror separation of 30 m and 90 beams a light path of 2.7 km is realized. So far a vacuum of only 10^{-2} to 10^{-3} mbar was sufficient; the planned large interferometers will need a pressure below 10^{-6} mbar in order to avoid spurious signals caused by the fluctuating number of gas molecules hit by the light beam.

11.10   Conclusion

So far, it is not quite clear whether optical delay lines or Fabry–Perot resonators are finally best suited to realize the very long paths necessary for a full scale interferometric gravitational wave detector.

A Fabry–Perot system allows the mirrors to be relatively small; this eases the implementation of more than one detector inside the same vacuum housing and helps to keep the mechanical resonance frequencies high. On the other hand an interferometer with Fabry–Perots is more difficult to operate – in addition to the path-difference control, the cavity in each arm has separately to be kept on resonance. The demands to the servos for the mirror alignment are also much higher than for a delay line system.

Compared to Fabry–Perot resonators, delay lines also show less distortion of the wavefronts due to absorption of light power. For these reasons delay lines will be implemented in a future detector run by the German group.

Acknowledgement

I want to thank the Garching Gravitational Wave group, particularly Roland Schilling, for the good cooperation that was a precondition for preparing this article.

References

Billing, H., Maischberger, K., Rüdiger, A., Schilling, R., Schnupp, L. and Winkler, W. (1978). Ninth Texas Symposium on Relativistic Astrophysics, Munich 1978, Internal Report MPI-PAE/Astro 175.

Billing, H., Winkler, W., Schilling, R., Rüdiger, A., Maischberger, K. and Schnupp, L. (1983). In *Quantum Optics, Experimental Gravitation, and Measurement Theory*, eds. P. Meystre and M. O. Scully, pp. 525–66, Plenum Press, New York.

Caves, C. M. (1981). *Phys. Rev. D* **23**, 1693–708.

Decher, R., Randall, J. L., Bender, P. L. and Faller, J. E. (1980). *SPIE* **228**, Active Optical Devices and Applications, 149–53.

Del Fabbro, R., Di Virgilio, A., Giazotto, A., Kautzky, H., Montelatici, V. and Passuello, D. (1987). *Phys. Lett. A (Netherlands)* **124**, 253–7.

Drever, R. W. P., Ford, G. M., Hough, J., Kerr, I. M., Munley, A. J., Pugh, J. P., Robertson, N. A. and Ward, H. (1983a). *Ninth Intern. Conf. on General Relativity and Gravitation* (GR9), Jena 1980, ed. E. Schmutzer, pp. 265–7, Cambridge University Press.

Drever, R. W. P., Hough, J., Munley, A. J., Lee, S.-A., Spero, R., Whitcomb, S. E., Ward, H., Ford, G. M., Hereld, M., Robertson, N. A., Kerr, I., Pugh, J. R., Newton, G. P., Meers, B., Brooks, E. D. III and Gursel, Y. (1983b). In *Quantum Optics, Experimental Gravitation, and Measurement Theory*, eds. P. Meystre and M. O. Scully, p. 503, Plenum Press, New York.

Fattaccioli, D., Boulharts, A., Brillet, A. and Man, C. N. (1986). *J. Opt. (Paris)* **17**, 115–28.

Forward, R. L. (1978). *Phys. Rev. D* **17**, 379–90.

Gertsenshtein, M. E. and Pustovoit, V. I. (1963). *Sov. Phys.-JETP* **16**, 433–5.

Herriot, D., Kogelnik, H. and Kompfner, R. (1964). *Appl. Opt.* **3**, 523–6.

Hough, J., Hoggan, S., Kerr, G. A., Mangan, J. B., Meers, B. J., Newton, G. P., Robertson, N. A., Ward, H. and Drever, R. W. P. (1984). Journées Relativistes, Aussois, 1984; Proceedings in: 'Gravitation, Geometry, and Relativistic Physics', *Lecture Notes in Physics* **212**, 204–12.

Hutchinson, J. R. (1979). *ASME J. Appl. Mech.* **46**, 139–44.

Jacobs, S. F., Johnston, S. C. and Hansen, G. A. (1984). *Appl. Opt.* **23**, 3014–16.

Jacobs, S. F., Shough, D. and Connors, C. (1984). *Appl. Opt.* **23**, 4237–44.

Kerr, G. A. and Hough, J. (1989). *Appl. Phys. B*, in press.

Kogelnik, H. (1965). *Bell Syst. Tech. J.* **44**, 455–94.

Leuchs, G., Maischberger, K., Rüdiger, A., Schilling, R., Schnupp, L. and Winkler, W. (1987). June 1987, Internal Report MPQ 129.

Maischberger, K., Rüdiger, A., Schilling, R., Schnupp, L., Winkler, W. and Leuchs, G. (1988). In *Experimental Gravitational Physics*, pp. 316–21, World Scientific, Singapore.

Man, C. N. and Brillet, A. (1984). Journées Relativistes, Aussois, 1984; Proceedings in 'Gravitation, geometry, and relativistic physics', *Lecture Notes in Physics* **212**, 22–5.

Man, C. N., Brillet, A., Cregut, O., Marraud, A., Shoemaker, D., Vinet, J. Y., Boulanger, J. L., Bradaschia, C., Del Fabbro, R., Di Virgilio, A., Giazotto, A., Kautzky, H., Montelatici, V. and Passuello, D. (1989). *Proc. Fifth Marcel Grossman Meeting (Perth, 1988)*. eds. D. G. Blair and M. J. Buckingham, pp. 1787–1801, World Scientific, Singapore.

Meers, B. J. (1988). *Phys. Rev. D* **38**, 2317–26.

Pierce, J. R. (1954). *Theory and Design of Electron Beams*, Van Nostrand, New York.

Rüdiger, A., Schilling, R., Schnupp, L., Winkler, W., Billing, H. and Maischberger, K. (1981). *Optica Acta* **28**, 641–58.

Schilling, R., Schnupp, L., Winkler, W., Billing, H., Maischberger, K. and Rüdiger, A. (1981). *J. Phys. E: Sci. Instrum.* **14**, 65–70.

Schnupp, L., Winkler, W., Maischberger, K., Rüdiger, A. and Schilling, R. (1985). *J. Phys. E: Sci. Instrum.* **18**, 482–5.

Schutz, B. F. and Tinto, M. (1987). *Mon. Not. R. Astron. Soc. (GB)* **224**, 131–54.

Schutz, B. F. and Tinto, M. (1989). *Mon. Not. R. Astron. Soc. (GB)* (in press).

Shoemaker, D. H., Winkler, W., Maischberger, K., Rüdiger, A., Schilling, R. and Schnupp, L. (1986). *Fourth Marcel Grossmann Meeting, Rome* 1985, ed. R. Ruffini, pp. 605–14, Elsevier, Amsterdam.

Shoemaker, D., Schilling, R., Schnupp, L., Winkler, W., Maischberger, K. and Rüdiger, A. (1988). *Phys. Rev. D* **38**, 423–32.

Thorne, K. S. (1989). Caltech preprint GRP-220.

Weiss, R. (1972). Progress Report, Research Laboratory of Electronics, MIT **105**, 54–76.

Winkler, W. (1983). Dissertation, München, 1983, Internal Report MPQ 74.

12
Fabry–Perot cavity gravity-wave detectors

R. W. P. DREVER

12.1 Introduction

The gravitational wave detection technique discussed here is a long-baseline nearly-free-mass technique, devised initially with the aim of obtaining high gravity-wave sensitivity with minimum practicable cost. The distinctive part of the technique is the use as sensors of a pair of optical cavities formed between mirrors attached to test masses defining two perpendicular baselines, illuminated by an external laser source. To introduce the basic concept it may be useful to summarize the train of ideas which led up to it.

Experience and analyses in the early 1970s of resonant-bar gravity-wave detectors indicated that, although it is in principle possible to achieve by this technique the high sensitivity likely to be required for detection of expected astronomical sources, the small energy exchange with the gravitational wave leads to increasingly difficult experimental problems as sensitivity is improved. Alternative techniques using free test masses at large separations, monitored by optical or microwave methods, can sample much larger baselines and make relatively less serious any thermal, seismic, and amplifier noise, as well as the uncertainty-principle quantum limit for the test masses. Measurement of the small relative displacements involved, which might correspond at 1 kHz to strains of order one part in 10^{21} or less in a 1 kHz bandwidth, is however a serious challenge for interferometers of any kind. If a simple Michelson interferometer were used the photon shot noise limit would demand an impractically high light flux. One way of improving sensitivity was proposed by R. Weiss (1972): the use of an optical delay line to cause the beam in each arm of a Michelson interferometer to pass many times between the test masses, so that the changes in total optical path lengths are increased. This is an effective technique, but it does require large mirrors and vacuum pipes of correspondingly large diameter to accommodate the many folded beams. Further, early experimental work with optical delay lines at the Max Planck Institute in Munich, and with a White cell multireflection system at the University of Glasgow, showed up a scattering problem. Scattering of light from one reflection spot to another can give competing optical paths which are not identical in the two arms. This can give phase noise if the light wavelength fluctuates, unless special precautions are taken.

The 'Fabry–Perot' technique was conceived as a means of minimizing mirror and pipe diameters, and at the same time avoiding the scattering noise problem (Drever *et al.*, 1983a). The test masses incorporate relatively small mirrors, one of which is partially transmitting, arranged to cause the light to reflect many times between the same pair of spots and give a sharp resonance condition analogous to that in the classical Fabry–Perot etalon. The system is then very sensitive to changes in either wavelength or in separation between the mirrors. If two similar cavities are illuminated via a beamsplitter by light from a single source, then the effect of wavelength changes can be discriminated against, and the system made sensitive to differential changes in cavity length which may be caused by gravitational radiation.

An arrangement in which, in principle, this might be done is indicated schematically in figure 12.1. Long optical cavities are formed between suitably curved mirrors attached to each pair of test masses, spanning the baselines between them. Light from a laser passes via a beamsplitter to both cavities, and it

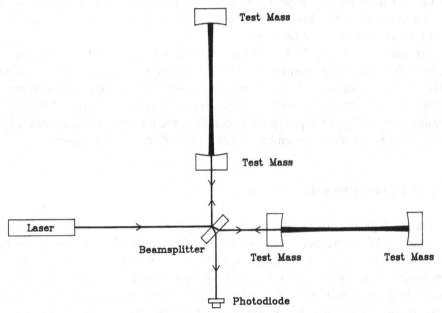

Figure 12.1. Simplified diagram to illustrate the principle of a basic Fabry–Perot cavity gravity-wave detector. With the cavities formed between each pair of test masses in resonance with the laser light, it is arranged that the output beams from the cavities arrive at the photodiode with almost exactly opposite phase to one another, giving near minimum intensity. A small differential change in the length of each cavity gives a change in phase of output light which is magnified by the number of reflections in the cavity, leading to a change in light intensity at the photodiode. (In this and following diagrams the thickness of lines representing light beams is intended to suggest the resonance modes in which light builds up inside the optical cavities, and should not be taken as a real representation of beam diameter or intensity.)

is arranged that resonance takes place in each cavity, giving effectively many superimposed bounces of the light between a single spot on each mirror. A change in length of either cavity causes a change in phase of the light leaving it. The length changes due to gravity waves would give differential phase changes in the two arms, which in principle may be measured with sensitivity similar to that of a delay-line system. However, the diameter required of the mirrors and the space required for the light beam is significantly less for the Fabry–Perot method.

For this technique to be practicable, it is necessary to have ways of obtaining a suitably stable light frequency, and of monitoring small changes in phase in the cavities. Techniques for carrying out these tasks were an important part of the original Fabry–Perot gravity-wave detector concept, and will be discussed below.

It may be noted here that if a sufficiently stable light source were available, a single optical cavity which monitored the distance between one pair of test masses might in principle detect gravity waves. Unfortunately at present there does not seem to be any light source which has adequate frequency stability over the relevant time scales, apart from a laser stabilized to a cavity similar to that used for the test mass measurement itself. The differential use of at least two similar cavities, oriented to be affected differently by the gravitational wave, is necessary in the current Fabry–Perot technique.

Since the basic Fabry–Perot detector system was conceived, several further methods for enhancing sensitivity, and for exploiting the small diameter of the cavity beams to make possible operation of several different interferometers within the same vacuum system, have been proposed (Drever, 1983). The possibility of making multiple interferometers share one set of beam pipes seems likely in the long term to be an important aspect of this optical system.

12.2   Principle of basic interferometer

To achieve high sensitivity in a Fabry–Perot cavity it is desirable to reflect the light back and forth between the mirrors for a time approaching the period of the gravitational wave; so such a cavity will have an extremely narrow optical bandwidth, typically less than 1 kHz. It is clear that the light source used to illuminate the system must have a bandwidth no larger than this, and a key part of the original idea for a Fabry–Perot gravity-wave detector was a new way of stabilizing the frequency of the light from a laser, to a cavity of this type (Drever et al., 1983b). Previous laser stabilizing systems usually involved operating the laser slightly down one side of the resonance peak in the response of a cavity, and then using changes in the intensity of light transmitted by the cavity as a measure of wavelength changes. The concept here involves measurement of phase difference – and not intensity – between light within the cavity and light from the laser, and use of this phase difference as a measure of deviation from resonance. A high gain servo system can use the phase signal to bring the laser into precise

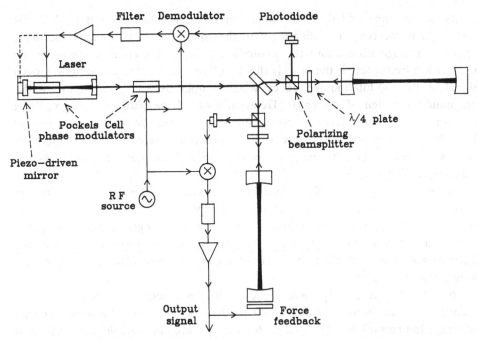

Figure 12.2. A more practical arrangement for a Fabry–Perot cavity gravity-wave detector, in which the laser wavelength is defined by the cavity on the right of the diagram, and the force required to maintain the second (lower) cavity in resonance with this is monitored. The diagram is simplified – auxiliary components are required to condition the beams and control the test masses in a real system.

resonance. A radiofrequency phase modulation technique can be employed to reduce effects of low frequency laser intensity noise; the principle is illustrated in a simplified diagram of one version of a Fabry–Perot gravity-wave interferometer system given in figure 12.2. In this diagram the optical cavity on the right-hand side is used to stabilize the wavelength of the laser. We discuss this part of the system first.

Plane polarized light from the laser is phase modulated by passage through a Pockels cell, at a frequency in the range of 10 to 20 MHz, and passes to a 50% beamsplitter. Half of the light goes to the right-hand optical cavity, through a polarizing beamsplitter and a quarter-wave plate. The axes of the quarter-wave plate are oriented at 45 degrees to the polarization of the input light, so that circularly polarized light enters the cavity. Light coming back from the input mirror of the cavity is circularly polarized in the opposite sense, is transformed into plane polarized light with polarization orthogonal to that of the input beam, and is reflected by the polarizing beamsplitter to the photodetector at the top of the diagram. The light arriving at this photodiode can be considered to have two components: the phase-modulated laser light directly reflected by the input cavity mirror; and light emerging from within the cavity – which has built up over the cavity storage time and thus has had its modulation sidebands removed. If the laser light is precisely in resonance with the cavity these two components have opposite average phase, and the photodiode output has no component at the modulation frequency. If the laser is slightly off resonance, the photodiode gives a signal at the modulation frequency whose amplitude and phase indicate the magnitude and sign of the error. Demodulation of the photodiode signal by a coherent demodulator gives a voltage signal which may be applied to a second Pockels cell within the laser cavity itself, so that the wavelength of the light from the laser is driven closer to the cavity resonance, and the laser becomes locked in wavelength to the cavity.

The bandwidth achievable with this stabilization system is not limited in principle by the storage time of the optical cavity, and a very high loop gain can be obtained with a suitable amplifier system. The wavelength of the laser light is then tightly locked to the spacing between the cavity mirrors, which in the diagram are shown attached to a pair of test masses which form one arm of a gravity-wave detector.

To complete the gravity-wave detector the same technique may be used a second time – but in this case it is employed to lock the length of an optical cavity spanning the second arm of the detector to the wavelength of the now-stabilized laser light. In figure 12.2 the light leaving the beamsplitter in the downwards direction enters the second optical cavity, and, as before, demodulation of the light coming back from the input cavity mirror gives a signal which measures the deviation from resonance. In this case the signal is amplified and used to apply an electrostatic force to the lower test mass, in a direction to bring the cavity closer to resonance. If the loop gains in the servo loops for both cavities are high, then the spacing between the two pairs of mirrors will be locked together, and the electrostatic force required to maintain this condition can become a measure of differential forces induced by gravitational radiation or other phenomena.

The interferometer configuration just described is in principle slightly less sensitive at its photon shot noise limit than the arrangement indicated in figure 12.1, since the output beams from the two cavities are not recombined together at the beamsplitter, and separate measurement of phase is made in each arm instead of a single differential measurement. With available mirrors in a laboratory-scale interferometer the difference is small, although in a large system the recombined arrangement may give a strain sensitivity which is better by a factor of nearly $2\sqrt{2}$, and it facilitates high power operation. In designs for practical versions of Fabry–Perot interferometers with recombined output beams, the laser may be stabilized to one of the main cavities, or to the average of both of them, using only a small part of the returned light; or a separate cavity to which the main arms are indirectly locked may be used as a primary reference for the laser stabilization.

The sensitivity of an interferometer of this type may be limited by many factors, but an important fundamental one comes from the fluctuations in light intensity detected by the photodiode arising from 'photon shot noise', the quantum fluctuations in the number of detected photons. For a given input light power, the photon shot noise sets a limit to the accuracy of measurement of optical phase difference, and thus to change in optical path within the cavities. A small motion of one of the test masses will cause a change in optical path which is increased by the number of times the light traverses the distance between the mirrors forming the cavity. Early in this work it was expected that losses in the mirrors would limit the number of reflections which could be used; and in this regime the sensitivity would improve with the effective number of reflections. However, the development of low-loss mirrors for other applications has led to the potential availability of multilayer dielectric mirrors having reflective losses less than one part in 10^4, and with such mirrors the total time that the light spends within the cavity becomes significant. If the effective storage time of the light is longer than the period, or the time scale, of the gravity wave there may be reversals in the mirror motion, and some resultant cancellation of output signal. The cavity behaves in some respects like an integrator, and as light storage time increases, displacement sensitivity improves up to a limiting value which is approached when the storage time matches the time scale of the gravity wave. In the configuration of figure 12.1, this limiting sensitivity, for detection at unity signal-to-noise ratio of a pulse of strain amplitude h, is given approximately by $h = [(\lambda f^3 \pi \hbar)/(I\eta c)]^{1/2}$.

Here λ is the wavelength of the light; f is the main frequency in the spectrum of the gravity-wave pulse: the amplitude h is assumed to be measured over a bandwidth equal to this; $\hbar = $ Planck's constant$/(2\pi)$; $I = $ input light power, $\eta = $ photodiode quantum efficiency; and $c = $ velocity of light. For $f = 1000$ Hz, $\lambda = 514$ nm, $I = 10$ W, and a photodiode of near unity quantum efficiency, this indicates a potential gravity-wave amplitude sensitivity of order 8×10^{-21}.

It may be noted that in this case the shot noise limit to sensitivity is independent of arm length – but in practice limits to sensitivity set by other factors, such as stochastic forces, thermal noise, or the uncertainty-principle quantum limit for the test masses, make a long baseline essential.

The diagrams shown are highly simplified, and in practice it is necessary to take precautions against many phenomena which may disturb sensitive optical measurements, such as lateral and angular motions of the laser beams, fluctuations in laser intensity, and spurious reflections and scattering between components. This makes the real system relatively complex. We will discuss some of these practical issues later; we concentrate on basic principles at this stage.

In an interferometer system in which the outputs from two arms are recombined at a beamsplitter, there are several advantages in adjusting the optical paths so that the output beam to the photodiode is near an intensity minimum – the photodiode is near a 'dark fringe'. It can be shown that if the measurement is limited by photon shot noise, and amplifier noise is unimportant, this gives in principle the maximum possible sensitivity. In practice a high frequency phase modulation technique may be employed to minimize effects of low frequency laser amplitude noise, and a feedback servo used to keep the interferometer centered on an intensity minimum. This arrangement is particularly useful when light power levels in the interferometer are high, for the photodiode has to handle only a small fraction of the total light flux, and a relatively small, high-sensitivity, photodiode may be used. With the photodiode near a dark fringe, most of the light from the interferometer passes out through the other side of the beamsplitter, in a direction towards the laser, and is available for other uses if required.

The potential sensitivity of this system is interesting, but higher performance would be desirable. Some improvement can be obtained by increasing the laser power, but there are practical limits to this. However, some further methods for improving sensitivity of interferometers of this type without increasing input power have been devised, and are likely to be important in large-scale gravity-wave detectors.

12.3   Enhancement of sensitivity by light recycling

When experiments with Fabry–Perot interferometers at Caltech showed that mirrors were potentially available which could give light storage times in laboratory-scale optical cavities comparable with the time scales of signals of interest, and could be expected to give much longer storage times in larger systems, it began to become evident that we might be entering a realm of new possibilities in optical experiments. The amount of light which has to be absorbed by a photodetector to give a sensitive null measurement is in principle very small, and much less than the circulating light power required in the interferometer arms. If scattering and dissipation can be kept small in the whole optical system there may be little loss of light in a period corresponding to the time scale of the gravity wave, and much of the light may still be present at the end of this measurement period. Thus there is a possibility for using most of the light again. In a suitable optical system the light may be 'recycled' many times, so that the light power effective for the measurement may be significantly larger than the output of the laser itself. The basic concepts (Drever, 1983) may be applied to both delay-line Michelson interferometers and to Fabry–Perot interferometers with beam recombining; we discuss only the latter case here.

With the basic interferometer system shown in figure 12.1, the main optical modification to achieve broadband recycling involves the introduction of an additional mirror between the laser and the beamsplitter to return light to the interferometer, with servo control to maintain correct phase of the recycled light. A schematic diagram of the resulting arrangement is given in figure 12.3. The recycling mirror forms what is effectively a large Fabry–Perot cavity encompassing the two main cavities and the beamsplitter, and its position and reflectivity are

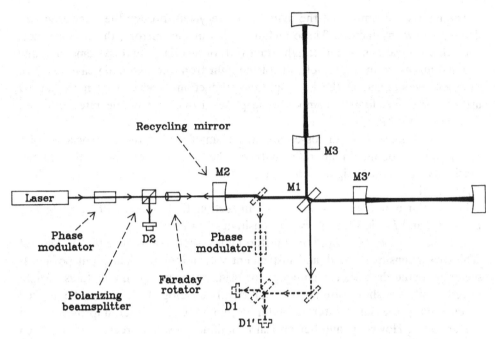

Figure 12.3. Basic arrangement of a Fabry–Perot interferometer with broadband light recycling, using mirror M2 to return light to the system. Effects of laser intensity noise may be reduced by applying high-frequency modulation: one way of doing this is shown by broken lines. A small part of the steady stored light in the recycling system is taken out to a side arm, phase modulated, and added to the differential output of the interferometer at an auxiliary beamsplitter. Phase changes in the main interferometer output can be determined from the signals from photodiodes D1 and D1'.

chosen to give maximum resonance build-up of light in the whole system. The method used to maintain resonance is similar to that described above for holding resonance between the laser in figure 12.2 and the cavity on the right-hand side of that figure. The laser beam is phase modulated at high frequency, and light returning from the recycling mirror is diverted to a photodiode D2, coherently demodulated, and the resulting phase error signal used to adjust either the laser frequency or the position of the recycling mirror. In the former case, the two interferometer cavities are maintained in resonance by separate servo systems using auxiliary photodiodes, not shown, to adjust the cavity lengths by applying feedback forces to the end masses in a way similar to that indicated for the lower cavity in figure 12.2.

For normal operation the reflectivities of the cavity input mirrors for each arm are chosen to give cavity storage times which approximately match the time scale of the gravity waves of interest. Maximum build-up of light is obtained with a transmission for the recycling mirror which suitably matches the total losses in the system.

The number of times that the light can be recycled through the interferometer depends on many factors. These include losses in the mirrors, the beamsplitter, and other optical components; light removed for auxiliary servo systems; and any residual mismatch in the wavefronts of the light from the two main cavities when they are recombined at the beamsplitter. Reduction of wavefront mismatch by use of a specially figured compensating plate in one arm of the interferometer may be practicable.

If it is assumed that all losses are made small except those associated with cavity mirrors of maximum reflectivity R, then the photon shot noise limited sensitivity at unity signal to noise ratio may approach $h = [(\lambda f^3 \hbar \{1 - R\})/(LI\eta)]^{1/2}$.

For similar parameters to those used in section 12.2 above, with $R = 0.99995$, $L = 4$ km, and $I = 50$ W, this gives a sensitivity $h = 1.2 \times 10^{-22}$.

This basic type of recycling interferometer seems a very efficient way of achieving a sensitive broadband instrument with minimum light input power. It seems probable that when techniques for using squeezed quantum states of light to reduce photon shot noise become sufficiently developed, these may be applied to essentially the same interferometer design to give further improvement in performance. However, another optical technique, 'resonant recycling', has been devised to obtain further improvement in shot noise limit to sensitivity in a narrow bandwidth (Drever, 1983). In the original version of this scheme, the cavities in the two arms of the interferometer are coupled together by a high-reflectivity mirror in such a way that the normal resonance modes are split, with a splitting which matches the period of the gravity waves of interest. One of the two resonances is then made to match the frequency of light from the laser, while the other matches the frequency of a sideband of this light produced by the mirror motions due to the gravity wave: both resonances enhance the output signal. Another variant of this basic idea has been proposed recently by B. J. Meers: here the second resonance is obtained by putting an additional recycling mirror at the output of the type of broadband recycling system shown in figure 12.3. The latter configuration has been called 'dual recycling'.

12.4   Resonant recycling and dual recycling

12.4.1   Resonant recycling

The original idea for resonant recycling arose while trying to find a way of detecting the very weak periodic gravitational radiation signals expected from pulsars. It grew out of the realization that if light could be stored in an arm of an interferometer for a time matching half a period of the gravity wave it might be made to exchange with light from the other arm at the same time as the phase of the gravity wave reverses, so that a signal might build up continuously (Drever, 1983). The first description of this concept, applied to both a delay line and a

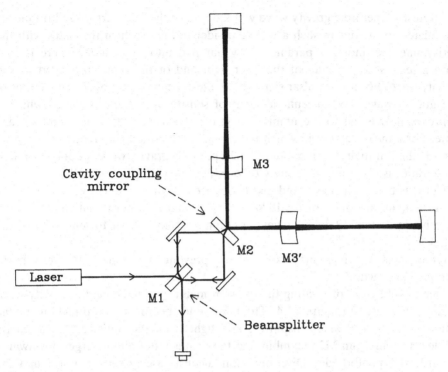

Figure 12.4. One configuration for a resonant recycling interferometer, with the main cavities coupled by mirror M2 to give one resonance at the frequency of the laser and a second resonance at a sideband frequency produced by modulation by the gravitational wave.

Fabry–Perot interferometer, was given in these terms; but in the case of the Fabry–Perot interferometer it is simpler to consider the mechanism as production of sidebands of the laser frequency by the gravity wave, and the system as providing a resonant enhancement of a sideband signal. A simple form of resonant recycling Fabry–Perot interferometer is shown in figure 12.4. Here the optical cavities in each arm are directly coupled to one another by a mirror M2. This has a high reflectivity, chosen so that its transmission approximately matches the total losses in the pair of cavities, giving maximum build-up of laser light in this part of the system.

   If we consider a single axial resonance mode of the same order in each cavity, then the coupling has the effect of giving the combined system two modes of oscillation, a symmetrical one – pumped by the laser – with the light in phase in the two cavities, and an antisymmetrical one in which it is out of phase. The degree of coupling between the two cavities is determined by the transmission of their input mirrors M3 and M3′ and also by interference effects in the optical path between these mirrors. To tune the interferometer, the mirrors M3 and M3′ are adjusted so that the frequency splitting matches the frequency of the gravity wave of interest. A periodic gravity wave will then move the end mirrors and frequency modulate the cavities in such a way that sideband light which resonates with the antisymmetrical mode is parametrically pumped into that mode. There is thus both a resonance build-up of the laser light and of the signal produced by the gravity wave. For a fast pulsar signal the build-up can extend over many cycles of the gravity wave, and an enhancement of sensitivity may be obtained which is approximately equal to the number of gravity-wave periods in the light storage time. For a measurement in which the signal is integrated coherently over a total time τ, the sensitivity for amplitude h of a periodic gravity wave, at unity signal to noise ratio, is given approximately by $h = [(\lambda \hbar c (1 - R)^2)/(L^2 I \eta \pi \tau)]^{1/2}$.

Here, as before, it is assumed that losses are determined only by the reflectivity R of the main mirrors. For $I = 10$ W, $\tau = 10^7$ s, and all other parameters as in the example in section 12.3, this gives a photon shot noise limit to sensitivity of the order of $h = 1 \times 10^{-28}$.

Other noise sources may, of course, prevent this sensitivity from being obtained in practice.

One possible way of exciting the interferometer and detecting the gravity-wave signal is indicated in figure 12.4. The laser sends beams in two directions to the mirror M2 via a 50% beamsplitter, and light from the antisymmetrical cavity mode returning from M2 recombines at the beamsplitter and emerges downwards towards the photodiode. Discrimination against laser intensity noise may be improved by introducing high frequency phase modulation, for example by a method like that shown in figure 12.3, in which some phase-modulated light is coherently added to the main output signal. Techniques of this type can be applied to most of these interferometer configurations, but are omitted here for simplicity.

12.4.2   Dual recycling

An alternative 'dual recycling' method (Meers, 1988) of obtaining the two resonances required for a resonant recycling interferometer is indicated in figure 12.5. Here a second recycling mirror M4 is added to the output of a standard broadband recycling interferometer. The first recycling mirror M2 acts just as in the broadband recycling system of section 12.3 to increase the laser light flux in the interferometer arms. The second recycling mirror M4 is adjusted to give a resonance with the two main cavities combined in antiphase with one another at a sideband of the laser frequency produced by the gravity wave. This enhances the gravity-wave signal just as in the resonant recycling system of section 12.4.1, and detection of the signal may be facilitated by a similar heterodyne system.

In a narrow-band operation the maximum sensitivity of this system is similar to that of the resonant recycling system of section 12.4.1, although there are differences. The presence of the two extra mirrors in the resonant part of the dual recycling system can potentially give it more flexibility in adjustment of overall response. By broadening the bandwidth of the output cavity, while keeping the

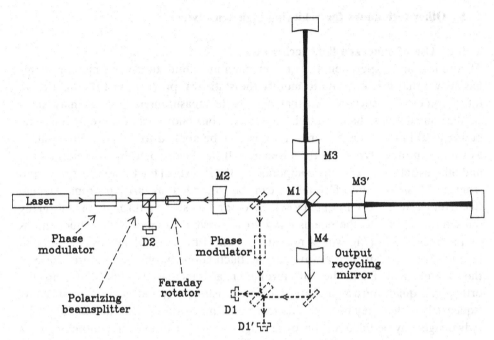

Figure 12.5. A dual recycling interferometer configuration. The output mirror M4 provides the second resonance required for resonant recycling, with the input mirror M2 giving the resonance at the laser frequency. The high-frequency side-arm modulation method outlined in figure 12.3, indicated by broken lines, may be used here also to reduce intensity noise.

input cavity narrow, a wider bandwidth can be obtained for a given peak sensitivity. However, losses in substrates may be higher. It may be noted that addition of a recycling mirror to the resonant recycling system of figure 12.4, in front of the beamsplitter M1, can give that system some similar characteristics.

12.4.3 Resonant recycling interferometers in general

The two resonant recycling systems discussed above are examples out of a wider range of configurations which can be envisaged. These systems are likely to have more general application than investigation of periodic signals alone. They can be adjusted for relatively broadband response, and then may be competitive with the broadband recycling systems of section 12.3 for pulse searches; while for particular pulse waveforms they may be more sensitive. Also in searches for a stochastic background of gravitational radiation by cross-correlation techniques they may give higher overall sensitivity. Further, it can be expected that these techniques will be relatively tolerant of optical distortions and figure errors, for the exchange of light between the interferometer arms tends to average out some imperfections. In general, resonant recycling systems of various kinds look promising for the future.

12.5   Other techniques for achieving high sensitivity

12.5.1   Use of squeezed light techniques

Theoretical and experimental work on quantum limits to optical measurements
has shown that it is possible to modify the statistical properties of the light in an
interferometer in such a way that a suitable measurement process may show
smaller fluctuations than would be obtained from photon shot noise at the same
power level in the usual system; and this may be applied to Fabry–Perot systems
as well as to other types of interferometer. It has been noted by C. Caves (1981)
and others that a fundamental component of the statistical fluctuations in the output
from a Michelson interferometer can be regarded as arising from vacuum
fluctuations at optical frequency which enter the system through the normally
unused direction into the beamsplitter. It is possible to modify these fluctuations
by parametric pumping with a non-linear optical medium which covers the open
beamsplitter port, in such a way that the fluctuations in one quadrature phase of
the output light from the interferometer are increased while those in the
orthogonal quadrature phase are decreased. Effectively the fluctuations become
'squeezed' so that they become smaller in one quadrature phase than in the other.
Advantage may be taken of this by arranging that detection of the output signal is
made using the quieter quadrature phase alone; which may be done with another
coherently pumped optical parametric amplifier. The ultimate noise in this, and
other, squeezed light measurement techniques can in principle be less than in
normal operation of an interferometer with the same light power.

Application of these techniques to Fabry–Perot interferometers shows promise
for the future, and may lead to improvements in sensitivity without increase in
light power. Some limitations should be mentioned, however. Optical losses at
mirrors and beamsplitters can introduce fluctuations which are not reduced by the
squeezing, and may obviate the potential gains in sensitivity. In particular, when
recycling techniques are pushed to a point where losses in the interferometer
arms prior to detection are the limiting factor, as in resonant recycling and dual
recycling for narrow-band operation, squeezing techniques may not give significant advantage. However, with broadband recycling and wide band dual
recycling where light is detected before losses have become large, squeezing may
in principle give a useful improvement in sensitivity.

12.5.2   Use of auxiliary interferometers to reduce seismic noise

Up to this point we have considered photon shot noise as the main source limiting
the sensitivity of an interferometric gravity-wave detector; but other noise sources
may also limit practical systems. Seismic motions of the ground, which may
produce stochastic forces on the test masses through the mass suspension systems,
are a potential noise source which may limit performance at low frequencies.
Passive isolation techniques can provide good seismic isolation at frequencies
above a few hundred hertz, but their effectiveness tends to decrease as frequency

falls. Various types of active feedback isolation systems have been proposed and developed with the aim of improving the low frequency performance of gravity-wave detection systems.

The relatively small diameter of the cavity mirrors and light beams in a Fabry–Perot gravity-wave detector makes it reasonable to consider putting several different interferometer beams within one set of vacuum pipes. This has made more practical a fairly simple technique for reducing low frequency seismic noise which was conceived earlier. An auxiliary interferometer is used to monitor relative motions between the suspension points of the test masses at the opposite ends of each interferometer arm. The output from this interferometer is applied to a transducer system which moves the suspension point at one end of the arm in such a way that relative motion along the direction of the arm is decreased. With the pair of suspension points locked together in this way, seismic motion components in the direction of the arm become effectively equal for the two test masses, and thus tend to cancel.

This active isolation technique is a fairly simple one for a Fabry–Perot interferometer, and can be expected to considerably reduce relative motions of the test masses. Experimental work on a system of this type has been done by Y. T. Chen. Such an arrangement is expected to usefully improve detector sensitivity at low frequencies. Further, as it reduces the dynamic range requirements for the servo systems controlling the overall lengths of the main interferometer arms it may contribute to improved performance at higher frequencies also.

12.6 Experimental strategies with Fabry–Perot systems

The Fabry–Perot interferometer system for gravity-wave detection was conceived initially mainly as a compact, and therefore relatively economical, optical system for achieving the required sensitivity. However, the compact nature of the mirrors and beams relative to those required for delay-line detectors has gradually led to new ideas about ways of using the interferometers which have influenced plans for experimental strategies for gravity-wave searches. For example, experience with searches for gravity-wave bursts using earlier broad-band bar detectors showed the major advantage to be gained in discrimination against spurious signals by having at least a pair of detectors operating in coincidence at one site, along with one or more at a distant location to verify possible detections (Drever et al., 1973). With laser detectors it was soon realized that if two interferometers could share a single set of beam pipes the cost of adding the second would be relatively small. Further, when planning for kilometre-scale interferometer facilities began about six years ago it became apparent that it would be wise to make any large facilities capable of accommodating delay-line interferometers as well as Fabry–Perot ones, and this implies that the beam pipes would be capable of accommodating several

Fabry–Perot interferometers. Gradually it became apparent that it could be economical to design large detection facilities to house several detectors, and this in turn has affected ideas for experimental search strategies. We summarize some of the concepts here, and their influence on the design of large detection facilities.

12.6.1 Use of interferometers of different length

The first proposed experimental strategy which became more viable with the possibility of sharing beam pipes was the concept of operating interferometers of different length side by side. This idea grew from attempts to find ways of reducing the cost of using pairs of interferometers at one site to improve the discrimination of gravity-wave signals from those due to other phenomena (Drever *et al.*, 1983c). One early arrangement proposed involved sharing of test masses and of some vacuum tanks, with separate beam pipes forming the sides of a square to give discrimination against non-gravity-wave disturbances. However, a performance nearly as good may be achieved at lower cost by sharing beam pipes, so that a single L-shaped system accommodates two interferometers which are made to respond differentially to non-gravity-wave effects by making their arm lengths significantly different. A convenient ratio of arm lengths is around 2:1, so that a vacuum tank half way along each main arm houses an end mass for the shorter interferometer.

In this case a gravitational wave will produce displacements (or forces) twice as large in the long interferometer as in the shorter one: very few other disturbances will produce displacements in this ratio. Thus a disturbance of any one of the test masses, a release of strain in any one suspension wire, or a change in optical path produced by a release of a burst of gas in any of the beam pipes, can be discriminated against. The discrimination is of course only effective with signals which are several times the mean noise – and these are likely to be the most important signals. For signals near the noise level the technique can give only marginal discrimination, but even in this case the addition of the two interfero-meter outputs will give slightly better sensitivity than that of a single interfero-meter alone.

This two-interferometer system seems capable of giving a sufficiently good signature for a gravity wave to reject many local spurious phenomena. For positive detection of a gravity-wave pulse, however, a coincidence with another detector at a distant site is still essential. The two-interferometer local system, giving signals in the required ratio, can nevertheless add greatly to the significance of a coincidence with a remote interferometer, and can reduce accidental coincidence rates by a large factor. The effective detection sensitivity for low rate burst events can thus be significantly improved.

12.6.2 Concurrent operation of interferometers for different purposes

The feasibility of operating several different interferometers within the same vacuum system can facilitate a comprehensive search for gravitational waves in other ways. In particular it can make it possible to develop new interferometer techniques without interrupting searches with existing interferometers; and it can make it viable to operate specialized interferometer systems designed to give maximum sensitivity for a specific kind of signal without stopping more general broad-range searches. We consider some aspects of these possibilities in turn.

(i) Gravity-wave searches and interferometer development

Fully realistic testing of long-baseline interferometers can only be carried out when the necessary large vacuum pipe systems become available, and it can be expected that the first interferometers to go into these systems will have far from the ultimate performance. Further development of the techniques will then be possible – and will be very important to achieve high sensitivity. Here the possibility of operating several interferometers within the same vacuum facilities can give another bonus – it can make it practicable to use one interferometer as a testbed for new developments without interrupting gravity-wave searches being made with other interferometers. A dilemma familiar to experimenters – how to divide instrument time between observation and technical development – may thus be avoided. This mode of operation – performance of continuing gravity-wave searches and observation concurrently with the development of new techniques – may be a very effective one, and is an important practical reason for attempting to accommodate additional interferometers in a large-scale facility.

(ii) Optimized searches for specific signals

It has been shown above how a suitable choice of optical system parameters can enhance the performance of an interferometer for certain types of signals; a particular example being a search for a continuing periodic wave, for which the optimum detector would have the narrowest bandwidth obtainable with available mirror losses, centred at the appropriate frequency. Even for broadband searches, the design of an interferometer for optimum performance at low frequencies differs from that for higher frequencies. In the latter case, for example, it is advantageous to make test masses relatively small so that frequencies of internal resonance modes are well above the gravity-wave frequency of interest. In this case also, the optimum cavity storage time is short; and the optimum light power on the test masses – where near the uncertainty-principle limit quantum fluctuations in radiation pressure may balance photodiode shot noise – is high. A complete search for gravity-wave signals is likely to involve a range of interferometer parameters and different types of interferometer, and here again it is clear that use of a number of interferometers in one vacuum system may make it practicable to carry out more sensitive searches in a given time than with a single one.

12.6.3 Detector and vacuum system arrangements to facilitate efficient experiments

The possibilities of running several interferometer beams within a single set of vacuum pipes can be used to the fullest extent only in vacuum systems designed for this mode of operation, and there has already been some development in concepts for doing this. In designing such a system using the typical interferometer layouts indicated earlier it can be particularly difficult to find enough space for those components which would be naturally located at the intersection of the two main arms, such as beamsplitters. To alleviate congestion in the intersection region, a concept in which separate vacuum tanks are used for test masses and beamsplitters was introduced by the writer in 1984. Tanks for test masses are located along the two main arms of the system, and beamsplitter tanks are placed along a line bisecting the angle between the two arms, with auxiliary evacuated pipes linking the tanks. One version of this arrangement is shown in figure 12.6. Auxiliary mirrors deflect the light from the beamsplitters to the appropriate main Fabry–Perot cavity. This type of system provides several virtual intersection regions, giving useful space for beamsplitters and other optical components outside the main interferometer arms. It was the basis for early plans for a system which could accommodate up to six separate Fabry–Perot interferometers. A further concept of practical importance was added more recently: the test masses on their suspensions are lowered into position on the main beam line through horizontal gate valves above the main vacuum pipe. A chamber above each gate valve can provide an evacuated housing for the test mass if it is necessary to withdraw it from the beam line, and can be used as an airlock to allow individual test masses to be removed or inserted without disturbing operation of other interferometers in the system. The beamsplitter tanks can be closed off from the main system by gate valves in the auxiliary linking pipes. This overall design thus provides a relatively economical way of accomodating several interferometers in a single pair of beam pipes, while making it possible to access the main components of each interferometer for replacement or adjustment without affecting the vacuum in the system or interrupting seriously the operation of the other interferometers.

In such a system it is important to avoid coupling of motion or vibration noise from one interferometer to another. To help ensure this, the main seismic isolation for each test mass is located in the airlock vacuum chamber above it, and is supported from the ground by a structure isolated from the main vacuum system walls by flexible metal bellows. (The latter concept is similar to one introduced in a vacuum system designed at the Rutherford Laboratory.) Additional vacuum tanks to house auxiliary optical components such as beam-conditioning cavities may be attached to the beamsplitter tanks as necessary.

There are many ways of arranging a multiple-interferometer system of this type, and the arrangement shown is just one example of a general concept which was stimulated by the compact nature of Fabry–Perot and similar optical

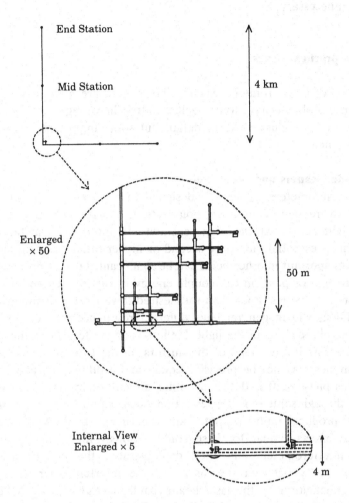

End Station

Mid Station

4 km

Enlarged
× 50

50 m

Internal View
Enlarged × 5

4 m

Figure 12.6. Schematic arrangement for a gravitational wave detector system which enables several Fabry–Perot interferometers to operate independently within a single pair of main vacuum pipes. Provision is shown for interferometers which monitor half the maximum arm length as well as the full arm length, to provide a gravitational wave 'signature' from the dependence of relative test mass displacement on length of baseline. The test masses are housed in vacuum chambers, with access by air-locks, along the line of each arm; while beamsplitters and other optical components are housed in separate chambers near the corner of the L, located along the bisector of the angle between the arms. The diagram is intended to indicate concepts only: layout and dimensions suggested are illustrative only of some possibilities.

configurations. The basic concept can clearly be extended to other types of optical system when necessary.

12.7 Some practical issues

The diagrams of interferometer systems shown here are greatly simplified, and many practical problems have to be dealt with to achieve high sensitivity. There is not space here to discuss these in detail, but some important aspects may be briefly mentioned.

12.7.1 Mode cleaners and fibre filters

Practical interferometers are usually designed to be insensitive to first order to fluctuations in frequency, intensity, polarization, and geometrical parameters of the input light beam. At the high sensitivity required for gravitational wave detection this is essential, and here second or higher order coupling mechanisms which are unimportant in other fields can be significant. For example, fluctuations in beam direction or position can couple energy in varying degree into modes of the main optical cavities other than the desired lowest order symmetrical mode. As these modes will in general have different resonance frequencies, they will cause varying phase shifts in the light. If there is any difference in the alignment, or in figure errors in the shape of the mirrors, in the two arms, the coupling to undesired modes may not be precisely equal, and a differential phase shift may result. If this phase shift is detected, either directly or by interference with the main mode through some other imperfection, such as photodiode non-uniformity, noise can be produced on the output. Early experiments by the Max Planck group showed that effects similar to these could seriously limit the sensitivity of a Michelson interferometer, and they demonstrated the use of an auxiliary Fabry–Perot cavity between the laser and the interferometer as a filter for geometrical fluctuations in the laser beam (Rudiger et al., 1981). A filter cavity like this, used as a 'mode cleaner', is also an effective way of reducing input beam fluctuations for a Fabry–Perot interferometer.

Several other methods of reducing geometrical fluctuations in an input beam have been used or proposed. A single-mode optical fibre, used first with a Michelson interferometer by the MIT group, has been employed with several Fabry–Perot detectors. This technique has the advantage of not requiring any frequency locking to the input light; although with small-diameter fibres there is a power limitation, and a possibility of non-linear phenomena at the high power density required in the fibre. Use of one or more spatial filters for reducing beam fluctuations has been analysed by B. J. Meers, but is in general less effective for a given power loss. Most current gravity-wave interferometers employ optical fibres or mode cleaning cavities to condition the input beam.

A relatively large mode cleaner has been proposed as an intrinsic part of the optical system for a Fabry–Perot interferometer for operation with kilometre-scale arms, in a design developed two years ago by the writer. In this case the mode cleaning cavity has some additional functions and special features. The cavity is designed as the main frequency reference for the input light in the gravity-wave frequency region; it defines the input beam direction at these frequencies; and it also acts as a high-frequency amplitude, phase, and frequency filter. To make this cavity quiet and relatively free from thermal noise it is about 12 m long, and is formed between mirrors suspended by wires from seismically isolated points, in a way similar to that for the test masses of the main interferometer. The D.C. stability of the light wavelength is determined by a separate smaller rigid cavity, through a feedback system to the mode cleaning cavity which is effective only at frequencies much lower than the gravity-wave frequency. In this interferometer design a second similar cavity is also proposed as a filter between the interferometer output and the main photodetector, to help reduce effects of scattered light and other light in undesired modes. To allow passage of the modulation sidebands, the free spectral range of the cavity is arranged to match the modulation frequency for the main interferometer.

Some further points about mode cleaning cavities may be mentioned here. In a system operating with high laser power it may be necessary to limit the finesse of a mode cleaner to minimize heating effects in the cavity mirrors, and it may then be appropriate to use more than one mode cleaner in series to achieve sufficient filtering. In this context it is worth noting that useful mode cleaning action may be provided in a recycling interferometer by the effective cavity formed by the whole system.

It may be remarked that a mode cleaning cavity system may provide a more general benefit than the control of specific geometrical fluctuations. At the levels of sensitivity required for gravitational wave detection unforeseen phenomena of various kinds may disturb operation. If, however, everything involved in the sensitivity region of the system is stationary for times of order of the period of the gravity wave the chance of noise appearing near this periodicity is reduced. A mode cleaner can help ensure that all the parameters describing the light as it enters the interferometer are stationary, and thus may reduce risks of disturbance by even unpredicted optical phenomena.

12.7.2   Beam heating effects in mirrors and other components and techniques for reducing it

In laboratory-scale optical cavities typical beam diameters at the mirrors are of order a millimetre or less, and it is possible to get high flux densities in the beam spot with a fairly low power laser. Losses in the mirror can then produce large temperature gradients, leading to distortion of the mirror and changes in refractive index in the substrate. Effects of this kind have been observed in many laboratories working with optical cavities. In the interferometer arms of kilometre-scale gravitational wave detectors, the spot diameters will be larger and flux densities lower for the same input power. However in spite of the relatively small temperature gradients these can be significant over a greater distance in a thick substrate, and it was noted by W. Winkler (private communication, 1988) that first order estimates of the optical effects suggest that they may be largely independent of beam diameter for a given total beam power.

Initial observations of heating effects have been made in several laboratories, and at stored power levels of order 1 kW of green light some mirrors have been free from significant heating effects, while other mirrors, with different coating losses, have shown significant distortion of the cavity modes. It is not yet clear whether with the most suitable mirrors absorption losses are large enough to cause heating effects to set limits to power levels in practical gravity-wave interferometers.

The magnitude of the effects on an interferometer of the heat produced by a given amount of mirror absorption loss does depend on the optical configuration used, and it is possible to design optical cavity interferometers which are less vulnerable to heating effects than those illustrated earlier here. In a typical cavity mirror there are two main effects: heat produced in the coating causes a change in refractive index of the substrate, giving distortion of the input and output beams, and temperature rise in the substrate causes thermal expansion and distortion of the mirror shape. The situation may be ameliorated by arranging the cavity so that light does not have to be transmitted by the main mirror substrates. Then thermal lensing effects in the substrate are avoided, and, further, the substrate material may be chosen for high thermal conductivity and low expansion rather than for good optical properties, so the thermal effects are further reduced. One way of realizing a cavity of this type is to introduce a thin wedged plate of low-loss material, such as fused silica, into the cavity, and use this as a coupling device. One surface of the plate may be arranged to be at Brewster's angle to the beam and thus free of reflection, and the other surface oriented to give by reflection the required input and output coupling to the cavity. No coatings are required on the plate, so thermal effects in it arise only through its intrinsic absorption, which may be significantly less than in typical mirror coatings.

If cavity heating effects are sufficiently reduced by these or other means, then heating in the beamsplitter may become a limiting factor. This may be reduced by avoiding mirror coatings in the beamsplitter, and achieving the beamsplitting action by evanescent wave coupling between two prisms. If heating in the main cavities is still a limiting factor, each Fabry–Perot cavity may be folded using a delay-line mirror configuration so that the heat is distributed over several reflection spots. Such an arrangement would use up more space in the main vacuum pipes than an unfolded system. In some respects systems of this type exhibit some of the characteristics of delay line Michelson interferometers when these are used with dual recycling to reduce the number of reflection spots and economize in mirror size. With equal mirror areas the two systems will be limited at about the same power level, with a possible slight advantage to the folded Fabry–Perot due to the lower flux through the beamsplitter in that configuration. Such systems would represent some compromise between economy of beam-pipe space and power handling, and may be rendered unnecessary by improvement in mirror and substrate performance.

12.8   Conclusion

This overview of Fabry–Perot cavity gravity-wave detectors is far from complete, and has had to omit discussion of many interesting and important aspects. The aim has been to give an understanding of the basic principles and of development of some of the ideas. The technique itself is far from fully developed, and large improvements in sensitivity are to be expected. Currently, the fields of gravity-wave detection, and of the interferometers themselves, are in a state of transition. Detailed plans and designs for scaling up from laboratory instruments to kilometre-baseline detectors are being prepared, while at the same time it is becoming apparent that there is still much room for originality and variety in optical configurations. The problems of handling high light flux in the interferometer arms are likely to lead to new designs, as already indicated. Earlier systems are likely to be superseded by configurations which may combine features of both Fabry–Perot and delay line interferometers. Currently, some recycled Fabry–Perot configurations which may be folded if required seem promising and flexible, but the choices depend on properties of mirror materials, and are likely to change. In any case, the research already carried out on Fabry–Perot and other systems is likely to significantly influence the instruments which will eventually be of practical importance for investigating gravitational radiation.

Acknowledgements

Much of the work described here has been supported by the California Institute of Technology, and by the National Science Foundation in a series of grants (including in large part grant PHY-8504136, superseded by PHV-8803557). This support, and the stimulation and collaboration of many colleagues, is gratefully acknowledged.

References

Caves, C. M. (1981). *Phys. Rev. D* **23**, 1693.
Drever, R. W. P., Hough, J., Bland, R. and Lessnoff, G. W. (1973). *Nature* **246**, 340.
Drever, R. W. P. (1983). In *Gravitational Radiation*, NATO Advanced Physics Institute, Les Houches, June 1982, eds. N. Deruelle and T. Piran, p. 321, North Holland Publishing, Amsterdam.

Drever, R. W. P., Ford, G. M., Hough, J., Kerr, I. M., Munley, A. J., Pugh, J. R., Robertson, N. A. and Ward, H. (1983a). *Proceedings of the Ninth International Conference on General Relativity and Gravitation (Jena 1980)*, ed. E. Schmutzer, p. 265, VEB Deutscher Verlag der Wissenschaften, Berlin.

Drever, R. W. P., Hall, J. L., Kowalski, F. V., Hough, J., Ford, G. M., Munley, A. J. and Ward, H. (1983b). *Appl. Phys.* **B31,** 97.

Drever, R. W. P., Hough, J., Munley, A. J., Lee, S.-A., Spero, R., Whitcomb, S. E., Pugh, J., Newton, G., Meers, B. J., Brooks III, E. and Gursel, Y. (1983c). In *Quantum Optics, Experimental Gravity, and Measurement Theory*, eds. S. Meystere and M. O. Scully, p. 503, Plenum Publishing, New York.

Meers, B. J. (1988). *Phys. Rev. D* **38,** 2317.

Rudiger, A., Schilling, R., Schnupp, L., Winkler, W., Billing, H. and Maischberger, K. (1981). *Optica Acta* **28,** 641.

Weiss, R. (1972). *Quartr. Progr. Rep. Res. Lab. Electr. MIT*, **105,** 54.

13

The stabilisation of lasers for interferometric gravitational wave detectors

J. HOUGH, H. WARD, G. A. KERR, N. L. MACKENZIE, B. J. MEERS, G. P. NEWTON, D. I. ROBERTSON, N. A. ROBERTSON, AND R. SCHILLING

13.1 Introduction

All laser interferometers rely on measuring the strain in space caused by a gravitational wave, sensitivities of the order of 10^{-22} over millisecond timescales being required to allow a good probability of detection.

In principle the strain as monitored by the change in separation of two test masses hung as pendulums can be measured against the wavelength of light from a stable source, but the degree of wavelength or frequency stability required of the source is unreasonably high. It is much more conceivable to measure the distance between test masses along an arm with respect to the distance between similar masses along a perpendicular arm. This is particularly appropriate since the interaction of a gravitational wave is quadrupole in nature and so can cause an opposite sign of length change in the two arms. The measurement of a differential length change of this type when performed by interferometry puts much less demand in principle on the frequency stability of the illuminating laser light – since a Michelson interferometer is insensitive to changes in the wavelength of the light used if the path lengths are equal. However, in practice a fairly high degree of frequency stability is required. In the case of optical delay lines in the arms of a Michelson interferometer this is a result of the difficulty in achieving equal path lengths and of some light being scattered back early without completing the full number of reflections (Billing *et al.*, 1983). In the case of resonant cavities in the arms it is a result of needing to keep the cavities on resonance and of practical imbalances in the responses of the two cavities to frequency fluctuations of the laser light (Hough *et al.*, 1983).

Frequency stability, however, is not the only parameter to be considered, as the interferometers also have some sensitivity to both power and geometry fluctuations of the illuminating light, again mainly due to small asymmetries. To a large extent intensity fluctuations are reduced in importance by the use of nulling techniques, and geometry fluctuations are reduced by the use of single mode optical fibres and mode cleaners. However, other suppression techniques will probably be required in addition. To some extent the stability required for the laser source depends on the scale of the interferometer and on whether optical delay lines or resonant cavities are used; the levels required will be discussed where appropriate.

13.2   Laser frequency stability

13.2.1   Delay line systems
As was mentioned earlier there are two ways in which laser frequency fluctuations can couple into the output signal of a Michelson interferometer with delay lines in the arms:

  (i)   via a static path difference between the two arms, and
  (ii)   as a result of the interference of scattered light with the main beam.

(i) Static path difference
For efficient operation of an interferometer system the fringe visibility has to be rather good. One of the most obvious requirements for high visibility is a proper overlap of the two interfering beams. For delay line systems this can only be achieved with identical ratios between mirror separation and radius of curvature for the two arms (see chapter 11 by W. Winkler in this book). Unavoidable differences in the curvatures of the delay line mirrors may result in operation where the lengths of the arms differ slightly. The resulting relationship between the limit to gravitational wave amplitude detectable and the fluctuations δv of the laser frequency v is given by

$$h \sim \frac{\Delta l}{l} \cdot \frac{\delta v}{v},$$

where Δl is the difference in length of the two arms of length l. Experience of the research group at MPQ, Garching (e.g. Rüdiger *et al.*, 1982) suggests that $\Delta l/l$ may be expected to be about 10^{-3} unless the curvature of the mirrors is made adjustable. Hence for proposed detectors where $h \sim 10^{-22}$ over millisecond timescales (i.e. linear spectral density $h \sim 3 \times 10^{-24}/\sqrt{(\text{Hz})}$), this requires $\delta v/v \sim 3 \times 10^{-21}/\sqrt{(\text{Hz})}$.

(ii) Light scattering
Light scattering at almost any point in the interferometer and recombining with the main beam will have travelled an optical path different by ΔL from that of the main beam; and typically ΔL may be of the order of the total optical path of one of the delay line arms. In this case, if the fraction of detected amplitude due to scattering effects is σ, the sensitivity to frequency fluctuations in terms of gravitational wave amplitude is

$$h \sim \sigma \frac{\delta v}{v}.$$

Measurements at MPQ (Rüdiger *et al.*, 1982) have suggested that $\sigma \sim 10^{-4}$ is presently a reasonable estimate, although this number may be significantly reduced for the new improved mirror coatings becoming available. Thus $\delta v/v \sim 3 \times 10^{-20}/\sqrt{(\text{Hz})}$ is required for standard coatings.

13.2.2 Cavity systems

For detectors using optical cavities, the laser and cavity systems must have high enough relative stability that the cavities are kept resonating at all times. However, this is not a very demanding constraint and is less important than the fact that any lack of balance in the phase response of the cavities may result in frequency fluctuations appearing as a noise source. If the required strain sensitivity of the detector is expressed as a linear amplitude spectral density $h/\sqrt{(\text{Hz})}$ the required level of stability of the light incident on the cavities is

$$\frac{\delta v}{v} \sim h \cdot D,$$

where D is a measure of the common mode rejection of the whole system which results from the comparison of the light from the two cavities. In fact D^{-1} is essentially the fractional fluctuation in the optical lengths of the cavities of length l and finesse F, i.e.

$$D^{-1} \sim \frac{\Delta[Fl]}{Fl} = \frac{\Delta F}{F} + \frac{\Delta l}{l}.$$

Experiment suggests that a value of D of at least 100 should be achievable. (And with the use of mirror systems of adjustable reflectivity to tune the finesse of the cavities D may reach $\sim 10^4$.) Hence $\delta v/v \sim 3 \times 10^{-22}/\sqrt{(\text{Hz})}$.

Thus we see that reduction of laser frequency fluctuations is probably more critical in the case of systems with Fabry–Perot cavities in the arms.

Light scattering of the type mentioned in section 13.2.1(ii) is probably less important for cavity systems as the beams in the cavities lie on top of each other; the required stability mentioned above should be sufficient to avoid noise due to this effect.

For the present prototype detectors where a linear amplitude spectral density of $10^{-20}/\sqrt{(\text{Hz})}$ is being sought over a frequency range from several hundred hertz to several kilohertz, the laser frequency stability required will be approximately $10^{-18}/\sqrt{(\text{Hz})}$ over this range for a common mode rejection ratio of 100.

13.2.3 Laser frequency stabilisation

The stabilisation of laser frequency has been an area of steady progress (see, for example, the reviews of Hall, 1978; Hamilton, 1989; Salomon, Hils and Hall, 1988). There has been a considerable amount of work done on the development of short term frequency stabilisation techniques for argon lasers (Camy, 1979; Drever *et al.*, 1983a; Hackel, Hackel and Ezekiel, 1977; Kerr *et al.*, 1985; Rüdiger *et al.*, 1982; Shoemaker *et al.*, 1985), dye lasers (Helmcke, Lee and Hall, 1982; Hough *et al.*, 1984) and YAG lasers (Shoemaker *et al.*, 1989). In the case of argon lasers a significant improvement in performance can be obtained passively by carefully isolating the mirrors of the laser cavity from the vibrations introduced by the flow of cooling water.

Active methods of stabilisation require a frequency reference discriminator which only needs to have very high short term stability (this may be part of the gravitational wave detector itself), a suitable amplifying system and a transducer or transducers which can alter the frequency of the laser.

A suitable discriminator is a Fabry–Perot cavity, especially now, as finesses of up to 20 000 can be achieved with available laser gyro mirrors (Anderson, Frisch and Masser, 1984). In the case of the gravitational wave detectors the long arms, forming either delay lines or Fabry–Perot cavities, can be used as very efficient frequency discriminators.

Transducers to change the laser frequency can operate by changing the length of the laser cavity and may be piezoelectric devices moving one of the cavity mirrors (Helmcke, Lee and Hall, 1982; White, 1965), or electro optic phase modulators placed inside the laser cavity (Barger, West and English, 1975; Drever *et al.*, 1983a; Lee, Helmcke and Hall, 1979), or a combination of both. They may also operate by changing the frequency of the light after it has left the laser. Acousto optic frequency modulators (Camy *et al.*, 1982; Hall, Layer and Deslattes, 1977), electro optic phase modulators (Kerr *et al.*, 1985) and combinations of these can be used (Hall and Hansch, 1984).

For reasons of stability of a servo system it is generally the case that the larger the amount of stabilisation required, i.e. the higher the loop gain of the servo system, the wider is the bandwidth required for the system. Much of the experimental work carried out with regard to gravitational wave detectors has been to develop discriminator techniques, feedback transducers and servo electronics to allow a bandwidth of several megahertz. It should be noted however that if one of the long arms is used as the frequency discriminator the bandwidth will probably be limited approximately to 1/(one pass light transmit time in the arm) or ~150 kHz for a 1 km arm; in this situation, a more complicated arrangement involving an auxiliary reference cavity of wider bandwidth may be useful.

13.2.4 Optical cavity as a frequency discriminator

For many years optical cavities in transmission have been used as frequency discriminators (for a modern analysis see Hils and Hall, 1987). In order to eliminate conversion of laser power fluctuations into frequency noise while preserving a bipolar error signal it was usual to compare an independent sample of the laser power with the cavity transmission signal near its half maximum tuning point. A fundamental limit to the fluctuation in frequency from that defined by the cavity is set by the photon noise in the detected light. At frequencies much lower than that set by the bandwidth of the cavity the limiting linear spectral density of the frequency fluctuations is given by

$$\delta v \sim \Delta v_C \sqrt{\left(\frac{\hbar \omega}{P_0 T \eta}\right)} \quad \text{Hz}/\sqrt{(\text{Hz})},$$

where Δv_C is the reference cavity bandwidth (full width at half maximum), P_0 is the mode matched input laser power, T is the on resonance transmission efficiency of the optical cavity, and η is the photodetector quantum efficiency. However, there are two principal difficulties in reaching this level of performance. Firstly the laser medium and cavity fluctuations which produce the frequency fluctuations also produce fluctuations in the laser beam direction and size. These may change the mode matched cavity transmission and may couple into the stabilising loop. Secondly the finite cavity bandwidth and resulting transient response errors set a limit to servo bandwidth and loop gain achievable (Helmcke, Lee and Hall, 1982). Both of these difficulties can be overcome. For example, the effect of beam geometry fluctuations can be reduced significantly by the use of radiofrequency phase modulation on the input light which allows a laser to be locked on the peak of a cavity resonance curve. The servo bandwidth can be increased by using the cavity in reflection rather than transmission. A combination of these techniques – rf reflection locking – has proved very success-ful (Drever *et al.*, 1983b) and is similar to a method proposed by Pound (1946) for the stabilisation of microwave oscillators. This involves making use of the beating of the light which is reflected from the input cavity mirror with the light which leaks back out of the cavity, an rf phase modulation technique on the input light being used to allow a suitable discriminator signal to be obtained. A schematic diagram of a possible experimental arrangement is shown in figure 13.1.

In more detail, the light from the laser is phase modulated at a frequency considerably higher than the cavity linewidth. For a small modulation index ($\beta < 1$) the input light amplitude and hence the amplitude reflected directly back from the input mirror will be of the form

$$A_0 = [J_0(\beta) \sin(\omega t) + J_1(\beta) \sin(\omega t + \omega_m t) - J_1(\beta) \sin(\omega t - \omega_m t)]$$

where ω is the angular frequency of the laser light and ω_m is the modulation angular frequency. Since the sidebands at $(\omega \pm \omega_m)$ lie outside the resonant linewidth of the cavity, the light which leaks back out has amplitude proportional to $A_C J_0(\beta) \sin(\omega t + \phi)$ where ϕ is a measure of the relative phase between the input light and the light in the cavity. Thus the intensity of the light detected on a photodiode collecting the total light 'reflected' from the cavity has a component at the modulation frequency proportional to $A_0 A_C J_0(\beta) J_1(\beta) \phi$ for small ϕ. For laser fluctuations at Fourier frequencies below the linewidth of the cavity the signal from the photodiode is proportional to the frequency fluctuations of the laser; while for fluctuations at Fourier frequencies higher than the linewidth the signal is proportional to the integral of the frequency fluctuations. In particular it is found

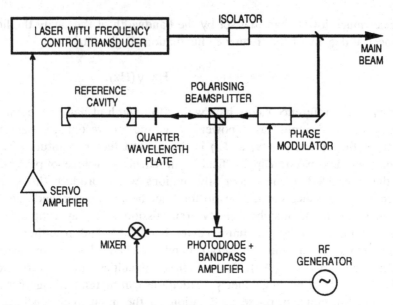

Figure 13.1. Schematic diagram of rf reflection locking technique.

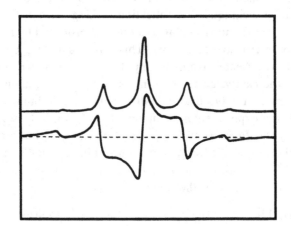

Figure 13.2. Signals from a Fabry–Perot cavity. *Upper curve*: transmitted laser intensity, as frequency of phase modulated laser beam scans through resonant frequency of the cavity. Positions of the modulation sidebands are clearly visible. *Lower curve*: output of the coherent demodulator. Locking point is the zero in the high slope region at the centre.

that the signal at v_L is proportional to

$$\frac{A_0 A_C J_0(\beta) J_1(\beta) \xi}{\sqrt{\left[1 + \left(\frac{2v_L}{\Delta v_C}\right)^2\right]}}$$

where v_L is the Fourier frequency of the laser frequency fluctuations of amplitude ζ and $\Delta \nu_C/2$ is the halfwidth of the cavity resonance. A full description of the cavity response is given by Ford (1981). On coherent detection at the modulation frequency the signal shape as the frequency of the laser drifts through resonance is as shown in figure 13.2, with the signal being zero when the cavity is exactly on resonance.

This system has a number of good points. For example, it provides a highly suitable discriminator curve for locking the frequency of the laser and when on resonance the signal, being approximately zero, has only a small dependence on low frequency fluctuations of the laser amplitude. Also the signal is not directly dependent on shape or direction changes of the input laser beam except through any phase changes these may cause (and of course there will be such effects).

The dependence of signal amplitude with Fourier frequency of the fluctuations has a simple low pass filter response and so is easy to use in a wideband servo system. In general for such a discriminator it can be shown that the limitation to the frequency fluctuations detectable set by photon shot noise is given by

$$\frac{\delta \nu}{\nu} = \frac{\Delta \nu_C}{\nu} \sqrt{\left[\frac{\hbar \omega (1 - J_0^2(\beta) M V_0)}{16 \eta P_0}\right]}$$

$$\cdot \frac{1}{M J_0(\beta) J_1(\beta)} \cdot \frac{1}{(1 \pm \sqrt{(1 - V_0)})} \cdot \sqrt{\left[1 + \left(\frac{2\nu_L}{\Delta \nu_C}\right)^2\right]} \bigg/ \sqrt{(\text{Hz})},$$

where

V_0 is the visibility in reflection of the cavity fringes $(P_{max} - P_{min})/P_{max}$ if the laser light is perfectly mode matched to the fundamental mode of the cavity and no phase modulation is present,

M is the fraction of the input light power P_0 actually matched to the fundamental mode of the cavity,

and the other symbols are as defined earlier. Relations between M, V_0, β and experimentally measurable quantities are given in box 13.1.

Note, however, the term $(1 \pm \sqrt{(1 - V_0)})$ in the denominator of the above. The negative sign applies when the amplitude of the light leaking back out of the cavity is smaller than that directly reflected back from the input mirror and is the situation normally met with for a typical high finesse cavity. The positive sign applies when the amplitude from the cavity is larger than that directly reflected from the input and this situation is met for a low loss cavity whose finesse is made low by choosing a high value for the transmission of the input mirror. (This case is relevant for the cavities in the arms of the kilometre scale gravitational wave detector.)

In a gravitational wave detector with cavities in the arms, one of these cavities can act as the discriminator; and for the prototype gravitational wave detector at

Box 13.1. *Useful relations*

Measure	Use with	To calculate
τ_s, l	—	$F = \dfrac{\pi c \tau_s}{2l}, \quad \Delta\nu_C = \dfrac{1}{\pi\tau_s}$ $\left[\text{note also that: } F = \dfrac{\pi\sqrt{(r_1 r_2)}}{1 - r_1 r_2} \right]$
t_1^2	F	$V_0 = \dfrac{t_1^2 F}{\pi}\left[2 - \dfrac{t_1^2 F}{\pi} \right]$
γ	—	$J_0(\beta) \sim \dfrac{1}{\sqrt{\left(1 + \dfrac{2}{\gamma}\right)}}$ (good approx. for $\gamma > 2$) $[\beta \sim 2[\sqrt{(\gamma + 2)} - \sqrt{\gamma}]]$ $J_1(\beta) = \dfrac{J_0(\beta)}{\sqrt{\gamma}}$
V_a	$V_0, J_0^2(\beta)$	$M = \dfrac{V_a}{V_0 J_0^2(\beta)}$

Key (refer also to earlier definitions of symbols):

τ_s cavity storage time; $1/e$ decay time of reflected intensity; $1/e^2$ decay time of transmitted intensity

l cavity length

F cavity finesse

r_1, r_2 amplitude reflectivities of cavity mirrors

$\Delta\nu_C$ cavity optical bandwidth – full width at half maximum; note that the integrator action of the cavity in a reflection locking scheme is equivalent to an RC filter with a corner frequency of $\Delta\nu_C/2$

t_1^2 power transmittance of cavity input mirror

V_0 theoretically optimum visibility – with no modulation and with perfect mode matching

γ ratio of power at the fundamental laser frequency to that in one of the first order phase modulation sidebands; typically ~ 6

V_a apparent measured visibility – with modulation and with possibly imperfect mode matching

M fraction of the input light power matched to the fundamental mode of the cavity

$J_i(\beta)$ Bessel's function of the first kind of order i

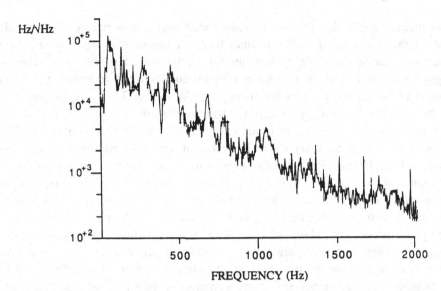

Figure 13.3. Linear spectral density of frequency noise in a typical argon ion laser. The vertical scale is in Hz/√(Hz), the noise present in a 1 Hz bandwidth. (Laser cavity mirrors have been mounted on a separate resonator to reduce vibrational effects introduced by flowing cooling water.)

Glasgow, with the following typical parameters:

$\Delta\nu_C \sim 4 \times 10^3$ Hz (cavity length = 10 m, finesse $\sim$ 4000)
$V_0 \sim 0.8$, $\eta \sim 0.5$, $\beta \sim 0.7$, $\nu = 5.8 \times 10^{14}$ Hz,
$P_0 \sim 30$ mW, $M \sim 0.8$,

the photon noise limited sensitivity to frequency fluctuations is given by

$$\frac{\delta\nu}{\nu} \sim 5 \times 10^{-20} \sqrt{\left[1 + \left(\frac{2\nu_L}{4 \times 10^3}\right)^2\right]} \bigg/ \sqrt{\text{(Hz)}}$$

or at 1 kHz, $\delta\nu \sim 3 \times 10^{-5}$ Hz/√(Hz).

The practical frequency noise of a typical single-line single-mode argon laser (Spectra-Physics model 170) is as shown in figure 13.3.

Thus it can be seen that a loop gain of $\sim 10^8$ at 1 kHz is required currently to reach photon noise limited performance. It is less difficult to achieve such performance if a servo system of relatively wide bandwidth (>1 MHz) is used.

13.2.5 Transducers for laser frequency control

Piezoelectric elements driving one or both laser cavity mirrors make relatively simple control elements. However, due to mechanical resonances and propagation delays they tend to be limited in use to frequencies below 100–200 kHz. For higher frequency operation intracavity electro optic modulators of ADP or AD*P are very suitable as these can be operated resonance free to frequencies of many megahertz. However, there is some optical loss associated with such devices, and experiments suggest that the loss increases with optical power level. This results in less light power being available than from an unmodified argon laser and in some cases may even result in short modulator lifetime. Experience at Glasgow suggests that the use of such an intracavity modulator tends to reduce the single mode laser output power available from 5 or 6 W to 2 or 3 W. The loss in power can be avoided by using piezoelectric transducers on the laser cavity at low frequencies in combination with an extracavity modulator operating at high frequencies. This extracavity device may be an acousto optic modulator (Hall, Layer and Deslattes, 1977), double passed to avoid causing fluctuations in the direction of the transmitted laser beam. However, time delay in the acousto optic device tends to limit its useful servo bandwidth to about 100 kHz. This may be extended by using a combination of such a modulator with an extracavity electro optic phase modulator (Hall and Hansch, 1984) or, if the frequency corrections required at high frequency are small (as with an argon laser) the extracavity device may be a single electro optic phase modulator (Kerr *et al.*, 1985). Up till the present most frequency stabilising systems for gravitational wave detectors with resonant Fabry–Perot cavities in the arms have used the rf reflection locking system with piezoelectric transducers and an intracavity phase modulator in the laser. However, recent work at Glasgow has been directed at achieving the required loop gain with only piezoelectric devices.

13.2.6   Feedback amplifying systems

The aim of the laser stabilising systems for gravitational wave detectors is to achieve the required level of frequency stability for the illuminating laser over the operating frequency range of a few hundred hertz to a few kilohertz. This requires very high servo loop gain, and to achieve this a combination of fast amplifier response and settling time with relatively large dynamic range at low frequency (to cope with the large low frequency laser fluctuations) is needed. Further, in a gravitational wave detector, if the stabilising cavity is formed between test masses suspended as pendulums, acquisition of lock of the servo system is more difficult than when a fixed static cavity is used. Movement of the pendulum masses causes a high rate of passage of fringes through the system and thus requires significant dynamic range of the servo system at high frequencies. In order to combine fast operation with large low frequency dynamic range a bypass amplifier system is used (Helmcke, Lee and Hall, 1982) in which the high frequency and low frequency signals are separated after the coherent detector and are individually amplified before application to the laser frequency control transducers. An important aspect of such a system is that for acquisition of lock, the high frequency part of the signal should be able to reach the laser even when the low frequency amplifiers are saturated.

13.2.7   Design of the servo system

For servo system stability it is important that the slope of the open loop gain with frequency should be less than 12 dB/octave at the frequency at which the loop

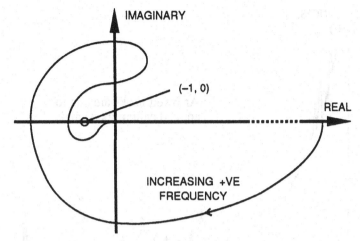

Figure 13.4. Schematic diagram of the positive frequency part of the Nyquist plot of open loop gain for a typical laser frequency stabilising system (not to scale). Real part can be $\sim 10^{12}$ at zero frequency.

gain is unity. However, below this frequency the slope of the curve can become increasingly higher, and slopes as high as 24 dB/octave (M. Zucker, 1986, Caltech, private communication) have been reported to give satisfactory operation. As a result of the accompanying phase changes, which can exceed a lag of 270° at some frequencies, the servo loop is usually only conditionally stable – i.e. only stable over a definite range of gain – but this causes no difficulty in practice. A typical Nyquist plot for the open loop system is shown in figure 13.4 and the reason for the conditional stability can be clearly seen if it is remembered that the contour must not enclose the point $(-1, 0)$.

It should be noted that one of the frequency rolloffs in the system which provides 6 dB/octave at the unit gain point can be due to the finite bandwidth of the discriminator cavity if the light reflected from the cavity is used. The rest of the filtering is carried out in the amplifiers and in the transducer drive circuits.

13.2.8 Typical performance of such a system
The reflection rf sideband method has been used extensively for the stabilisation of dye lasers for spectroscopy (Hough *et al.*, 1984), of argon lasers for the gravitational wave experiments at Glasgow (Ward *et al.*, 1985) and Caltech (Spero, 1985) and of YAG lasers being developed at Orsay (Shoemaker *et al.*, 1988) to replace argon lasers in future detectors. A general idea of the level of stabilisation achievable can be obtained from some typical results with the argon laser on the prototype interferometer at Glasgow. These results were obtained using rf phase modulation at 12 MHz on single mode argon light at 514.5 nm with one of the Fabry–Perot cavities forming the arms of the gravitational wave detector used as the frequency reference. The finesse of this cavity was approximately 4000, the visibility of the fringes with modulation was about 60%,

FREQUENCY NOISE
(Hz/√Hz)

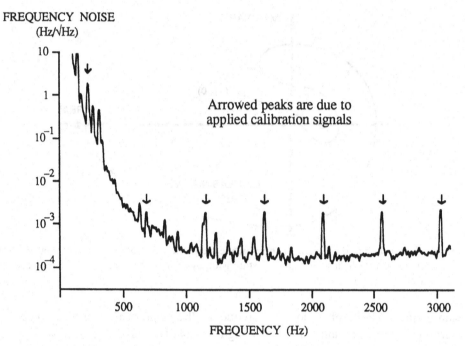

Figure 13.5. Glasgow 10 m prototype interferometer. Upper limit to the spectral density of the frequency noise of the stabilised laser.

the modulation index was 0.7, and the detected power in reflection with the cavity not on resonance was approximately 30 mW. Two spectra are shown. Figure 13.3 shows the correction signal applied to the phase modulator in the laser in terms of the linear spectral density of frequency fluctuations about the resonant frequency of the cavity; and this corresponds to the frequency noise of the unstabilised laser. An upper limit to the residual frequency fluctuations can be obtained by locking a second cavity, e.g. the Fabry–Perot in the other arm, to be resonant with the stabilised light from the laser, the reflection locking technique again being used. This time the feedback transducer may be a piezoelectric device driving one of the mirrors in the cavity or a coil/magnet arrangement driving the position of the mass holding one of the cavity mirrors. The feedback signal (up to the unity gain point of the loop) is a measure of the difference between the resonant frequency of the cavity and the frequency of the laser, and thus its linear spectral density as shown in figure 13.5 gives the required information. This is an upper limit to residual frequency fluctuations as there may be other noise sources present in the second cavity. Note that the signals out of the phase detector in the stabilising loop were equivalent to an apparent frequency noise of $<5 \times 10^{-5}$ Hz/$\sqrt{(\text{Hz})}$ at 1 kHz, consistent with a loop gain of $>10^8$ at this frequency.

The delay line detector at the MPQ in Garching does not require the same level of frequency stabilisation as those using cavities in the arms. However, the

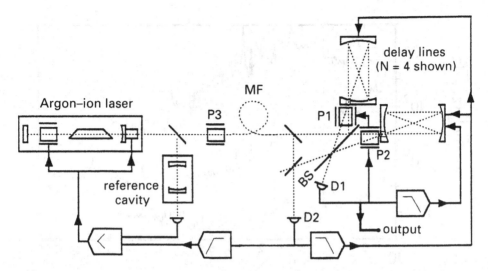

Figure 13.6. Schematic diagram of the MPQ 30 m laser interferometer. The illumination is provided by an argon laser shown at the left. The two arms of the Michelson interferometer are on the right. Frequency stabilisation and interferometer servo systems are shown at the bottom.

requirements are still considerable and R. Schilling and the MPQ group have developed a novel two loop stabilising scheme with each loop unconditionally stable (Shoemaker *et al.*, 1985). This system allows easy acquisition of lock, and considerable flexibility of use, and has a loop gain which is the product of the two loops. A schematic diagram of the system is shown in figure 13.6.

The laser is first locked to a 25 cm long fixed Fabry–Perot cavity in a vacuum chamber with feedback being applied to both a piezoelectrically driven mirror and an electro optic modulator in the laser cavity. This servo system is equipped with a particular input which allows wideband control of the laser frequency for stabilised operation. The stabilised light from the laser is interfered with the rejected light from the whole interferometer to give a further correction signal which is added back into the first feedback loop to provide further stabilisation. The performance of this system with one loop and then with both loops operating is shown in figure 13.7.

13.2.9   Current developments

In experiments at Glasgow there has been some experimental evidence that the presence of an electro optic modulator inside the cavity of an argon laser causes small amplitudes of unwanted laser modes and can significantly limit the output power available. Kerr *et al.* (1985) demonstrated that reasonably high loop gain could be achieved without an intracavity electro optic modulator. Piezoelectric drive of the cavity mirrors was used for low frequencies and an extracavity electro optic phase corrector used for high frequencies. The loop gain achieved with this

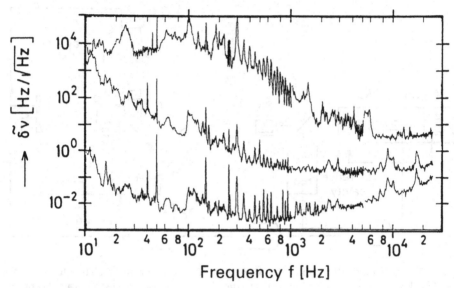

Figure 13.7. MPQ system. Frequency noise in the illuminating laser beam. The top curve is the frequency noise present in the uncontrolled laser output; the middle curve is the noise level when the Fabry–Perot cavity is used as the reference, and the bottom curve is the error signal for the system including the interferometer as the 'second reference'.

system compared favourably with that achieved in contemporary systems using intracavity devices. Since then considerably higher performance has been required for the Glasgow prototype and has been obtained with intracavity devices. But recently in Glasgow very high loop gains have been achieved with only piezoelectric mirror drives (loop gains of $>10^8$ at 1 kHz) and the system looks promising for gravitational wave detectors. It should be noted that a two loop stabilisation scheme is being used, the laser first being stabilised to an auxiliary cavity and then the combination being stabilised to a long cavity with suspended test masses. The two loop scheme is not essential in principle but allows easier acquisition of the locked condition with the very high gain loops involved; it follows to some extent the two loop scheme developed at the Max-Planck-Institute for delay line type detectors (Sheomaker *et al.*, 1985).

An extracavity phase corrector element has recently been added to this arrangement to reduce frequency noise in the region of 100–200 kHz as this noise can cause dynamic range problems in the locking of the cavity in the other detector arm. A schematic of the two loop scheme is shown in figure 13.8. Typical results are shown in figure 13.9.

13.2.10 Future prospects

Present methods of reducing the short term frequency fluctuations of lasers seem to be providing very satisfactory results for the present interferometric gravitational wave detectors. However, several orders of magnitude improvement are

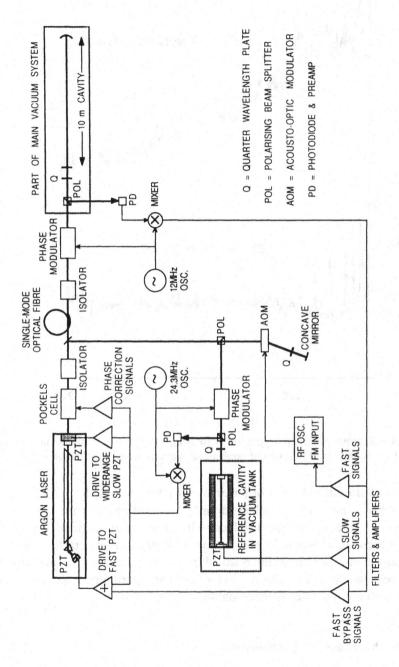

Figure 13.8. Schematic diagram of two loop laser frequency stabilising system being developed in Glasgow.

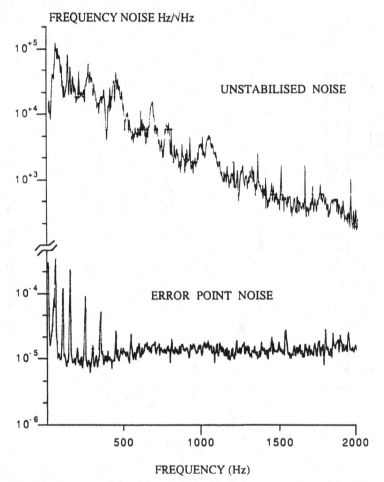

Figure 13.9. Glasgow two loop frequency stabilising system. Error signal in the laser stabilising loop is expressed in terms of equivalent fluctuations in laser frequency.

still required before the desired order of stability for the long baseline detectors will be reached. This could be achieved partly by splitting off more light for the stabilising loops with a consequent reduction in the photon noise limited level and a necessary increase in overall loop gain; partly by electronic subtraction from the detector output of a portion of the residual frequency fluctuations as measured at the error point of the laser stabilising servo loop; and partly, perhaps, by the use of lasers which are intrinsically more stable.

13.3   Laser beam geometry stabilisation

In any interferometer fluctuations in the geometry of the input laser beam will give rise to spurious output signals if the interfering phasefronts are not perfectly matched in alignment and shape. Such mismatching is always present to some degree, and in practice it has been found necessary to reduce the natural beam geometry fluctuations to keep them from being a serious source of noise (Billing *et al.*, 1979). Various techniques have been investigated in the course of the development of the prototype interferometers, and their main features are outlined in the following sections.

Obvious types of geometry fluctuations are those in lateral and angular position of the beam, and in its size and phasefront curvature. These deviations from an ideal beam are normally the most significant in practice and hence the most important to reduce.

13.3.1   Passive suppression of geometry fluctuations
Two main passive approaches have been used to improve beam quality:

    (i)   a resonant cavity used as a mode cleaner, and
   (ii)   a single-mode optical fibre.

(i)   The mode cleaner
The use of a mode cleaner cavity was first proposed in the context of gravity wave laser interferometers by the Max-Planck group, initially as a method for reducing lateral beam jitter (the beam symmetriser – Maischberger *et al.*, 1979), and later in a more general form for controlling all types of geometry fluctuations (Rüdiger *et al.*, 1981).

The mode cleaner is a high-Q optical resonator through which the laser beam propagates on its way to the interferometer. The incident field distribution from the laser can be expanded into a series of eigenmodes of the resonator with the fundamental mode being the one desired for transmission to the interferometer. In operation the cavity is adjusted so that the fundamental mode is resonant. The higher order modes, describing the unwanted geometry fluctuations of the input laser beam, will in general not be in resonance and will thus be considerably suppressed.

For a symmetrical resonator of length d, with mirrors of equal radius of curvature R, and power reflectivity ρ^2, the transmission suppression factor, S_{mn}, of the T_{mn} mode relative to that of the fundamental T_{00} mode is given by (Rüdiger *et al.*, 1981)

$$S_{mn} = \sqrt{\left(1 + \left(\frac{2\rho}{1 - \rho^2}\sin[(n + m)\Psi]\right)^2\right)},$$

where $\cos\Psi = 1 - (d/R)$. By suitable choice of resonator length and mirror curvature it is possible to achieve significant simultaneous suppression factors for the important low order modes. A cavity geometry with $d/R \sim 0.6$ or ~ 1.4 gives about 90% of the maximum suppression for positional fluctuations and about 75% of the maximum for size changes. It also gives reasonable suppression for modes at least up to sixth order.

In early experiments at the Max-Planck-Institute (Rüdiger *et al.*, 1981), suppression factors of ~20 in field amplitude have been demonstrated for modes corresponding to lateral beam jitter together with significant reduction of beam size fluctuations. The availability now of mirrors of much lower loss should allow even more attenuation of higher order modes to be achieved.

Care has to be taken to avoid damage to the mirrors in the mode cleaner due to excessive optical power density, and it may be appropriate in large scale interferometers to use a mode cleaner cavity of several metres length. It would probably be impracticable to have a rigid resonator of this length which is sufficiently stable, but the mode cleaner cavity could be formed between suspended masses in a manner similar to that used for the actual interferometer.

(ii) Single-mode optical fibre

A single-mode optical fibre, as suggested by R. Weiss of MIT, improves the beam quality because any imperfections in the beam geometry are equivalent to higher order fibre modes which are not propagated in the fibre. In order to see how fast these modes decay with distance along the fibre, it is helpful to view the fibre as a waveguide in which the modes may be regarded as the superposition of plane waves travelling at an angle to the axis of

$$\theta \approx \frac{(m+1)\lambda}{4d},$$

where λ is the wavelength, d is the fibre diameter and m is the mode number; $m = 0$ for the fundamental, $m = 1$ for the first-order mode corresponding to a positional change of the beam with respect to the fibre, and $m = 2$ for the mode corresponding to a size change. The plane waves corresponding to the fundamental mode will be totally internally reflected at the core boundary and so will propagate essentially unattenuated; those corresponding to higher order modes will encounter the boundary at too steep an angle to be totally internally reflected and will lose energy on successive bounces. This gives rise to an exponential decay of the mode amplitude with a characteristic $1/e$ length, l_{decay}, given approximately by,

$$l_{\text{decay}} \approx \frac{8d^2}{\lambda \pi m^2 \sqrt{(2m^2 - 1)}}.$$

As an example, the fibre currently used at Glasgow has $d \sim 2.5\,\mu$m, giving a decay length of $\sim 3 \times 10^{-5}$ m for $m = 1$. If a fibre with $d = 10\,\mu$m were available, which would be more suitable for higher power levels, its decay length would be ~ 0.5 mm. It is clear from these numbers that it is in principle possible to obtain a large attenuation of the beam geometry fluctuations using a relatively short piece of fibre. In practice care has to be taken to avoid reintroducing positional noise after the fibre. One approach, proposed and used by the Max-Planck group (Shoemaker *et al.*, 1985), is to mount the output end of the fibre on a separately suspended test mass inside the main vacuum tank for the interferometer.

13.3.2 Active control of beam pointing

An active feedback system to control lateral and angular fluctuations has been developed at Glasgow (Meers, Newton and Drever, 1983). The system uses quadrant photodiodes to monitor the location of the laser beam with respect to a suitable reference at two separated points. In initial experiments the photodiodes were mounted on the interferometer test mass carrying the main beamsplitter. To control both lateral and angular beam position two transducers are required separated by a suitable distance, each capable of beam deflection in two orthogonal dimensions. Signals derived from the quadrant detectors are fed back to the transducers and cause the input laser beam's position in space to be defined relative to that of the photodiodes. In practice it proved convenient to split each transducer into two elements; mirrors mounted on moving coil transducers were used for low frequencies ($<\sim 10$ Hz) where the deviations to be corrected were large, and piezo driven mirrors with much wider bandwidth but more restricted dynamic range were used for faster signals. Servo bandwidths of order 30 kHz were obtained using the piezo transducers, with loop gains of ~ 100 achievable up to 1 kHz. With the relatively small amount of power used of ~ 1 mW per photodiode, the noise level of the sensing system corresponded to a beam movement of $\sim 4 \times 10^{-11}$ m/$\sqrt{}$(Hz), approximately 100 times smaller than the natural positional fluctuations occurring at 1 kHz of the commercial large frame argon ion laser used.

A significant advantage of the active method is that very little optical power need be lost from the main beam; apart from the small amount extracted for the quadrant detectors, the only losses are those occurring on reflection at the transducers.

Extension of this technique to control the size and convergence of the laser beam is clearly possible using two concentric diodes to measure beam size in two locations together with servo controlled lenses in the beam path.

It should be noted that it is also possible to use position sensitive measurements of the relative phase of the interfering beams in an interferometer to produce signals suitable for driving the aligning and focussing transducers, rather than using references that are stable but independent of the interferometer itself.

13.4 Laser intensity stabilisation

Laser power fluctuations in the frequency band of interest can couple into an interferometric gravitational wave detector in two ways – firstly in the fringe detection process, and secondly due to possible unbalanced radiation pressure effects on the test masses in the cavities (Hough *et al.*, 1986; Weiss, 1972).

13.4.1 Fringe detection process

In principle the use of radiofrequency modulation/demodulation techniques at frequencies where the laser is photon noise limited makes the measurement of phase changes independent of low frequency power changes. However, any offsets or errors in the length stabilising systems such that the detected output lies away from the null locking point on the fringe can introduce a sensitivity to them.

In fact the required power stability is given by

$$\frac{\delta P}{P} \sim h \left(\frac{\Delta l}{l}\right)^{-1},$$

where Δl is the offset of an arm of length l from its correct locking point.

Typically for a detector of arm length of a few kilometres, $\Delta l/l$ might be about 10^{-16} (depending on the loop gain of the servo system), resulting in $\delta P/P \sim 3 \times 10^{-8}/\sqrt{(\text{Hz})}$. This requires a modest degree of stabilisation of the laser power.

13.4.2   Radiation pressure effects

These only become important in the case of high laser power and the use of optical recycling. Given reasonable fractional imbalance of power in the two arms of the interferometer and imbalance in the test masses the power stability required seems moderate (Hough et al., 1986). For the present short baseline prototype detectors typical values of $\Delta l/l \sim 10^{-14}$ lead to a required $\delta P/P$ of $10^{-6}/\sqrt{(\text{Hz})}$.

However, there is an additional possibility of chaotic problems in Fabry–Perot cavities at high power level and this is currently being investigated theoretically in Paris (Aguirregabiria and Bell, 1987; Deruelle and Tourrenc, 1984) and in Glasgow (Meers and MacDonald, 1989). The work in Glasgow suggests that there should be no difficulties for kilometre scale detectors. The effect is also predicted to be negligible for the present prototypes with their relatively low input laser powers. Some experimental work related to this issue has been performed by Dorsel et al. (1983) and some directly relevant experiments are planned by Brillet and colleagues at Orsay.

13.4.3   Methods of intensity stabilisation

Most commercial argon lasers are relatively noisy in their intensity in the region of a few hundred hertz to a few kilohertz. Typical performances of argon ion lasers are shown in figures 13.10 and 13.11.

Such lasers are fitted as standard with light control in which a small fraction of the output light has its intensity measured by a photodiode. The output from this diode is compared with a reference level and then amplified and fed back to the power supply to control the tube current. The commercial systems typically work well at zero and low frequency but give little or no improvement in the kilohertz part of the spectrum. This method can be improved either by modifications to the circuitry driving the current amplifiers of the power supply (R. Schilling, 1986, MPQ Garching, private communication) or by adding an extra current drive (A. Brillet, 1986, CNRS Orsay, private communication). With the extra current drive, suppression of intensity fluctuations of greater than 40 dB at 1 kHz have

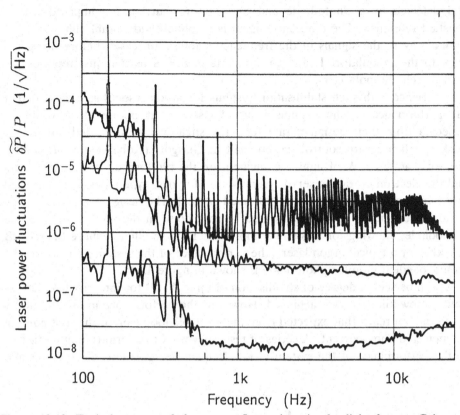

Figure 13.10. Typical spectra of the power fluctuations in the light from a Coherent Innova 90-5 laser (MPQ). The upper curve shows the natural power fluctuations; the middle curve shows the fluctuations after some passive improvement to the power supply; and the lower curve shows the residual fluctuations with the stabilisation system in operation.

been achieved by Brillet and colleagues using a servo loop of unity gain point ~100 kHz; and with modifications to the commercial circuitry of a Coherent Innova 90-5 laser, MPQ obtained the results shown in figure 13.10.

An alternative method of stabilisation which has been fairly widely adopted is the active control of the intensity by an electro optic or acousto optic modulator in the laser output beam (J. L. Hall and J. J. Snyder pioneered this technique in an unpublished manuscript; see also Hall, Layer and Deslattes, 1977, and Layer, 1979.) Several such commercial systems exist. In the electro optic case, a Pockels cell modulator with two pairs of transverse cut ADP crystals with axes at right angles to each other is placed between two crossed polarisers such that the axes of the modulator are at 45° to the axes of the polarisers. The intensity of light transmitted through this combination can be adjusted by varying the voltage applied to the modulator. Typically the applied voltage is such that approximately 50% to 80% of the light is transmitted. A small fraction of this light is split off and detected by a photodiode and the resulting current is compared with a constant reference. The difference signal is amplified and suitably filtered – in a similar way to the signals in the frequency stabilisation case – before being fed back to the modulator. If an acousto optic device is used it just replaces the Pockels cell and polarisers.

Experience with such stabilisation systems for argon lasers has indicated that the performance of the systems is usually somewhat poorer than would be expected from their design parameters. Investigations were carried out with a Pockels cell intensity control system built at Glasgow (Robertson *et al.*, 1986). Approximately 1 mW of light was incident on the photodiode in the servo loop and the electronics for the servo system were of bypass type (similar to that used in the frequency stabilising loops described earlier). The design was such that the performance should have been limited by shot noise in the photocurrent of the photodiode, allowing for a possible improvement in stability of more than 40 dB at 1 kHz for a typical argon laser. The performance of the system as measured by an independent monitor photodiode is shown in figure 13.11.

As can be seen a degree of stabilisation of up to 40 dB for intensity fluctuations below a few kilohertz was achieved. However, the performance in this frequency range did not reach that expected from the available loop gain or the shot noise in the detected photocurrent. A series of tests suggested two serious limiting factors to the performance of the system at low frequencies: geometry fluctuations and

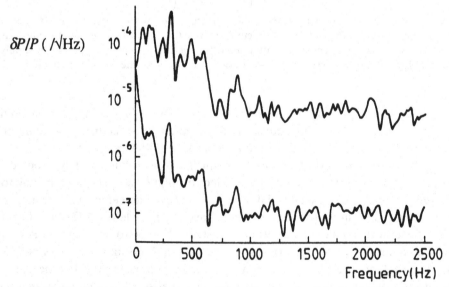

Figure 13.11. Typical spectra of the power fluctuations in the light from a Spectra-Physics Model 165 laser. The upper curve shows the natural power fluctuations, and the lower curve shows the residual fluctuations with the electro optic stabilisation system in operation.

frequency fluctuations of the laser light. The use of an optical fibre to remove geometry fluctuations, together with frequency stabilisation of the laser, greatly improved the situation allowing the design performance to be achieved over the frequency range of interest.

The level of laser intensity stability achieved is more than adequate for the current prototype detectors; and it is believed that the improvement needed for the longer baseline detectors proposed is well within the bounds of present technology.

13.5 Conclusion

It is clear that the operation of all the laser interferometric detectors – both the present prototypes and the proposed longer baseline instruments – depends on a high level of stabilisation of the laser light used for illumination.

Experimental development has reached such a stage that it is generally believed that the level of stability required to allow design sensitivity to be obtained should be achievable, and work is progressing at a number of laboratories to provide further demonstration of this.

References

Aguirregabiria, J. M. and Bell, L. L. (1978). *Phys. Rev. A* **36**, 3768.
Anderson, D. Z., Frisch, J. C. and Masser, C. S. (1984). *Appl. Opt.* **23**, 1238–45.
Barger, R. L., West, J. B. and English, T. C. (1975). *Appl. Phys. Lett.* **27**, (1), 31–3.
Billing, H., Maischberger, K., Rüdiger, A., Schilling, R., Schnupp, L. and Winkler, W. (1979). *J. Phys. E., Sci. Instrum.* **12**, 1043–50.
Billing, H., Winkler, W., Schilling, R., Rüdiger, A., Maischberger, K. and Schnupp, L. (1983). *Quantum Optics, Experimental Gravity, and Measurement Theory*, eds. P. Meystre and M. O. Scully, pp. 525–66, Plenum Press, New York.
Camy, G. (1979). Thèse de docteur Ingenieur, Université Paris 13.
Camy, G., Pinaud, D., Courtier, N. and Hu Chi Chu (1982). *Rev. Physique Appl.* **17**, 357–63.
Deruelle, N. and Tourrenc, P. (1984). *Lecture Notes in Physics*, vol. **212**, pp. 232–7, Springer, Berlin.
Dorsel, A., McCullen, J. D., Meystre, P., Vignes, E. and Walther, H. (1983). *Phys. Rev. Lett.* **51**, 1550–3.
Drever, R. W. P., Hall, J. L., Kowalski, F. V., Hough, J., Ford, G. M., Munley, A. J. and Ward, H. (1983a). *Appl. Phys. B* **31**, 97–105.
Ford, G. M. (1981). Ph.D. Thesis, University of Glasgow.
Hackel, L. A., Hackel, R. P. and Ezekiel, S. (1977). *Metrologia* **13**, 141–3.
Hall, J. L. (1978). *Science* **202**, 147–56.
Hall, J. L. and Hansch, T. W. (1984). *Opt. Lett.* **9**, 333–4.
Hall, J. L., Layer, H. P. and Deslattes, R. D. (1977). *IEEE J. Quantum Electronics* **QE-13**, 45D–46D.

Hamilton, M. (1989). *Contemporary Physics*, **30**(1), 21–33.

Helmcke, J., Lee, S. A. and Hall, J. L. (1982). *Appl. Opt.* **21**, 1686–94.

Hils, D. and Hall, J. L. (1987). *Rev. Sci. Instrum.* **58**, 1406.

Hough, J., Drever, R. W. P., Munley, A. J., Lee, S.-A., Spero, R., Whitcomb, S. E., Ward, H., Ford, G. M., Hereld, M., Roberston, N. A., Kerr, I., Pugh, J. R., Newton, G. P., Meers, B., Brooks, E. D. III and Gursel, Y. (1983a). In *Quantum Optics, Experimental Gravity, and Measurement Theory*, eds. P. Meystre and M. O. Scully, pp. 515–24, Plenum Press, New York.

Hough, J., Hils, D., Rayman, M. D., Ma, L.-S., Hollberg, L. and Hall, J. L. (1984). *Appl. Phys. B* **33**, 179–85.

Hough, J., Meers, B. J., Newton, G. P., Robertson, N. A., Ward, H., Schutz, B. F., Drever, R. W. P., Tolcher, R. and Corbett, I. F. (1986). *British Design Study Report* GWD/RAL/86-001.

Kerr, G. A., Robertson, N. A., Hough, J. and Man, C. N. (1985). *Appl. Phys. B* **37**, 11–16.

Layer, H. P. (1979). *Appl. Opt.* **18**, 2947–9.

Lee, S. A., Helmcke, J. and Hall, J. L. (1979). *Proc. Laser Spectroscopy IV*, Rottach-Egern, June 1979, p. 130.

Maischberger, K., Rüdiger, A., Schilling, R., Schnupp, L., Winkler, W. and Billing, H. (1979). *Proc. of 2nd Marcel Grossman Meeting on General Relativity, Trieste 1979*, ed. R. Ruffini, pp. 1083–100, North Holland, Amsterdam.

Meers, B. J. and MacDonald, N. (1989). *Phys. Rev. A* **40**, (7), 3754.

Meers, B. J., Newton, G. P. and Drever, R. W. P. (1983). Unpublished report, University of Glasgow.

Pound, R. V. (1946). *Rev. Sci. Instrum.* **17**, 490.

Robertson, N. A., Hoggan, S., Mangan, J. B. and Hough, J. (1986). *Appl. Phys. B* **39**, 149–53.

Rüdiger, A., Schilling, R., Schnupp, L., Winkler, W., Billing, H. and Maischberger, K. (1981). *Optica Acta* **28**, 641.

Rüdiger, A., Schilling, R., Schnupp, L., Winkler, W., Billing, H. and Maischberger, K. (1982). Internal Report MPQ 68, November 1982, Max-Planck-Institut für Quantenoptik, Garching.

Salomon, C., Hils, D. and Hall, J. L. (1988). *J. Opt. Soc. B* **5**, (8), 1576.

Shoemaker, D., Winkler, W., Maischberger, K., Rüdiger, A., Schilling, R. and Schnupp, L. (1985). Internal Report MPQ 100, August 1985, Max-Planck-Institut für Quantenoptik, Garching, and in *Proc. of 4th Marcel Grossmann Meeting on Recent Developments in Relativity and Gravitation*, ed. R. Ruffini, pp. 605–14, Elsevier Science Publishers, Amsterdam.

Shoemaker, D., Brillet, A., Man, C. N., Cregut, O. and Kerr, G. A. (1989). *Opt. Lett.* **14**, (12), 609.

Spero, R. (1985). *Proc. of 4th Marcel Grossman Meeting on Recent Developments in Relativity and Gravitation*, ed. R. Ruffini, p. 615, Elsevier Science Publishers, Amsterdam.

Ward, H., Hough, J., Newton, G. P., Meers, B. J., Robertson, N. A., Hoggan, S., Kerr, G. A., Mangan, J. B. and Drever, R. W. P. (1985). *IEEE Trans. Instrum. Measurement* **IM-34**, no. 2, 261–5.

Weiss, R. (1972). MIT Quarterly Progress Report (Research Laboratory of Electronics) **105**, 54.

White, A. D. (1965). *IEEE J. Quantum Electronics* **QE-1**, no. 8, 349–57.

14

Vibration isolation for the test masses in interferometric gravitational wave detectors

N. A. ROBERTSON

14.1 Introduction

14.1.1 Why good broadband seismic isolation is an essential design feature for laser interferometric antennas

One of the key features of laser interferometric detectors is the potential wideband nature of their operation. Proposed long baseline detectors are intended to achieve sensitivities in the region $h \sim 10^{-21}$ to 10^{-22} or better over a range of frequencies f from a few tens of hertz (possibly as low as 10 Hz) to a few kilohertz in a bandwidth $\Delta f \approx f/2$. If the performance of such detectors is limited by photon shot noise in the output light, for constant light power the effect of this noise source decreases towards lower frequencies for a constant light intensity, when the detectors are operated in searches for burst sources or a stochastic background. However, other sources of noise have spectra which rise towards lower frequencies. These include thermal noise from the pendulum suspensions of the masses, and, more particularly, seismic noise. In fact it is likely that the extent to which these detectors can be operated with reasonable sensitivity at the lower end of the frequency spectrum will depend crucially on the level of seismic and mechanical isolation achievable. Since there are interesting sources of gravitational waves in the region of ten to a few hundred hertz, such as fast pulsars and coalescing compact binary systems, it is advantageous to incorporate as much seismic isolation as practicably possible into the design of these detectors. However, exactly how much isolation is required as a function of frequency will depend not only on the sensitivity being sought, but also on the intrinsic spectrum of the seismic noise at the site chosen for the detector. It is therefore instructive firstly to consider the general spectrum of seismic noise to be expected before going on to consider the question of isolation.

14.1.2 The spectrum of seismic noise

Sources of seismic noise, both natural and artificial, are numerous and varied (see e.g. Richter, 1958). Natural phenomena such as tectonic motions of the earth's crust, storms, wind and water in motion (including ocean waves) and artificial sources such as traffic, machinery and general industrial activity combine to produce a general continuous background level of seismic motion which we term seismic noise. Obviously with such a wide variety of potential noise sources, the level of seismic noise can vary greatly from place to place. In particular in the geophysical 'high frequency' regime (>0.5 Hz) seismic noise is mostly generated by local disturbances such as wind, moving water and so-called 'cultural noise'. The level of seismic noise between a very quiet site and a noisy site can vary by at least two or even three orders of magnitude. The quietest sites are those away from man-made sources and the coastline, and below the earth's surface (e.g. down a mine). In the absence of mines with suitable long tunnels at right angles, research groups intending to build long baseline detectors are planning to site them on or near the surface of the ground at sites reasonably removed from man-made noise.

Measurements of the spectrum of seismic noise at frequencies >0.5 Hz have been carried out by many workers, and it seems that the levels of seismic noise in the vertical and horizontal directions are essentially of the same order of magnitude. One fairly consistent finding at reasonably quiet sites is that the linear spectral density of displacement in each dimension appears to vary to a good approximation as $1/f^2$ (corresponding to a spectral density of acceleration independent of frequency). Measurements taken at some gravitational wave laboratories give figures such as $(3 \times 10^{-7}/f^2)$ m/$\sqrt{}$(Hz) at Garching (Shoemaker *et al.*, 1988), $(10^{-6}/f^2)$ m/$\sqrt{}$(Hz) at Pisa (Giazotto, 1987), and $(3 \times 10^{-6}/f^2)$ m/$\sqrt{}$(Hz) at Glasgow, all in bandwidths between approximately 1 Hz and 1 kHz. At a potential site for a long baseline detector in Scotland the linear spectral density approximated to $(10^{-7}/f^2)$ m/$\sqrt{}$(Hz). For comparison, Weiss (1972) has reported measurements at a very quiet site, namely down a mine at a depth of about 0.5 km, which approximated to $(2 \times 10^{-9}/f^2)$ m/$\sqrt{}$(Hz).

From these figures, one can see the large variation between a typical laboratory and a 'very quiet' site underground. To try to estimate the isolation necessary we shall assume a figure of $(10^{-7}/f^2)$ m/$\sqrt{}$(Hz) for horizontal motions, over the range 1 Hz to 1 kHz, but one should appreciate that this figure could vary depending on the site.

To calculate how such a spectrum of seismic noise would affect an interferometric detector, we recall that such a detector consists in its simplest form of three test masses suspended at three corners of a square to form two horizontal arms at right angles. The relative displacements of the two arm lengths are monitored to look for the effect of a gravitational wave interacting with the detector. Thus it is mainly the horizontal component of ground noise which is of importance. Some sensitivity to vertical motions will be present if the detector arms are not level or if there is coupling between horizontal and vertical modes. In the discussions which follow in this chapter it will be assumed that it is mainly horizontal isolation which has to be provided.

The simple pendulum suspension of each test mass has a transfer function in the horizontal direction between the point of suspension (assumed to be rigidly

Table 14.1. *Estimation of the extra seismic isolation required for the test masses of an interferometric gravitational wave detector. Each test mass is assumed to be suspended as a simple pendulum of resonant frequency 1 Hz.*

	$f = 10$ Hz	$f = 100$ Hz	$f = 1$ kHz
Sensitivity required (h)	10^{-22}	10^{-22}	10^{-22}
Noise level due to seismic motions	$6 \times 10^{-14}\left(\dfrac{1\,\text{km}}{l}\right)$	$2 \times 10^{-17}\left(\dfrac{1\,\text{km}}{l}\right)$	$6 \times 10^{-21}\left(\dfrac{1\,km}{l}\right)$
Extra isolation required	$6 \times 10^{8}\left(\dfrac{1\,\text{km}}{l}\right)$	$2 \times 10^{5}\left(\dfrac{1\,\text{km}}{l}\right)$	$60\left(\dfrac{1\,\text{km}}{l}\right)$

attached to the ground for the present) and the mass itself, which is given by $(f_0/f)^2$ at frequencies above f_0 (the resonant frequency of the pendulum) for a high Q suspension. (We shall ignore at present the effect of the finite mass of the suspension wire which introduces other resonances of the suspension – the so-called 'violin' modes of the suspending wires.) Thus each mass will have an r.m.s. displacement of

$$\frac{10^{-7}f_0^2}{f^4}\frac{\text{m}}{\sqrt{(\text{Hz})}}.$$

Assuming uncorrelated motions of the masses in the detector, the r.m.s. noise introduced into the relative displacement of the two arms of the detector will be

$$\bar{x}(f) = \frac{2 \times 10^{-7}f_0^2}{f^4}\frac{\text{m}}{\sqrt{(\text{Hz})}}, \qquad (14.1)$$

and thus the sensitivity limit $\bar{h} = \bar{x}(f)/l$ due to seismic motions for a detector of arm length l will be

$$\bar{h}(f) \sim 2 \times 10^{-10}\left(\frac{1\,\text{Hz}}{f}\right)^4\left(\frac{f_0}{1\,\text{Hz}}\right)^2\left(\frac{1\,\text{km}}{l}\right)\cdot\frac{1}{\sqrt{(\text{Hz})}} \qquad (14.2)$$

Considering as an example the detection of burst sources of duration $\tau \approx 1/f$, and bandwidth $\Delta f \approx f/2$, and aiming for a detector sensitivity of $h \approx 10^{-22}$, we can compute the necessary isolation required *over and above* that achieved with the simple pendulum suspensions assuming $f_0 = 1$ Hz.

From the figures in table 14.1, the magnitude of the problem may now be appreciated. The extra isolation required at 1 kHz should be achieved with simple passive isolation techniques such as have been used with success for resonant bar detectors and for the prototype interferometric detectors. However, the problem increases dramatically towards lower frequencies. The isolation required at 100 Hz should be achievable using combinations of several well known and relatively simple passive isolation techniques. For operation below this frequency one will require to extend the simple passive techniques or introduce active isolation systems.

14.2   Methods of isolation

In this section, methods of isolation, with particular application to the free test masses of laser interferometric detectors, will be reviewed. For the purposes of presentation, the methods are subdivided into passive and active. However, as will be made clear, this division is often somewhat arbitrary, since some systems to be described incorporate both passive and active means of isolation.

14.2.1   Passive techniques

(i) Simple pendulum suspensions
Reference has been made in the previous section to the fact that the test masses are suspended as pendulums. A simple pendulum suspension is probably the simplest isolation system one can envisage. For a point mass, m, suspended on a wire of length L and negligible mass, we may calculate the transfer function between the displacement of the point of suspension, x_0, and the displacement of the mass, x_1, in the horizontal direction, as a function of angular frequency ω. Assuming a damping force given by $F_b = b(\dot{x}_1 - \dot{x}_0)$, the transfer function (for small displacements) is

$$\frac{x_1}{x_0} = \frac{\gamma s + \omega_0^2}{s^2 + \gamma s + \omega_0^2} \tag{14.3}$$

where $s = j\omega$, $\omega_0^2 = g/L$, $\gamma = b/m$.

This ratio is also known as the transmissibility. It should be noted that another damping term proportional only to the velocity of the mass itself (so-called internal damping) may in some cases be present. This will not be considered here.

It can immediately be seen that above the resonant frequency, given by $f_0 = \sqrt{(g/L)}/2\pi$, the transmissibility or transfer function is essentially ω_0^2/ω^2 or f_0^2/f^2 until a frequency given approximately by $f = f_0 Q$ (where Q is the quality factor, $Q = \omega_0/\gamma$). Above this frequency the function tends to f_0/Qf. Since it is planned to use suspensions of high quality factor ($Q > 10^6$) to avoid introducing excess motion due to thermal noise of the suspensions, for all possible working frequencies of the detector (say 10 Hz to 10 kHz) the isolation will be $(f/f_0)^2$. Thus for $f_0 = 1$ Hz ($L = 0.25$ m) the isolation achieved at 1 kHz, for example, would be 10^6 with the above model. Increasing the length L would lead to increased isolation, but the choice of length is restricted by practical considerations such as the size of vacuum tank required and the violin mode resonances treated below.

Such a suspension forms the basic isolation for the test masses of an interferometric detector. In practice, however, the assumptions of a point mass and suspending wire(s) of negligible mass are not valid. The masses will have finite size, and if not suspended exactly at their centres of mass, there will be tilting modes of oscillation which may introduce excess noise into the longitudinal motion of the mass, especially around the resonant frequencies of these tilting modes. In practice such tilting modes (and rotational modes also) of the masses may be controlled and damped by electronic feedback at low frequencies (<10 Hz). This, combined with suspending the masses close to their centres of mass to reduce cross-coupling of modes and to make the unwanted modes as low frequency as possible, should be enough to remove any significant noise due to these modes.

A potentially more serious problem is likely to arise from the finite mass of the suspending wires which produces other resonances of the wires – the violin mode resonances. The transfer function for the pendulum suspension in this case (assuming at present no damping losses) is given by (Robertson, 1981; Shoemaker et al., 1988)

$$\frac{x_1}{x_0} = \frac{1}{\cos(\omega L/c) - \dfrac{\omega c}{g}\sin(\omega L/c)}, \tag{14.4}$$

where the velocity of propagation of waves along the wire, $c = \sqrt{(T/\rho_L)}$, the tension in the wire $T = mg$, and ρ_L is the linear density of the wire.

The lowest resonance can easily be derived from this equation under the condition $\omega L/c \ll 1$, and one obtains the simple isolation function. This is equivalent to equation (14.3) without damping ($\gamma = 0$):

$$\frac{x_1}{x_0} = \frac{\omega_0^2}{\omega_0^2 - \omega^2}, \tag{14.5}$$

where $\omega_0^2 = g/L$. All the other resonances can be approximated to a series of frequencies f_n such that

$$f_n = \frac{c}{\lambda_n} = \sqrt{\left(\frac{mg}{\rho_L}\right)} \cdot \left(\frac{n}{2L}\right), \tag{14.6}$$

where $\lambda_n = 2L/n$. In between these resonances the isolation is at best $\omega c/g$, and so increases only as f, not as f^2 as for the ideal simple pendulum. The validity of this simple analysis has been demonstrated by the group at Garching, who have compared theoretical and experimental data for the transfer function of the suspension system of a mirror in their 30 m prototype detector (Shoemaker et al., 1988). At the resonant frequencies f_n, the isolation will obviously decrease, and depending on the Q's of these resonances, the pendulum mass motion could in fact be enhanced (e.g. for the case of no damping x_1/x_0 becomes infinite at these frequencies). Thus the presence of the violin modes can in principle cause two undesirable features in the isolation achieved. The first of these is the decrease in attenuation or even the amplification of noise at the frequencies f_n. The second is the change in the dependence of the isolation with frequency from f^2 to f. For completeness one should mention a third problem which is not strictly concerned with isolation, namely that such resonances may cause difficulties in any servo system used to control the overall path length of the arms of an interferometric detector by introducing undesirable phase changes in the transfer function. However, it is beyond the scope of this chapter to consider this effect.

The problems introduced by the violin modes could in principle be ameliorated by shifting their frequencies to be above the frequency band in which the detector operates. However, this is not in practice very easy. For example, noting that $c = \sqrt{(\text{breaking stress}/N\rho)}$, where N is the factor by which the breaking stress exceeds the stress in the wire and ρ is the density of the wire, and using characteristics for piano wire, which has a very high breaking stress to density ratio, we find

$$f_1 = 880 \sqrt{\left(\frac{2}{N}\right)\left(\frac{f_0}{1\,\text{Hz}}\right)^2} \text{ Hz.} \qquad (14.7)$$

f_1 could be increased by increasing f_0 but this would be at the expense of decreasing the passive isolation ($\propto 1/f_0^2$). Thus it is likely that at least the first two or three violin mode resonances will fall within a typical detector's working frequency band. If the first of these is made sufficiently high, as in the example above, the deterioration of the rate of increase of isolation from f^2 to f above this frequency should not be a serious problem. However, at the specific resonant frequencies themselves the decrease in isolation could be very significant. The magnitudes of the peaks in the transmissivities will depend on the actual Q's of these resonances, and it may prove necessary to provide notches at these frequencies in the detector signal output, or to electronically damp the resonances in some manner.

(ii) Compound pendulums

An obvious way to increase isolation is to use two or more pendulum suspensions in series (see, for example, Lorrain, 1977). The resulting transfer function between the top point of suspension and the lowest mass has the form $f_1^2 \cdot f_2^2 \ldots f_n^2/f^{2n}$ for an n-stage suspension system at frequencies above the highest resonant frequency of the coupled system. Here the f_n's are the normal modes and the effects of damping and violin mode resonances have been neglected. It can be seen that the isolation achievable increases as f^{2n}, but its actual value at any particular frequency also depends on the normal mode frequencies, which will be a function of the resonant frequencies of each stage separately and of the ratios of the masses of each stage. For example, for a two-stage system with upper and lower resonant frequencies of each stage separately being f_1 and f_2 and masses of m_1 and m_2, respectively, the transfer function is given by

$(1 + m_2/m_1)f_1^2 \cdot f_2^2/f^4$ at frequencies f, above the highest normal mode. Thus to achieve the best isolation one would like to have $m_2 \ll m_1$, which also has the beneficial effect of keeping the highest normal mode low. In practice, however, it may be difficult to achieve this, especially for low frequency operation of an interferometric detector, when the test masses will be required to be of the order of 1000 kg to reduce the effects of thermal noise.

A two-stage pendulum system is being used successfully by the group at Garching in their 30 m prototype detector, and its measured isolation agrees well with the expected transfer function (Shoemaker et al., 1988). Here the upper pendulum is in fact a massive plate suspended by coil springs, and the lower pendulum is a mirror (forming one of the test masses of the detector) suspended in a wire sling. The use of coil springs as the upper pendulum suspension gives some isolation against vertical seismic and mechanical motions of the point of support which may couple weakly into horizontal and rotational motions of the suspended optical component through cross-coupling effects.

A more ambitious seven-stage pendulum with isolation in both horizontal and vertical directions is being developed by a group at Pisa (Del Fabbro et al., 1987, 1988). The idea here is to design a passive isolation system which will allow operation of an interferometric detector in the frequency range down to 10 Hz. The system being investigated consists of six masses of 100 kg and a final test mass of 400 kg, each separated by 0.7 m of steel wire. Thus the total length is approximately 5 m. Each of the masses except the lowest incorporates a gas spring supporting the next mass to provide isolation in the vertical direction. The highest resonant frequencies in the horizontal and vertical directions are calculated to be approximately 3 Hz and 5 Hz, respectively, and such a scheme should allow an isolation of the test masses of 10^{11} and 10^8 at 10 Hz for horizontal and vertical motions, respectively. Initial investigation of this system looks promising.

It should be mentioned here that care must be taken in designing any multiple suspension for the masses of an interferometric detector to ensure that horizontal displacements due to thermal noise from the upper stages do not exceed those due to the thermal noise level in the final pendulum suspension of the test mass, which will necessarily be of high Q ($>10^6$), at frequencies of operation of the detector. In practice this means that the Q of any upper stage must be larger than some minimum value which will depend on the masses and resonant frequencies of the system. This criterion is not likely to be difficult to achieve for the type of n-stage pendulum systems described above. However, this may not necessarily be the case when considering isolation stacks used in conjunction with a single pendulum suspension, and we shall return to this point in the next section.

Another interesting point to note in this respect is the potential contribution to the horizontal displacements of a suspended mass introduced by thermal noise associated with vertical motions of the suspension. These may be converted into horizontal motions via cross-coupling. This idea has been addressed by Del

Fabbro *et al.* (1987, 1988) who have measured the level of cross-coupling in one stage of their seven-stage pendulum. They conclude that the required Q at the suspension in the vertical direction should be achievable.

One final topic which should be noted here is that the use of multiple stage suspensions may increase the complexity of any feedback systems used to control the separation of two such suspended test masses. This will be particularly true if one wishes to provide electronic damping of some or all the normal modes of the system. Such damping is desirable to avoid large motions at the resonant frequencies of the system.

(iii) Acoustic isolation stacks

A method of isolation which has been applied with success in gravitational wave detectors, both of resonant bar and interferometric design, is the use of isolation stacks consisting of alternating layers of a heavy metal (e.g. lead) and a soft elastic material (e.g. rubber). These stacks are normally situated between the 'ground' and the point of suspension of the resonant bar or test mass. Each stage consisting of a layer of metal and rubber has a behaviour in each dimension similar to a mass connected to a spring. In the horizontal direction one is considering shear motions inside the rubber, and in the vertical direction compressional motions. Thus the transfer function for one layer is the same as for a simple pendulum suspension (equation (14.3)), where the resonant frequency is related to the dimensions and elastic modulus of the rubber, and the mass of the metal block. In general the resonant frequency of each layer will be higher in the vertical direction than in the horizontal.

For a stack of n layers, each of resonant frequency f_0 in the appropriate direction, the transfer function from bottom to top is given by f_0^{2n}/f^{2n}, where it is assumed that each stage is identical, the damping is low, and that we are considering frequencies above the normal modes. For high damping the isolation will increase only as f^n and not f^{2n}. Thus for example a five-stack system close to being critically damped could in theory give an isolation factor of 10^5 at 100 Hz for $f_0 = 10$ Hz.

It is interesting to note here that an n-stage stack does not have the same transfer function as an n-stage pendulum of identical masses, with the same natural frequencies of each stage in the two cases. This is due to a fundamental difference in the dynamics of the two systems. In the pendulum case, unlike the stack case, the restoring force on each mass in the chain depends on all the masses below it. However in a stack, the restoring force on each mass only depends to first order on the properties of the rubber, assumed identical in each stage. An n-stage pendulum of equal masses and resonant frequency of each stage alone of f_0 has a transfer function of $n: f_0^{2n}/f^{2n}$ under the same conditions as above, and thus its isolation is worse by a factor $n!$. An n-stage pendulum also has the disadvantage of a higher maximum normal mode frequency, which approaches $2\sqrt{(n)}f_0$ for large n. In comparison the highest normal mode frequency of a stack approaches

$2f_0$ for large n. However, it is possible that loading effects on the dynamics of the rubber in the stacks may result in the behaviour becoming closer to that of a multiple pendulum system.

Thus stacks would appear to have good isolation properties and indeed, as has been mentioned above, they have been in use with success in past and present detectors. However, they can have one or two disadvantages. For example lead and rubber would not be compatible materials for use in the ultra-high vacuum systems ultimately required for interferometric detectors, and other materials would have to be found. Secondly the resonant frequency typically achieved by one layer of a stack is in the region of several hertz, so that isolation is only starting to increase above frequencies of approximately 10 Hz. Thirdly achieving high Q's for each stage is not as easy as for pendulum suspensions. This has an advantage when considering the undesirable enhancement of motion at the normal mode frequencies (motion proportional to Q), but has two disadvantages, namely the decreased isolation, and the possible increase in noise due to thermal noise in the stack. Also in practice measured isolation achieved may not equal the theoretical isolation but may flatten out at some level due to the effects of internal resonances of the masses of the stacks and cross-coupling of modes, and a stack may couple in tilts more strongly than a multiple pendulum system. However, given all these provisos, stacks are still likely to be a useful tool for isolation, used in conjunction with one or more stages of pendulum suspension.

(iv) Pneumatic air mounts

Widely used and commercially available systems for providing isolation platforms on which sensitive experiments may be carried out make use of pneumatic air mounts. These typically consist of pressurised cylinders with an internal piston supporting the load. The gas under the piston acts as a spring of low spring constant in both horizontal and vertical dimensions giving isolation in principle of the same form as a damped mechanical spring or a simple pendulum. In practice the form of the isolation achieved can be such that there is low amplification at the resonant frequency as if the system were reasonably damped, but the isolation increases above this frequency as f^2 and not f as would be expected. A transfer function of this characteristic shape can be achieved by providing damping which is only of a significant value at frequencies around the resonant frequency of the air mount (see, for example, the Newport Catalog no. 100). An example of a similar transfer function in a pendulum isolation system has been demonstrated by Lorrain (1977). In his system reasonable damping of the resonance of a test mass on its pendulum suspension is achieved through a small auxiliary mass suspended below it, thus maintaining the f^2 isolation factor for the test mass.

The commercial pneumatic isolation mounts available act in both horizontal and vertical directions, with resonant frequencies typically in the 0.5 to 2 Hz region (the vertical frequency being lower than the horizontal), and quoted isolation figures of better than 100 above a few hertz are normal. Isolation in the vertical is usually better, and may reach 1000 by a few tens of hertz. However, it should be noted that the isolation of such systems has not been extensively investigated at high frequencies (hundreds of hertz), and there is some evidence for high frequency structural resonances which may have the effect of flattening off the isolation level achieved at these higher frequencies.

An attractive feature of such systems is their load bearing capacity, which can easily be several tonnes. Thus such isolation mounts could be used to support a massive platform from which the test masses in an interferometric detector would be suspended, either directly or via a multi-layer stack situated on the platform. Obviously such a scheme would have to be incorporated separately at the centre and the two ends of the interferometer.

As described in the section on compound pendulums, gas springs form an integral part of the seven-stage pendulums which are under investigation at Pisa (Del Fabbro *et al.*, 1988). Unlike the commercial pneumatic air mounts discussed above, these gas springs are intended to provide vertical isolation only. A gas spring has advantages over a helical steel spring for application to the suspension of heavy loads, since the static extension is more easily controlled, and hysteresis effects are reduced.

14.2.2   Active techniques

(i) Simple pendulum with feedback to lower its resonant frequency
It has already been seen that the isolation achieved using a simple pendulum can be increased by increasing the length of the suspension, which is equivalent to lowering the natural resonant frequency of the pendulum. There are obviously practical limits to the length which can easily be utilised. Geologists have overcome the problem of making very low frequency horizontal seismometer suspensions by using either a construction similar to a slightly tilted hinged gate, or an inverted pendulum held vertical by weak springs. These would appear to have some disadvantages in the case of the suspension of the test masses of a gravitational wave detector. Firstly these suspensions will strongly couple in tilting motions of the ground (whereas a simple pendulum suspension to first order does not couple in tilts about its point of suspension). Secondly the achievement of a very high Q suspension with these techniques may be significantly more difficult than for a simple pendulum. However, the principle of using a lower natural resonant frequency suspension in a gravitational wave detector may be achieved with a simple pendulum by increasing the apparent length using active feedback (R. W. P. Drever, private communication). In practice the idea is to monitor the relative displacement of the point of suspension of a pendulum mass with respect to the mass itself, and feed a suitably amplified signal back to the point of suspension in such a way as to reduce the motion of the suspension point and hence also of the mass itself. The effect of such a feedback scheme is to lower the apparent resonant frequency of the pendulum, as will be shown shortly.

A system for achieving improved isolation against horizontal disturbances (which are probably the most important for interferometric detectors as currently conceived) using the idea expressed above has been extensively investigated at Glasgow (Robertson, 1981; Robertson *et al*, 1982) and subsequently at Pisa (Campani, Giazotto and Passuello, 1986; Giazotto, Passuello and Stefanini, 1986). The basic concept is similar in principle to that of Erath *et al.*, reviewed by Melton (1957), for lengthening the period of a mass and spring seismometer by electronic means. Improved isolation against vertical motions has also been developed, notably by Rinker (1983) and Rinker and Faller (1984). They have constructed a vertical 'anti-seismic' system using a test mass suspended from a helical spring. The resultant 'super-spring' had its natural frequency of 0.9 Hz reduced to an effective value of 3×10^{-3} Hz by feedback. Work on a prototype vertical isolation system for gravitational wave antennas has also been carried out by Saulson (1984) at MIT, and a reduction in natural resonant frequency from 4.5 Hz to 0.04 Hz was achieved.

To understand how one such system works for horizontal motions, consider the following simplified analysis, and refer to figure 14.1. The equations of motion for the mass (assuming a perfect simple pendulum and no natural damping) is

$$\ddot{x}_1 = \omega_0^2(x_0' - x_1), \tag{14.8}$$

where x_0' is the position of the point of suspension.

Now assume a signal is fed back to a motion transducer at the point of suspension (for example a piezo-electric transducer), such that $x_0' = x_0 + A(x_1 - x_0')$, where A is the gain of the feedback loop.

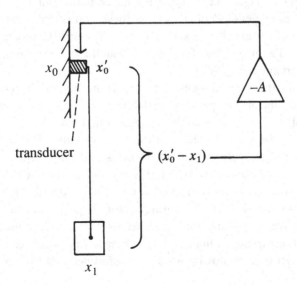

Figure 14.1. Schematic diagram of an active anti-seismic system.

Thus it can be shown that

$$\frac{x_1}{x_0} = \frac{\omega_0^2}{s^2(1+A) + \omega_0^2} = \frac{\omega_0^2/(1+A)}{s^2 + \omega_0^2/(1+A)}. \tag{14.9}$$

Comparing this with equation (14.3) we see that this active system behaves as a pendulum of new resonant angular frequency $\omega_0' = \omega_0/\sqrt{(1+A)}$, and that above this frequency the transfer function tends to $\omega_0^2/[\omega^2(1+A)]$, i.e. the isolation has been improved by the factor $1+A$. Further if the feedback function takes the form $A(1+a/s)$ instead of A alone, where A and a are constants and $s = j\omega$, the effect is to electronically damp the resonance, thus avoiding significant amplification of motion at the new resonant frequency without compromising the isolation achieved above that frequency. It is interesting to note that the damping is achieved by using the *integral* of the sensing signal, and this may be understood by recalling (c.f. equation (14.8)) that the sensing signal is proportional to the acceleration of the mass. This method of achieving damping of the motion of the mass is different from the form of feedback required to stabilise a suspended mass to the ground, where a differentiated sensing signal is normally required.

The above theory explains the principle of operation of an active anti-seismic system. However, there are several practical points which need to be considered. These include: (i) the achievement of D.C. stability in the loop, since the transfer function from the driving point at the top of the suspension to the differential sensing signal tends to zero at very low frequencies and this may cause difficulties in avoiding saturation in electronic elements in the loop, (ii) the effects of additional resonances such as violin modes of the suspension wires, (iii) the use of a practical integration function with finite gain at D.C., and (iv) the method of sensing the required signal $(x_1 - x_0')$. This latter point will be discussed briefly here, but for a more detailed discussion of these and other points the reader its referred to the various articles on such feedback systems (Campani, Giazotto and Passuello, 1986; Giazotto, 1987; Giazotto, Passuello and Stefanini, 1986; Robertson, 1981; Robertson *et al.*, 1982).

One can obtain the desired signal for vertical motions by directly sensing the extension of the supporting spring as in Rinker and Faller's *Superspring* (1984), or by the use of a vertical accelerometer (Saulson, 1984). For a system designed to work in the horizontal direction with a mass suspended as a pendulum, the problem is somewhat different. One could sense the required signal using accelerometers whose outputs were suitably combined to remove any signals due to tilts of the ground. However, thermal noise associated with the small masses in the accelerometer tends to be a limiting factor to performance. In principle it seems better to use the suspended mass as its own accelerometer and deduce $(x_1 - x_0')$ by measuring the inclination of the suspending wires. A structure rigidly attached to the ground cannot be used as a reference against which to measure this inclination if the system is to be unaffected by ground tilts, and thus a freely suspended arm of very low resonant frequency is required as a reference (Drever et al., 1983; Hough et al., 1983; Robertson et al., 1982). This arm effectively provides an inertial reference direction which allows the desired signal to be sensed, as shown in figure 14.2. The sensing signal may be measured capacitively, or optically using an interferometric scheme. The optical scheme has an important practical advantage in requiring less critical positioning of the test mass with respect to the reference arm.

A prototype horizontal system in Glasgow, incorporating a reference arm and utilising capacitive sensing, achieved a loop gain A of approximately 60 over a bandwidth of 30 Hz for a pendulum of mass 0.5 kg and natural resonant frequency 1.7 Hz, giving a new resonant frequency of ~0.2 Hz (Robertson et al., 1982). Some work was also carried out on an optical sensing scheme, and the potential of such a system was clearly established (Robertson, 1981). At Pisa a system based on this optical scheme but with some important new additions, was used to achieve isolation for a 100 kg test mass. One of the key additions was the use of a D.C. motor of large dynamic range which acted as a supplementary integrating feedback element used in conjunction with a piezo-electric transducer at the point of suspension. With the feedback in operation, the natural pendulum frequency of 0.5 Hz (pendulum length = 1 m) was reduced to 0.012 Hz (virtual pendulum length of 1.7 km), at frequencies up to at least 10 Hz (Giazotto, Passuello and Stefanini, 1986). There were difficulties with the operation of the system due to air motions when the reference arm was free to move and not damped in any way. However this may not be a serious problem when such systems are operated under vacuum.

The potential of these isolation systems has been clearly demonstrated by

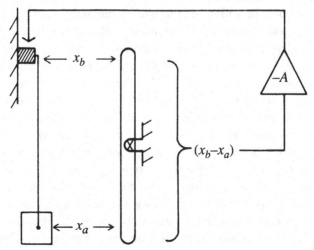

Figure 14.2. Active anti-seismic system with a reference arm.

several workers. However there are still some practical problems which will need more investigation before such systems can be incorporated into the design of an interferometric detector.

(ii) Feedback to control the relative separation of the points of suspension of two or more test masses

In an interferometric detector one is interested in looking at the relative separations of test masses. Thus, rather than attempting to improve the isolation of each test mass separately, one could monitor the separation of the points of suspension in each of the two arms of the detector, and feed back a suitable signal to these points to minimise changes in their separation. Such a scheme is at present under development at Caltech, where the relevant motions are sensed using Michelson interferometers, with lengths of optical fibre acting as the references (Chen and Drever, 1987).

(iii) Active measurement and correction for seismic noise

If one can measure the motions of the free masses in an interferometric detector due to ground vibrations (for example by using reference arms), instead of using the signal in a feedback loop to reduce such motions, one could subtract these signals from the main interferometric output signal. Alternatively, by monitoring the relative separations of the points of suspension of the test masses in the two arms of the interferometer, one could deduce the effects on the relative separation of the test masses and subtract this resulting signal from the main interferometer output. To implement such systems will require a very good model of the detailed dynamics of the suspensions, including all resonances.

(iv) Actively controlled platforms

In the section on passive isolation, the use of pneumatic air mounts to isolate massive platforms was discussed. An extension of this idea is to incorporate active feedback into a passive inertial platform system to increase the isolation. Such systems have been described by Weinstock (1968) and Lorenzini (1972) among others, and a commercial system (EVIS) has recently been announced by Newport (Catalog no. 100). These platforms are typically based on modified conventional pneumatic isolation systems. For example, Weinstock describes the addition of two-dimensional active tilt isolation achieved by sensing tilts using accelerometers and gyroscopes, and using the amplified signal to drive a servomechanism to restore the platform to level. Lorenzini describes a more ambitious scheme with a combination of tiltmeters, angular accelerometers, seismometers and a gyrocompass measuring the six degrees of freedom of the system, and with pressure transducers and electromagnetic shakers compensating for motions induced by ground vibrations in three dimensions. The commercial Newport device incorporates accelerometers as sensors and force transducers as the feedback element.

It is instructive to note here that these isolation systems work on a principle different from the active anti-seismic schemes described in section (i) above. Here the motion which is sensed is that of the mass to be isolated itself, and the feedback is applied between the mass and the ground. In the active systems outlined in (i), the signal sensed is the relative motion of the point of suspension of the mass with respect to the mass itself, and the feedback is applied between the ground and the point of suspension. However, the form of the isolation achieved is very similar.

One could envisage using actively controlled isolation platforms in the same manner as described in the section on pneumatic air mounts, where the platforms are used to provide a stable isolated base from which to suspend the test masses of the detector.

14.3   Conclusions

In the previous section we have seen that there are several potential solutions to the problem of seismic isolation for free masses. In section 14.1 we have seen how dramatically the magnitude of the problem of isolation increases towards lower frequencies in an interferometric gravitational wave detector. To achieve the required level of isolation at frequencies above approximately 100 Hz one could combine the isolation properties of pneumatic mounts and a stack of several stages with the simple pendulum suspension. For performance below 100 Hz it is likely that either one of the active systems described above, or a many-stage pendulum such as that being investigated at Pisa, will be required.

While there is still much work to be done in investigating and developing the more novel systems described above, the prospects are encouraging that some or all of these schemes will provide enough isolation from seismic disturbances to allow the optimum low frequency performance of laser interferometric gravitational wave detectors to be achieved.

Acknowledgements

The author wishes to thank J. Hough and A. Rüdiger for help and advice in writing this article.

References

Campani, E., Giazotto, A. and Passuello, D. (1986). *Rev. Sci. Instrum.* **57**, 79–81.
Chen, Y. T. and Drever, R. W. P. (1989). In *Proceedings of International Symposium on Experimental Gravitational Physics, Guangzhou, August 1987*, World Scientific, Singapore.

Del Fabbro, R., Di Virgilio, A., Giazotto, A., Kautzky, H., Montelatici, V. and Passuello, D. (1987). *Phys. Lett. A* **124,** 5.

Del Fabbro, R., Di Virgilio, A., Giazotto, A., Kautzky, H., Montelatici, V. and Passuello, D. (1988). *Rev. Sci. Instrum.* **59,** (2), 292–7.

Drever, R. W. P., Hough, J., Munley, A. J., Lee, S. A., Spero, R., Whitcomb, S. E., Ward, H., Ford, G. M., Hereld, M., Robertson, N. A., Kerr, I., Pugh, J. R., Newton, G. P., Meers, B. J., Boorks, E. D. III, and Gursel, Y. (1983). In *Proceedings of NASI on Quantum Optics and Experimental General Relativity*, Bad Windsheim, August, 1981, eds. P. Meystre and M. O. Scully (Plenum, New York), pp. 503–14.

Giazotto, A. (1987). *Proceedings of 7th Conference on General Relativity and Gravitational Physics*, eds. U. Bruzzo, R. Cianci and E. Massa, pp. 449–63, World Scientific, Singapore.

Giazotto, A., Passuello, D. and Stefanini, A. (1986). *Rev. Sci. Instrum.* **57,** 1145–51.

Hough, J., Drever, R. W. P., Munley, A. J., Lee, S. A., Spero, R., Whitcomb, S. E., Ward, H., Ford, G. M., Hereld, M., Robertson, N. A., Kerr, I., Pugh, J.R., Newton, G. P., Meers, B. J., Brooks, E. D. III, and Gursel, Y. (1983). In *Proceedings of NASI on Quantum Optics and Experimental General Relativity*, Bad Windsheim, August, 1981, eds. P. Meystre and M. O. Scully (Plenum, New York), pp. 515–24.

Lorenzini, D. A. (1972). *Proc. AIAA Guidance and Control Conference*, Stanford, California, paper no. 72–843.

Lorrain, P. (1977). *Rev. Sci. Instrum.* **48,** 1397–9.

Melton, B. S. (1957). *Advances in Electronics and Electron Physics,* vol. IX, pp. 307–11, Academic Press, New York.

Richter, C. F. (1958). *Elementary Seismology,* W. H. Freeman & Co., San Francisco.

Rinker III, R. L. (1983). Ph.D. Thesis, University of Colorado.

Rinker III, R. L. and Faller, J. E. (1984). In *Precision Measurements and Fundamental Constants II,* eds. B. N. Taylor and W. D. Phillips, Natl. Bur. Stand. (U.S.), Spec. Publ., **617,** pp. 411–17.

Robertson, N. A. (1981). Ph.D. Thesis, University of Glasgow.

Robertson, N. A., Drever, R. W. P., Kerr, I. and Hough, J. (1982). *J. Phys. E. Sci. Instrum.* **15,** 1101–5.

Saulson, P. R. (1984). *Rev. Sci. Instrum.* **55,** 1315–20.

Shoemaker, D., Schilling, R., Schnupp, L., Winkler, W., Maischberger, K. and Rüdiger, A. (1988). *Phys.Rev. D* **38,** 423–32.

Weinstock, H. (1968). NASA Technical Report TRR-281.

Weiss, R. (1972). MIT Quart. Progress Report No. 105.

15
Advanced techniques: recycling and squeezing

A. BRILLET*, J. GEA-BANACLOCHE†, G. LEUCHS†, C. N. MAN*
AND J. Y. VINET*

15.1 Introduction to recycling

All the long baseline interferometers for the detection of gravitational radiation which are presently being studied are based on the construction of a large, Michelson-like interferometer with an armlength of 1 to 4 km, containing some kind of gravito-optic transducer in each arm. In order to improve the shot-noise limited sensitivity, all these interferometers will use high-power lasers, in conjunction with so-called light recycling techniques. The basic idea of recycling was proposed by R. W. P. Drever (1983): it consists in building a resonant optical cavity which contains the interferometer, so that, if the losses are low and if the cavity is kept on resonance with the incoming monochromatic light, there is a power build-up which results in a reduction of the shot noise. This can be realized in different ways, depending on the geometry of the gravito-optic transducer (delay line or Fabry–Perot).

A general theory of recycling interferometers was recently developed and published (Vinet *et al.*, 1988) and the Garching (Schnupp, 1987) and Orsay (Man, 1987) groups have obtained the first experimental verifications of the efficiency of this technique. In this chapter, we first remind the reader of the main ideas and results of the theory, which is fully expressed in Vinet *et al.* (1988). We then describe today's experimental achievements, and we end up with a short discussion of possible future improvements.

15.2 Theory of recycling interferometers

The aim of this section is to establish several simple models and associated formulas giving the ultimate photon-noise limited sensitivities of some presently experimented interferometer configurations. We deliberately choose to discuss in some detail the most simple recycling configurations, at the cost of not describing the most exotic ones. The interested reader is welcome to consult Meers (1988)

---

* Sections 15.1–15.4. † Sections 15.5–15.8.

and Vinet *et al.* (1988). In order to carry out this program, some special tools are useful: first, a set of standard parameters built from values of optical elements allowing comparisons. Secondly, a common formalism allowing a straightforward derivation of the properties of arbitrary optical configurations.

The cases of non-recycling delay-line and Fabry–Perot Michelson interferometers are treated first in order to develop the formalism. Then we apply these results to various recycling configurations and discuss the relative merits of each configuration according to the frequency range, to the bandwidth of the signal and to the value and the localization of the optical losses which limit the power build-up.

15.2.1   Optics in a weakly modulating medium

(i) General principles

Consider a plane, transverse, traceless, monochromatic gravitational wave of frequency v_g, which propagates perpendicularly to the interferometer plane $(z = 0)$, and is linearly polarized along the directions of the (orthogonal) interferometer arms $(x = 0$ and $y = 0$ respectively):

$$[h_{ij}(x, y, z, t)]_{z=0} = h_{ij} \cdot \cos(\Omega t + \Phi) \tag{15.1}$$

with

$$h_{ij} = \mathrm{diag}(h, -h, 0); \ \Omega = 2\pi v_g.$$

At every point of the optical path, the light frequency spectrum will resolve in a carrier frequency

$$v_{\mathrm{opt}}$$

and two sidebands

$$v_{\mathrm{opt}} \pm v_g.$$

Considering only first order effects in h, the optical amplitudes at an arbitrary point of the interferometer are of the form

$$\mathbf{A}(t) = \left[A_0 + \frac{1}{2} h A_1 e^{i(\Omega t + \Phi)} + \frac{1}{2} h A_2 e^{-i(\Omega t + \Phi)} \right] e^{-i\omega t} \tag{15.2}$$

$$\omega = 2\pi v_{\mathrm{opt}}.$$

We represent the action of gravitational transducers upon already modulated light by linear operators $\mathbf{S}$ acting upon generalized amplitudes

$$\mathbf{A} = (A_0, A_1, A_2)$$

as

$$\mathbf{A}' = \mathbf{S} \cdot \mathbf{A}.$$

According to the formalism developed in Vinet *et al.* (1988), these operators have the general form

$$S = \begin{pmatrix} S_{00} & 0 & 0 \\ S_{10} & S_{11} & 0 \\ S_{20} & 0 & S_{22} \end{pmatrix}. \tag{15.3}$$

In this formalism, the diagonal elements S_{ii} represent the ordinary reflectance (or transmittance) of the transducer for each frequency (carrier and sidebands), whereas S_{10} and S_{20} characterize the power transfer from the carrier to the sidebands, i.e. the sensitivity to the gravitational wave. Optical elements with dispersion and no gravitational sensitivity will be represented by diagonal matrices, elements without dispersion or G sensitivity by scalar matrices (mirrors, splitters etc.).

The whole interferometer is itself a gravitational transducer and has therefore an associated global operator **S**. If the limiting noise reduces to the shot noise, the signal to noise ratio (SNR) is:

$$\text{SNR} = h \sqrt{\left(\frac{\eta P \tau_i}{\hbar \omega_{\text{opt}}}\right)} S; \qquad S = |S_{10}| + |S_{20}|, \tag{15.4}$$

where τ_i and η are, respectively, the integration time and quantum efficiency of the photodetector, and P is the power of the source. In other words, the minimum, photon-noise limited, detectable h is:

$$h_{\text{PN}} = \sqrt{\left(\frac{\hbar \omega_{\text{opt}}}{\eta P}\right)} \frac{1}{S} \sqrt{(\delta v)}, \tag{15.5}$$

where δv is the bandwidth of the detector. In what follows, we shall consider the quantity S, that we call normalized signal to noise ratio (NSNR), as the quantity to be optimized.

(ii) Standard gravito-optic transducers

Gravito-optic transducers are optical devices in which the gravitational wave (GW) is supposed to have a detectable perturbing effect, such as delay lines and Fabry–Perot cavities. Both have associated operators (respectively **D** and **F**), which can be related to the elementary propagation operator **X** corresponding to a round trip in the perturbed vacuum.

We have $\mathbf{A}' = \mathbf{X} \cdot \mathbf{A}$ with

$$\mathbf{X} = \left\{ \begin{array}{ccc} e^{2i\omega L/c} & 0 & 0 \\ i\varepsilon \dfrac{\omega L}{c} \dfrac{\sin(\Omega L/c)}{\Omega L/c} e^{i(2\omega - \Omega)L/c} & e^{2i(\omega - \Omega)L/c} & 0 \\ i\varepsilon \dfrac{\omega L}{c} \dfrac{\sin(\Omega L/c)}{\Omega L/c} e^{i(2\omega + \Omega)L/c} & 0 & e^{2i(\omega + \Omega)L/c} \end{array} \right\} \tag{15.6}$$

The value of ε is $+1$ for a round trip along the x-axis, and -1 along the y-axis. Consider now an n-fold delay line with two mirrors of intensity reflection

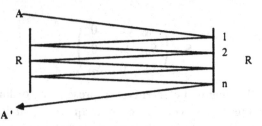

Figure 15.1. n-fold delay line (notations).

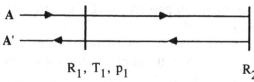

Figure 15.2. Reflecting Fabry–Perot cavity (notations).

coefficients R (figure 15.1). It consists in n iterations of the $\mathbf{X}$ operator and $2n - 1$ iterations of the operator $i\sqrt{R}$. Its associated operator is thus $i\mathbf{D}$, where:

$$\mathbf{D} = (-1)^{n-1}\sqrt{(R)}^{2n-1}\mathbf{X}^n. \tag{15.7}$$

Consider a Fabry–Perot cavity (see figure 15.2) with a front mirror of intensity reflection and transmission coefficients R_1, T_1, with losses p_1, and with a rear mirror of intensity reflection coefficient R_2. The associated operator $i\mathbf{F}$ looks like the ordinary reflectance of a Fabry–Perot cavity but with the ordinary phase factor replaced by $\mathbf{X}$:

$$\mathbf{F} = [\sqrt{(R_1)} + (1 - p_1)\sqrt{(R_2)}\mathbf{X}] \cdot [1 + \sqrt{(R_1 R_2)}\mathbf{X}]^{-1}. \tag{15.8}$$

Response of delay-line type detectors   In more detail, the delay-line operator involves the three following elements:

$$\begin{cases} D_{00} = (-1)^n\sqrt{(R)}^{2n-1}e^{4in\pi v_{\mathrm{opt}}L/c} \\[2ex] D_{10} = (-1)^n\sqrt{(R)}^{2n-1}i\varepsilon\dfrac{v_{\mathrm{opt}}}{v_g}\sin\left(\dfrac{2n\pi v_g L}{c}\right)e^{2i\pi(2v_{\mathrm{opt}}-v_g)nL/c} \\[2ex] D_{20} = (-1)^n\sqrt{(R)}^{2n-1}i\varepsilon\dfrac{v_{\mathrm{opt}}}{v_g}\sin\left(\dfrac{2n\pi v_g L}{c}\right)e^{2i\pi(2v_{\mathrm{opt}}+v_g)nL/c}. \end{cases} \tag{15.9}$$

The action of this operator on an unmodulated wave is therefore a pure phase modulation. It is convenient to introduce some parameters which have their counterparts in the case of cavities:
By introducing the storage time

$$\tau_s = \frac{2nL}{c}$$

the time constant

$$\tau'' = \frac{2L}{cR^*}, \qquad R^* = 1 - R,$$

and the ratio $t = \tau_s/\tau''$ which has the minimum value $t_m = R^*$, we get:

$$R^{n-1/2} = e^{-(t-t_m/2)}.$$

In what follows, we shall consider kilometric interferometers ($L \approx 3$ km) so that the minimum value of τ_s, i.e. $2L/c$, is about 2×10^{-5} s, and high reflectivity coatings ($R^* \approx 10^{-4}$) so that τ'' is about 0.2 s. This set of parameters will be referred to as the reference antenna. As will be shown later, the best τ_s for a given gravitational frequency $v_g^{(0)}$ is of order $1/2v_g^{(0)}$. Insofar as we consider gravitational frequencies smaller than a few kilohertz, we can assume $t \gg t_m$. Two more parameters are useful: the normalized gravitational frequency $f = 2\pi v_g \tau''$ and the maximum quality factor $Q = 2\pi v_{opt} \tau''$. (In the reference antenna, when visible light is used, the quality factor Q is about 7.5×10^{14} and $f \approx 1.26 v_g/\text{Hz}$).

The approximate form of the operator **D** simplifies now to:

$$\begin{cases} |D_{00}| = e^{-t} \\ |D_{10}| = |D_{20}| = \dfrac{Q}{f} e^{-t} \left| \sin\left(\dfrac{ft}{2}\right) \right| \end{cases} \tag{15.10}$$

If the detection system involves two delay lines, it has a NSNR

$$S(f) = \frac{2Q}{f} e^{-t} \left| \sin\left(\frac{ft}{2}\right) \right|. \tag{15.11}$$

For a given gravitational frequency corresponding to f_0, there exists an optimal normalized storage time:

$$t_0 = \frac{2}{f_0} \tan^{-1}\left(\frac{f_0}{2}\right). \tag{15.12}$$

Note that $f_0 \to 0$ yields $\tau_s^{(0)} \to \tau''$, and thus τ'' may be interpreted as the maximum value of the optimal storage time.

The optimized NSNR is then:

$$S(f) = \frac{2Q}{f} \left| \sin\left[\frac{f}{f_0} \tan^{-1}\left(\frac{f_0}{2}\right)\right] \right| \exp\left[-\frac{2}{f_0} \tan^{-1}\left(\frac{f_0}{2}\right)\right]. \tag{15.13}$$

We have at $f = f_0$:

$$S(f) = \frac{2Q}{\sqrt{(f_0^2 + 4)}} \exp\left[-\frac{2}{f_0} \tan^{-1}\left(\frac{f_0}{2}\right)\right]. \tag{15.14}$$

Therefore, we can give the limiting value of S when $f_0 \to 0$:

$$S(0) = Q/e.$$

For $f_0 \gg 1$, a good approximation of the optimal storage time is given by $t_0 = \pi/f_0$, i.e.

$$\tau_s^{(0)} = \frac{1}{2v_g^{(0)}}$$

and the NSNR becomes simply:

$$S(f) = \frac{2Q}{f} \left| \sin\left(\frac{\pi f}{2 f_0}\right) \right|$$

i.e.

$$S(v_g) = \frac{2v_{opt}}{v_g} \left| \sin\left(\frac{\pi v_g}{2v_g^{(0)}}\right) \right|. \tag{15.15}$$

Response of Fabry–Perot type detectors   For the Fabry–Perot cavity operator, the relevant elements are

$$
\begin{cases}
F_{00} = \dfrac{(1-p_1)\sqrt{(R_2)}e^{2i\omega L/c} + \sqrt{(R_1)}}{1 + \sqrt{(R_1 R_2)}e^{2i\omega L/c}} \\[2mm]
F_{10} = i\varepsilon T_1 \sqrt{(R_2)}\,\dfrac{v_{opt}}{v_g} \sin\left(\dfrac{\Omega L}{c}\right) \dfrac{e^{i(2\omega-\Omega)L/c}}{(1 + \sqrt{(R_1 R_2)}e^{2i\omega L/c})(1 + \sqrt{(R_1 R_2)}e^{2i(\omega-\Omega)L/c})} \\[2mm]
F_{20} = i\varepsilon T_1 \sqrt{(R_2)}\,\dfrac{v_{opt}}{v_g} \sin\left(\dfrac{\Omega L}{c}\right) \dfrac{e^{i(2\omega+\Omega)L/c}}{(1 + \sqrt{(R_1 R_2)}e^{2i\omega L/c})(1 + \sqrt{(R_1 R_2)}e^{2i(\omega+\Omega)L/c})}.
\end{cases}
\tag{15.16}
$$

The eigenfrequencies of the cavity are determined by the condition $\exp(-2i\omega_0 L/c) = -1$ and consequently, when the optical source is resonant, the preceding operator denotes pure phase modulation. We need now some dimensionless parameters analogous to the delay line's. We may define the time constant of the cavity by

$$\tau_s' = \frac{2L}{c[1 - \sqrt{(R_1 R_2)}]}. \tag{15.17}$$

Owing to the constraint $0 < R_1 < 1 - p_1$, we have:

$$\frac{2L}{c} < \tau_s' < \tau'' \equiv \frac{2L}{c[1 - \sqrt{((1-p_1)R_2)}]}. \tag{15.18}$$

The ratio of the time constant to its maximum value will be called the normalized time constant t; it obeys

$$t_m < t < 1 \quad \text{with} \quad t_m \equiv 1 - \sqrt{((1-p_1)R_2)}.$$

In the general case, the source is eventually detuned by Δv_{opt} from a resonance v_0 which leads us to introduce the normalized detuning defined by $\Delta f = 2\pi \Delta v_{opt} \tau''$. With these notations, the ordinary reflectance of the cavity has the exact expression:

$$|F_{00}|^2 = \frac{1}{R_2} \frac{(1 - 2t + tt_m)^2 + (1 - t_m)^2(1 - t_m/t)\Delta f^2 t^2 \operatorname{sinc}^2(\Delta f t_m/2)}{1 + \Delta f^2 t^2 (1 - t_m/t) \operatorname{sinc}^2(\Delta f t_m/2)}, \tag{15.19}$$

where the notation $\text{sinc}(x)$ denotes the function $\sin(x)/x$. Fortunately a simple approximate form can be given when Δf is much smaller than the free spectral range and when $t_m \ll 1$, which is satisfied when gravitational frequencies are restricted to a range of values less than a few kilohertz:

$$|F_{00}| = \sqrt{\left[1 - \frac{4t(1-t)}{1 + \Delta f^2 t^2}\right]}. \tag{15.20}$$

The minimum value of $|F_{00}|$ is reached at resonance where

$$|F_{00}|_{\text{res}} = |1 - 2t|.$$

Within the same approximation, we have

$$\text{Arg } F_{00} = \pi + \tan^{-1}\left[\frac{2\Delta ft(1-t)}{1 - 2t - \Delta f^2 t^2}\right] \tag{15.21}$$

and

$$\begin{cases} |F_{10}| = \dfrac{Qt(1-t)}{\sqrt{(1 + \Delta f^2 t^2)}\sqrt{(1 + (\Delta f + f)^2 t^2)}} \\[4mm] |F_{20}| = \dfrac{Qt(1-t)}{\sqrt{(1 + \Delta f^2 t^2)}\sqrt{(1 + (\Delta f + f^2)t^2)}}. \end{cases} \tag{15.22}$$

In particular when the source is resonant ($\Delta f = 0$), we have the special case:

$$\begin{cases} |F_{00}| = |1 - 2t| \\[4mm] |F_{10}| = |F_{20}| = \dfrac{Qt(1-t)}{\sqrt{(1 + f^2 t^2)}} \end{cases} \tag{15.23}$$

We have consequently for the NSNR:

$$S(f) = \frac{2Qt(1-t)}{\sqrt{(1 + f^2 t^2)}}. \tag{15.24}$$

When the normalized gravitational frequency $f_0 = 2\pi v_{\text{g}}^{(0)}\tau''$ tends to zero, the optimal value of τ_{s}' has the limiting value $\tau''/2$ and the limiting value of the SNR turns out to be $Q/2$. When $f_0 \gg 1$, the function $S(t)$ begins to saturate as soon as

$$t_0 = \frac{2}{f_0},$$

i.e.

$$\tau_{\text{s}}'^{(0)} = \frac{1}{\pi v_{\text{g}}^{(0)}}.$$

This value will be taken as a reasonable choice, for the true optimal value is much higher but irrelevant, giving only a slightly better value of S. In this case, we have for the NSNR:

$$S(f) = \frac{4Q}{\sqrt{(f_0^2 + 4f^2)}}; \tag{15.25}$$

in particular,

$$S(f_0) = 1.78 \frac{Q}{f_0} . \tag{15.26}$$

In dimensional expression, this is

$$S(v_g) = \frac{4v_{opt}}{v_g^{(0)}} \frac{1}{\sqrt{[1 + (2v_g/v_g^{(0)})^2]}} \tag{15.27}$$

and

$$S(v_g^{(0)}) = 1.78 \frac{v_{opt}}{v_g^{(0)}} . \tag{15.28}$$

Let us emphasise an important point: with Fabry–Perot cavities, it is possible to use a detuned source with respect to the cavity eigenfrequency of an amount $\Delta f = f_0$ so that the sideband generated by the gravitational wave is resonant:

$$v_{opt} = v_0 + v_g^{(0)},$$

leading to

$$|F_{00}| = \sqrt{\left(1 - \frac{4t(1 - t)}{1 + f_0^2 t^2}\right)}$$

and

$$|F_{10}| = \frac{Qt(1 - t)}{\sqrt{(1 + f_0^2 t^2)}\sqrt{(1 + (f - f_0)^2 t^2)}} .$$

When $f_0 \gg 1$, a reasonable choice of τ_s' is again

$$\tau_s' = \frac{1}{\pi v_g^{(0)}}$$

and with only one resonant sideband, the optimized NSNR becomes

$$S(f) = 0.89 \frac{Q}{f_0} \frac{1}{\sqrt{(1 + 4(1 - f/f_0)^2)}} \tag{15.29}$$

or

$$S(v_g) = 0.89 \frac{v_{opt}}{v_g^{(0)}} \frac{1}{\sqrt{(1 + 4(1 - v_g/v_g^{(0)})^2)}} . \tag{15.30}$$

Figure 15.3 gives a comparison of the sensitivities versus v_g for a delay line, for a Fabry–Perot both at resonance and with detuning, in the conditions we have described above. The detuned Fabry–Perot is less sensitive than the other configurations, but the fact that it brings a higher reflectance makes it interesting when recycling is applied, as we will see in the next part.

15.2.2   Standard recycling

(i) Principles of the standard recycling
A classical Michelson interferometer tuned at a dark fringe behaves just like a mirror: most of the power incoming from the source is reflected back. We can use

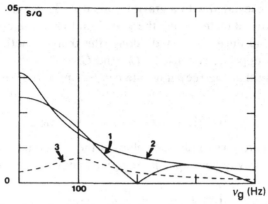

Figure 15.3. Transfer function of a Michelson interferometer with multipass arms (optimized at 100 Hz): 1, delay lines; 2, resonant FP cavities; 3, detuned FP cavities.

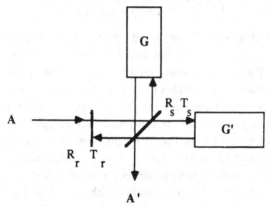

Figure 15.4. Sketch of the standard recycling setup.

it as the second mirror of a cavity; the front mirror of it is called the recycling mirror. It will be shown that this configuration increases the SNR. Figure 15.4 gives the principle of operation. Let R_r, T_r, p_r be the parameters (reflectivity and transmittivity coefficients, losses) of the recycling mirror, and R_s, T_s, p_s those of the splitter.

It is easy to show that at a dark fringe we have an operator **S** for the whole system:

$$\mathbf{A'} = \mathbf{S} \cdot \mathbf{A},$$

where the relevant coefficients of **S** are:

$$\begin{cases} |S_{10}| = (1-p_s)\sqrt{(T_r)}\,\dfrac{|G_{10}|}{1-(1-p_s)\sqrt{(R_r)}\,|G_{00}|} \\[4mm] |S_{20}| = (1-p_s)\sqrt{(T_r)}\,\dfrac{|G_{20}|}{1-(1-p_s)\sqrt{(R_r)}\,|G_{00}|}\,. \end{cases} \qquad (15.31)$$

G is the operator associated with a gravito-optic transducer, either a delay line or a Fabry–Perot cavity, directed along the y-axis, and **G′** is the operator associated with the same transducer, directed along the x-axis; both have the same coefficients G_{ii} but opposite coefficients G_{01} and G_{02}.

One sees already that the recycling rate can be optimized for given losses and **G**, we find:

$$[\surd(R_r)]_{\text{optim}} = (1 - p_r)(1 - p_s)\,|G_{00}|.$$

So that it is possible to give optimized values of the components of the NSNR. By assuming low extra cavity losses $p = p_r + 2p_s$ we may write simply:

$$
\begin{cases}
|S_{10}| = \dfrac{|G_{10}|}{\surd(1 - (1-p)\,|G_{00}|^2)} \\[4mm]
|S_{20}| = \dfrac{|G_{20}|}{\surd(1 - (1-p)\,|G_{00}|^2)}.
\end{cases}
\tag{15.32}
$$

The problem to be discussed below is the optimization of either the storage time for delay lines, or the decay time for cavities, when recycling is applied. In the general situation, the optimal value of that time constant will depend on the gravitational frequency f_0 at which one wants to optimize, and on the recycling losses denoted by p. Two frequency ranges will appear: the low frequency range and the high frequency range. In the low frequency range, long storage times are required, the recycling losses are therefore dominated by the reflectivity losses in the arms, and the optimal storage time is almost independent of p. In the high frequency range, the required storage times are relatively short, so that the losses in the arms may happen to be comparable with the recycling losses p, and the optimal storage time will depend on both p and f_0. The effective value of p will be determined not only by the losses of the recycling mirror and of the beamsplitter but also by the fact that the interference on the beamsplitter may be affected by small misalignments or by a slight asymmetry between the two arms of the Michelson interferometer.

(ii) Standard recycling with delay lines
In the case when **G** represents a delay line, as shown earlier, we have

$$
\begin{cases}
|G_{00}| = e^{-t} \\[4mm]
|G_{10}| = |G_{20}| = \dfrac{Q}{f}\left|\sin\!\left(\dfrac{ft}{2}\right)\right|e^{-t}.
\end{cases}
\tag{15.33}
$$

After some algebra, one finds the optimized frequency response of the recycling setup:

$$S(f) = \frac{Q}{f}\,\surd\!\left(\frac{2f_0}{x_0}\right)\left|\sin\!\left(\frac{x_0 f}{2f_0}\right)\right|.
\tag{15.34}$$

In particular, if the frequency f_0 is within the especially interesting band $1 \ll f_0 \ll 1/p$, say 50 Hz to 500 Hz in the reference antenna, we can write:

$$S(f) = 0.92 \frac{Q}{f} \sqrt{(f_0)} \left| \sin\left(\frac{1.17f}{f_0}\right) \right|,$$

in particular

$$S(f_0) = 0.85 \frac{Q}{\sqrt{(f_0)}},$$

or, in ordinary notation:

$$S(v_g) = 0.92 \frac{v_{opt}}{v_g} \sqrt{(2\pi v_g^{(0)} \tau'')} \, |\sin(1.17 v_g / v_g^{(0)})|. \tag{15.35}$$

Such an optimized transfer function (for 100 Hz) is plotted in figure 15.5.

(iii) Case of resonant Fabry–Perot cavities

The relevant operator **G** is now **F**:

$$\begin{cases} |F_{00}| = |1 - 2t| \\ \\ |F_{10}| = |F_{20}| = Qt(1-t)/\sqrt{(1+f^2t^2)} \end{cases} \tag{15.36}$$

and, within the band $50 \to 500$ Hz in the reference antenna, we can take $t_0 = 1/f_0$, i.e.

$$\tau_s^{\prime(0)} = \frac{1}{2\pi v_g^{(0)}}.$$

The optimized frequency response of the recycling setup is now:

$$S(f) = Q \sqrt{\left(\frac{f_0}{f^2 + f_0^2}\right)},$$

in particular

$$S(f_0) = \frac{Q}{\sqrt{(2f_0)}}.$$

In ordinary notation we have:

$$S(v_g) = v_{opt} \sqrt{\left(\frac{2\pi \tau''}{v_g^{(0)}}\right)} \frac{1}{\sqrt{[1 + (v_g / v_g^{(0)})]}}. \tag{15.37}$$

An optimized transfer function (for 100 Hz) is represented on figure 15.5 so as to be compared with the case of delay lines.

(iv) Case of detuned Fabry–Perot cavities

As already noted, Fabry–Perot cavities can be driven out of resonance, leading to a different response to gravitational frequencies, to a slightly worse signal amplitude, but a higher reflectivity. One can expect this mode of operation to give interesting results in a recycling configuration. Let us discuss this idea.

Assume the optical frequency to be detuned with respect to an eigenfrequency of the cavity by an amount equal to the gravitational frequency to be detected;

$$\nu_{opt} = \nu_0 + \nu_g^{(0)}. \tag{15.38}$$

The detuned cavity operator, as shown earlier, contains the following elements;

$$\begin{cases} |F_{00}|^2 = 1 - \dfrac{4t(1-t)}{1+\Delta f^2 t^2} \\[4mm] |F_{10}| = \dfrac{Qt(1-t)}{\sqrt{(1+\Delta f^2 t^2)}\sqrt{(1+(\Delta f - f)^2 t^2)}}. \end{cases} \tag{15.39}$$

With an optimal recycling rate for $\nu_g = \nu_g^{(0)}$ and with $\Delta f = f_0$, we find the NSNR as

$$S(f) = \frac{S(f_0)}{\sqrt{(1+(f-f_0)^2 t^2)}}. \tag{15.40}$$

The peak value, $S(f_0)$ is

$$S(f_0) = \frac{Qt(1-t)}{\sqrt{(p(1+f_0^2 t^2) + 4t(1-t)(1-p))}}. \tag{15.41}$$

We shall consider that $p \ll t$, for it will be seen that this approximation holds even for relatively high frequencies (up to kilohertz in the reference antenna) due to the fact that high values of the optimal time constant τ_s' are required in the detuned system. The resulting expression for the peak value of the NSNR is

$$S(f_0) = Q\sqrt{\left[\frac{t(1-t)^2}{(pf_0^2 - 4)t + 4}\right]}. \tag{15.42}$$

The optimal value of t is

$$t_0 = \frac{2}{3+\sqrt{(1+2pf_0^2)}} \Rightarrow S(f_0) = Q\sqrt{\left[\frac{2(1+\sqrt{(1+2pf_0^2)})}{(3+\sqrt{(1+2pf_0^2)})^3}\right]}. \tag{15.43}$$

The NSNR has a narrow band type behavior characterized by a bandwidth of

$$\delta f = (3 + \sqrt{(1+2pf_0^2)})\sqrt{3}. \tag{15.44}$$

The transfer functions of delay-line and resonant FP interferometers are not essentially changed by standard recycling apart from a gain factor, but the transfer function of the detuned FP interferometer becomes resonant, due to the long time constant required (see figure 15.5). It is easily seen that, in the limit $pf_0 \to 0$, $S(f_0) \to Q/4$ and $\delta f \to 4\sqrt{3}$.

With a choice of x_0 two or three times less than the optimal value, the linewidth is seen to increase while the peak value is only slightly decreased, giving interesting transfer functions, with a finite band response localized in the gravitational spectrum. Examples of transfer functions corresponding to that mode of operation are plotted in figure 15.6.

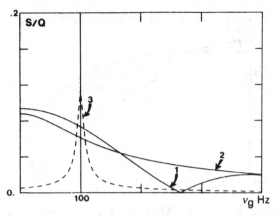

Figure 15.5. Transfer function of a Michelson interferometer with multipass arms and standard recycling (optimized at 100 Hz): 1, delay lines; 2, resonant FP cavities; 3, detuned FP cavities.

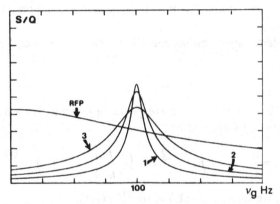

Figure 15.6. Transfer functions for standard recycling with detuned FP cavities and non-optimal time constants: 1, $t = t_{opt}$; 2, $t = t_{opt}/2$; 3, $t = t_{opt}/4$. RFP: standard recycling interferometer with resonant FP cavities (for comparison).

Figure 15.7 resumes the discussion of standard recycling systems by a plot of the optimal NSNR value for the different systems. The sensitivity gain provided by recycling techniques is of the order of 10 above 50 Hz, and becomes negligible below 10 Hz (where the sensitivity is unlikely to be shot-noise limited anyway).

15.2.3 Numerical estimations

We give below numerical estimations of the shot-noise limited sensitivity based on the reference antenna ($L = 3$ km, mirror losses $p_r = 10^{-4}$, recombination losses $p_s = 10^{-3}$, effective laser power $\eta P = 10$ W at a wavelength of 0.5 μm).

Apart from the ordinary non-recycling interferometers involving delay lines or FP cavities, the standard recycling scheme provides us with three new wideband

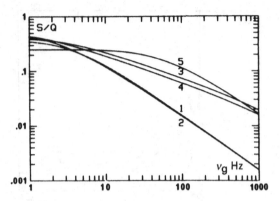

Figure 15.7. Optimal values of the NSNR versus gravitational frequency for standard recycling systems. 1, delay lines – no recycling; 2, resonant FP – no recycling; 3, delay lines – standard recycling; 4, resonant FP – standard recycling; 5, detuned FP – standard recycling (peak value).

systems that are to be compared. Recall the essential features of each:

Michelson with delay lines and no recycling   Zero frequency limit of the NSNR amplitude:

$$S(0) = Q/e. \tag{15.45}$$

Optimal storage time for $f_0 \gg 1$, corresponding optimal response and minimum detectable, photon-noise limited h:

$$t_0 = \frac{\pi}{f_0}, \quad S(f) = \frac{2Q}{f} \left| \sin\left(\frac{\pi f}{2 f_0}\right) \right|, \quad S(f_0) = \frac{2Q}{f_0}. \tag{15.46}$$

At $v_g^{(0)} = 100$ Hz, we have $h_{PN} = 1.7 \times 10^{-23}/\sqrt{}(Hz)$.

Michelson with FP cavities and no recycling   Zero frequency limit of the NSNR amplitude: $S(0) = Q/2$. Optimal time constant, and optimized response;

$$t_0 = \frac{2}{f_0}, \quad S(f) = \frac{4Q}{f_0} \frac{1}{\sqrt{(1 + 4f^2/f_0^2)}}, \quad S(f_0) = \frac{4}{\sqrt{5}} \frac{Q}{f_0}. \tag{15.47}$$

For $v_g^{(0)} = 100$ Hz we obtain $h_{PN} = 1.9 \times 10^{-23}/\sqrt{}(Hz)$.

Michelson with delay line and standard recycling   Zero frequency limit of the NSNR amplitude: $S(0) = 0.4Q$. Optimal time constant, optimized response in the band $1 \ll f_0$, $pf_0 \ll 1$:

$$t_0 = \frac{2}{f_0}, \quad S(f) = 0.92 \frac{Q}{f} \sqrt{}(f_0) \left| \sin\left(\frac{1.17f}{f_0}\right) \right|, \quad S(f_0) = 0.85 \frac{Q}{\sqrt{}(f_0)}. \tag{15.48}$$

For $v_g^{(0)} = 100$ Hz, we have $h_{PN} = 3.5 \times 10^{-24}/\sqrt{}(Hz)$.

Michelson with resonant FP cavities and standard recycling Zero frequency limit of the NSNR: $S(0) = Q/2$. Optimal time constant, optimized response in the band $f_0 \gg 1$, $pf_0 \ll 1$:

$$t_0 = \frac{1}{f_0}, \quad S(f) = \frac{Q}{\sqrt{(f_0)}} \frac{1}{\sqrt{(1 + f^2/f_0^2)}}, \quad S(f_0) = \frac{Q}{\sqrt{(2f_0)}}. \quad (15.49)$$

For $v_g^{(0)} = 100\,\text{Hz}$, we have $h_{\text{PN}} = 4.2 \times 10^{-24}/\sqrt{(\text{Hz})}$.

We can conclude that delay lines and Fabry–Perot systems are almost equivalent from this theoretical point of view, either in conventional or recycling antennas. Furthermore we note that standard recycling provides a gain of $0.4\sqrt{f_0}$ within the preceding range. When no recycling is applied, the optimal NSNR is proportional (for $f_0 \gg 1$) to Q/f_0, i.e. to $v_{\text{opt}}/v_g^{(0)}$ and thus is independent of the interferometer armlength, provided that a suitable time constant is achieved: whether it has been obtained by many reflections over a short distance or by few reflections over a long distance does not matter. On the other hand, when recycling is applied, we see that the NSNR becomes proportional to $Q/\sqrt{f_0}$ and that the interferometer size is now important: a larger size allows fewer reflections in achieving the optimal time constant, and thus lowers the reflectivity losses of the arms, which permits a higher power build-up in the system and finally a better SNR. If now we examine the very low frequency limit, all systems are limited by their upper bound on the possible storage times: this is why the zero frequency limits for all wideband systems are a fraction of Q. In that very low frequency part of the gravitational spectrum, the photon-noise limited sensitivity will improve linearly with the length of the detector (the fact that they will more likely be limited by thermal or seismic noise does not change this linear dependence).

15.3 Experimental results

15.3.1 Internal modulation

The first experiments on a standard recycling were performed in Garching and in Orsay, in 1987, on simple delay-line systems with $n = 1$. Their primary goal was to demonstrate the validity of the recycling concept, by a direct measurement of the 'noise equivalent power' (NEP) (Schilling, 1988) defined as the laser power which would produce the same phase sensitivity in a perfect, non-recycling, interferometer. Then,

$$P_{\text{ne}} = \frac{h v}{\Delta \Phi^2},$$

where $\Delta \Phi^2$, the spectral density of phase fluctuations of the detected signal, is the quantity which is directly measured. From the viewpoint of these experiments, this parameter is more useful than the sensitivity h, because it characterizes specifically the efficiency of the interferometric technique, and is independent of the storage time of the arms. Actually, if the detection technique is perfect, the NEP is simply related to the NSNR, by the relation: $P_{ne} = \eta \cdot P \cdot S$.

The other goal of these experiments was to evaluate the experimental difficulties and limitations which can be encountered when using recycling. Light from an argon ion laser is brought to the interferometer through a monomode optical fiber. A first electro-optic phase modulator excited by a high frequency oscillator is used to produce the error signal allowing the laser frequency to be locked in reflexion to a resonance in the recycling cavity, in the usual way. (See chapter 13 for details.) When this lock operates, the carrier frequency is stored in the cavity while the phase modulation sidebands are reflected. A second (internal) servo-loop is needed in order to lock the Michelson interferometer to a dark fringe in transmission, so that most of the incident light is reflected towards the recycling mirror. Its error signal is produced by the phase modulation generated by the two electro-optic (E-O) phase modulators situated in the arms of the Michelson and excited by a common oscillator with opposite phases. (See chapter 11 by W. Winkler for the description of the standard phase modulation technique of a Michelson interferometer.) The main difference between the Garching and Orsay experiments is that the first one was performed with separately suspended optical components, while the second one was made on a semi-rigid interferometer. In Garching, the noise measurements were made at 1 kHz, well within the bandpass of the internal servo-loop, while in Orsay the measurements were made at higher frequency (50 kHz), in order to avoid the low frequency region, which is spoiled by the mechanical resonances of the semi-rigid structure. Except for that, the results were quite similar: we both observed the improvement of the noise equivalent power by a factor of the order of 10 to 15. This result was simultaneously a success, a disappointment, and a lesson, because: (i) it is the first experimental proof of the validity of the recycling concept, (ii) it is lower than the factor of 100 we hoped for, and (iii) it taught us that the internal E-O components were the main cause of losses, and that they should be removed: not only do they introduce some absorption, but they are also slightly birefringent, and there is evidence that the total losses increase at high optical power, which is not promising for the future high power interferometers.

15.3.2   External modulation

(i) Theoretical considerations

In Orsay, we decided to try a different interferometric technique where the light is not modulated inside the interferometer. It remains necessary to apply somewhere a high frequency phase or frequency modulation, in order to maintain the frequency transposition which frees us from the excess laser amplitude noise,

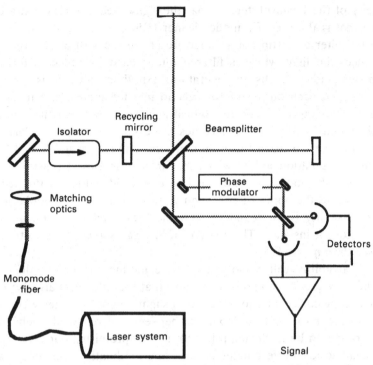

Figure 15.8. Experimental setup for a recycling interferometer with external modulation.

but this modulation is now imposed on a local-oscillator beam which serves to make a kind of heterodyne detection of the signal beam. All this is better explained by figure 15.8, which shows the block diagram of experiment realized in Orsay in 1988: the local-oscillator beam is taken from a spurious reflexion of the light reflected by one of the arms on the (A. R. coated) side of the beamsplitter, and it is phase modulated before being recombined in a second beamsplitter with the signal beam transmitted by the Michelson interferometer. This arrangement guarantees that both the local oscillator and the signal beam have traveled nearly the same length, so that the laser frequency fluctuations do not reintroduce too much noise.

Ideally, if the contrast of this second (external) interferometer is high, i.e. if the local-oscillator geometry and power are the same as those of the signal beam, the best sensitivity will be obtained by locking the external interferometer to a dark fringe on one of the outputs and using one single detector on this output: then, the modulation can be kept low and the theoretical sensitivity is the same as in the case of an internally modulated interferometer, $P_{ne} = \eta \cdot P \cdot S$.

In practice we observed that the contrast of the external interferometer is always very bad, because the lack of contrast of the internal interferometer is generally not due to a difference in the reflectivities of its arms, but to their imperfect matching, resulting from the residual birefringence and bulk inhomogeneity of the beamsplitter, so that the signal beam is very distorted. The solution to that is the use of a mode cleaner (Rüdiger *et al.*, 1981). We tried a simple spatial filter consisting of an aperture at the focus of a telescope, but, in such a device, the light which is filtered out is partially reflected towards the interferometer and perturbs its operation. Another solution is to put two detectors, one on each output of the second interferometer, and to detect the difference of their outputs. The theoretical analysis shows that the sensitivity is slightly degraded, and that it becomes necessary to apply a large phase modulation in order to reach this optimum sensitivity: $P_{he} = \eta \cdot P \cdot S \cdot (2/3)$, and the optimum modulation index, which maximizes the Bessel function $J_1(m)$, is $m = 1.8$. This high modulation index is one of the difficulties of this technique, because it requires the application of high voltages to the E-O modulator.

A third possibility is to frequency shift the local-oscillator, with an acousto-optic modulator for instance. The analysis of this case shows that sensitivity is still a bit worse: $P_{ne} = \eta \cdot P \cdot S/2$.

A fourth possibility will be to use a phase modulation before the recycling mirror, whose frequency should be chosen so that the sidebands are also resonant with the recycling cavity, and to introduce a slight length asymmetry between the two arms, in order to create a dephasing between the reflected sidebands. This technique, proposed by L. Schnupp (1988), is very promising but very difficult to try on a small scale interferometer because the modulation frequency, which is inversely proportional to the length difference between the two arms, must be very high in a small system.

(ii) Experimental results

In our experimental conditions, the best results were obtained with the second solution. We observed a recycling factor of 40, and could operate the interferometer without damage with an input power of 300 mW, yielding a stored power of 12 W. Above this value we observed a saturation of the stored power, and a slight but irreversible degradation of the beamsplitter. This result must be considered as very encouraging because, even if the *power* levels are much smaller than those we need to reach in the large interferometers, the *intensity* we reached is already high ($>3\,\text{kW/cm}^2$), and even larger than it will need to be. Furthermore, the coatings of the beamsplitter in use were not of supermirror quality, so there is room for large improvements.

As concerns the sensitivity measurements, they were done in less favorable conditions, after the damage had occurred. Another limitation came from the fact that it was not possible to drive the phase modulator with the optimum modulation index, $m = 1.8$, without saturating the driver amplifier. For a low modulation index (up to $m = 1$) we found, nevertheless, that the measured sensitivity was in very good agreement with the theoretical expectation. The best P_{ne} we have reached up to now is 2 W.

15.4 Recycling: the current status

We have presented a unified formalism for the study of the simplest kinds of recycling interferometers which have been proposed for the detection of gravitational waves. This allowed us to compare the relative shot-noise limited sensitivities of these interferometers. The important results are that:

The sensitivity gain brought by the use of recycling techniques varies with the gravitational frequency: for the reference antenna, in the frequency range between 50 and 500 Hz, it is roughly equal to the square root of this frequency (expressed in Hz) in the case of a wideband antenna (standard recycling).

The use of a recycling technique calls for very long armlengths: the sensitivity is proportional to the length in the case of a narrowband recycling system, and to the square root of the length, in the case of standard recycling.

Delay-line or Fabry–Perot gravito-optic transducers show essentially the same sensitivity in all cases, but the Fabry–Perot systems are much more versatile: while any modification of the transfer function of a delay-line system requires a major change of the apparatus (moving or changing mirrors), the response of a Fabry–Perot system can be adapted rapidly with just a slight change of the laser frequency or a micrometric movement of one mirror.

The technique of standard recycling with detuned cavities gives the possibility of finding a compromise between bandwidth and peak sensitivity, which should prove to be very useful, especially at the time of the detection of the first signals, when the sensitivity of wideband systems will still be marginal.

The smallest detectable gravitational wave amplitude, h, obtainable with a realistic laser ($\eta P = 10$ W) and the use of recycling techniques should guarantee the observation of a few events per year, since the present theoretical estimations for strong extragalactic sources in the VIRGO cluster give amplitudes h around $3 \times 10^{-23}/\sqrt{(\text{Hz})}$.

The first experiments (at Garching and Orsay) have confirmed the theoretical analysis and brought the highest NEP recorded up to now. They have also uncovered where the experimental difficulties are, and have triggered the use of new interferometric techniques which eliminate these difficulties. In addition they have provided us with the understanding which allows us now to generate reliable computer simulations of the interferometer. These studies are now sufficiently well advanced that one can be confident in the feasibility of reasonably simple interferometers having the desired sensitivity of $h = 10^{-23}/\sqrt{(\text{Hz})}$ where many events per year should already be observable. Some still more advanced recycling techniques (synchronous recycling, dual recycling) have been proposed, and theoretically analyzed. They are very difficult to test experimentally on a small scale interferometer, but they could be tried in a second generation of antennas, together with the squeezing techniques, to improve the sensitivities by maybe an order of magnitude. We expect this field to show a very fast evolution once the first real large detectors have been built.

15.5   Use of squeezed states in interferometric gravitational-wave detectors

In the effort to detect the extremely weak signals associated with gravitational waves, many sources of noise have to be identified and reduced to as low a level as possible. When the more usual mechanical and thermal noises have thus been taken care of (see, e.g., chapter 11 by W. Winkler), there remain fundamental limits arising from the quantum-mechanical nature of the apparatus. In a detector of the interferometric type (Michelson or Fabry–Perot) it was shown by C. M. Caves (1980, 1981) that the shot-noise (also called photon-counting) error and the radiation-pressure error were complementary in the sense of the uncertainty principle: that is, a decrease in one of them must be accompanied by an increase in the other, just as a reduction in the uncertainty associated with the position of a particle must result in an increasing uncertainty regarding its momentum. Caves identified an optimum operating point as a trade-off between the two error sources and called the minimum detectable mirror displacement at this optimum point the 'standard quantum limit' (SQL). (Interestingly, the SQL could also be obtained without any explicit reference to the light in the interferometer, from a straightforward application of the position–momentum uncertainty principle to the mirrors.)

The SQL has been discussed repeatedly over the years (see, for instance, Bondurant and Shapiro, 1984; Caves, 1985; Ni, 1986; and Yuen, 1984), and its status as a truly fundamental limit has been questioned occasionally. (Perhaps unfortunately, the term 'standard quantum limit' has also come to be used, outside the gravitational-wave detection community, to refer to the ordinary shot-noise limit to the detection of weak light signals.) As originally shown by Caves, given the interferometer length and the mirror mass, the SQL is reached for a certain laser power. Roughly speaking, we can see this as follows: the shot-noise limit is related to the fluctuations in the number of detected photons, n, for very weak signals. These fluctuations are approximately Poissonian for an ideal laser beam, which means that the mean Δn scales as $\sqrt{n}$; hence the *relative* size of the fluctuations, and with it the shot-noise limit, decreases as the laser power is increased. On the other hand, as the laser power increases the error due to radiation pressure on the mirrors increases: hence the existence of a trade-off point, for some optimum laser power, where both error sources are equal in magnitude, which is where the SQL is realized.

For low laser power, the shot-noise error is much larger than the radiation pressure error, and thus much larger also than the ideal 'standard quantum limit'. While the latter could in principle be reached just by increasing the laser power appropriately, this is not really quite so simple. At high powers the lasers are likely to be more noisy by themselves, and high light intensities may also damage the optical elements of the interferometer and the photodiodes used for signal detection, or the mirrors may be heat deformed.

Caves realized that the use of the so-called 'squeezed states' of the electromagnetic field, which at the time had been studied (under a different name) mostly in the context of optical communications (Yuen, 1976; Yuen and Shapiro, 1980), could make it possible to reach the SQL for lower laser powers than those predicted for ordinary laser light ('coherent state' light); in any event, it would reduce the shot-noise error for a given laser power, at the expense of increasing the (in practice negligible) radiation pressure error. He devised a very ingenious method for doing so, the essence of which will be presented in the following section. It is clear, in retrospect, that the possibility of such an application for squeezed states has been since then one of the major motivations for theoretical and experimental work in this subfield of quantum optics.

In the years since Caves's original proposal, several prototype interferometric detectors of gravitational waves have been developed to the point where they operate at the shot-noise limit for the laser powers employed (e.g. Shoemaker *et al.*, 1988), and even though increases in laser power are planned for the next generation of 'real' detectors, the SQL is still far away in the future. At the same time, squeezed states of light have been experimentally generated*, with large amount of squeezing in some cases, and moreover the feasibility of Caves's scheme for noise reduction has been demonstrated experimentally as well (Xiao, Wu and Kimble, 1987) in a conventional (table-top size) Mach–Zender interfero-meter, by H. J. Kimble's group at the University of Texas. These developments make the possibility of using squeezed light in interferometric gravitational-wave detectors particularly interesting at the moment.

Some of the aspects of the actual implementation of these ideas have been studied in a series of papers (Gea-Banacloche and Leuchs, 1987a,b, 1989); in particular the ways in which several imperfections in the interferometer (such as losses, wavefront distortion, or beam misalignment) may affect adversely the squeezed-state performance have been considered. It has also been studied which features are desirable in the spectrum of squeezing of the light for use with the well-established experimental method to stabilize the interferometer at a 'dark fringe' using the modulation technique. The results, especially those concerning interferometer imperfections, will be summarized in section 15.7. We may anticipate here that even small departures of the fringe pattern from ideal unit visibility are found to limit seriously the usefulness of squeezed light for noise reduction, which may be taken as an additional motivation to strive for the highest fringe visibility in these devices.

Another scheme to increase the sensitivity without increasing the laser power was suggested by R. W. P. Drever (Drever *et al.*, 1983), and is known as 'light recycling'. As explained in sections 15.1–15.4, this method consists in using an additional mirror in an interferometer operating at a dark fringe, to add coherently the light that would otherwise escape on the other side of the

---

* For a recent survey of the field, see the special issues of *J. Mod. Opt.* **34** (1987), nos. 6/7, and *J. Opt. Soc. Am. B* **4** (1987), no. 10.

beamsplitter (the bright-fringe side) onto the incoming beam. The result is that the whole interferometer becomes one large optical cavity where the light intensity builds up to a level which may be many times higher than that of the incident light (i.e., the laser output), just as in an ordinary resonant Fabry–Perot cavity. In this way one achieves a larger signal-to-noise ratio by increasing the light intensity (since in the shot-noise limited regime, the noise increases as the square root of the intensity only), but without actually requiring a more powerful laser.

It is possible in principle to combine *both* squeezing and light recycling to increase the signal-to-noise ratio even further. We shall show how this may be done, in some detail, in section 15.8.

15.6 The principles of noise reduction using squeezed states

In this section we shall illustrate the essence of Caves's method for squeezed-state noise reduction by considering the specific case of a Michelson interferometer (figure 15.9) operating at a 'null' or dark fringe. More general arrangements are discussed in Caves's original paper; here we shall concentrate on the dark-fringe case because it helps to visualize how the method works, and also because it is especially suited for the recycling technique to be discussed later.

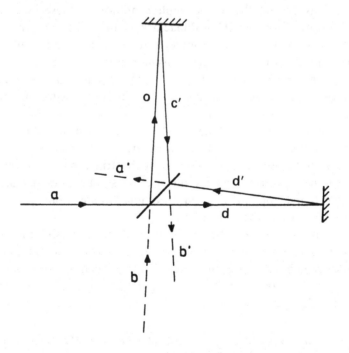

Figure 15.9.

The letters shown on figure 15.9 for each ray will be used to designate the field amplitude at that point in the light path and, quantum mechanically, the annihilation operators for the modes of the field corresponding to propagation in the direction given. The interferometer, as shown on the figure, has two input ports, corresponding to rays a and b; light in these two beams couples to the same internal modes c and d. We shall take a to be the input laser beam (which we shall always assume to be in a coherent state), while, for the time being, we assume that nothing is coming in the direction of b. Also, we shall take b' to be the output mode whose intensity is measured by a photodetector. (The other output mode, a', will be ignored for now; it could be used in a light recycling scheme, as we shall show later.)

Under these circumstances, operation at a dark fringe means that the path length is adjusted so that no light comes out at b' in the absence of a signal (in fact, a signal, such as a gravitational wave, would manifest itself by changing the optical path difference between the two arms, so that some light would then exit at b'). By conservation of energy, this means that all the light must come out at a'. This is ensured by the fact that the beamsplitter introduces a π phase difference between the waves reflected on each of the two sides. What happens, then, is that, in the absence of a signal, all the a light exits at a' and all the b light (were there any) exits at b'. If a small signal is present, the path length difference is altered, with the result that at b' we would get some amount of the a light, plus still most of b.

We may express all the foregoing in formulas assuming, for simplicity, that the beamsplitter introduces no phase shift upon reflection for a wave incident on the side of a, and a π phase shift for a wave reflected on the side of b. Then (the beamsplitter reflection and transmission coefficients being equal),

$$c = (a + b)/\sqrt{2} \tag{15.50a}$$

$$d = (a - b)/\sqrt{2} \tag{15.50b}$$

If the total phase shift along arm $c(d)$ is $\phi_c(\phi_d)$, we have

$$b' = \frac{1}{\sqrt{2}}\left[\frac{1}{\sqrt{2}}(a + b)e^{i\phi_c} - \frac{1}{\sqrt{2}}(a - b)e^{i\phi_d}\right]$$
$$= e^{i(\phi_c + \phi_d)/2}[i\sin(\phi/2)a + \cos(\phi/2)b], \tag{15.51}$$

where

$$\phi \equiv \phi_c - \phi_d \tag{15.52}$$

is equal to $2n\pi$ for operation at a dark fringe, and departs slightly from this value in the presence of a signal. (Actually, ϕ will typically be chosen to depart from $2n\pi$ by some constant, or variable (modulated), phase offset. The purpose of this phase offset is discussed in Caves (1980, 1981) and Gea-Banacloche and Leuchs (1987a); here we shall not be concerned with it.)

Equation (15.51) is valid quantum-mechanically if a and b are the annihilation operators for the incoming light in modes a and b, and b' the annihilation operator for the outgoing mode; it is easy to verify, for instance, that

$$[b'^{\dagger}, b'] = 1. \tag{15.53}$$

Note that, quantum-mechanically, the operator b, satisfying $[b^{\dagger}, b] = 1$, must be included in equation (15.51) even if, classically, *no* light is coming in from the direction of b (or otherwise equation (15.53) would not hold). We say then that b is in the vacuum state, or that it corresponds to a vacuum field. Caves (1980, 1981) argued that the shot-noise limit could be understood as arising from this vacuum field.

The vacuum field contains exactly zero photons, although formally it does appear to have non-zero electric and magnetic fields, for the expectation values of E^2 and B^2 in the vacuum state are non-zero (actually, they are infinite when all the modes of the radiation field are added, although for each single mode they have a finite value). The first property means that it is, strictly speaking, unobservable; the second property, however, appears to suggest that it might interfere with another field and thus give rise to observable effects. This may in fact be inferred from the following expression, which follows from equation (15.51), for the photon number operator (the intensity, roughly speaking) of the mode b':

$$b'^{\dagger}b' = \sin^2(\phi/2)a^{\dagger}a + \cos^2(\phi/2)b^{\dagger}b - i\sin(\phi/2)\cos(\phi/2)(a^{\dagger}b - b^{\dagger}a). \tag{15.54}$$

The last term in equation (15.54) is the interference term one would expect in the presence of a signal (i.e., if $\phi \neq 2n\pi$), since in that case, as we mentioned earlier, some light from mode a exits at b' along with most of b. When b is in the vacuum state the expectation value of this term is, of course, zero. However, the expectation value of its *square* is not zero, not even in the vacuum state, because of the presence in it of a term

$$\langle 0| bb^{\dagger} |0\rangle = 1. \tag{15.55}$$

To be specific, assume that a is in a coherent state $|\alpha\rangle$, such that

$$a|\alpha\rangle = \alpha|\alpha\rangle \tag{15.56}$$

with

$$\alpha = \sqrt{(\bar{n}_a)}e^{i\delta}. \tag{15.57}$$

Here $\bar{n}_a$ is the average number of photons in mode a and δ is the phase of the field a. Then, the expectation value of the square of the last term in equation (15.54) is proportional to

$$-\langle(a^{\dagger}b - b^{\dagger}a)^2\rangle = -\bar{n}_a\langle(be^{-i\delta} - b^{\dagger}e^{i\delta})^2\rangle + \langle b^{\dagger}b\rangle$$
$$= 4\bar{n}_a\langle b_2^2\rangle + \bar{n}_b. \tag{15.58}$$

Here the 'phase quadrature', b_2, of the field b, relative to the phase δ, has been introduced; the amplitude and phase quadratures (b_1 and b_2, respectively) are defined by

$$b_1 = (b^\dagger e^{-i\delta} + b e^{i\delta})/2 \tag{15.59a}$$

$$b_2 = (b^\dagger e^{-i\delta} - b e^{i\delta})/2i. \tag{15.59b}$$

(Their physical meaning will be considered shortly.) In the vacuum state,

$$\langle b_1 \rangle = \langle b_2 \rangle = 0 \tag{15.60a}$$

$$\langle b_1^2 \rangle = \langle b_2^2 \rangle = \tfrac{1}{4}. \tag{15.60b}$$

In this case, $\bar{n}_b = 0$, and the first term on the r.h.s. of equation (15.58) is equal to $\bar{n}_a$. In general, equation (15.58) shows that the expectation value of the square of the last term in equation (15.54) is (apart from the usually negligible term $\bar{n}_b$) proportional to the expectation value of the square of the phase quadrature, b_2, of the field b, and to $\bar{n}_a$. It is important to realize that the remainder of this chapter has only been written because of this fact: that the noise in the output mode b' depends only on one of the quadratures (b_2) of the input mode and not on the other one (b_1).

Indeed, the term we are discussing, with zero average value but non-zero root-mean-square value, may properly be labeled a noise term. Formally it looks like an interference term between the field a and the vacuum field b. Further, its r.m.s. value is precisely that of shot noise, i.e. $\sqrt{(\bar{n}_a)}$, and b is in the vacuum state. In fact, if the fluctuations in $n_{b'} = b'^\dagger b'$ are calculated from equation (15.54), it is found that the largest contribution comes from precisely this interference term, provided that $\sin(\phi/2) \ll 1$ (as would be the case for a very weak signal). Letting $\delta\phi$ be equal to the departure of ϕ from the value $2n\pi$, we find from equation (15.54) (expanding the trigonometric functions)

$$\langle n_{b'} \rangle \simeq \bar{n}_a \, \delta\phi^2/4 \tag{15.61}$$

$$\sqrt{(\langle \Delta n_{b'}^2 \rangle)} \simeq \sqrt{(\bar{n}_a \langle b_2^2 \rangle)} \delta\phi. \tag{15.62}$$

Equation (15.61) gives the signal, and equation (15.62) the r.m.s. noise, at a photodetector in the b' path. Equating them we find the minimum detectable $\delta\phi$

$$\delta\phi \gtrsim 2/\sqrt{(\bar{n}_a)} \tag{15.63}$$

if b is in the vacuum state, so that equation (15.60b) holds.

Equation (15.63) is the 'shot-noise limit'. (Actually, the factor of two on the r.h.s. of equation (15.63) may be made to disappear if a small, fixed, controllable phase offset is introduced, so that the interferometer does not operate exactly at a dark fringe, but very close to it; see Caves (1980, 1981) or Gea-Banacloche and Leuchs (1987a) for details.)

Caves realized that the shot-noise limit could be interpreted in this way as arising from the interference of the laser field with the 'vacuum entering the interferometer through the unused port' (that is, b in figure 15.9). He further realized that, as equation (15.62) shows, this limit could be brought down if one injected at port b some kind of light for which the value of $\langle b_2^2 \rangle$ were smaller than for the vacuum. Such is, precisely, squeezed light.

To understand what the quadratures b_1 and b_2 mean, and to gain some insight into the way in which the noise reduction is achieved, it is convenient to introduce at this point a pictorial representation of the optical fields involved which is a generalization of the phasor representation, familiar from classical optics, where the (complex) electric field amplitude is represented by a vector ('phasor') in the complex plane.

The field a, for instance, could be represented by figure 15.10: the large arrow indicates the average value, whereas the small ones represent the fluctuations in the amplitude and phase quadratures, defined in a way analogous to those of b (equation 15.59b):

$$a_1 = (a^\dagger e^{-i\delta} + a e^{i\delta})/2 \tag{15.64a}$$

$$a_2 = (a^\dagger e^{-i\delta} - a e^{i\delta})/2i, \tag{15.64b}$$

from which we see

$$a = (a_1 + ia_2)e^{i\delta}. \tag{15.65}$$

We have $\langle a_1 \rangle = \sqrt{(\bar{n}_a)}$, $\langle a_2 \rangle = 0$, and, just as for the vacuum, for an arbitrary coherent state $\langle \Delta a_1^2 \rangle = \langle \Delta a_2^2 \rangle = \frac{1}{4}$. Fluctuations in a_1 may be pictured as adding to the average value of a a small phasor along the direction of $\langle a \rangle$ (amplitude fluctuations), while fluctuations in a_2 add a phasor perpendicular to it: if $\langle a_1 \rangle \gg \Delta a_2$, this results, to first order, in a change in the direction, but not the amplitude, of a (phase fluctuation).

The quadrature operators have the commutator

$$[a_1, a_2] = \tfrac{1}{4} \tag{15.66}$$

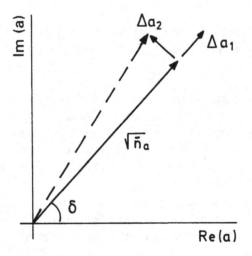

Figure 15.10.

from which follows the uncertainty relation

$$\Delta a_1 \Delta a_2 \geqq \tfrac{1}{4}. \tag{15.67}$$

This means that, e.g., the fluctuations in the phase quadrature can be reduced ('squeezed') below the coherent-state value $\tfrac{1}{2}$ at the expense of increasing the amplitude fluctuations.

For the field b, consider the 'squeezed vacuum' state shown in figure 15.11. It still has zero mean, so it is represented conventionally as an 'error ellipse', enclosing the possible phasors pointing in all possible directions; but the fluctuations in the phase quadrature b_2 are smaller than for the true vacuum (the fluctuations in b_1 being correspondingly larger). That this is the way to minimize the interference term in equation (15.54) may be understood by looking at equation (15.51), which shows the fields a and b adding out of phase by $\pi/2$ (because of the factor of i); that is, to obtain the field b' the phasor representing a in figure 15.10 is to be rotated by $\pi/2$, then added to b. The amplitude of a is thus aligned along the phase quadrature, b_2, of b, and this gives rise to the largest contribution to the interference term, namely, the first term on the r.h.s. of equation (15.58).

An even more graphical way to see why squeezing the phase quadrature of b helps is presented in figure 15.12, which shows the difference between the two phasors representing the fields c' and d' (for simplicity, assume that $(\phi_c + \phi_d)/2 = 2n\pi$, so that c' and d' lie almost along the direction of a, except for the small angle ϕ). This difference is in fact proportional to b' with our sign convention for the beamsplitter,

$$b' = (c' - d')/\sqrt{2} \tag{15.68}$$

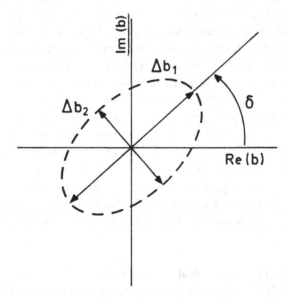

Figure 15.11.

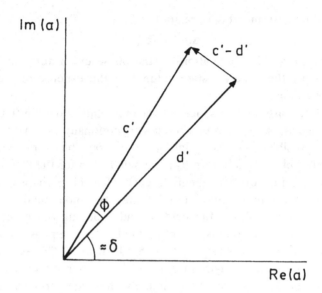

Figure 15.12.

and, as the figure shows, it lies mostly in the direction perpendicular to a; that is, the direction of the b_2 quadrature. A measurement of the intensity of b' is a measurement of the (square of the) length of the small arrow in figure 15.12.

Now, c' and d' are just c and d slightly rotated, and from equations (15.50) we can see that the fluctuations in a cancel in the difference $c - d$, but the fluctuations in b *add* when $c - d$ is formed. Thus the main source of uncertainty in the length of the small arrow in figure 15.12 is the fluctuations of b, and especially those fluctuations that lie along the arrow's length (the ones orthogonal to it do not change the length to first order, if they are sufficiently small). These are the b_2 fluctuations, and hence the advantage in squeezing them.

We may draw other tentative conclusions from figure 15.12 and equations (15.50), (15.68). For instance, we may expect that the fluctuations in the input beam a are going to be irrelevant in any case, since they cancel exactly when $\phi = 0$ and will be proportional to at least $\delta\phi$ for small $\delta\phi$. Thus there would be nothing to be gained by squeezing the input laser light. This conjecture is correct, although, as we shall see in the following section, it is no longer true in the presence of asymmetric losses in the two arms of the interferometer. At any rate, a detailed analysis is necessary to ascertain the relative size of all the noise terms in the presence of both a signal and a squeezed input (see Caves, 1980, 1981, or, for a simplified treatment, Gea-Banacloche and Leuchs, 1987a). One of the results of this analysis is the need for the phase-offset (constant or variable) mentioned earlier.

15.7   Squeezed states for non-ideal interferometers
The original analysis of Caves (1980, 1981) assumed an ideal interferometer, with no losses, and with unity fringe visibility. In practice, of course, any real

interferometer will have losses in both arms, and if they are not exactly equal the light cancellation at the output port cannot be perfect (that is, the dark fringe is not totally dark anymore). This reduces the visibility of the fringes, which is defined in the conventional way

$$V = \frac{I_{max} - I_{min}}{I_{max} + I_{min}} \tag{15.69}$$

Here $I_{max}(I_{min})$ is the intensity at a bright (dark) fringe; V has a maximum value of unity, which is reached only when $I_{min} = 0$.

It was already remarked by Caves (1980, 1981) that losses of any type would have a detrimental effect on his scheme, because losses in general tend to destroy squeezing*: that is, they tend to equalize the fluctuations in both quadratures, at the vacuum noise level (or higher, at finite temperature). In addition to this, however, unbalanced losses in the two arms, leading to a non-zero I_{min}, are especially harmful, because the background light leaving the b' port fluctuates, even in the absence of a signal, introducing additional noise. The amplitude of this background light is proportional to that of the input laser light, and fluctuates with it. Thus, unlike for the ideal interferometer of section 15.6, the intensity fluctuations of the laser light cannot be neglected. This is in fact the case whenever there is any background light at a dark fringe (i.e., whenever the fringe visibility is less than unity), regardless of whether it is due to unbalanced losses or to other causes, such as wavefront distortion or beam misalignment.

Figure 15.13 is the equivalent of figure 15.12 for the case when there are unequal losses in both arms: the phasor for field c' is assumed to be reduced by a factor r_1 and that for d' by a factor r_2. Fluctuations in the length of the input

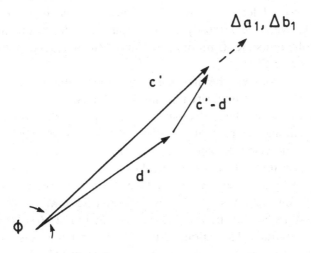

Figure 15.13.

---

* Noise in linear amplifiers (or attenuators) with special emphasis on the applications to non-classical light has been discussed, for instance, by Caves (1982). See also Loudon and Shepherd (1984).

phasor a (the a_1 quadrature) are also going to appear in c' and d' multiplied by those two different factors, so they will not exactly cancel in the difference $c' - d'$, unlike in section 15.6.

In addition to this, figure 15.13 also suggests that fluctuations in b_1, the *amplitude* quadrature of b, would now affect the length of $c' - d'$ to zeroth order in $\delta\phi$, since they have now a significant projection along the direction of this phasor. This conjecture is also correct. All this may be easily seen from the following expression for b', which follows from equations (15.50) and (15.60) and the simple assumption $c' = r_1 c$, $d' = r_2 d$:

$$b' = \tfrac{1}{2}(r_1 - r_2)a + \tfrac{1}{2}(r_1 + r_2)b + tv. \tag{15.70}$$

Here v is the annihilation operator for a field mode in the vacuum state; its presence in equation (15.70) is necessary to satisfy the commutation relations. (It can be viewed as a 'vacuum fluctuation' coupled in through the lossy mirror, e.g., by transmission or scattering.) The coefficient t is given by $t = \surd(1 - (r_1^2 + r_2^2)/2)$. The main effect of the term tv is to destroy the squeezing, as we said earlier.

From equation (15.70) we see that $b'^\dagger b'$ contains terms of the form

$$\frac{1}{4}(r_1 - r_2)^2 a^\dagger a \tag{15.71}$$

as well as of the form

$$\frac{1}{4}(r_1^2 - r_2^2)(a^\dagger b + b^\dagger a). \tag{15.72}$$

The fluctuations of terms (15.71) and (15.72) are the two new noise sources introduced by unequal losses (clearly, both terms vanish if $r_1 = r_2$). The term (15.71) fluctuates with the intensity of the input laser field; the term (15.72), when a is in a coherent state, is proportional to b_1, the amplitude quadrature of b (compare equation 15.59a).

Of these two new sources, the first one might conceivably be reduced by using a laser with sub-Poissonian intensity fluctuations (that is, $\Delta n_a < \surd(\bar{n}_a)$). Operation of such a laser has, in fact, been demonstrated experimentally by Yamamoto and his coworkers, although at present it seems unlikely that it might be used in a gravitational-wave detector (Machida, Yamamoto and Itaya, 1987).

The second noise term (15.72) is particularly harmful from the point of view of using squeezed light to reduce the shot-noise limit, since, as we saw in the previous section, this requires the use of squeezed light with reduced noise in the *phase* quadrature b_2; but, by the uncertainty relation (15.67), such light must have increased amplitude fluctuations, Δb_1, which form the noise term (15.72). Thus, as the squeezing of Δb_2 increases, the noise term (15.72) grows, until, at some point, it becomes dominant, and further squeezing can only degrade the signal-to-noise ratio.

A detailed study of an interferometer with unequal losses, presented in Gea-Banacloche and Leuchs (1987b) yielded the following expression for the minimum detectable $\delta\phi$;

$$\delta\phi_{min} = \sqrt{\left(\frac{2}{\bar{n}_a}\right)}\sqrt{[\sqrt{(C)} + 1/r_1 r_2 + e^{-2s} - 1]}, \qquad (15.73)$$

where

$$C = \frac{3}{2}(1 - V)\frac{r_1^2 + r_2^2}{r_1^3 r_2^3}\left(1 - \frac{1}{4}(r_1 + r_2)^2(1 - e^{2s})\right). \qquad (15.74)$$

In equation (15.74), V is the fringe visibility, which for this model is easily seen to be given by $V = 2r_1 r_2/(r_1^2 + r_2^2)$. (Equation (15.73) is actually an approximation, valid when V is very close to unity.) The parameter $s \geq 0$ determines the amount of squeezing ($s = 0$ means no squeezing, that is, b is a true vacuum in that case). As we anticipated, equations (15.73) and (15.74) contain both terms which decrease and terms which increase as the squeezing increases, the latter arising from term (15.72).

A particularly noteworthy feature of equations (15.73), (15.74) is their dependence on the visibility, which is like the square root of $1 - V$ inside the square root in equation (15.74). This means that very small deviations of V from unity have a large adverse effect (the derivative of $\sqrt{(1 - V)}$ is infinite at $V = 1$). By the same token, however, this means that there is a substantial payoff to be gained from bringing V as close to unity as possible, and that efforts in this direction are of great consequence for any noise-reduction schemes in interferometers.

We mentioned earlier other possible causes, besides unequal losses, for a visibility less than unity: these include wavefront distortions (aberrations) affecting the two beams differently before recombination, as well as simple beam misalignment at the beamsplitter exit (and perhaps others such as partial depolarization of the beams). These differ from the losses in that they are 'dispersive' in nature, rather than absorptive: that is, the energy in the light beams is not lost, merely transferred to other spatial (or polarization) modes. This has an interesting consequence: while any non-zero background intensity must introduce fluctuations proportional to the intensity fluctuations of the laser beam, and therefore terms like (15.71) ('dispersive' effects not being an exception), we have found (Gea-Banacloche and Leuchs, 1989) that in a purely dispersive case terms like (15.72), proportional to the amplitude fluctuations of b, need not arise.

The reason for the different behavior of term (15.72) lies in the fact that it is an interference term, linear in a and b: when an aberration, for instance, removes some of the amplitude of $a + b$, say, from a certain spatial mode, causing a term like (15.72) to appear when the light intensity in that mode is computed, a term identical but with the opposite sign appears in the intensity of another spatial mode (in the general case, it may actually be distributed over an, in principle, infinite set of modes). Then, if the detector is sensitive to the total intensity in all the spatial modes, a cancellation is in principle possible.

A model example to illustrate this point is a model of a dispersive element in one of the beam paths as a simple beamsplitter, transmitting, say, mode c with a coefficient ε, and therefore reflecting an amplitude $\sqrt{(1 - \varepsilon^2)}$ into another mode which we may call v'. We have then

$$c' = \varepsilon(a + b)/\sqrt{2} + \sqrt{(1 - \varepsilon^2)}v \tag{15.75a}$$

$$v' = \varepsilon v - \sqrt{(1 - \varepsilon^2)}(a + b)/\sqrt{2}. \tag{15.75b}$$

Suppose, for simplicity, that d is not altered, i.e. $d' = d$. Then we have (using equations 15.75a and 15.50b in equation 15.68)

$$b' = \frac{1}{2}(\varepsilon - 1)a + \frac{1}{2}(\varepsilon + 1)b + \sqrt{(1 - \varepsilon^2)}v/\sqrt{2}. \tag{15.76}$$

This looks very much like equation (15.70), and in fact if only $b'^\dagger/b'$ is measured, a term like (15.72) is obtained (with $r_1 = \varepsilon$, $r_2 = 1$). The difference is that now we could, at least in principle, think of measuring the intensity of mode v' and adding it to that of b'. If we do that, we find in $v'^\dagger v'$, from equation (15.75b), a term proportional to

$$(1 - \varepsilon^2)(a^\dagger b + b^\dagger a), \tag{15.77}$$

which exactly cancels the corresponding interference term in $b'^\dagger b'$. (We must assume that v' goes through the 50/50 beamsplitter in figure 15.9, just as c' does, so that its amplitude, too, is reduced by a further factor of $1/\sqrt{2}$ from what is shown in equation (15.75b).) The more complete analysis of Gea-Banacloche and Leuchs (1989) supports this conclusion, when arbitrary aberrations in both beams, and an infinite set of spatial modes, are considered.

We thus see that, at least in this highly idealized example where the detector collects all the energy spread over the relevant spatial modes, aberrations and other dispersive imperfections do not necessarily couple in noise proportional to the fluctuations in the amplitude quadrature of b; thus, unlike for the case of unbalanced losses, no harm is done by squeezing the phase fluctuations of b as much as one can, to reduce the shot-noise limit. Nonetheless, the inevitable presence of fluctuations proportional to the intensity fluctuations of the laser beam (the term (15.71)), which is always the consequence of a fringe visibility less than unity, is also highly detrimental in this case. The formula obtained in Gea-Banacloche and Leuchs (1989) for the minimum detectable $\delta\phi$ in the dispersive case reads

$$\delta\phi_{min} = \sqrt{\left(\frac{1}{n_a}\right)}\sqrt{[\sqrt{2}V(1 - V) + e^{-2s}]}, \tag{15.78}$$

and, like equations (15.73), (15.74), it exhibits a dependence on $\sqrt{(1 - V)}$ inside a square root.

The conclusion of our analysis is that squeezing can substantially improve the sensitivity of an interferometer only if the fringe visibility is very close to unity.

Other issues of interest addressed in Gea-Banacloche and Leuchs (1987a,b) concern the 'ideal' spectrum of squeezing for certain specific measurement schemes and interferometer imperfections. For example, for the case when the measurement technique used is phase modulation at a dark fringe, we find that for squeezing to be useful the so-called 'spectrum of squeezing' must extend at least out to twice the modulation frequency on each side of the cavity resonance. The interested reader is referred to Gea-Banacloche and Leuchs (1987a,b) for further details.

15.8 Squeezing and light recycling

As discussed in sections 15.1–15.4, light recycling is a technique to increase the sensitivity of a shot-noise-limited interferometer by sending the light that would otherwise escape at port a' of figure 15.9 back to the beamsplitter, using a partially transmitting mirror of high reflectivity. This may be done as shown in figure 15.14, with the laser light entering now through mirror M_1.

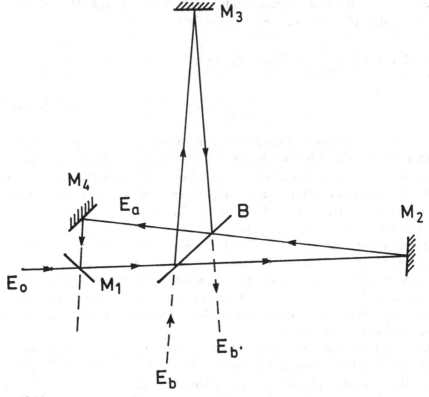

Figure 15.14.

Assume that the interferometer is set to operate exactly at a dark fringe. Then, as discussed in section 15.6, all the light entering at port a exits at port a'. As far as the a light is concerned, the whole arrangement of beamsplitter B and mirrors M_2 and M_3 is equivalent to one perfectly reflecting mirror. This 'mirror', together with the perfect reflector M_4 and the partially transmitting M_1, forms then an optical 'ring' cavity where the intensity of the light may build up to many times the incident laser intensity. When this more intense light probes the lengths of the arms of the interferometer, a larger signal results, as we shall see below.

On the other hand, consider the b light. With the interferometer set at a dark fringe, it exits entirely at b': none of it exits at a', which means that none of it really enters the 'cavity' where the laser light is 'recycled'. As far as the b light is concerned, the arrangement B, M_2 and M_3 is equivalent to a perfect reflector, sending it back to b' and effectively keeping it out of the recycling cavity. Thus, at least when no signal is present, the addition of M_1 and M_4 has no effect on the b light, and we may therefore expect squeezing of b to work independently of, and in addition to, the recycling of a.

To study what happens when there *is* a signal present, so that some a light leaks out of the cavity and some b light leaks in, consider the following equation for the slowly varying amplitude of the electric field operator inside the cavity, $E_a(t)$, as a function of E_0, the laser electric field amplitude just outside M_1, and E_b, the electric field amplitude associated with the field b:

$$E_a(t + L/c) = e^{i\omega l/c}\left(\frac{1}{\sqrt{2}}\left[\frac{1}{\sqrt{2}}(tE_0 + rE_a) + \frac{1}{\sqrt{2}}E_b e^{i\phi_b}\right]e^{i\phi_c}\right.$$

$$\left. + \frac{1}{\sqrt{2}}\left[\frac{1}{\sqrt{2}}(tE_0 + rE_a) - \frac{1}{\sqrt{2}}E_b e^{i\phi_b}\right]e^{i\phi_d}\right). \quad (15.79)$$

Here L is the total length of the light path in the cavity, including one arm of the interferometer, while l is only the length of the light path in the external cavity M_1–B–M_4; r and t are the amplitude reflection and transmission coefficients of mirror M_1; and all the slowly varying amplitudes on the r.h.s. of equation (15.79) are evaluated at time t. (Strictly speaking, E_0 and E_b should be evaluated at different times, if E_b is the field just outside the beamsplitter, since light has to propagate a finite distance from M_1 to B; we have accounted for that in the phase factor $\exp(\phi_b)$ and ignored the difference in the slowly varying amplitudes, assuming that the distance from M_1 to B is sufficiently small. Likewise we have ignored possible small differences in length between the arms of the interfero-meter, in the slowly-varying amplitudes (they contribute, of course, to the phase factors $\exp(\phi_c)$ and $\exp(\phi_d)$).)

Equation (15.79) is easily obtained by following the light through one cavity round trip. We have chosen to write it in terms of the electric field operators, rather than directly in terms of annihilation operators, because the cavity field operators would require a different normalization. The equation for the field leaving at port b' is also easily obtained:

$$E_{b'}(t + l/c) = \left(\frac{1}{\sqrt{2}}\left[\frac{1}{\sqrt{2}}(tE_0 + rE_a)e^{-i\phi_b} + \frac{1}{\sqrt{2}}E_b\right]e^{i\phi_c}\right.$$

$$\left. - \frac{1}{\sqrt{2}}\left[\frac{1}{\sqrt{2}}(tE_0 + rE_a)e^{-i\phi_b} - \frac{1}{\sqrt{2}}E_be^{i\phi_b}\right]e^{i\phi_d}\right)$$

$$= e^{i(\phi_c + \phi_d)/2}(ie^{-i\phi_b}(tE_0 + rE_a)\sin(\phi/2) + E_b\cos(\phi/2)) \quad (15.80)$$

with ϕ given by equation (15.52). Introducing the angle ψ_0, $\psi_0 \equiv \omega l/c + (\phi_c + \phi_d)/2$ we may rewrite equation (15.79) as

$$E_a(t + L/c) = e^{i\psi_0}((tE_0 + rE_a)\cos(\phi/2) + ie^{i\phi_b}E_b\sin(\phi/2)). \quad (15.81)$$

Equation (15.81) is a difference equation, embodying the cavity boundary conditions, which might be converted into a differential equation through the approximation $E_a(t + L/c) \approx (1 + (L/c)\,d/dt)E_a(t)$, to follow the time evolution of the slowly varying amplitude $E_a(t)$. Instead, we shall expand the slowly varying amplitudes into monochromatic components, as in

$$E_a(t) = \int \tilde{E}_a(\Omega)e^{-i\Omega t}\,d\Omega \quad (15.82)$$

and solve (the Fourier transform of) equation (15.81) for $\tilde{E}_a(\Omega)$, which is easily done if the angles ψ_0 and ϕ do not change in time. This is, of course, only an approximation, since any gravitational-wave signal would yield a time-dependent ϕ, but this may be neglected to lowest order. In practice, actually, the largest part of ϕ would be either a constant (the constant phase offset to which we have alluded earlier) or a well-defined sinusoidal function of time (in the external phase-modulation stabilization technique); the latter would introduce a coupling between the components $\tilde{E}_a(\Omega)$ which could also be treated without much difficulty, although we shall not do so here.

The solution for $\tilde{E}_a(\Omega)$ has the simple form

$$\tilde{E}_a(\Omega) = e^{i\psi}\frac{t\tilde{E}_0\cos(\phi/2) + ie^{i\phi_b}\tilde{E}_b\sin(\phi/2)}{1 - re^{i\psi}\cos(\phi/2)} \quad (15.83)$$

where ψ depends on Ω,

$$\psi = \psi_0 + \Omega L/c. \quad (15.84)$$

This may be inserted in the Fourier transform of equation (15.80) to yield the output at b' as a function of the two input fields, E_0 and E_b, only. Apart from an overall phase factor, the result is

$$\tilde{E}_{b'} = ie^{i\phi_b}\frac{t\tilde{E}_0}{1 - re^{i\psi}\cos(\phi/2)}\sin(\phi/2)$$

$$+ \left[\cos(\phi/2) - \frac{re^{i\psi}\sin^2(\phi/2)}{1 - re^{i\psi}\cos(\phi/2)}\right]\tilde{E}_b. \quad (15.85)$$

The limit $r = 0$ corresponds to no recycling and is easily seen to reduce to equation (15.51). In fact, now that the intracavity field has been eliminated, equation (15.85) would hold as a relationship between annihilation operators for the monochromatic components of the input and output fields (dividing through by an appropriate normalization constant, which would be the same for all the fields appearing in equation (15.85)). We may then think of $\bar{E}_0$, $\bar{E}_b$ and $\bar{E}_{b'}$ as being essentially the same as a, b and b', respectively, in equation (15.51).

Equation (15.85) shows that the amplitude of a may be enhanced appreciably when the recycling cavity is near resonance and ϕ is not too large. In fact, some enhancement will take place as long as $\cos(\phi/2) > (1 - t)/r$. This leads naturally to an increase in the signal, which is proportional to the light intensity. The shot-noise limit also follows from equation (15.85) as it did from equation (15.51), and by reasoning along the same lines as in section 15.6 one can see that it involves again a specific quadrature of the b field (here shifted by the phase ϕ_b in addition to the phase δ).

The only possible concern is whether the amplitude of the b field is also increased by recycling, which would raise the shot-noise level above what the change in the intensity of a alone would do. This is actually not the case. On resonance, that is, when $\exp(i\psi) = 1$ (which yields maximum enhancement of the a light), the coefficient of $\bar{E}_b$ in equation (15.85) becomes

$$\frac{\cos(\phi/2) - r}{1 - r\cos(\phi/2)}, \tag{15.86}$$

which, as a function of ϕ, is easily seen to be always bounded between -1 and 1.

We may conclude, then, recycling is perfectly compatible with squeezing, and both methods can be used simultaneously to improve the signal-to-noise ratio: recycling does it by increasing the light intensity (the largest possible enhancement factor in equation (15.85) is $t/(1 - r) \simeq 2/t$ when $\phi \ll t$, the approximation holding when $t \ll 1$), thus increasing the signal, whereas squeezing an appropriate quadrature of the b light reduces the noise itself.

Acknowledgments

Some parts of sections 15.1–15.4 have been reprinted from Vinet et al. (1988), with the permission of the American Physical Society.

References

Bondurant, R. S. and Shapiro, J. H. (1984). *Phys. Rev. D* **30,** 2548.
Caves, C. M. (1980). *Phys. Rev. Lett.* **45,** 75.
Caves, C. M. (1981). *Phys. Rev. D* **23,** 1693.
Caves, C. M. (1982). *Phys. Rev. D* **26,** 1817.
Caves, C. M. (1985). *Phys. Rev. Lett.* **54,** 2465.

Drever, R. W. P. (1983). In *Gravitational Radiation,* eds. N. Deruelle and T. Piran, p. 321, North Holland, Amsterdam.

Drever, R. W. P., Hough, J., Munley, A. J., Lee, S.-A., Spero, R., Whitcomb, S. E., Ward, H., Ford, G. M., Hereld, M., Robertson, N. A., Kerr, I., Pugh, J. R., Newton, G. P., Meers, B., Brooks, E. D. III and Gursel, Y. (1983). In *Quantum Optics, Experimental Gravitation, and Measurement Theory,* eds. P. Meystre and M. O. Scully, p. 503, Plenum Press, New York.

Gea-Banacloche, J. and Leuchs, G. (1987a). *J. Mod. Opt.* **34,** 793.

Gea-Banacloche, J. and Leuchs, G. (1987b). *J. Opt. Soc. Am. B* **4,** 1667.

Gea-Banacloche, J. and Leuchs, G. (1989). *J. Mod. Opt.* **36,** (10), 1277.

Loudon, R. and Shepherd, T. J. (1984). *Optica Acta* **31,** 1243.

Machida, S., Yamamoto, Y. and Itaya, Y. (1987). *Phys. Rev. Lett.* **58,** 1000.

Man, C. N. (1987). Communication at 'Gravitational wave data analysis', NATO Advanced Research Workshop, Cardiff.

Meers, B. (1988). *Phys. Rev. D* **38,** 2317.

Ni, W.-T. (1986). *Phys. Rev. A* **33,** 2225.

Rüdiger, A., Schilling, R., Schnupp, L., Winkler, W., Billing, H. and Maischberger, K. (1981). *Optica Acta* **28,** 641.

Schilling, R. (1988). Communication at the Workshop on Gravitational Wave Detection, Munich, May 1988.

Schnupp, L. (1987). Communication at 'Gravitational wave data analysis', NATO Advanced Research Workshop, Cardiff.

Schnupp, L. (1988). Communication at the European Collaboration Meeting on Interferometric Detection of Gravitational Waves, Sorrento, September 1988.

Shoemaker, D., Schilling, R., Schnupp, L., Winkler, W., Maischberger, K. and Rüdiger, A. (1988). *Phys. Rev. D* **38,** 423.

Vinet, J. Y., Meers, B., Man, C. N. and Brillet, A. (1988). *Phys. Rev. D* **38,** 433.

Xiao, M., Wu, L.-A. and Kimble, H. J. (1987). *Phys. Rev. Lett.* **59,** 278.

Yuen, H. P. (1976). *Phys. Rev. A* **13,** 2226.

Yuen, H. P. (1984). *Phys. Rev. Lett.* **52,** 1730.

Yuen, H. P. and Shapiro, J. H. (1980). *IEEE Trans. Inf. Theory* **IT-26,** 78.

16

Data processing, analysis, and storage for interferometric antennas

BERNARD F. SCHUTZ

16.1 Introduction

Laser-interferometric gravitational wave antennas face one of the most formidable data handling problems in all of physics. The problem is compounded of several parts: the data will be taken at reasonably high data rates (of the order of 20 kHz of 16 bit data); they may be accompanied by twice as much 'housekeeping' data to ensure that the system is working appropriately; the data will be collected 24 hours a day for many years; the data need to be searched in real time for a variety of rare, weak events of short duration (one second or less); the data need to be searched for pulsar signals; the data from two or more detectors should be cross-correlated with each other; and the data need to be archived in searchable form in case later information makes a re-analysis desirable. One detector might generate 400 Mbytes of data each hour. Even using optical discs or digital magnetic tapes with a capacity of 3 Gbytes, a network of four interferometers would generate almost 5000 discs or tapes per year. The gathering, exchange, analysis, and storage of these data will require international agreements on standards and protocols. The object of all of this effort will of course be to make astronomical observations. Because the detectors are nearly omni-directional, a network of at least three and preferably more detectors will be necessary to reconstruct a gravitational wave event completely, from which the astronomical information can be inferred.

In this chapter I will discuss the mathematical techniques for analysing the data and reconstructing the waves, the technical problems of handling the data, and the possibilities for international cooperation, as they appear in mid-1989. This discussion can only be a snapshot in time, and a personal one at that. The subject is one that can be expected to develop considerably in the next decade. I will orient the discussion toward ground-based interferometers, with the sensitivity and spectral range expected of the instruments that are planned to be built in the next decade. Much of the discussion naturally is equally applicable to present prototypes, but it is important to look ahead towards future detectors so that their data problems can be anticipated in their design. A large part of the section on data analysis also applies to space-based interferometers or to the analysis of ranging data for interplanetary spacecraft, although in these cases the volume of data is much lower because they operate as low-frequency detectors. I will also assume that the interferometers will operate with a bandwidth greater than that of the signal, even when they are configured in a resonant mode. In the extreme narrow-banding case, in which the detectors have a bandwidth smaller than that of the waves, the data analysis problem resembles that for bar detectors, as discussed by Pallottino and Pizzella in chapter 10.

16.1.1 Signals to look for

The likely sources of gravitational radiation are described by David Blair in part I of this book. If a source is strong enough to stand out above the noise in the time-series of data coming off the machine, then simple threshold-crossing criteria can be used to isolate candidate events. If the event is too weak to be seen immediately, it may still be picked up by pattern-matching techniques, but the sensitivity to such events will depend upon how much information we have about the expected waveform. At the present time, we have little idea of what waveform to expect from bursts of radiation from gravitational collapse (supernovae or electromagnetically quiet collapses), so their detectability depends upon their being strong enough to stand up above the broad-band noise. (Future detailed numerical calculations of gravitational collapse may change this, of course.) On the other hand, we have detailed predictions for the waveforms from binary coalescence and from continuous-wave sources such as pulsars; these can be extracted from noisy data by various techniques, such as matched filtering. Pulsars with a known position may be found from the output of a single detector by sampling techniques. An all-sky search for unknown pulsars will be performed at a sensitivity that will ultimately be limited by the available computing power. Cross-correlation techniques between detectors can search for a stochastic background of radiation and detect weak, unpredicted signals.

16.2 Analysis of the data from individual detectors

Bursts and continuous-wave signals can in principle be detected by looking at the output of one instrument. Of course, one must have coincident observations of the same waves in different detectors, for several reasons: to increase one's confidence that the event is real, to improve the signal-to-noise ratio of the detection, and to gain extra information with which to reconstruct the wave. The simplest detection strategy splits into two parts: first find the events in single detectors, then correlate them between detectors. In most cases this is likely to work, but in some cases it will only be possible to detect signals in the first place by cross-correlating the output of different detectors. In this section I will address the problem of finding candidate events in single detectors. Cross-correlation will be treated later.

16.2.1   Finding broad-band bursts

A broad-band burst is an event whose energy is spread across the whole of the bandwidth of the detector, which I will take to be something like 100–5000 Hz (although considerable efforts are now being devoted to techniques for extending the bandwidth down to 40 Hz or less). To be detected it has to compete against all of the detector's noise, and the only way to identify it is to see it across a pre-determined amplitude threshold in the time-series of data coming from the detector. The main burst of radiation from stellar core collapse may be like this. Numerical simulations of axisymmetric collapse (Evans, 1986; Piran and Stark, 1986) reveal, among other things, that after the main burst there is – at least if a black hole is formed – a 'ringdown phase' in which the radiation is dominated by the fundamental quasi-normal mode of the black hole. This phase lends itself to some degree of pattern-recognition, such as that which I will describe for coalescing binaries in the next section. But it is unlikely that ringdown radiation will substantially improve the signal-to-noise ratio of a collapse burst, since it is damped out very quickly. Some simplified models of non-axisymmetric collapse (e.g. Ipser and Managan, 1984) suggest that if angular momentum dominates and non-axisymmetric instabilities deform the collapsing object into a tumbling tri-axial shape, then a considerable part of the radiation will come out at a single slowly changing frequency. If future three-dimensional numerical simulations of collapse bear this out, then this would also be a candidate for pattern-matching. But one must bear in mind that even if we have good predictions of waveforms from simulations, there will be an intrinsic uncertainty due to our complete lack of knowledge of the initial conditions we might expect in a collapse, particularly regarding the angular momentum of the core. So it is not yet clear whether collapses will ever be easier to see than the time–series threshold criteria described next would indicate.

(i)   Simple threshold criteria

The idea of setting thresholds is to exclude 'false alarms' – apparent events that are generated by the detector noise. Thresholds are set at a level which will guarantee that any collection of events above the threshold will be free from contamination from false alarms at some level. The 'guarantee' is of course only statistical, and it relies on understanding the noise characteristics of the detector. I will assume here that the noise is Gaussian and white over the observing bandwidth.

This should be a good first approximation, but there are at least two important refinements: first, detector noise is frequency-dependent, and when we consider coalescing binaries this will be important; and second, we must allow for unmodelled sources of noise that will occasionally produce large-amplitude 'events' in individual detectors.

This latter noise can be eliminated by demanding coincident observations in other detectors, provided we assume that it is independent of noise in the other detectors and that it is not Gaussian, in particular that there are fewer low-amplitude noise events for a given number of large-amplitude ones than we would expect of a Gaussian distribution. This implies that the cross-correlated noise between detectors will be dominated by the Gaussian component. These assumptions are usually made in data analysis, but it is important to check them as far as possible in a given set of data.

Thresholds for single detectors Assuming that the noise amplitude n in any sampled point has a Gaussian distribution with zero mean and standard deviation σ, the probability that its absolute value will exceed a threshold T (an event that we call a 'false alarm' relative to the threshold T) is

$$p(|n| > T) = \left(\frac{2}{\pi}\right)^{1/2} \frac{1}{\sigma} \int_T^\infty e^{-n^2/2\sigma^2} \, dn = \left(\frac{2}{\pi}\right)^{1/2} \left(\frac{\sigma}{T} - \frac{\sigma^3}{T^3} + \cdots\right) e^{-T^2/2\sigma^2}. \quad (16.1)$$

In the asymptotic approximation given by the second equality, the first term gives 10% accuracy for $T > 3.2\sigma$, and the first two terms give similar accuracy for $T > 2.5\sigma$. If we want the expected number of false alarms to be one in N_{obs} data points, then we must choose T such that

$$p(|n| > T) = 1/N_{\text{obs}}. \quad (16.2)$$

This is a straightforward transcendental equation to solve. For example, if we imagine looking for supernova bursts of a typical duration of 1 ms, then we might be sampling the noise in the output effectively 1000 times per second. (If we want to reconstruct the waveform we might use the data at its raw sampled rate, say 4 kHz; but this would require a larger signal-to-noise ratio than simple detection, for which we could use the data sampled at or averaged over 1 ms intervals.) If we wish no more than one false alarm per year, then we must choose $T = 6.6\sigma$.

Thresholds for multiple detectors If we have two detectors, with independent noise but located on the same site, then we can dig deeper into the noise by accepting only *coincidences*, which occur when both detectors simultaneously cross their respective thresholds T_1 and T_2. Given noise levels σ_1 and σ_2, respectively, the criterion for the threshold is

$$p(|n| > T_1)p(|n| > T_2) = 1/N_{\text{obs}}, \quad (16.3)$$

For two identical detectors ($\sigma_1 = \sigma_2$), each making 1000 observations per second, the threshold T needs to be set at only $4.5\sigma_1$ to give one false alarm per year. Similarly, three identical detectors on the same site require $T = 3.6\sigma$ and four can be set at $T = 3.0\sigma$. The improvement from two to four detectors is a factor of 1.5 in sensitivity, or a factor of three in the volume of space that can be surveyed, and hence a similar improvement in the expected event rate. This favourable cost/benefit ratio – in this case, a factor of three improvement in event rate for a

Table 16.1. *Thresholds (in units of σ) for various arrays and false-alarm probabilities.*

| Number of detectors | False-alarm probability | | | |
	$1/3 \times 10^{10}$	$1/1.5 \times 10^{12}$	$1/6 \times 10^{12}$	$1/3 \times 10^{14}$
1	6.63	7.19	7.37	7.88
2	4.53	4.93	5.06	5.43
3	3.59	3.92	4.03	4.33
4	3.03	3.31	3.41	3.67

factor of two increase in expenditure – is characteristic of networks of gravitational wave detectors, and indeed of any astronomical detector network whose sensitivity is limited by internal noise uncorrelated between instruments. In table 16.1 appropriate thresholds for a number of possible computer arrays and interesting false-alarm probabilities are given. (The last two columns are relevant to coalescing binaries, as discussed later.) The detectors are assumed to be identical. Notice that the thresholds are relatively insensitive to the false-alarm probability, since we are far out on the Gaussian tail. Thresholds are given in units of σ, the r.m.s. noise amplitude.

(ii) Threshold criteria with time delays

I have qualified the discussion of multiple detectors so far by demanding that they be on the same site; the reason is that if they are separated, then allowing for the possible time delay between the arrival of a true signal in different detectors opens up a larger window of time in which noise can masquerade as signal. Suppose that two detectors are separated by such a distance that the maximum time delay between them is W measurement intervals. (For example, Glasgow and California are separated by about 25 ms, which we take to be effectively ±25 measurement intervals for collapse events. This gives a total window size of 50 measurements.) Then in equation (16.3), the appropriate probability to use on the right-hand side is $1/N_{obs}/W$, since each possible 'event' in one detector must be compared with W possible coincident ones in the other.

In table 16.1, the second and fourth columns of thresholds correspond to false-alarm probabilities that are one-fiftieth of the first and third columns, respectively. For two identical detectors, this 'typical' window $W = 50$ raises the threshold T from 4.53σ to 4.93σ. This is a 9% decrease in sensitivity, or a 29% decrease in the volume of space that can be surveyed.

For three detectors, the situation begins to get more complex: as we will see later, if three detectors see an event that lasts considerably longer than their resolution time, there is a self-consistency check which may be used to reject spurious coincidences. (The check is that three detectors can determine the direction to the source, which must of course remain constant during the event.) For four detectors, even a few resolution times are enough to apply a self-consistency check. In principle, the quantitative effect of these corrections will depend on the signal-to-noise ratio of the event, since strong events can be checked for consistency more rigorously than weak events. But the level of the threshold in turn will determine the minimum signal-to-noise ratio. A full study of this problem has not yet been made, and can probably only be undertaken in the light of a more thorough investigation of the signal-reconstruction problem (see section 16.5).

16.2.2   Extracting coalescing binary signals

Coalescing binaries are good examples of the type of signal that will probably only be seen by applying pattern-matching techniques: the raw amplitude from even the nearest likely source will be below the level of broad-band noise in the detector. Nevertheless, the signal is so predictable that interferometers should be able to see such systems ten times or more as distant as collapsed sources. We will see that the signal depends on two parameters, so when we discuss the coincidence problem from the point of view of pattern-matching, we will have to consider the added uncertainty caused by this.

(i)   The coalescing binary waveform

The amplitude of the radiation from a coalescing binary depends on the masses of the stars and the frequency f of the radiation, which together determine how far apart the stars are. It is usual to assume that the stars are in circular orbits. This is a safe assumption if the binary system has existed in its present form long enough for its orbit to have shrunk substantially, since the timescale for the loss of eccentricity, $e/\dot{e}$, is 2/3 of the similar timescale for the decrease of the semimajor axis a. If the binary has only recently been formed, e.g. by tidal capture in a dense star cluster, then more general waveforms can be expected. This complication will not be treated here.

Amplitude   The model assumes point particles in a Newtonian orbit, with energy dissipation due to quadrupolar gravitational radiation reaction; corrections to this are discussed briefly below. The radiation amplitude when the radiation frequency is f is given by the function:

$$A_h(f) = 2.6 \times 10^{-23} \left(\frac{\mathcal{M}}{M_\odot} \right)^{5/3} \left(\frac{f}{100 \text{ Hz}} \right)^{2/3} \left(\frac{100 \text{ Mpc}}{r} \right), \qquad (16.4)$$

where $\mathcal{M}$ is what I shall call the *mass parameter* of the binary system, defined for a system consisting of stars of masses m_1 and m_2 by the equation

$$\mathcal{M} = m_1^{3/5} m_2^{3/5} / (m_1 + m_2)^{1/5}, \qquad (16.5)$$

or equivalently by the more transparent formula,

$$\mathcal{M}^{5/3} = \mu M_T^{2/3}, \qquad (16.6)$$

where μ is the usual reduced mass and M_T the total mass of the system. A system consisting of two $1.4M_\odot$ stars has $\mathcal{M} = 1.22M_\odot$.

The numerical value of $A_h(f)$ is actually the *maximum* observable value of the amplitude which one obtains when the system is viewed down the axis of its angular momentum. One must insert angular factors in front of the expression to get the wave amplitude in other directions. If one averages over these angular factors *and* over the angular factors that describe the antenna pattern of an interferometer, one obtains an effective *mean amplitude* only $2/5$ of the maximum (Krolak, 1989; Thorne, 1987).

Frequency The binary's orbital period changes as gravitational waves extract energy from the system. The frequency of the radiation is twice the orbital frequency, and its rate of change is

$$\frac{df}{dt} = 13\left(\frac{\mathcal{M}}{M_\odot}\right)^{5/3}\left(\frac{f}{100\ \text{Hz}}\right)^{11/3}\ \text{Hz s}^{-1}. \tag{16.7}$$

The maximum wave amplitude we expect, therefore, has the time-dependence

$$h_{\text{max}}(t) = A_h[f(t)]\cos\left(2\pi\int_{t_a}^{t} f(t')\,dt' + \Phi\right), \tag{16.8}$$

where t_a is an arbitrarily defined 'arrival time', at which the signal reaches the frequency f_a, and Φ is the signal's phase at time t_a. This depends on where in their orbits the stars are when the frequency reaches f_a. The amplitude increases slowly with the frequency-dependence of A_h.

Doing the frequency integral explicitly gives

$$f(t) = 100\ \text{Hz} \times \left[\left(\frac{f_a}{100\ \text{Hz}}\right)^{-8/3} - 0.33\left(\frac{\mathcal{M}}{M_\odot}\right)^{5/3}\left(\frac{t-t_a}{1\ \text{s}}\right)\right]^{-3/8}. \tag{16.9}$$

The phase integral is then

$$2\pi\int_{t_a}^{t} f(t')\,dt' = 3000\left(\frac{\mathcal{M}}{M_\odot}\right)^{-5/3}$$
$$\times\left\{\left(\frac{f_a}{100\ \text{Hz}}\right)^{-5/3} - \left[\left(\frac{f_a}{100\ \text{Hz}}\right)^{-8/3} - 0.33\left(\frac{\mathcal{M}}{M_\odot}\right)^{5/3}\left(\frac{t-t_a}{1\ \text{s}}\right)\right]^{5/8}\right\}. \tag{16.10}$$

Putting this into equation (16.8) for $h_{\text{max}}(t)$ gives the desired formula, which we will use in the next section.

Notice that coalescence in the two-point-particle model occurs when $f = \infty$. For a system whose radiation is at frequency f, the remaining lifetime until this occurs is

$$T_{\text{coal}}(f) = 3.0\left(\frac{\mathcal{M}}{M_\odot}\right)^{-5/3}\left(\frac{f}{100\ \text{Hz}}\right)^{-8/3}\ \text{s}. \tag{16.11}$$

This is $3/8$ of the formal timescale $f/\dot{f}$ deducible from equation (16.7). Of course, for realistic stars the Newtonian point-particle approximation breaks down before this time, but if the stars are neutron stars or solar-mass black holes, corrections need be made only in the last second or less. Corrections due to post-Newtonian effects are the first to become important in this case, followed by tidal and mass-transfer effects. These have been considered in detail by Krolak (1989) and Krolak and Schutz (1987). If at least one of the stars is a white dwarf, tidal corrections will become important when T_{coal} is still 1000 years or so, and f is tens of millihertz; the system would only be observable from space (Evans, Iben and Smarr, 1987).

Fourier transform of the coalescing binary signal   We shall need below not only the waveform $h(f)$, but also its Fourier transform. We shall denote the Fourier transform of any function $g(t)$ by $\tilde{g}(f)$, given by

$$\tilde{g}(f) = \int_{-\infty}^{\infty} g(t)e^{-2\pi i f t}\, dt. \tag{16.12}$$

Provided that the frequency of the coalescing binary signal is changing relatively slowly (i.e., that $T_{coal} \gg 1/f$), the method of stationary phase can be used to approximate the transform of $h_{max}(t)$, $\tilde{h}_{max}(f)$ (Dhurandhar, Schutz and Watkins, 1990; Thorne, 1987). We shall only need its magnitude,

$$|\tilde{h}_{max}(f)| \approx 3.7 \times 10^{-24} \left(\frac{\mathcal{M}}{M_\odot}\right)^{5/6} \left(\frac{f}{100\,\text{Hz}}\right)^{-7/6} \left(\frac{100\,\text{Mpc}}{r}\right) \text{Hz}^{-1}. \tag{16.13}$$

This gives good agreement with the results of some numerical integrations performed by Schutz (1986). We shall use it in the following sections.

(ii) The mathematics of matched filtering: finding the signal
Matched filtering is a linear pattern-matching technique designed to extract signals from noise. For references on the theory outlined in this and subsequent sections, the reader may consult a number of books on signal analysis, such as Srinath and Rajasekaran (1979).

Describing the noise   To use matched filtering we have first to define some properties of the noise, $n(t)$. We expect that $n(t)$ will be a random variable, and we use angle brackets $\langle\ \rangle$ to denote expectation values of functions of this noise. It is usually more convenient to deal with the noise as a function of frequency, as described by its Fourier transform $\tilde{n}(f)$. We shall assume that the noise has zero mean,

$$\langle n(t)\rangle = \langle \tilde{n}(f)\rangle = 0.$$

We shall also assume that the noise is *stationary*, i.e. that its statistical properties are independent of time. Then the *spectral density* of (amplitude) noise $S(f)$ is defined by the equation

$$\langle \tilde{n}(f)\tilde{n}^*(f')\rangle = S(f)\delta(f - f'), \tag{16.14}$$

where a * denotes complex conjugation. This says two things: (i) the noise at different frequencies is uncorrelated; and (ii) the autocorrelation of the noise at a single frequency has variances $S(f)$, apart from the normalization provided by the delta function, which arises essentially because our formalism assumes that the noise stream is infinite in duration. (Texts on signal processing often define $S(f)$ in terms of a normalized Fourier transform of the autocorrelation function of a discretely sampled time-series of noise $n_j(t)$. The continuous limit of this definition is equivalent to ours.) Since $n(t)$ is real, $S(f)$ is real and an even function of f.

Noise in an interferometer   *White* noise has a constant spectrum, which means that $S(f)$ is independent of f. Interferometers have many sources of noise, as described in chapter 11 by W. Winkler in this volume or by Thorne (1987). In this treatment we will consider only two: shot noise, which limits the sensitivity of a detector at most frequencies; and seismic noise, which is idealized as a 'barrier' that makes a lower cutoff on the sensitivity of the detector at a frequency f_s.

The shot noise is intrinsically white (that is, as a noise on the photodetector), but — depending on the configuration of the detector — the detector's sensitivity to gravitational waves depends on frequency, so the relevant noise is the photon white noise divided by the frequency response of the detector (called its *transfer function*). We denote this 'gravitational wave' spectral density by $S_h(f)$. I will assume that the detector is in the standard recycling configuration, so that (allowing for the seismic cutoff) we have

$$S_h(f) = \begin{cases} \dfrac{1}{2}\sigma_f^2(f_k)[1 + (f/f_k)^2] & \text{for } f > f_s, \\ \infty & \text{for } f < f_s. \end{cases} \tag{16.15}$$

Here f_k is the so-called 'knee' frequency, which may be chosen by the experimenter when recycling is implemented, and $\sigma_f(f_k)$ is the standard deviation of the frequency-domain noise at f_k.

In the usual discussions of source strength vs. detector noise (e.g. Thorne, 1987), what is taken to be the detector noise as a function of frequency f is $\sigma_f(f)$, not $[S_h(f)]^{1/2}$, because it is assumed in those discussions that the knee frequency f_k will be optimized by the experimenter for the particular range of frequencies being studied, so that σ_f is representative of the noise that the experimenter would encounter. Later in this section we will see that the optimum value of f_k for observing coalescing binaries is $1.44f_s$.

The matched filtering theorem   Now, the fundamental theorem we need in order to extract the signal from the noise is the matched filtering theorem. If we have a signal $h(t)$ buried in noise $n(t)$, so that the output of our detector is

$$o(t) = h(t) + n(t),$$

and if the Fourier transform of the signal is $\bar{h}(f)$, then any stationary, linear operation on the output can be expressed as a correlation with a *filter* $q(t)$:

$$c(t) = (o \circ q)(t)$$

$$= \int_{-\infty}^{\infty} o(t')q(t' + t)\,dt' \tag{16.16}$$

$$= \int_{-\infty}^{\infty} \bar{o}(f)\bar{q}^*(f)e^{2\pi ift}\,df \tag{16.17}$$

The expectation value of the output $c(t)$ of the filter is the filter's signal,

$$\langle c(t) \rangle = (h \circ q)(t). \tag{16.18}$$

The noise that passes through the filter is Gaussian if $n(t)$ is Gaussian, and its variance is

$$\langle [c(t) - \langle c(t) \rangle]^2 \rangle = \int_{-\infty}^{\infty} S(f)\,|\bar{q}(f)|^2\,df. \tag{16.19}$$

This gives a 'raw' signal-to-noise ratio of

$$\frac{S}{N}(t) = \frac{(h \circ q)(t)}{\left[\int_{-\infty}^{\infty} S(f)\,|\bar{q}(f)|^2\,df\right]^{1/2}}. \tag{16.20}$$

The idea of matching the filter to the signal comes from finding the filter $q(t)$ that maximizes this signal-to-noise ratio. It is not difficult to show that the optimal choice of filter for detecting the signal $h(t)$ is

$$\bar{q}(f) = k\bar{h}(f)/S_h(f), \tag{16.21}$$

where k is any constant. With this filter, if the output contains a signal, then $c(t)$ will reach a maximum at a time t that corresponds to the time in the output stream at which the signal reaches the point $t' = 0$ in the waveform $h(t')$. Of course, noise will distort the form of $c(t)$, but the expected amplitude signal-to-noise ratio S/N in $c(t)$ (ratio of maximum value to the standard deviation of the noise) is given by the key equation

$$\left(\frac{S}{N}\right)^2_{\text{opt}} = 2\int_0^{\infty} \frac{|\bar{h}(f)|^2}{S_h(f)}\,df. \tag{16.22}$$

This is the largest S/N achievable with a linear filter. Moreover, given a waveform $h(t)$ that one wants to look for, and given a seismic cutoff frequency f_s, one can ask what value of the knee frequency f_k one should take in $S_h(f)$ in equation (16.22) to maximize S/N. For coalescing binaries, one can use the explicit expression for $\bar{h}(f)$ given in equation (16.13) to show that this value, as mentioned earlier, is (Krolak, 1989; Thorne, 1987)

$$(f_k)_{\text{opt}} = 1.44 f_s.$$

Thresholds for the detection of coalescing binaries   Naturally, in a real experiment one does not know if a signal is present or not. One then uses the size of S/N to decide on the likelihood of the correlation being the result of noise. A widely used criterion is the Neyman–Pearson test of significance (Davis, 1989), based on the *likelihood ratio*, defined as the ratio of the probability that the signal is present to the probability that the signal is absent (false alarm). If the noise is Gaussian, then the Neyman–Pearson 'best' criterion is just to calculate the chance of false alarm in the matched filter given by equation (16.21), exactly as described in section 16.2.1(i) with x/σ replaced by S/N.

Searches for coalescing binaries can therefore be carried out by applying threshold criteria to the correlations produced by filtering. The false-alarm probabilities for detecting a coalescing binary have to be calculated with some care, however, because we must allow for the fact that we have in general to apply many independent filters, for different values of the mass parameter $\mathcal{M}$, and this increases the chance of a false alarm. I will consider the necessary corrections in section 16.2.2(iii) below.

Determining the time-of-arrival of the signal   It is important for gravitational wave experiments that, by filtering the data stream, one not only determines the presence of a signal, but one also fixes its 'time-of-arrival', defined as the time t_{arr} at which the signal reaches the $t' = 0$ point in the filter $h(t')$. The standard deviation in the measurement of t_{arr} is δt_{arr}, which is given by an equation similar to equation (16.22) (Dhurandhar, Schutz and Watkins, 1990; Srinath and Rajasekaran, 1979):

$$\frac{1}{\delta t_{\mathrm{arr}}^2} = 2 \int_0^\infty \frac{|\tilde{h}(f)|^2}{S_h(f)}\, df = 8\pi^2 \int_0^\infty \frac{f^2\, |\tilde{h}(f)|^2}{S_h(f)}\, df, \qquad (16.23)$$

where $\tilde{h}(f)$ is the Fourier transform of the time derivative of $h(t)$. If either the signal or the detector's sensitivity is narrow-band about a frequency f_0, then a reasonable approximation to equation (16.23) is

$$\delta t_{\mathrm{arr}} = \frac{1}{2\pi f_0} \frac{1}{S/N}, \qquad (16.24)$$

where S/N is the optimum signal-to-noise ratio as computed from equation (16.22). This is a good approximation as long as S/N is reasonably large compared to unity. If we use equation (16.13) for $\tilde{h}(f)$ then it is not hard to show that, for coalescing binaries (Dhurandhar, Schutz and Watkins, 1990)

$$\delta\tau_{\mathrm{arr}} = 0.84\left(\frac{100\,\mathrm{Hz}}{f_{\mathrm{s}}}\right) \frac{1}{S/N}\,\mathrm{ms}. \qquad (16.25)$$

For example, if the signal-to-noise ratio is 7 (the smallest for detection by a single detector) and the seismic limit is 100 Hz, then the timing accuracy would be 0.1 ms. If the signal-to-noise is as high as 30, which could occur a few times per year (see below), then the signal could be timed to $30\,\mu s$. Considering that the time it takes the wave to travel from one detector to another will typically be 15–20 ms, this timing accuracy would translate into good directional information. I will explain below how this can be done.

However, in practice it will turn out that these numbers are too optimistic, perhaps by a factor of two. The reason is that one needs to determine other parameters as well from the signal, such as the mass parameter $\mathcal{M}$ and the phase. The errors in these parameters correlate, with the result that δt_{arr} is affected by, for example, $\delta\mathcal{M}$. Schutz (1986) has shown numerically that a small change in the mass parameter can masquerade as a displacement in the time-of-arrival of the signal. This effect will have to be quantified before realistic estimates of the timing accuracy can be made.

Another serious source of error in timing has been stressed by Alberto Lobo (private communication). As is apparent in the calculations of Schutz (1986), when a waveform has a frequency that changes only slowly with time, there can be an ambiguity in the identification of the peak in the correlation that gives the correct time-of-arrival. This is because a shift of the filter by one cycle relative to the waveform will not degrade the correlation much if the frequency is roughly constant. Our timing accuracy formula gives in some sense the width of the correlation peak, but the spacing between peaks is much larger, of order $1/f_0$ for coalescing binaries. Unless the signal-to-noise ratio is high enough to permit reliable discrimination between peaks, this may be the dominant timing error. It is possible that cross-correlation between detectors will still be able to give correct time delays, as in section 16.4.2 below, but this remains to be investigated.

It may seem paradoxical that, if detector physicists succeed in lowering the seismic barrier to, say, 50 Hz, the arrival-time-resolution given by equation (16.24) appears to get worse as f_s^{-1}! This is not a real worsening, of course: the increase in S/N due to the lower seismic cutoff (gaining as $f_s^{-7/6}$ if f_k remains optimized to f_s) more than compensates the $1/f_s$ factor, and the timing accuracy improves.

Implications for the sampling rate   In practice, one only samples the data stream at a finite rate, not continuously. It is clear from equation (16.22) that one must sample at least as fast as is required to determine $\tilde{h}(f)$ at all frequencies that contribute significantly to the integral for the optimum signal-to-noise ratio: at least twice as fast as the largest required frequency in $\tilde{h}(f)$. For the coalescing binary, whose transform is given approximately by equation (16.13), the power spectrum $|\tilde{h}(f)|^2$ falls off as $f^{-7/3}$, and the recycling shot noise multiplies a further factor of f^{-2} into this. Thus, when f rises to, say, four times f_s, the integrand in equation (16.22) will have fallen off to about 0.005 of its value at f_s. Truncating the integration here should be enough to guarantee that the filter comes within 1% of the optimum signal-to-noise ratio. This would require a sampling rate of $8f_s$, or 800 Hz if we take $f_s = 100$ Hz.

Similar but more stringent requirements apply if one wants good timing. If the sampling rate is smaller than twice the largest frequency at which the integrand in equation (16.23) contributes significantly, then in the numerical calculation the arrival time accuracy will be worse than optimum. This is an important lesson: *in choosing one's sampling speed one should ensure that one can get good accuracy in equation* (16.23), *whose integrand falls off less rapidly with frequency than that of equation* (16.22). If one does sample at an adequate rate, then it is possible to determine the time-of-arrival of a signal to much greater precision than the sampling time, provided the signal-to-noise ratio is much greater than unity. (See, for example, the numerical experiments reported by Gursel and Tinto, 1989.) For a coalescing binary, taking timing accuracy into account does not significantly increase the sampling rate over that required for a good signal-to-noise ratio.

Determining the parameters of the waveform   Naively, one might expect that by performing filtering of the incoming data stream with many independent filters, one would just identify the filter that gives the best correlation with the signal and then infer the mass parameter, phase, amplitude, and time-of-arrival from that. It is possible to do better than this, however, using these values as a starting point. This is called non-linear filtering, and there are many possible ways to proceed. For our problem, one of the most attractive is the Kallianpur–Striebel (KS) filter, described by Davis (1989). Rather than reproduce Davis's clear discussion of this method, I will simply refer the reader to his article and to the M.Sc. thesis of Pasetti (1987), which is the first attempt to design a numerical system capable of detecting coalescing binary signals and estimating their parameters. Pasetti gives listings of his computer programs and tests them on simulated data.

(iii) Threshold criteria for filtered signals

Number of filters needed   When searching a data stream for coalescing binary signals, we cannot presume ahead of time that we know what the mass parameter $\mathcal{M}$ will be: not all neutron stars may have mass $1.4M_\odot$, and some binaries may contain black holes of mass 15 or $20M_\odot$. We therefore will have to filter the data with a family of filters with $\mathcal{M}$ running through the range, say, 0.25–$30M_\odot$.

How many filters should there be? This question has not yet received enough study. The calculations of Dhurandhar, Schutz and Watkins (1991) show that two filters with mass parameters differing by a few per cent have significantly reduced correlation, so the filters in the family should not be more widely spaced than this. However, it is not known whether they should be more closely spaced, to avoid missing weak signals. If we take successive filters to have mass parameters that increase by 1% at each step, then we need about 500 filters to span the range $(0.25, 30)$ in $\mathcal{M}$.

However, there is also another parameter in the filter, equation (16.8): the phase Φ, about which I have so far said little. When the wave arrives at the detector with frequency f_s, so that it is just becoming detectable, its phase may be anything: this depends on the binary's history. Filters with different phases must therefore be used. Inspection of equation (16.8) reveals that the phase is a constant within the cosine term for the duration of the signal. It follows that only two filters with different phases will suffice to determine the phase and amplitude of the signal on the assumption of a given mass parameter. For convenience one might choose $\Phi = 0$ and $\Phi = \pi/2$. This increases the number of filters to about 1000. In section 16.2.2(v) we will look at the computing demands that this filtering makes on the data analysis system. In the present section we shall consider the signal-to-noise implications.

Effective sampling rate First it will be necessary to establish what the filtering equivalent of the sampling rate is, so that we can calculate the probability of, say, one false alarm per year. In our original calculation of the false-alarm probability, the sampling rate told us how many independent data points there were per year, on the assumption of white noise, which meant that each data point was statistically independent, no matter how rapidly samples were taken. In the present case, the output of the filter is the correlation given in equation (16.16). It has noise in it, but the noise is no longer white, having been filtered. The key number that we want here is the 'decorrelation time', defined as the time interval τ_s between successive applications of the filter that will ensure that the outputs of the two filters are statistically independent. The analogue here of the sampling rate in the burst problem is $1/\tau_s$, which I will call the *effective sampling rate*. This is the rate at which successive independent data points arrive from each filter.

To develop a criterion for statistical independence, we consider the autocorrelation function of the filter output when the detector output $o(t)$ is pure noise $n(t)$:

$$a(\tau) = \int_{-\infty}^{\infty} c(t)c(t + \tau)\,\mathrm{d}t. \tag{16.26}$$

We shall take the decorrelation time to be the time τ_s such that $a(\tau)$ is small for all $\tau > \tau_s$. We can learn what this is by noting that it is not hard to show that the Fourier transform of $a(\tau)$ is, when the optimal filter given in equation (16.21) is used,

$$\tilde{a}(f) = \frac{|\tilde{h}(f)|^2}{S_h(f)}. \tag{16.27}$$

For coalescing binaries, we have already discussed some of the properties of this function in section 16.2.2(ii). It is strongly peaked near f_s, and in particular the seismic barrier cuts it off rapidly below f_s. It follows that for times $\tau \gg 1/f_s$ the autocorrelation function is nearly zero: the effective sampling rate is about f_s. To play it safe, we will work with a rate twice this large, or an effective sampling time of 0.005 s. This gives effectively 6×10^9 samples – statistically independent filter outputs – per year.

Thresholds for coalescing binary filters   Now, assuming that the noise is Gaussian, the calculation of the false-alarm probability for any size network looks similar to our earlier one in section 16.2.1(ii). What we have to allow for is that there will be some 1000 independent filters, each of which could give a false alarm. Of course, the false alarm occurs only if each detector registers an event in the *same* filter, so it is like doing 1000 independent experiments with no filter at all and a sampling time of 0.005 s, or one experiment with no filter and a sampling time of 5×10^{-6} s. This increases the number of points by a factor of 200 over the number we used in section 16.2.1(i), but this factor makes only a modest difference in the level of the thresholds. For example, for one false alarm per year, and no correction for time-delay windows, the thresholds are: for one detector, 7.4; for two, 5.1; for three, 4.0; and for four, 3.4. For example, the three-detector threshold is 12% higher than for unfiltered data taken at 1 kHz. For further details see table 16.1.

These figures should not be taken as graven in stone: they illustrate the consequences of a particular set of assumptions. A better calculation of the noise properties of the filters is needed, and in any case one will have to ensure that the detector noise really obeys the statistics we have assumed.

(iv) Two ways of looking at the improvement matched filtering brings

The discussion of matched filtering so far has been fairly technical, with the emphasis on making reliable and precise estimates of the achievable signal-to-noise ratios and timing accuracy. In this section I will change the approach and try to develop approximate but instructive ways of looking at the business of matched filtering. The idea is to understand how matched filtering improves the sensitivity of an interferometer beyond its sensitivity to wide-band bursts. We will look at two points of view: comparing the sensitivity of the detector to broad-band and narrow-band signals that have either (i) the same amplitude or (ii) the same total energy.

Improving the visibility of signals of a given amplitude   Let us consider two signals of the same amplitude h, one of which is a broad-band burst of radiation centred at f_0 and the other of which is a relatively narrow-band signal with n cycles at roughly the frequency f_1. The signals are observed with different recycling detectors optimized at their respective frequencies, f_0 and f_1, possibly contained in the same detector system, as is envisioned in some present designs. The broad-band signal has

$$\left(\frac{S}{N}\right)^2_{bb} = 2 \int_0^\infty \frac{|\tilde{h}(f)|^2}{S_h(f)} \, df.$$

$$\approx \frac{2}{\sigma_f^2(f_0)} \int_0^\infty |\tilde{h}(f)|^2 \, df$$

$$\approx \frac{1}{\sigma_f^2(f_0)} \int_{-\infty}^\infty |h(t)|^2 \, dt. \tag{16.28}$$

Now, the integrand in equation (16.28) for a burst lasts typically only for a time $1/f_0$, so we have

$$\left(\frac{S}{N}\right)_{bb} \approx \frac{h}{\sigma_f(f_0)f_0^{1/2}}. \tag{16.29}$$

For the narrow-band signal, we obtain again equation (16.28), but with f_0 replaced by f_1. Now, however, the signal lasts n cycles, a time n/f_1. This leads immediately to

$$\left(\frac{S}{N}\right)_{nb} \approx \frac{hn^{1/2}}{\sigma_f(f_1)f_1^{1/2}}. \tag{16.30}$$

Comparing equations (16.29) and (16.30), we see that a narrow-band signal has an advantage of $n^{1/2}$ over a burst of the same amplitude and frequency, provided we have enough understanding of the signal to use matched filtering*.

For the coalescing binary one may approximate n by $f^2/\dot{f}$, and this can be large (of order 200). Coalescing binaries gain further when compared to supernova bursts because of their lower frequency: because σ_f depends on f as $f^{1/2}$, there is a further gain of a factor of f_0/f_1, which can be 7 or so. Therefore, a coalescing binary signal might have something like 100 times the S/N of a supernova burst *of the same amplitude*! This exaggerates somewhat the advantage that coalescing binaries have as a potential source of gravitational waves, since their intrinsic amplitudes may be smaller than those from supernovae, but it does show why they are such interesting sources.

Improving the visibility of signals of a given energy   The other way of looking at filtering is in terms of energy. This is very instructive, because it shows 'why' matched filtering works. We have just seen that a narrow-band signal with n cycles has a higher S/N than a broad-band burst of one cycle that has the same amplitude and frequency, by a factor of $n^{1/2}$. But the *energy* in the narrow-band signal is n times that in the burst. This is because the energy flux in a gravitational wave is

$$\mathcal{F}_{gw} \approx \frac{4c^3}{\pi G}h^2f^2, \tag{16.31}$$

and thus the total energy E in a signal passing through a detector during the time n/f that the burst lasts is given by the proportionality

$$E \propto h^2f^2(n/f) = nfh^2.$$

If we solve this expression for nh^2 and put it into equation (16.30), we find

$$\frac{S}{N} \propto \frac{E^{1/2}}{f\sigma_f(f)}. \tag{16.32}$$

---

* For this reason, plots of burst sensitivity for broadband detectors, such as one finds in Thorne (1987), typically plot the *effective amplitude* $hn^{1/2}$ of a signal, rather than just h. This allows one to compare supernova bursts and coalescing binary signals on the same graph.

Since this is independent of n, it applies to broad-band and narrow-band signals equally. It shows that if two signals send the same total energy through an interferometric detector, and if they have the same frequency, then they will have the same signal-to-noise ratio, again provided we have enough information to do the matched filtering where necessary.

This provides a somewhat more realistic comparison of coalescing binaries and supernovae, since a coalescing binary radiates a substantial amount of energy in gravitational waves, of the order of $0.01M_\odot$. This is similar to the energy one might expect from a moderate to strong gravitational collapse. The advantage that coalescing binaries have is that they emit their energy at a lower frequency. The factor of $f\sigma_f \propto f^{3/2}$ in equation (16.32) gives them an advantage of a factor of roughly 20 over a collapse generating the same energy at the same distance. If laser interferometric detectors achieve a broad-band sensitivity of 10^{-22}, as current designs suggest will be possible, then they will be able to see moderate supernovae as far away as 50 Mpc. This volume includes several starburst galaxies, where the supernova rate may be much higher than average. They will therefore also be able to see coalescing binaries at distances approaching 1 Gpc.

(v) The technology of real-time filtering

Basic requirements In this section I will discuss the technical feasibility of performing matched filtering on a data stream in 'real time', i.e. keeping up with the data as it comes out of a detector. Since coalescing binaries seem to make the most stringent demands, I will take them as fixing the requirements of the computing system. We have seen that we need a data stream sampled at a rate of about 1 kHz in order to obtain the best S/N and timing information, so I will use this data rate to discover the minimum requirements. It is likely that the actual sampling rates used in the experiments will be much higher, but they can easily be filtered down to 1 kHz before being analysed. If the seismic cutoff is 100 Hz, then the duration of the signal, at least until tidal or post-Newtonian effects become important, will be less than 2 s in almost all cases. This means that a filter need have no more than about 2000 2-byte data points.

The quickest way of doing the correlations necessary for filtering is to use fast Fourier transforms (FFTs) to transform the filter and signal, multiply the signal transform by the complex conjugate of the filter transform, and invert the product to find the correlation. The correlation can then be tested for places where it exceeds pre-set thresholds, and the resulting candidate events can be subjected to further analysis later. This further analysis might involve: finding the best value of the mass parameter and phase parameter; filtering with filters matched to the post-Newtonian waveform to find other parameters that could determine the individual masses of the stars; looking for unmodelled effects, such as tides or mass transfer; looking for the final burst of gravitational radiation as the two stars coalesce; and of course processing lists of these events for comparison with the outputs of other detectors. Since the number of significant events is likely to be relatively small, the most demanding aspect of this scenario is likely to be the initial correlation with 1000 coalescing binary filters.

Discrete correlations   One way the processing might be done is as follows. The discrete correlation between a data set containing the N values $\{d_j,\ j = 0, \ldots, N-1\}$ and a filter containing the N values $\{h_k, k = 0, \ldots, N-1\}$ is usually given by the *circular* correlation formula:

$$c_k = (d \circ h)_k = \sum_{j=0}^{N-1} d_j h_{j+k}, \qquad k = 0, \ldots, N-1, \tag{16.33}$$

where we extend the filter by making it periodic:

$$h_{j+N} = h_j \ \forall j.$$

The circular correlation formula has a danger, because the data set and filter are not really periodic. In practice, this means that we should make the data set much longer than the (non-zero part of the) filter, so that only when the filter is 'split' between the beginning and the end of the data set does the circular correlation give the wrong answer. Thus, even if each filter requires only $N_h \le 2000$ points, it is more efficient to split the data set up into segments of length $N \gg N_h$ points, and to use a filter which has formally the same length, but the first $N - N_h$ of whose elements are zero. (I am grateful to Harry Ward for stressing the need to pay attention to this point.) The 'padding' by zeros ensures that the periodicity of h corrupts only the last N_h elements of the correlation. This can be rectified by forgetting these elements and beginning the next data segment N_h elements before the end of the previous one: this overlap ensures that the first N_h elements of the next correlation replace the corrupt elements of the previous one with correct values. Since this procedure involves filtering some parts of the data set twice, it is desirable to make it a small fraction of the set, namely to make N_h small compared to N. This efficiency consideration is, however, balanced by the extra numerical work required to calculate long correlations, increasing as $\ln N$. This arises as follows.

Correlation by FFT   The fastest way to do long correlations on a general-purpose computer is to use Fourier transforms (or related Hartley transforms). For a discrete data set $\{d_j, j = 1, \ldots, N-1\}$ the discrete (circular) Fourier transform (DFT) is the set $\{\tilde{d}_k, k = 1, \ldots, N-1\}$ given by

$$\tilde{d}_k = \sum_{j=0}^{N-1} d_j e^{-2\pi ijk/N}, \tag{16.34}$$

whose inverse is

$$d_j = \frac{1}{N} \sum_{k=0}^{N-1} \tilde{d}_k e^{2\pi ijk/N}. \tag{16.35}$$

Then the discrete version of the convolution theorem equation (16.17) is as follows. Given the (circular) correlation $\{c_j\}$ of two sets $\{d_j\}$ and $\{h_j\}$ as in equation (16.33), its DFT is

$$\tilde{c}_k = (\tilde{d}_k)^* \tilde{h}_k, \tag{16.36}$$

where an asterisk denotes complex conjugation.

Fast Fourier transform (FFT) algorithms may require typically $3N \log_2 N$ real floating-point operations (additions and multiplications) to compute the transform of a set of N real elements, provided N is an integer power of two (which can usually be arranged). (I neglect overheads due to integer arithmetic concerned with the index manipulations in such routines and, possibly significantly, memory access overheads.) To compute the correlation of two such sets, then, would require three transforms – two to produce $\tilde{d}_k$ and $\tilde{h}_k$ and a third to invert the product $\tilde{c}_k$ – and the multiplication of the two original transforms, giving a total of $9N \log_2 N + 4N$ real floating-point operations. This is to be compared with the $2N^2 - N$ operations required to calculate the correlation directly from equation (16.33). As long as $N \geq 16$ it will be quicker to use FFTs.

In practice, one would compute once and store the DFT of all M filters, so that in real time the data would have to be transformed only once, and then M products of data and filter calculated and inverse-transformed. This would require $3N(M + 1) \log_2 N + 4NM$ floating-point operations.

Optimal length of a data set We must now remind ourselves that in order to achieve the economies of the FFT algorithm, we must use the circular correlation, which has an extra cost associated with the overlaps we are required to take in successive data sets. For a given filter length (say $N_f < N$ non-zero points in the filter time-series), we can reduce the fractional size of these overlaps by making N larger, but this increases the cost of the FFT logarithmically in N. Is there an optimum ratio N_f/N? The total cost of analysing a data set containing a very large number $N_{tot} \gg N$ of elements, split up into segments of length N is

$$N_{\text{fl·pt ops}} = \frac{N_{tot}}{N - N_f} [3N(M + 1) \log_2 N + 4NM].$$

We want to minimize this with respect to variations in N holding N_f and M (the number of filters) fixed. It is more convenient to introduce the variable $x = N_f/N$, which measures the fractional overlap of successive data sets. In terms of x the expression is:

$$N_{\text{fl·pt ops}}(x) = \frac{N_{tot}}{1 - x} \left[3(M + 1) \log_2 \frac{N_f}{x} + 4M \right]. \tag{16.37}$$

As long as the number of filters M is large, the optimum x will be independent of M: it will depend only on N_f, the 'true' length of the filter. This is illustrated in table 16.2, which gives x and $N_{\text{fl·pt ops}}/N_{tot}M$, the number of floating-point

Table 16.2. *The consequences of various strategies for applying filters of 'true' length* N_f, *padded out with zeros to a length* N, *to very long data sets. See text, especially equation (16.37), for details.*

N_f	N	x	$N_{\text{fl-pt ops}}/N_{\text{tot}}M$
1000	2^{11}	0.488	72
1000	2^{12}	0.244	53
1000	2^{13}	0.122	49
1000	2^{14}	0.061	49
1000	2^{15}	0.031	50
2000	2^{12}	0.488	78
2000	2^{13}	0.244	57
2000	2^{14}	0.122	52
2000	2^{15}	0.061	52
2000	2^{16}	0.031	54

operations per data point per filter, as required by various strategies, always taking N_{tot} to be an integer power of two.

If we take N_f to be 2000, then the optimum x is 0.057; if $N_f = 1000$ then the best x is 0.061. But the minimum in $N_{\text{fl-pt ops}}$ is a flat one, and one can increase the value of x quite a bit without compromising speed. This is important, because each stored filter transform must contain N points, so the larger we make x, the smaller will be our core memory requirements. From this it is clear that choosing an overlap between successive data sets of around 25% gives a CPU demand that is only slightly higher than optimum and reduces storage requirements to a minimum.

Demands on computing power   Based on these calculations, and assuming a data rate of 1000 2-byte samples per second with a 2 s filter length ($N_f = 2000$), it follows that doing 1000 filters in real time requires a computer capable of 60 Mflops (where 1 Mflop is 10^6 floating-point operations per second), and storage for 1000 filters, each of length 16 kbytes. This is within the capabilities of present-day inexpensive (<$100k) workstations with add-on array-processors, or of stand-alone arrays of transputers or other fast microprocessors. In five years it should be trivial.

There are many possible ways to speed up the calculation if CPU rates are a problem. It may be that special-purpose digital-signal-processing chips would be faster than general-purpose microprocessors for this problem. It might be possible to do the calculation in block-integer format rather than floating-point, with filters that consist of crude steps rather than accurate representations of the waveform (Dewey, 1986). These should be analysed further. Another possible CPU-saver is described in the next section.

(vi) Smith's interpolation method for coalescing binaries

A different way of looking at coalescing binary signals An alternative strategy for coalescing binaries has been proposed and implemented by Smith (1987). This interesting idea is based upon the following observation: if two coalescing systems of different mass parameters happen to have the same time of coalescence, then their signals' frequencies will remain strictly proportional to one another right up to the moment of coalescence. This follows from the fact that df/dt is proportional to a power of f, so that, as remarked after equation (16.11), there is a constant α independent of the masses such that $T_{coal} = \alpha f / \dot{f}$. If two signals with present frequencies f_1 and f_2 have the same T_{coal}, then it follows that

$$\frac{df_1}{df_2} = \frac{\dot{f}_1}{\dot{f}_2} = \frac{f_1}{f_2} .$$

Since if their times to coalescence are equal at one time then they are necessarily equal for all later times, this equation can be integrated to give $f_1/f_2 = \text{const}$.

Now suppose that the data stream is sampled at constant increments of the phase of signal 1, i.e. it is sampled at a rate that accelerates with the frequency f_1. Then if a Fourier transform is performed on the sampled points, the signal will appear just as pure sinusoid, allowing it to be identified without sophisticated filtering. Moreover, and this is the key point, every other signal with the same time to coalescence will have been sampled at constant increments of its phase as well, since its frequency has been a constant times the first signal's frequency. So signals from any binary coalescing at the same time, no matter what its mass parameter, will be exposed by the single Fourier transform. Thus, one Fourier transform would seem to have done the work of all 1000 filters!

How much work is required? The situation is not quite that good, however, because a signal with a different coalescence time will not be visible in the transform of the points sampled in the manner just described. Therefore, data must be sampled over again at the increasing rate *ending at each possible time of coalescence of the binary*. If this is done, then every possible signal will be picked up.

One way of implementing this method would be to sample the detector output at a constant rate (e.g. 1000 Hz) and then interpolate to form the data sets that are given to the FFT routine. (Livas, 1987, used this method to search for pulsars in a particular direction.) If we compare this interpolation method with the filtering described earlier, one trades the work of doing 1000 Fourier transforms on a stretch of data for the work of interpolating many times. The actual comparison depends on the number of operations required by the interpolation algorithm, but in general Smith's method with interpolation becomes more attractive as the number of filters one must use increases.

Stroboscopic sampling Another way of implementing Smith's method – and the way she herself used – would be to sample the detector output very fast, say at

10 kHz, and then to extract a data set at a slower rate (perhaps 500–1000 Hz) by selecting from the sampled points those points closest in time to the places one ideally would wish to sample. This is a far faster procedure than interpolating, and it seems to me that it would not necessarily be less accurate than a simple interpolation algorithm. I will call this *stroboscopic sampling*; we will meet it again when we discuss searches for pulsars. I do not know of any detailed theoretical analysis of it; in particular, one would like to understand what it does to the noise background. One also has to be careful about aliasing problems. The idea, at least in astronomy, seems to go back to Horowitz (1969), who devised it for optical searches for pulsars.

Comparison with matched filtering   It may well be that for 1000 filters Smith's method will be more efficient than filtering. However, it has at least two significant disadvantages over filtering:

(1) It is restricted only to looking for the Newtonian coalescing binary signal: even any corrections (such as for post-Newtonian effects) will have to be searched for by filtering the sampled data sets, and the sets are essentially useless in searches for other kinds of signals that we may wish to filter from the data.

(2) Signals with the same coalescence time but different mass parameters will enter the observing window (say, $f > 100$ Hz) at different times, and this presents a possible problem that was first pointed out by Harry Ward. If one decides to break the data stream into sets of length, say, 2–3 s, appropriate to coalescing $1.4M_\odot$ neutron stars starting at 100 Hz, then the set will be much too long for a signal from a binary system of two $14M_\odot$ black holes that will coalesce at the same time. The black hole system will have frequency 24 Hz when the data set begins, and will be buried in the low-frequency detector noise. When the data are transformed, this noise will be included in the transform, and the signal-to-noise ratio will accordingly be reduced. The matched filtering method does not suffer from this drawback, since it filters out the low-frequency noise. It might be possible to avoid this problem by pre-filtering the data stream before it is sampled or interpolated, removing the low-frequency noise (and signal).

Given our present uncertainties about sources, my own prejudice is to use filtering because of its inherent flexibility; but Smith's method may become important if filtering places too great demands on the computing system.

16.2.3   Looking for pulsars and other fixed-frequency sources

(i) Why the data-analysis problem is difficult
There are many possible sources of gravitational radiation that essentially radiate at a fixed frequency. Pulsars, unstable accreting neutron stars (the Wagoner

mechanism), and the possible long-term spindown of a newly formed neutron star are examples. In some cases, such as nearby known pulsars, we will know ahead of time the frequency to look for and the position of the source. But most continuous sources may have unknown frequencies; indeed they will only be discovered through their gravitational waves. I will first discuss the detection problem for sources of known frequency, and then consider searches for unknown sources. Throughout this discussion, the word 'pulsar' will stand for any continuous source with a stable frequency. The most complete discussion of this problem of which I am aware is the Ph.D. thesis of Livas (1987).

If we were on an observing platform that had a fixed velocity relative to the stars, and therefore to any pulsar we might be looking for, then finding the signal would be just a matter of taking the Fourier transform of the data and looking for a peak at the known frequency. This is a special case of matched filtering, since the Fourier integral is the same as the correlation integral in equation (16.17) with the filter equal to a sinusoid with the frequency of the incoming wave. Therefore, the signal-to-noise ratio for an observation that lasts a time T_{obs} would increase as $T_{obs}^{1/2}$, just as in equation (16.30). However, the Earth rotates on its axis and moves about the Sun and Moon, and these motions would Doppler-spread the frequency and reduce its visibility against the noise.

How long do we have to look at a source before it becomes necessary to correct for the Earth's motion? If we consider only the Earth's rotation for the moment, then in a time T_{obs} the detector's velocity relative to the source changes by an amount $\Delta v = \Omega_\oplus^2 R_\oplus T_{obs}$, where $R_\oplus$ is the Earth's radius and $\Omega_\oplus$ its angular velocity of rotation. In a source of frequency f, this produces a change $\Delta f_{Dop} = vf/c$. But the frequency resolution of an observation is $\Delta f_{obs} = 2/T_{obs}$. The Doppler effect begins to be important if $\Delta f_{Dop} = \Delta f_{obs}$. Solving this for T_{obs} gives T_{max}, the maximum uncorrected observing time:

$$T_{max} = \left(\frac{2c}{\Omega_\oplus^2 f R_\oplus}\right)^{1/2} \approx 70\left(\frac{f}{1\,\text{kHz}}\right)^{-1/2}\ \text{min}. \tag{16.38}$$

Using the same formula for the effects of the Earth's orbit around the Sun gives a time roughly 2.8 times as long. The Earth's motion about the Earth–Moon barycentre also has a significant effect. Since any serious observation is likely to last days or longer, the Doppler effects of all these motions must be removed, even in searches for very low-frequency signals (10 Hz).

(ii) Angular resolution of a pulsar observation
The Doppler corrections one has to apply depend on the location of the source in the sky. Since the spin axis of the Earth is not parallel to orbital angular momentum vectors of its motion about the Sun or Moon, there is no symmetry in the Doppler problem, and every location on the sky needs its own correction.

It is of interest to ask how close two points on the sky may be in order to have the same correction; this is the same as asking what the angular resolution of an observation might be. Let us first imagine for simplicity that our detector participates in only one rotational motion, with angular velocity Ω and radius R. If two sources are separated on the sky by an angle $\Delta\theta$ (in either azimuth or altitude), then the difference between the Doppler corrections for the two sources depends on the *difference* between the changes in the detector's velocities relative to the two sources. For small $\Delta\theta$ this is $\Delta v = \Delta\theta\Omega^2 R T_{obs}$. Its maximum value is $2\Omega R\Delta\theta$. Using this velocity change, the argument is otherwise identical to that given in the previous section, provided that we keep Δv no larger than $2\Omega R$. The result is that

$$\Delta\theta = T_{max}^2 \, \max\left(\frac{\Omega^2}{4}, \frac{1}{T_{obs}^2}\right). \tag{16.39}$$

The dependence of this expression on T_{obs} will be significant when we come to discuss all-sky searches for pulsars in section 16.2.2(v) below, so it is well to remind ourselves how it comes about. There are two factors of T_{obs} because, as T_{obs} increases, (i) our frequency resolution increases, so we are more sensitive to the Doppler effect; and (ii) the Doppler velocity change over the observing period becomes larger.

When looking at a source with a frequency of 1 kHz, then for the Earth's rotation, and an observation lasting longer than half a day, this gives

$$\Delta\theta_{rot} = 0.02\left(\frac{f}{1\,\text{kHz}}\right)^{-1} \text{rad}, \tag{16.40}$$

which is about half a degree for a millisecond pulsar. The Earth's motion about the Earth–Moon barycentre can have a greater effect, falling to a minimum of 0.002 rad at two weeks. But this is swamped by the effect of the Earth's motion about the Sun, which gives

$$\Delta\theta_{orbit} = 1\times 10^{-6}\left(\frac{f}{1\,\text{kHz}}\right)^{-1}\left(\frac{T_{obs}}{10^7\,\text{s}}\right)^{-2} \text{rad}, \qquad \text{for } T_{obs} < 1\times 10^7\,\text{s.} \tag{16.41}$$

This reaches a minimum of about 0.2 arcsec for a millisecond pulsar observed for four months. Even at two weeks this motion gives a resolution of 2×10^{-5} rad, much finer than the Earth–Moon motion gives. So the orbital motion of the Earth always dominates the Earth–Moon motion. But it does not dominate the Earth's rotation for short times: up to about 20 hours the limit is given by equation (16.40).

For observations longer than about a day, the Earth's orbital motion therefore affords the better angular resolution, but it also makes the most stringent demands on applying the corrections. In particular, uncertainties in the position of the pulsar being searched, for orbital motion of the pulsar in a binary system, proper motion of the pulsar (e.g., a transverse velocity of 150 km s^{-1} at 100 pc), or unpredicted changes in the period (anything larger than an accumulated fractional change $\Delta f/f$ of $10^{-10}(f/1\,\text{kHz})^{-1}$) will all require special techniques to compensate for the way they spread the frequency out over more than the frequency resolution of the observation.

(iii) The technology of performing long Fourier transforms

We shall see that there are several different strategies one can adopt to search for pulsars, whether known ahead of time or not, but all of them can involve performing Fourier transforms of large data sets. It will help us compare the efficiencies of different strategies if we first look at how this might be done.

If one imagines that the observation lasts 10^7 s with a sampling rate of 1 kHz, then one must perform an FFT with roughly 10^{10} data points. This requires roughly $3N \log_2 N$ operations for $N = 2^{34} = 1.7 \times 10^{10}$. This evaluates to 1.7×10^{12} operations per FFT. Given the 50 Mflops computer we required earlier for filtering for coalescing binaries, this would take about 10 hours. This is not unreasonable: over 200 FFTs could be computed in the time it took to do the observation.

The real difficulty with this is the memory requirement: FFT algorithms require access to the whole data set at once. To achieve these processing speeds, the whole data set would have to be held in fast memory, all 20 Gbytes of it. Unless there is a revolution in fast memory technology, it does not seem likely that this will be possible, at least not at an affordable level. One could imagine being able to store the data on a couple of 10-Gbyte read/write optical discs, and then using a mass-store-FFT algorithm, which uses clever paging of data in and out of store. This would still be very slow, but its exact speed would depend on the computer system.

One method of calculating the Fourier transform would be to split the data set up into M chunks of length L, each chunk being small enough to fit into core. by performing FFTs on data sets of length L it is possible to calculate the contribution of each subset to the total transform. It is not hard to show that the work needed to construct the full transform from these individual sets is about M times the work needed to do it as a single set (see, e.g., Hocking, 1989). With a memory limit of 200 Mbytes and a machine capable of 50 Mflops, it might be possible to do one or two Fourier transforms in the time it takes to do the observation. With the same memory in a machine capable of 1 Gflop, one could do 40 Fourier transforms in the same time. These are big numbers for memory and performance, but they may be within reach of the interferometer projects by the time they go on-line. The numbers become even more tractable if we are looking for a pulsar under 100 Hz: with a data rate of only 100 Hz, say, the work for a given number M of subsets goes down by a factor of about 11. It is clear that it is possible to trade-off memory against CPU speed; the technology of the time will dictate how this trade-off is to be made.

If it proves impossible to compute the full transform exactly, there are approximate methods available, such as to subdivide the full set into M subsets as above, but then only to compute the power spectrum of each subset and to add the power spectra together. This reduces the frequency resolution by a factor of M, with a proportionate decrease in the spatial resolution and in the number of different positions that an observation might need to search. It also reduces the signal-to-noise ratio of the observation. it is likely that techniques developed for radio pulsar searches (Lyne, 1989) will be useful here as well.

(iv) Detecting known pulsars

The earliest example of using a wide-band detector to search for a known pulsar is the experiment of Hough *et al.* (1983), which set an upper limit of $h < 8 \times 10^{-21}$ on radiation from the millisecond pulsar, PSR 1937 + 214. Future interferometers could better this limit by many orders of magnitude, but they will have to do long observing runs (some 10^7 s) to achieve maximum sensitivity. The analysis of the vast amount of data such experiments will generate poses greater problems for analysis than those we addressed for coalescing binaries.

Let us assume that we know the location and frequency of a pulsar, and we wish to detect its radiation. We need to make a correction for the Doppler effects from the known position, or from several contiguous positions if the position is not known accurately enough ahead of time. One might be tempted to approach this problem by filtering, as for coalescing binaries. But because of the computational demands, this is not the best method. Much better is a numerical version of the standard radio technique called *heterodyning**, followed by stroboscopic sampling.

Difficulties with filtering for pulsars Let us consider first why filtering is unsuitable. In this context a filter is just a sinusoidal signal Doppler-shifted to give the expected arrival time of any phase at our detector. If only one rotational motion of our detectors were present, and if the observation were to last several rotation periods, then only points separated in the polar direction would need separate filters: points separated in azimuth have waveforms that are simply shifted in time relative to one another, and so correlating the data in time with only one filter would take care of all such points. This might be useful even for a pulsar of known position, since it might not be known to the accuracy of equations (16.40) and (16.41).

However, our detectors participate in at least *three* rotational motions about different centres, and the observations will probably last only a fraction of a period of the most demanding motion, the solar orbital one. This means that filters lose one of their principal advantages: searching whole data sets for similar signals arriving at different times.

Filtering requires that at least three FFTs of long data sets must be performed: of the filter, of the sampled data, and of their product to find the correlation.

---

* I am indebted to Jim Hough and Harry Ward for suggesting this method. The details in this section are based on conversations with them and with Norman MacKenzie, Tim Niebauer, and Roland Schilling.

Even for a well-known source, there will have to be several filters, because the phase of the wave as it arrives will not be predictable, nor will its polarization. The phase of the wave depends on exactly where the radiating 'lump' on the pulsar is. A given detector will respond to the two independent polarizations differently as it moves in orbit around the Sun; the polarization will generally be elliptical, but the proportion of the two independent polarizations and the orientation of the spin axis are unknown. Each of these variables must be filtered for, and each filter needs two more FFTs (the data set needs to be transformed only once). If the source's position and/or frequency are not known accurately, then even more filters will be required, each adding two further FFTs. Given the problems we saw we might have with FFTs, this could be a costly procedure.

Heterodyne detection Suppose the frequency of the pulsar is f_p in the barycentric frame (Solar System rest frame). Then Doppler effects of the Earth's motion plus uncertainties in the pulsar's frequency and its rate of change will require us to look in a narrow range of frequencies $(f_0, f_0 + \Delta f)$ containing f_p. The idea underlying heterodyning is that if the data contain a sinusoidal signal of frequency f,

$$s(t) = \sin(ft + \phi),$$

where ϕ is a possible phase, then if we *multiply* the signal by a 'carrier' sinusoid of frequency f_c in the bandwidth, the result can be written as

$$\sin(f_c t)s(t) = \frac{1}{2}\cos[(f - f_c)t + \phi] + \frac{1}{2}\cos[(f + f_c)t + \phi].$$

We may choose f_c so that the difference frequency $f - f_c$ is within a bandwidth Δf about zero, and yet it contains all the information (amplitude and phase) of the original signal. By filtering the resultant data set down to that bandwidth about the origin, and then re-sampling it at its (now much lower) Nyquist frequency, one can produce a data set containing many fewer points that will still contain all the information in the original band of frequencies. This set will be easier to apply Fourier transforms to than the original.

The saving in size is of order $\Delta f / f$, or 1×10^{-4} for the Doppler broadening due to the Earth's orbital motion. This would reduce the typical data set discussed in the previous section down from 10^{10} points to 10^6. This is of a size that can reasonably be handled on our 50 Mflops computer: an FFT can be done in a matter of seconds, so that complicated filtering and searches for signals become practical without expensive computing machinery.

When one looks at the details of how to implement heterodyning, one has to worry about how the noise is affected and how the procedure can be done with minimum cost. Much more work needs to be done on this question, but two possible implementations might be as follows. The first step in both is to filter the data stream with a narrow band-pass filter that allows only the required bandwidth through. This is to ensure that subsequent steps do not introduce noise (or signals) from other regions of the spectrum into our bandwidth.

In the first implementation, the next step would be to multiply by the heterodyne carrier with frequency $f_c = f_0$, i.e. at the lower edge of the bandwidth. This will ensure that noise from outside the bandwidth is not heterodyned. This allows the band-pass filter to be imperfect, as it must be if it is not to involve prohibitive amounts of computing: it will perhaps need to fall off by a factor of ten within a distance of $\Delta f/2$ of the edges of the band. Then a low-pass filter needs to be applied to get rid of the *sum* frequencies $f_c + f$. The resulting data set is still running at the rate of 10 kHz or so, but all we want is a narrow band, perhaps less than 1 Hz, about zero frequency. By *stroboscopically sampling* (defined earlier) this set at a rate equal to the appropriate Nyquist frequency ($2\Delta f$) in the barycentric frame for signals arriving from the pulsar's direction, one can produce a data set that is at once small and Doppler-corrected. This sampling involves accepting only one point in every 10^4 or so.

The alternative implementation, which might be even faster, is based on a suggestion of Norman Mackenzie. This is to apply stroboscopic sampling (at a slow rate f_s near the Nyquist rate) directly to the data set after it has been put through the band-pass filter but before heterodyning. This may be thought of as heterodyning by aliasing: what appears in the low-frequency spectrum of the sampled data set is the aliased signal. The aliasing condition is that an original frequency f will appear in the sampled set at a frequency $f - nf_s$, where n is an integer. By choosing n and f_s appropriately, it should be possible to alias the required range of frequencies into a range near zero, without introducing extraneous noise. If the sampling is done at a rate equal to the phase arrival rate for a constant frequency in the barycentric frame at the pulsar's position, it will make all the necessary Doppler corrections automatically. Because this is potentially a very fast method, it deserves more study.

Further refinements can be made. For instance, in the first heterodyning implementation, one should multiply independently by two carriers 90 degrees apart in phase, and then add the resultant difference signals with a similar 90 degree phase shift. This reinforces the signal but adds the two independent quadratures of noise together incoherently, so that the noise is reduced by $\sqrt{2}$ relative to the signal.

Moreover, once a 'slow' data set (near zero frequency) is produced, it may still be necessary to do quite a lot of work on it to extract a pulsar signal. One will have to correct for uncertainties in the pulsar position (and hence in the stroboscopic sampling rate), for changes in the pulsar's intrinsic frequency during the observation period, for possible proper motion or binary motion effects, for the changing orientation of the detector relative to the pulsar direction and so on.

However, regardless of which of the two types of heterodyning implementations turns out to be best, the general principle is clear: if we are only interested in a bandwidth Δf about a frequency f_p, then we should be able to deal with a data set sampled at an effective rate $2\Delta f$ rather than $2f_p$. The resultant savings in computing effort make it possible to contemplate on-line searches for a few selected pulsars with computing resources that are no larger than are needed for filtering for coalescing binaries.

(v) Searching for unknown pulsars

One of the most interesting and important observations that interferometers could make is to discover old nearby pulsars or other continuous wave sources. There may be thousands of spinning neutron stars – old dead pulsars – for each currently active one. The nearest may be only tens of parsecs away. But we would have to conduct an all-sky, all-frequency search to find them. We shall see in this section that the sensitivity we can achieve in such a search is limited by computer technology.

The central problem is the number of independent points on the sky that have to be searched. As we saw in equation (16.39), the angular resolution increases as the square of the observing time, so the number of patches on the sky increases as the fourth power. For observations longer than 20 hours, equation (16.41) implies

$$N_{\text{patches}} = 4\pi/(\Delta\theta)^2 = 1.3 \times 10^{13} \left(\frac{f}{1\,\text{kHz}}\right)^2 \left(\frac{T_{\text{obs}}}{10^7\,\text{s}}\right)^4. \tag{16.42}$$

We will now look at what seems to me to be the most efficient method of searching these patches.

The barycentric Fourier transform   The signal from a simple pulsar (i.e. one that does not have added complications like a binary orbit, a rapid spindown, or a large proper motion) would stand out as a strong peak if we were to compute its Fourier transform with respect to the time-of-arrival of the waves at the barycentre of the Solar System, which we take to be a convenient inertial frame. In this section I shall look at the relationship between this transform and the raw-data transform with respect to time at the detector, which relationship depends on the direction we assume for the pulsar. I also look at the relationship between the barycentric transforms of the same signal on two different assumptions for the pulsar position.

We shall need some notation. Let t_d be the time that a given part of the pulsar signal arrives at the detector. Let $t_b(\theta, \phi, t_d)$ be the time that the same signal would arrive at the barycentre if it comes from a pulsar at angular position (θ, ϕ). Let $s_d(t_d)$ be the signal itself at the detector and $s_b(t_b)$ the signal at the barycentre. Note that

$$s_b[t_b(\theta, \phi, t_d)] = s_d(t_d),$$

by definition. The relation between the two timescales is given by

$$t_b = t_d + k(\theta, \phi, t_d), \tag{16.43}$$

where the function k is slowly varying in time for our problem,

$$\left|\frac{\partial k}{\partial t_d}\right| \ll 1$$

due to the slow velocities that the Earth participates in. The inverse of equation (16.43) is

$$t_d = t_b + g(\theta, \phi, t_b) \tag{16.44}$$

Again the derivative of g is small. From the definition it is evident that

$$g(\theta, \phi, t_b) = -k[\theta, \phi, t_b + g(\theta, \phi, t_b)]. \tag{16.45}$$

The exact forms of the functions g and k are complicated, but they need not concern us here.

Now we wish to find the relation between the Fourier transform of s_b and that of s_a with respect to their respective local times. For a given set of detector data, we have

$$\tilde{s}_b(f_b, \theta, \phi) = \int_{-\infty}^{\infty} s_b[t_b(\theta, \phi)]e^{-2\pi i f_b t_b}\, dt_b,$$

$$= \int_{-\infty}^{\infty} s_d(t_d)e^{-2\pi i f_b t_b}\, dt_b, \tag{16.46}$$

$$= \int_{-\infty}^{\infty} \left[\int_{-\infty}^{\infty} \tilde{s}_d(f_d)e^{2\pi i f_d t_d}\, df_d\right]e^{-2\pi i f_b t_b}\, dt_b,$$

$$= \int_{-\infty}^{\infty} \tilde{s}_d(f_d)m(\theta, \phi, f_d, f_b)\, df_d, \tag{16.47}$$

where we define

$$m(\theta, \phi, f_d, f_b) = \int_{-\infty}^{\infty} e^{2\pi i f_d t_d(t_b)}\, e^{-2\pi i f_b t_b}\, dt_b. \tag{16.48}$$

The inverse of this relation is obtained by a simple permutation of indices:

$$\tilde{s}_d(f_d) = \int_{-\infty}^{\infty} \tilde{s}_b(f_b)n(\theta, \phi, f_d, f_b)\, df_b, \tag{16.49}$$

where the kernel here is

$$n(\theta, \phi, f_d, f_b) = \int_{-\infty}^{\infty} e^{2\pi i f_b t_b(t_d)}\, e^{-2\pi i f_d t_d}\, dt_d. \tag{16.50}$$

These equations allow us to find the barycentric transform from the detector transform, and vice versa. In principle, by applying equation (16.47) to the Fourier transform of the detector data one produces a transform in which the signal from a pulsar at a given position should stand out much more strongly. In practice, if one only wants to do this for a few cases, it is much more efficient to use stroboscopic sampling, which effectively computes equation (16.46) by selecting the appropriate values of the integrand. However, when searching the whole sky for pulsars this would involve more work than the method of the next section.

Barycentric transforms for nearby locations   If one has computed the barycentric transform $\tilde{s}_b$ for some location on the sky, the quickest way to find the transform for a nearby location is to find a direct transformation of $\tilde{s}_b$, rather than to start again with s_d or $\tilde{s}_d$. In this manner one can compute $\tilde{s}_b$ for one location and then 'step' around the sky from there. We derive in this section the appropriate equations.

Consider two locations (θ, ϕ) and (θ', ϕ'). We want $\tilde{s}_b$ at (θ', ϕ') in terms of that at (θ, ϕ). From equations (16.47) and (16.49) we have

$$\tilde{s}_b(f'_b, \theta', \phi') = \int_{-\infty}^{\infty} \tilde{s}_d(f_d) m(\theta', \phi', f_d, f'_b) \, df_d,$$

$$= \int_{-\infty}^{\infty} \int_{-\infty}^{\infty} \tilde{s}_b(f_b, \theta, \phi) n(\theta, \phi, f_d, f_b) m(\theta', \phi', f_d, f'_b) \, df_d \, df_b,$$

$$= \int_{-\infty}^{\infty} \tilde{s}_b(f_b, \theta, \phi) q(\theta', \phi', f'_b; \theta, \phi, f_b) \, df_b, \tag{16.51}$$

where we define the 'stepping' kernel q by

$$q(\theta', \phi', f'; \theta, \phi, f) = \int_{-\infty}^{\infty} m(\theta', \phi', f'', f') n(\theta, \phi, f'', f) \, df''. \tag{16.52}$$

If (θ', ϕ') is close to (θ, ϕ), then the kernel q should be sharply peaked in frequency near $f = f'$. In fact, it is easy to show from the inverse properties that

$$q(\theta, \phi, f'; \theta, \phi, f) = \delta(f - f') \quad \forall \theta, \phi.$$

The peaking of this function is in fact the mathematically precise way of doing the calculation we did roughly earlier, namely seeing how many independent patches on the sky one would have to search. Two angles are independent if q is wider than the frequency resolution of the observation.

The stepping method   The way to do an all-sky search uses in fact the converse of the last statement. In order to convert the barycentric Fourier transform for a source at one position to that at another, one must do the integral given in equation (16.51). If the two positions are adjacent patches on the sky, then by definition the function q will be only (at least on average) two frequency bins wide, so that one can produce the barycentric transform for the second patch from that for the first by a calculation taking of order N operations, where N is the number of data points. This can represent a significant saving over doing stroboscopic sampling and an FFT for each patch. This is particularly true for large data sets that exceed the core memory capacity of the computer, because FFT algorithms on such data sets will be very much slower. The present method does not suffer from this drawback because – after the first barycentric transform has been computed – it does not require the whole transform to be held in memory at once. I shall refer to this method as *stepping around the sky*.

Depth of a search as a function of computing power   We can now assemble what we know and make an assessment of the computing power required to make a search of a given sensitivity, at least by the method of stepping described here. From equation (16.42) the number of patches on the sky is

$$N_{\text{patches}} = 1.3 \times 10^{13} \left(\frac{f}{1 \text{ kHz}}\right)^2 \left(\frac{T_{\text{obs}}}{10^7 \text{ s}}\right)^4.$$

The data set will have a length

$$N_{\text{pts}} = 2 \times 10^{10} \left(\frac{f}{1 \text{ kHz}}\right) \left(\frac{T_{\text{obs}}}{10^7 \text{ s}}\right) \text{ points,}$$

provided we interpret f as the highest observable frequency, so we sample at a rate $2f$. If the stepping operation between adjacent patches requires ten real floating-point operations per data point, then we need to perform

$$N_{\text{fl-pt ops}} = 2.5 \times 10^{24} \left(\frac{f}{1 \text{ kHz}}\right)^3 \left(\frac{T_{\text{obs}}}{10^7 \text{ s}}\right)^5$$

floating-point operations to search the whole sky.

In order to do repeatable searches, it must be possible to analyse the data in roughly the time it takes to take it. If the computer speed is called $\mathscr{S}$, measured in floating-point operations performed per second, then the time to perform $N_{\text{fl-pt ops}}$ operations is $N_{\text{fl-pt ops}}/\mathscr{S}$ s. Ignoring overheads due to other factors, we therefore find that the time to analyse the data is

$$T_{\text{anal}} = 2.5 \times 10^{16} \left(\frac{f}{1 \text{ kHz}}\right)^3 \left(\frac{T_{\text{obs}}}{10^7 \text{ s}}\right)^5 \left(\frac{\mathscr{S}}{100 \text{ Mflops}}\right)^{-1} \text{ s.}$$

By equating T_{anal} and T_{obs}, we obtain the maximum observation time allowed by a computer of a given speed:

$$T_{\text{max}} = 4.4 \times 10^4 \left(\frac{f}{1 \text{ kHz}}\right)^{-3/4} \left(\frac{\mathscr{S}}{100 \text{ Mflops}}\right)^{1/4} \text{ s.} \tag{16.53}$$

This is about 12 hours for a 100 Mflops computer analysing data for millisecond pulsars (up to 1 kHz). If we lower our sights and try to search for pulsars under 100 Hz (still very interesting), we can run for about three days. Another improvement comes from making a narrow-band search. This is attractive anyway, since narrow-banding enhances the detector's sensitivity in the band-width. In a narrow-band search one would use heterodyning to reduce the size of the data set. For a bandwidth B, the analogue of equation (16.53) is

$$T_{\text{max}} = 2.1 \times 10^5 \left(\frac{f}{1\,\text{kHz}}\right)^{-1/2} \left(\frac{B}{2\,\text{Hz}}\right)^{-1/4} \left(\frac{\mathscr{S}}{100\,\text{Mflops}}\right)^{1/4} \quad \text{s}. \qquad (16.54)$$

This is better, but still permits only about 2.4 days of observing in a narrow bandwidth at 1 kHz.

The actual figures given here may change with the invention of more efficient algorithms, but what is not likely to change is that the minimum number of operations per patch on the sky scales linearly with the number of data points. This means in turn that the permissible observation time will grow only as the fourth root of the computer speed. Even worse, since the sensitivity one can reach in h scales as the square root of the observation time, the limits on h will scale as the *eighth* root of the computer speed! Changing from a desktop computer capable of 0.1 Mflops to a supercomputer capable of 10 Gflops improves one's limits on h by only a factor of four.

This is the central problem of the all-sky search for pulsars: it is quite possible to run detectors for several months gathering data, and this will probably be done to search for known pulsars, but computing power limits any all-sky, all-frequency search for unknown pulsars to periods of the order of days.

16.3   Combining lists of candidate events from different detectors

Until now I have kept the discussion to the analysis of one detector's data, but it is clear that for the best signal-to-noise ratio and for the extraction of complete astrophysical information, detectors must operate in coincidence. I will consider in this section the simplest method of coordinated observation: exchanging lists of events detected in individual detectors. I have elsewhere (Schutz, 1989) called this the 'threshold mode' of network data analysis, because each detector's criterion for an 'event' is that its amplitude crosses a pre-set threshold.

16.3.1   Threshold mode of data analysis

We have seen in section 16.2.1(ii) how the thresholds can be determined. Once events have been identified by the on-line computer – either in the time-series of data directly or by filtering – it is important that the data from these events be brought together and analysed as quickly as possible. If the event is a supernova, we have considerably less than a day before it might become bright enough to be seen optically, and optical astronomers need to be told of it as quickly as possible. If the event is a coalescing binary, there may be even more urgency: the absence of an envelope around a neutron star means that any radiation emitted may come out with much less delay than in a supernova. Since we know so little about what such events look like, it would be valuable to have optical telescopes and orbiting X-ray telescopes observe the region of the event as quickly as possible.

The rapid exchange of data is certainly possible: with modern computer networks, it would be easy to arrange that the on-line computers could automatically circulate lists of events and associated data periodically, such as every hour. We should bear in mind that, if the threshold is set so that a network would have a four-way false alarm only once per year at a data rate of 1 kHz, then each detector will see a spurious noise-generated event three times per second! It will be impossible to distinguish the real events from the spurious until the lists of events from the various detectors are compared. The initial lists need not contain much data, so links over the usual data networks will be fast enough at this stage.

What sort of data must be exchanged? If the event is seen in a filter, the list should include the amplitude of the event, the parameters of the best-fit filter, and an agreed measure of the time the signal arrived at the detector (such as when a coalescing binary signal reached some fiducial frequency, e.g. 100 Hz). It will probably also be necessary to include calibration data, as the sensitivity of interferometers will probably change from time to time. If the signal has a high signal-to-noise ratio, then it may be desirable to include other information, such as its correlation amplitude with other filters, or even the raw unfiltered data containing the signal. The feasibility of this will depend upon the bandwidth of available communication channels.

If the event is a broad-band burst seen in the time-series, then it will be even more important to exchange the raw data, along with timing and calibration information. If raw-data exchange is impossible, then at least some description of the event will be needed, such as when it first crossed the threshold, when it reaches its maximum, and when it went below threshold.

Once likely coincidences among detectors have been identified, it will then be useful to request the on-line computers to send out more detailed information about the selected candidate coincidences. Since these requests will be rarer, it will not overburden the communications networks to exchange raw data and more complete calibration information for the times in question. If the events then still seem significant, they should be broadcast to other astronomers and analysed more thoroughly at leisure.

16.3.2 Deciding that a gravitational wave has been detected

The question that underlies all of the present article is, how do we decide that a gravitational wave has actually arrived? Various of our topics, such as the construction and use of filters and the setting of appropriate thresholds, are important components of such a decision. What we want to stress in this section is that the laser interferometer community must make sure that it has well-defined criteria for accepting a gravitational wave event as real, and a well-defined procedure for modifying and updating these criteria, *before* it begins observing in earnest.

The first detection of a gravitational wave will be such a momentous event that – if it occurs in an interferometer network – those who operate the network should leave no room for doubt that the event was well above the threshold expected of known noise sources during the time of observation. If criteria are established ahead of time, there can be no question that they have been 'adapted' to the data; conversely, if criteria for gravitational wave detection are formulated after looking at the data, there is always doubt that the events that are then identified really have the significance that might be claimed for them.

In this connection, one should not naively believe that because an unexpected event has a signal-to-noise ratio that would give it a small probability p of arising by chance, then that automatically means that the probability of its being real is correspondingly high. It is very hard to make an accurate calculation of p, since it involves not only the modelled noise but also unmodelled noise and even the circumstances of an experiment. Some Bayesian-type criterion, which involves an *a priori* estimate of the probability that the candidate event would be real, should also be used in such circumstances.

This is not to say that there should be no criteria for accepting unexpected events or unpredicted waveforms. Provided the signal-to-noise ratio of such events is high enough and they have been processed in the same way as all other data, there should be no problem accepting them. But when the signal-to-noise ratio is relatively small and/or the data have been processed in a way that had not been agreed ahead of time, there is considerable danger of accepting false alarms as real.

What can and should be done, however, if unusual events with marginal signal-to-noise ratio are seen, is that new criteria can be adopted to look for them in subsequent data. If they continue to turn up – or if re-analysis of archived data show them – then they can be accepted as real. Similarly, if new theoretical models of gravitational wave sources are evolved, they can be incorporated into the criteria. But the community should not claim detections before this second stage of verification. In particular, if there are marginal and unexpected one-off events apparently associated with rare astronomical phenomena, then it may not be possible to call them real until they have been seen again, however long that may take.

16.4   Using cross-correlation to discover unpredicted sources

The threshold mode of analysis is unsuitable for some sources, such as continuous waves or weak events that we have not predicted well enough ahead of time to construct filters for. In these cases, the 'correlation mode' is appropriate: using cross-correlations between the data streams of different detectors.

Cross-correlation has its own problems, however: its signal-to-noise relations are rather different from filtering, and the different polarizations of different detectors mean that signals in two different detectors from the same gravitational wave may not exactly correlate. In the next section I will give a general discussion of cross-correlation, addressing the behaviour of noise and assuming that the two data streams contain the same signal. One solution to the problem of polarization has been given by Gursel and Tinto (1989). Their approach will be discussed in section 16.4.2.

16.4.1   The mathematics of cross-correlation: enhancing unexpected signals

It is useful to think of cross-correlation as the use of one data stream as a filter to find things in the other data stream. Thus, if the first stream contains a signal that hasn't been predicted, one can still find it in the second. If we adopt this point of view, then we must face two important differences between matched filtering and cross-correlation as a means of enhancing signal-to-noise ratios. These are:

(1)   The 'filter' is *noisy*. In fact, in the case of most interest, the signal is below the broad-band noise and the power in the filter is dominated by the noise. If we really had an instrument with an infinite bandwidth, then the noise power would be infinite and we would never see the signal. In practice, we will see below that we must filter the data down to a finite bandwidth before performing the correlation in order to achieve an acceptable signal-to-noise ratio.

(2)   The 'filter' also contains the signal we wish to find, of course, but the amplitude of this part of the filter is not known *a priori*: it is the amplitude of the incoming signal. This means that if the incoming signal is reduced by half, the response of the filter to it will go down by a factor of *four*. We shall see that this leads to the biggest difference between matched filtering and cross-correlation when they are applied to long wavetrains: the enhancement of signal-to-noise in cross-correlation increases only as the fourth root of the observing time or the number of cycles in the signal, not as the square root we found in equation (16.30).

If we have two data streams o_1 and o_2 containing the same signal h but independent noise amplitudes n_1 and n_2,

$$o_1(t) = h(t) + n_1(t), \qquad o_2(t) = h(t) + n_2(t), \tag{16.55}$$

their cross-correlation is

$$o_1 \circ o_2 = h \circ h + n_1 \circ h + h \circ n_2 + n_1 \circ n_2. \tag{16.56}$$

The 'signal' is the expectation of this (averaged over both noise amplitudes), which is just $h \circ h$. The variance of the correlation, however, is a problem. The final term contributes

$$\langle |n_1 \circ n_2|^2 \rangle = \left\langle \int \tilde{n}_1(f) \tilde{n}_1^*(f') \tilde{n}_2^*(f) \tilde{n}_2(f') e^{2\pi i (f - f')t} \, df \, df' \right\rangle$$

$$= \int S_1(f) S_2(f) \delta(f - f') \delta(f - f') e^{2\pi i (f - f')t} \, df \, df'.$$

The presence of *two* delta functions in the integrand makes this expression infinite: if we allow all the noise in the detectors to be cross-correlated, then the variance of the correlation will swamp the signal. The solution is (i) to *filter* the output down to a suitable bandwidth B before correlating, and (ii) to perform the correlation only over a finite stretch of data lasting a time T. If we use a superscript F to denote the filtered version of a quantity, then the analogue of $n_1 \circ n_2$ is

$$I_{12}(t) = \int_0^T n_1^F(t') n_2^{F'}(t' + t) \, dt'. \tag{16.57}$$

Its variance is

$$\langle |I_{12}(t)|^2 \rangle = \int_0^T \int_0^T \langle n_1^F(y) n_1^{F*}(y') n_2^F(y + t) n_2^{F*}(y' + t) \rangle \, dy \, dy'. \tag{16.58}$$

The key to evaluating this is the expectation

$$\langle n_1^F(t) n_1^F(t') \rangle = 2 \int_{f_1}^{f_2} S_1(f) \cos[2\pi f(t - t')] \, df, \tag{16.59}$$

where f_1 and f_2 are the lower and upper limits of the filtered frequency band $(f_2 = f_1 + B)$, and where the factor of two arises because negative frequencies must be included in the filtered data as well as positive ones. It is a straightforward calculation to show that, assuming for simplicity that $S_i(f)$ has the constant value σ_{if}^2 over the bandwidth, then for the most important case $2\pi f_1 T \gg 1$ and $2\pi BT \gg 1$,

$$\langle |I_{12}(t)|^2 \rangle \approx 2\sigma_{1f}^2 \sigma_{2f}^2 BT. \tag{16.60}$$

This part of the noise is proportional to the bandwidth of the data and the duration of the correlation. The duration will usually be chosen so that the above conditions on B and T are satisfied, for otherwise the experiment would be too brief to detect any signal that fits within the bandwidth B. The remaining contributions to the variance of the cross-correlation come from the second and third terms of equation (16.56) (strictly, from their filtered and finite-time analogues). These are just like equation (16.19), and add to equation (16.60) a term equal to $(\sigma_{1f}^2 + \sigma_{2f}^2) \int_0^T |h^F(t)|^2 \, dt$.

The case of most interest to us is where the 'raw' signal $h^F(t)$ is smaller than the time-series noise in the bandwidth B in each detector, $n_i^F(t)$. Then the variance is dominated by equation (16.60) and we have the following expression for the signal-to-noise ratio of the cross-correlation:

$$\frac{\text{correlation signal}}{\text{correlation noise}} = \frac{\int_0^T |h^F(t)|^2 \, dt}{[2\sigma_{1f}^2 \sigma_{2f}^2 BT]^{1/2}}. \tag{16.61}$$

This has considerable resemblance to the filtering signal-to-noise ratio given in equation (16.20), and this justifies and makes precise our notion that cross-correlation can be thought of as using a noisy data stream as the filter. To convert equation (16.20) into equation (16.61), we must (i) replace the filter in the numerator with the signal h^F that is in the noisy 'filter', and (ii) replace the filter power in the denominator with the noise power of the noise filter, since we have assumed this power is the largest contributor to the noise.

However, equation (16.61) does not give us the signal-to-noise ratio for the gravitational wave signal, since its numerator is proportional to the *square* of the wave amplitude. This is the effect that we noted at the beginning of this section, that the 'filter' amplitude is proportional to the signal amplitude. A better measure of the amplitude signal-to-noise ratio is the square root of the expression in equation (16.61):

$$\frac{S}{N} = \frac{\left[\int_0^T |h^F(t)|^2 \, dt \right]^{1/2}}{[2\sigma_{1f}^2 \sigma_{2f}^2 BT]^{1/4}} . \tag{16.62}$$

There are two cases to consider here: long wavetrains and short pulses.

(i) Long wavetrains

The best signal-to-noise is achieved if we match the observation time T to the duration of the signal or, in the case of pulsars, make T as long as possible. Let us assume for simplicity that the two detectors have the same noise amplitude, and let us denote by R the 'raw' signal-to-noise ratio of the signal (its amplitude relative to the full detector noise in the bandwidth B),

$$R = \frac{h}{(2B\sigma_f^2)^{1/2}} .$$

Then we find

$$\frac{S}{N} \approx \left(\frac{1}{2} BT \right)^{1/4} R. \tag{16.63}$$

The signal-to-noise ratio increases only as the fourth root of the observation time. If we are looking at, say, the spindown of a newly formed pulsar, lasting 1 s, and we filter to a bandwidth of 1 kHz because we don't know where to look for the signal, then the enhancement factor $(BT/2)^{1/4}$ is about five: short wavetrains are improved, but not dramatically. If we are looking at a pulsar, again in a broad-band search with 1 kHz bandwidth, but in an observation lasting 10^7 s, then the enhancement of signal-to-noise is a factor of about 250. This enhancement could be achieved by the $T_{obs}^{1/2}$ effect in a single-detector observation lasting only three minutes, for which the data could be trivially analysed. If the single detector is narrow-banded, the time would be even less. Therefore, cross-correlation is not a good way of finding pulsars.

There are other differences between filtering and cross-correlation. Since for signals below the broad-band noise ($R < 1$), we do not know where the signal is in the data stream used as a filter, it follows that we cannot determine the time-of-arrival of the signal from the correlation, apart from a relatively crude determination based upon the presence or absence of correlations between given data sets of length T. The correlation also does not tell us the waveform and therefore it cannot determine the true amplitude of the signal. It can, however, determine the time-delays between the arrival of brief events at different detectors.

(ii) Short pulses

Here one would set the bandwidth B equal to that of the pulse; if the pulse has duration roughly $T = 1/B$, and if again the two detectors have the same noise amplitude, then equation (16.62) gives a signal-to-noise ratio that is a factor of roughly $2^{1/4} \approx 1.2$ smaller than the optimum that filtering can achieve. For $TB \approx 1$ our approximations are breaking down, but it is reasonable that using this noisy filter would reduce the signal-to-noise by a factor of order two. Since in this case filtering does not enhance the signal-to-noise ratio, neither does cross-correlation: if a pulse is too weak to be seen above the broad-band (bandwidth B) noise in one detector, if will not be found by cross-correlation.

16.4.2   Cross-correlating differently polarized detectors

A more sophisticated approach to correlation has been devised by Gursel and Tinto (1989) in their approach to the signal-reconstruction problem, which I will describe in detail in section 16.5 below. It works if there are at least three detectors in the network. I shall neglect noise for simplicity in describing the method. If we let θ and ϕ be the angles describing the position of the source on the sky and we use α_i, β_i, and χ_i to represent the latitude, longitude, and orientation of the ith detector, respectively, and if we have some definition of polarization of the waves so that we can describe any wave by its amplitudes h_+ and $h_\times$, then the response $r = \delta l/l$ of the ith detector is a function of the form

$$r_i(t) = E_{+i}(\theta, \phi, \alpha_i, \beta_i, \chi_i)h_+[t - \tau_i(\theta, \phi)]$$
$$+ E_{\times i}(\theta, \phi, \alpha_i, \beta_i, \chi_i)h_\times[t - \tau_i(\theta, \phi)], \quad (16.64)$$

where $\tau_i(\theta, \phi)$ is the time-delay between receiving a wave coming from the direction (θ, ϕ) at some standard location and at the position of the detector. We shall define the 'standard location' by setting $\tau_1 = 0$. We need not be concerned here with the precise form of the functions E_{+i}, $E_{\times i}$, and τ_i, nor with the exact definitions of the various angles.

The response equations of the first two detectors may be solved for h_+ and $h_\times$ and substituted into the response equation for the third to predict its response, for an assumed direction to the source. Let this prediction be $r_{3\text{-pred}}$:

$$r_{3\text{-pred}}(t) = -[D_{23}r_1(t - \tau_3) + D_{31}r_2(t + \tau_2 - \tau_3)]/D_{12}, \quad (16.65)$$

where D_{ij} is the determinant

$$D_{ij} = E_{+i}E_{\times j} - E_{\times i}E_{+j}.$$

If there were no noise in the detectors, then for some choice of angles θ and ϕ there would be exact agreement between $p_{3\text{-pred}}$ and the actual data from detector 3, $r_{3\text{-obs}}$. Given the noise, the best one can do is to find the angles that minimize the squared difference $d(\theta, \phi)$ between the predicted and observed responses during the interval of observation:

$$d(\theta, \phi) = \int_0^T |r_{3\text{-obs}}(t) - r_{3\text{-pred}}(t)|^2 \, dt. \qquad (16.66)$$

Hidden in the integral for d are the correlation integrals we began with, e.g. $\int r_3(t) r_1(t - \tau_3) \, dt$. These will normally be the most time-consuming part of the computation of d for various angles, and should usually be done by FFTs. Once the correlations have been computed for all possible time-delays, they may be used to find the minimum of d over all angles; this will determine the position of the source. Notice that if the noise is small, this information can then be substituted back into equation (16.64) for the first two detectors to find $h_+(t)$ and $h_\times(t)$. This reconstructs the signal. But if the source is weaker than the noise, then this substitution will give mostly noise.

The information we have gained about the unpredicted source, even if it is weak, is that it is there: its position is known and its arrival time can be determined roughly by restricting the time-interval over which the correlation integrals are done and finding the interval during which one gets significant correlations.This is enough to alert other astronomers to look for something in the source's position.

The paper by Gursel and Tinto (1989) contains a more sophisticated treatment of the noise than we have described here, allowing for different detectors to have different levels of noise, and constructing almost optimal filters for the signals that weight given detector responses according to where in their antenna pattern the signal seems to be coming from. They also give the results of extensive simulations and estimate the signal-to-noise ratio that will be required to give good predicitons. This paper is an important advance towards a robust solution of the reconstruction problem.

16.4.3 Using cross-correlation to search for a stochastic background

Another very important observation that interferometers will make is to find or set limits upon a background of radiation. This is much easier to do than finding discrete sources of continuous radiation, because there is no direction-finding or frequency-searching to do. This problem has been discussed in detail by Michelson (1987).

The most sensitive search for a background would be with two detectors on the same site, with the same polarization. Current plans for some installations envision more than one interferometer in one vacuum system, which would permit a correlation search. One would have to take care that common external sources of noise are excluded, especially seismic and other ground disturbances, but if this can be done then the two detectors should respond identically to any random waves coming in, and should therefore have the maximum possible correlation for these waves. The correlation can be calculated either by direct multiplication of the sampled data points ($2N$ operations per time delay between the two data sets) or by Fourier transform methods as in section 16.2.3(iii) above. We are only interested in the zero-time-delay value of the correlation, but in order to test the reality of the observed correlation, one would have to compute points at other time delays, where the correlation is expected to fall off. (How rapidly it falls off with increasing time delay depends on the spectrum of the background.) The choice of technique – direct multiplication or Fourier transform – will depend on the number of time-delays one wishes to compute and the capacity of one's computer.

If separated detectors are used, the essential physical point is that two separated detectors will still respond to waves in the same way if the waves have a wavelength λ much longer than the separation between the detectors. Conversely, if the separation between detectors is greater than $\lambda/2\pi$, there is a significant loss of correlation. It is important as well to try to orient the detectors as nearly as possible in the same polarization state. In order to perform a search at 100 Hz, the maximum separation one would like to have is 500 km. This may be achievable within Europe, but it seems most unlikely that detectors in the USA will be built this close together. The data analysis is exactly the same as for two detectors on the same site.

16.5   Reconstructing the signal

The inverse problem is the problem of how to reconstruct the gravitational wave from the observations made by a network of detectors. A single detector produces limited information about the wave; in particular, on its own it cannot give directional information and therefore it cannot say what the intrinsic amplitude is. With three detectors, however, one can reconstruct the wave entirely. In the last two or three years there has been considerable progress in understanding the inverse problem: see Boulanger, le Denmant and Tourrenc (1988), Dhurandhar and Tinto (1988), Gursel and Tinto (1989), and Tinto and Dhurandhar (1989). I will summarize the main ideas as I understand them at present but this is an area in which much more development is likely soon. My thinking in this section has been shaped by conversations with Massimo Tinto and Kip Thorne.

16.5.1   Single bursts seen in several detectors

(i) Unfiltered signals
A gravitational wave is described by two constants – the position angles of its source, (θ, ϕ) – and two functions of time – the amplitudes of the two independ-

ent polarizations $h_+(t)$ and $h_\times(t)$. Simple counting arguments give us an idea of how much we can learn from any given number of detectors. I will assume here that we do not have an *a priori* model (filter) for the signal. For signals that stand out above the broad-band noise:

• A single detector gives its response $r(t)$ and nothing else. Nothing exact can be said about the waves unless non-gravitational data can be used, as from optical or neutrino detections of the same event.

• Two detectors yield two responses and one approximate time-delay between the arrival of the wave in one detector and its arrival in the other. Two functions of time and one constant should not be enough to solve the problem, and indeed they are not. The time-delay is only an approximate one, because the two detectors will generally be responding to different linear combinations of $h_+(t)$ and $h_\times(t)$, so there will not be a perfect match between the responses of the two detectors, from which the time-delay must be inferred. The time-delay will confine the source to an error-band about a circle on the sky in a plane perpendicular to the line joining the detectors. The antenna patterns of the detectors can then be used to make some places on this circle more likely than others, but the unknown polarization of the wave will not allow great precision here. If the location of the source can be determined by other means, and if noise is not too large, then the two responses can determine the two amplitudes of the waves.

• Three detectors cross the threshold into precision astronomy, at least when the signals stand out against the broad-band noise. Here we have three functions of time (the responses) and two constants (the time-delays) as data, and this should suffice. As described in section 16.4 above, correlations among the three detectors can pin down the location of the source and, if noise is not too important, the time-dependent amplitudes as well. In this case, there is redundant information in the data that effectively test Einstein's predictions about the polarization of gravitational waves: the waveforms constructed from any pair of detectors should agree with those from the other two pairs to within noise fluctuations.

(ii) Filtered signals

If noise is so important that filtering is necessary, there is a completely different way of doing the counting. A given filter yields only constants as outputs, such as the maximum value of the correlation and the time the signal arrives (i.e. when it best matches the filter). It does not give useful functions of time. We can only assume that the signal's waveform matches the 'best' filter, so instead of two unknown time-dependent amplitudes we will have the response of the filter, the time-of-arrival, and a certain number of parameter constants that distinguish the observed waveform from others in its family.

Let us concentrate on coalescing binaries. The signal from a coalescing binary is an elliptically polarized, roughly sinusoidal waveform. The filters form a two-parameter family, characterized by the mass parameter $\mathcal{M}$ and the phase of the signal Φ, as in equation (16.8). The parameters we want to deduce are: the amplitude h of the signal, the ellipticity e of its polarization ellipse (one minus the ratio of the minor and major axes), an orientation angle ψ of the ellipse on the sky, and the binary's mass parameter $\mathcal{M}$. From these data we can not only determine the distance to the system, but also the inclination angle of the binary orbit to the line of sight (from e) and the orientation of the orbital plane on the sky (ψ).

The mass parameter $\mathcal{M}$ will be determined independently in each detector, and of course they will all agree if the event is real. Each detector in addition contributes the response of the filter, the phase parameter, and the time-of-arrival; these data must be used to deduce the five constants $\{\theta, \phi, h, e, \psi\}$. Here is how various numbers of detectors can use their data*:

• One detector does not have enough data, so it can only make average statements about the amplitude.
• Two detectors provide four useful data: two responses, one phase difference, and one time-delay. (Only the *differences* between the phases and times-of-arrival matter: the phase and time-of-arrival at the first detector are functions of the history of the source.) If the two detectors were identically polarized, the phase difference would necessarily be zero. A non-zero phase difference arises because the two principal polarizations in an elliptically polarized wave are 90° out of phase, so if the detectors respond to different combinations of these two polarizations, they will have different phases. With four data chasing five unknowns, the solution will presumably be a one-dimensional curve on the sky, but the problem has not yet been studied from this perspective.
• Three detectors have seven data: three responses, two phase differences and two time-delays. The two time-delays are sufficient to place the source at either of the intersections of two circles on the sky. For either location, the three responses determine h, e, and ψ. Presumably the phase differences would be consistent only with one of these positions, thereby solving the problem uniquely and incidentally providing the phase differences as a test of general relativity's model for the polarization of gravitational waves.

16.6   Data storage and exchange

Although the amount of data generated by a four-detector network will be huge, I would argue strongly that our present ignorance of gravitational wave sources

* This discussion is very different from previous ones I have given, e.g. Schutz (1989). In these I had not yet appreciated the importance of being able to determine the phase parameter independently of the time-of-arrival. This extra information makes it possible to solve the inverse problem with fewer detectors than I had previously believed.

makes it important that the data should be archived in a form that is relatively unprocessed, and kept for as long a time as possible, certainly for several years. It may be that new and unexpected sources of gravitational waves will be found, which will make it desirable to go over old data and re-filter it. It may also be that new classes of events will be discovered by their electromagnetic radiation, possibly with some considerable delay after the event would have produced gravitational waves, and a retrospective search would be desirable. In any case, we have already seen that it will be important to exchange essentially raw data between sites for cross-correlation searches for unknown events. Once exchanged, it is presumably already in a form in which it can be stored.

16.6.1 Storage requirements
We have seen in the introduction that a network could generate 5000 optical discs or videotapes per year. Data compression techniques and especially the discarding of most of the housekeeping data at times when it merely indicated that the detector was working satisfactorily could reduce this substantially, perhaps by as much as a factor of four. The cost of the storage media is not necessarily trivial. While videotapes are inexpensive, optical discs of large capacity could cost $250k at present prices (which will, hopefully, come down). Added to this is the cost of providing a suitable storage room, personnel to supervise the store, and equipment to make access to the data easy.

16.6.2 Exchanges of data among sites
We have already seen how important it will be to cross-correlate the raw data streams. At a data rate of some 100 kbytes per second, or even at 30 kbytes per second if the data volume is reduced as described above, one would have difficulty using standard international data networks. But these networks are being constantly upgraded, and so in five years the situation may be considerably different: it may be possible, at reasonable cost, to exchange short high-bandwidth bursts of data regularly via optical-fibre-to-satellite-to-optical-fibre routes. Alternatively, a cheaper solution might be to exchange optical discs or videotapes physically, accepting the inevitable delay. If lists of filtered events were exchanged on electronic data networks, then there may be less urgency about exchanging the full data sets.

(i) Protocols, analysis and archiving
It will be clear from our discussion that exchanging and jointly analysing data will require careful planning and coordination among all the groups. Discussions to this end are in a rudimentary stage now, but could soon be formalized more. Besides decisions on compatible hardware, software, data formats and modes of exchange, there are a number of 'political' questions that need to be resolved before observations begin. We are dealing with data that the groups involved have spent literally decades of their scientific careers to be in a position to obtain,

and the scientific importance of actual observations of gravitational waves will be momentous. Questions of fairness and proprietary rights to the data could be a source of considerable friction if they are not clearly decided ahead of time. A model for some of these decisions could be the protocols adopted by the GRAVNET network of bar antennas, described elsewhere in this volume. Other models might be international VLBI, or large particle-physics collaborations.

Some of the questions that need to be addressed are:

- how much data needs to be exchanged;
- what groups have the right to see and analyse the data of other groups and what form of acknowledgement they need to give when they use it;
- what powers of veto groups have over the use of their data, for example in publications by other people;
- how long the proprietary veto would last before the data become 'public domain' (the funding agencies will presumably apply pressure to allow ready access to the data by other scientists after some reasonable interval of time);
- how long the data need to be archived.

Given the volume of data and the logistical complications of multi-way exchanges of it, it may be attractive to establish one or more joint data analysis and archiving centres. These could be particularly attractive as sites for any large computers dedicated to the pulsar-search problem. These would collect the data and store it, and perform the cross-correlations that can only be done with the full data sets on hand.

16.7   Conclusions

In this review I have set out what I understand about the data analysis problem as of September, 1989. Evidently, the field is covered very non-uniformly: coalescing binaries have received much more attention than pulsars or stochastic sources so far, and protocols for data exchange are something mainly for the future.

Nevertheless, it is clear that questions of the type we have discussed here will influence in an important way decisions about the detectors: how many there will be, where they will be located, what their orientations will be, what weights one should apply to the various important parameters affecting their sensitivity (e.g., length, seismic isolation, laser power) when deciding how to apportion limited budgets to attain the maximum sensitivity. Other questions that I have not addressed will also be important, particularly choosing the particular recycling configuration most suitable to searching for a given class of sources.

From the present perspective, it seems very likely that in ten years or so a number of large-scale interferometric detectors will be operating with a broadband sensitivity approaching 10^{-22}. The data should contain plenty of coalescing binaries and at least a few supernovae; but the most exciting thing that we can look forward to is the unexpected: will this sensitivity suffice to discover completely unanticipated sources? The best way to ensure that it does is to make sure that our data-analysis algorithms and data-exchange protocols are adequate to the task: given the enormous efforts being made by the hardware groups to develop the detectors, and the considerable amount of money that will be required to build them, it is important that development of the data-analysis tools not be left too late. Solutions to data-analysis problems must be developed in parallel with detector technology.

Acknowledgements

It is a pleasure to acknowledge the helpfulness of many colleagues, from whom I have learned much about data analysis and many of whom have contributed crucial insights into the problem. They include Mark Davis, Ron Drever, Jim Hough, Andrzej Krolak, Sean Lawrence, Alberto Lobo, Norman MacKenzie, Brian Meers, Eduardo Nahmad-Achar, Tim Niebauer, Norna Robertson, Roland Schilling, Kip Thorne, Massimo Tinto, Philippe Tourrenc, Harry Ward, Rai Weiss, and Sheryl Smith Woodward. I am grateful to Drs Hough, Robertson, Tinto, and Ware for correcting errors in earlier drafts of this chapter. Any remaining errors are, of course, my own.

References

Boulanger, J. L., le Denmant, G. and Tourrenc, Ph. (1988). *Phys. Lett. A* **126**, 213–18.
Davis, M. H. A. (1989). In *Gravitational Wave Data Analysis,* ed. B. F. Schutz, pp. 73–94, Kluwer, Dordrecht.
Dewey, D. (1986). In *Proceedings of the Fourth Marcel Grossmann Meeting on General Relativity,* ed. R. Ruffini, p. 581, Elsevier, Amsterdam.
Dhurandhar, S. V. and Tinto, M. (1988). *Mon. Not. R. Astr. Soc.* **234**, 663–76.
Dhurandhar, S. V., Schutz, B. F. and Watkins, W. J. (1991). In preparation.
Evans, C. R. (1986). In *Dynamical Spacetimes and Numerical Relativity,* ed. J. M. Centrella, pp. 3–39, Cambridge University Press.
Evans, C. R., Iben, Jr., I. and Smarr, L. (1987). *Astrophys. J.* **323**, 129–39.
Gursel, Y. and Tinto, M. (1989). *Phys. Rev. D* **40**, 3884–938.
Hocking, W. K. (1989). *Computers in Physics,* Jan/Feb 1989 issue, pp. 59–65.
Horowitz, P. (1969). *Rev. Sci. Instrum.* **40**, 369–70.
Hough, J., Drever, R. W. P., Ward, F., Munley, A. J., Newton, G. P., Meers, B. J., Hoggan, S. and Kerr, G. A. (1983). *Nature* **303**, 216.
Ipser, J. R. and Managan, R. A. (1984). *Astrophys. J.* **282**, 287.
Krolak, A. (1989). In *Gravitational Wave Data Analysis,* ed. B. F. Schutz, pp. 59–69, Kluwer, Dordrecht.
Krolak, A. and Schutz, B. F. (1987). *Gen. Rel. Gravit.* **19**, 1163–71.
Livas, J. C. (1987). Ph.D. Thesis, Massachusetts Institute of Technology, Cambridge, Mass.

Lyne, A. (1989). In *Gravitational Wave Data Analysis*, ed. B. F. Schutz, pp. 95–103, Kluwer, Dordrecht.

Michelson, P. F. (1987). *Mon. Not. R. Astr. Soc.* **227,** 933–41.

Pasetti, A. (1987). M.Sc. thesis, Imperial College, London.

Piran, T. and Stark, R. F. (1986). In *Dynamical Spacetimes and Numerical Relativity*, ed. J. M. Centrella, pp. 40–73, Cambridge University Press.

Schutz, B. F. (1986). In *Gravitational Collapse and Relativity*, H. Sato and T. Nakamura, pp. 350–68, World Scientic, Singapore.

Schutz, B. F. (1989). In *Gravitational Wave Data Analysis*, ed. B. F. Schutz, pp. 315–26, Kluwer, Dordrecht.

Smith, S. (1987). *Phys. Rev. D* **36,** 2901–4.

Srinath, M. D. and Rajasekaran, P. K. (1979). *An Introduction to Statistical Signal Processing with Applications*, Wiley, New York.

Thorne, K. S. (1987). In *300 Years of Gravitation*, eds. S. W. Hawking and W. Israel, pp. 330–458, Cambridge University Press.

Tinto, M. and Dhurandhar, S. V. (1989). *Mon. Not. R. Astr. Soc.* **236,** 621–7.

17

Gravitational wave detection at low and very low frequencies

RONALD W. HELLINGS

17.1 Introduction

The detection of gravitational waves with frequencies less than 1 kHz appears to be impossible on earth, due to the magnitude of the earth's seismic noise at these frequencies. These waves, therefore, will only be seen in space-based detectors.

A simple gravitational wave detector in space can be created by setting up two free masses and using an electromagnetic signal passing from one to the other as a probe of the spacetime curvature of the region between them. This is the fundamental idea involved in several gravitational wave detectors in space, including pulsar timing, two-way Doppler tracking of interplanetary spacecraft, and spaceborne interferometers. In this article we will discuss the theory and practice of such detectors.

The outline of this chapter is as follows. In section 17.2, we will briefly discuss the sources for the gravitational waves that are to be the targets of the space-based detectors. Then, in section 17.3, the effect of a plane gravitational wave on the arrival time of electromagnetic signals is derived. Our derivation follows that of Hellings (1983) and gives the same result as that first found by Estabrook and Wahlquist (1975). In sections 17.4 and 17.5, these results are used to discuss existing results from pulsar timing experiments and spacecraft Doppler tracking experiments. Finally, prospects for space interferometers are discussed in section 17.6.

17.2 LF and VLF gravitational waves

Current usage defines a gravitational wave 'low frequency' (LF) band from 1 Hz to about 10^{-5} Hz and a 'very low frequency' (VLF) band from 10^{-5} Hz to 10^{-9} Hz. These bands separate both as to the sources which are contained in each band and as to the methods used for detection.

Sources of gravitational radiation produce waves of three basic types: short broadband bursts from isolated violent astrophysical events, monochromatic signals of long duration typically produced by sources in fast rotation, and the

453

cosmic stochastic background produced by overlapping astrophysical sources or possibly still echoing from the big bang itself. Over the years, theorists have attempted to estimate the possible strengths of the waves from these sources (see, for example, Epstein and Clark, 1979; Press and Thorne, 1972; Ruffini and Wheeler, 1969). Generally speaking, the most likely astrophysical sources of gravitational waves will produce waves with frequencies in the range 10^{-5} to 10^4 Hz. Astrophysical sources in the LF part of this band (below 1 Hz) include active galactic nucleii, supermassive galactic core collapse, and close binary star orbits. Primordial cosmic sources, those that occur at or very shortly after the big bang, are just as likely to produce waves in one wavelength regime as in any other. A number of researchers have discussed the waves that would be created by inhomogeneity in the structure of the initial singularity (Adams *et al.*, 1982; Carr, 1980; Carr and Verdaguer, 1983; Grischuk, 1977; Krauss, 1985; Starobin- skii, 1979). Others have considered very early events such as primordial black hole formation (Barrow and Carr, 1978) or creation of macroscopic 'strings' or quark 'nuggets' in an early particle phase transition (Hogan and Rees, 1984; Vilenkin, 1981, 1985; Witten, 1984). LF band detectors might expect to see pulses and continuous monochromatic waves of astrophysical origin as well as the LF components of the stochastic gravitational wave background. VLF band detectors for the most part are looking only for the cosmic background.

Detection techniques also separate these two bands. In the LF band a complete cycle or pulse could be seen in a single typical one-day observing session while, in the VLF band, phase connection from several sessions must be done in order to allow an effect to be seen in a time series. The low end of the VLF band corresponds roughly to the professional lifetime of an observer and so provides the limit for waves that someone will actually observe. Waves in the ultra long wavelength (ULF) band, those with frequencies less than 10^{-9} Hz, are typically suggested to be observed via their instantaneous effect in things like anisotropy or inhomogeneity in the 3K background or as a nearly linear change in the period of a binary pulsar (Bertotti, Carr and Rees, 1983).

17.3   The effect of a gravitational wave on electromagnetically tracked free masses

The effect of a weak plane gravitational wave on an initially stationary free test mass, when expressed in the most natural coordinate system, is not to move the test mass away from its coordination position, but rather to flex the spacetime around the mass. Thus no accelerometer could detect the wave, but a signal traveling from one mass to another would be retarded or advanced in arrival time due to the changing curvature of the spacetime.

17.3.1   One-way tracking

A vacuum solution of Einstein's linearized field equations representing plane gravitational waves propagating along the z-axis has as line element

$$d\sigma^2 = c^2\,dt^2 - dx^2 - (1 + h\cos 2\psi)\,dx^2 - (1 - h\cos 2\psi)\,dy^2 - h\sin 2\psi\,dx\,dy,$$
(17.1)

where $h(t - z/c)$ is the amplitude of the wave and ψ is the angle between a principal polarization direction and the x-axis (figure 17.1). We choose the origin of the coordinate system at the earth, and orient the coordinate system so that the source is in the x–z-plane at an angle θ from the propagation direction of the gravitational wave (the z-axis) and at a distance l from the earth. In the derivation of the effect of gravitational waves on one-way timing, the earth and the source are assumed to be at rest (the motion of either one will only affect the results at order v/c).

A photon emitted at the source and moving toward the earth will follow a path in space that may be written to first order in h as

$$x = (l - s)\sin\theta; \qquad z = (l - s)\cos\theta,$$
(17.2)

where s is the distance parameter. The photon trajectories satisfy the null geodesic equation along this path

$$c^2\,dt^2 = (1 + h\cos 2\psi\,\sin^2\theta)\,ds^2.$$
(17.3)

This equation may be integrated along the path from emission at $t = t_1$ and

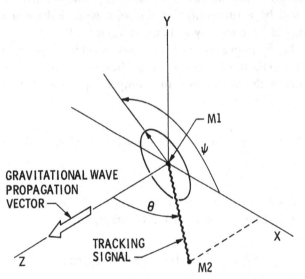

Figure 17.1. Geometry of the derivation of the gravitational wave effect on electromagnetic signals between two masses, M1 and M2. The wave propagation vector is along the z-axis. The x–z-plane contains the M1–M2 line-of-sight at an angle θ from the z-axis. The gravitational wave polarization vector makes an angle ψ with the x-axis in the x–y-plane.

$z = l \cos \theta / c$ to reception at $t = t_2$ and $z = 0$, giving

$$c(t_2 - t_1) = l + \frac{1}{2}c(1 - \cos \theta) \cos 2\psi[H(t_2) - H(t_1 - l \cos \theta / c)], \qquad (17.4)$$

where $H(u)$ is the indefinite integral of $h(u)$.

The quantity $c(t_2 - t_1)$ is the apparent distance to the source. Its derivative will be the apparent radial speed of the source and will be equal to its apparent Doppler shift. Taking the time derivative of equation (17.4), we have

$$z(t, \theta, \psi) \equiv \frac{\Omega_2 - \Omega_1}{\Omega_1} = \frac{1}{2}(1 - \cos \theta) \cos 2\psi[h(t) - h(t - l/c - l \cos \theta / c)], \qquad (17.5)$$

where Ω_1 is the frequency emitted by the source, Ω_2 is the frequency received at the earth, and where we have used the reception time $t \equiv t_2 = t_1 + l/c$ in the arguments of h.

Equation (17.5) gives the Doppler shift of a single photon passing from one body to another. The result depends on the value of the wave amplitude only at the times of emission and of reception. The response of a one-way tracking system to the passage of a gravitational wave is therefore to produce a two-feature signature in the tracking record, each feature proportional to the shape of the wave. For gravitational wave pulses, such as those derived from gravitational collapse, there will thus be two distinct pulses in the tracking record. For a sinusoidal gravitational wave, such as one would see from a binary system, the signature will be a superimposed sinusoid, with perhaps a slow phase and amplitude variation produced by a changing value of θ.

To derive the response of the one-way tracking system to a *stochastic background* of gravitational radiation (see Bertotti, Carr and Rees, 1983), we Fourier decompose the waves into the complex spectrum $\bar{h}$ and integrate over polarization and direction:

$$z(t) = \frac{1}{16\pi^3} \int d\psi \int d\mu \, d\phi \int d\omega \cos 2\psi(1 - \mu)\Re\{\bar{h}(\omega, \theta, \phi, \psi)[1 - e^{-i\omega(1+\mu)l/c}]e^{i\omega t}\},$$

$$(17.6)$$

where $\mu = \cos \theta$ and ω is the angular frequency. The power spectrum of z is the Fourier transform of the autocorrelation function,

$$S_z(\omega) = \frac{1}{\pi} \int_0^\infty d\tau \langle z(t)z(t + \tau)\rangle_t \cos \omega\tau, \qquad (17.7)$$

where the brackets, $\langle\rangle_t$, indicate an integral over t which gives in this case the autocorrelation function of z. We assume that the background is incoherent, so that

$$\int \bar{h}(\omega,\, \theta,\, \phi,\, \psi)\bar{h}(\omega,\, \theta',\, \phi',\, \psi')\, d\omega$$

$$= \delta(\theta - \theta')\delta(\phi - \phi')\delta(\psi - \psi') \int |\bar{h}(\omega,\, \theta,\, \phi,\, \psi)|^2\, d\omega, \quad (17.8)$$

and that it is isotropic and unpolarized, so that $|\bar{h}|^2$ will be independent of θ, ϕ, and ψ. The power spectrum of z may then be written in terms of the power spectrum of h or of the gravitational wave energy density $\rho_g(\omega)$:

$$S_z(\omega) = B(\omega l)S_h(\omega) = \frac{16\pi G}{c^2} \frac{B(\omega l)}{\omega^2} \rho_g(\omega), \quad (17.9a)$$

where the transfer function $B(\omega l)$ arises from the term in square brackets in equation (17.5) and is given by

$$B(u) = \frac{1}{6}\left[1 - \frac{3}{4}\left(\frac{2u - \sin 2u}{u^3}\right)\right], \quad (17.9b)$$

where $u \equiv \omega l/c$.

17.3.2 Two-way tracking

In tracking of a spacecraft or in similar two-way tracking of one free mass by another, a signal is generated at the first mass, coherently transponded or bounced off the second mass, and then received back at the first mass a round-trip light-time later. For such a system, equation (17.5) gives the Doppler shift on the second leg, the *downlink*, but the first leg remains to be computed.

The computation of the additional shift on *uplink* follows the same method as that which led to equation (17.5), with a few changes in variables. Using s as the approximate distance traveled by the photon from earth to spacecraft, we have

$$x = s \sin \theta \qquad z = s \cos \theta \quad (17.10)$$

The integral from emission at t_0 to reception at t_1 then gives a Doppler shift:

$$\frac{\Omega_1 - \Omega_0}{\Omega_0} = \frac{1}{2}(1 + \cos \theta)\cos 2\psi[h(t - l/c - l\cos\theta/c) - h(t - 2l/c)], \quad (17.11)$$

where everything is again written in terms of $t = t_1 + l/c$. Equations (17.5) and (17.11) combine to give a total two-way Doppler shift of

$$y(t,\, \theta,\, \psi) \equiv \frac{\Omega_2 - \Omega_0}{\Omega_0}$$

$$= \frac{1}{2}\cos 2\psi[(1 - \cos\theta)h(t) - 2\cos\theta h(t - l/c - l\cos\theta/c)$$

$$- (1 + \cos\theta)h(t - 2l/c)], \quad (17.12)$$

In equation (17.12), the result depends on the value of the wave amplitude only at the times and positions of emission, bounce, and reception. This gives rise to a three-pulse signature in the Doppler record, as equation (17.5) gave rise to a two-feature signature. This unique three-pulse signature will be very useful in increasing confidence in a possible gravitational wave detection by a spacecraft Doppler tracking experiment.

For the three-pulse signature to be apparent, however, it is necessary that the characteristic width of h be short compared to the light-time. If the pulse widths are long compared to l/c then the three terms in equation (17.12) will overlap and cancel to zeroth order, leaving only the first order residue as the signal in the frequency record. This may be seen by expanding h in a Taylor series about t so that equation (17.12) becomes

$$y(t, \theta, \psi) = -\frac{1}{c}\cos 2\psi \sin^2 \theta \dot{h}(t), \tag{17.13}$$

the integral of which gives an expression proportional to h:

$$\Delta\phi = \Omega_0 \frac{l}{c}\cos 2\psi \sin^2 \theta h(t), \tag{17.14}$$

where $\Delta\phi$ is the phase difference in cycles produced by the gravitational wave. This phase offset will be seen as a change in the apparent range between the masses (given by $\Delta l = c\Omega_0^{-1}\Delta\phi$) so equation (17.14) can also be written as the more familiar expression for spatial strain:

$$\frac{\Delta l}{l} = \cos 2\psi \sin^2 \theta h(t). \tag{17.15}$$

In the interaction of gravitational waves with the planets, the greatest sensitivity comes at gravitational wave periods comparable to the orbital periods, a period much longer than the light-time to the planet. Thus equation (17.15) would be the appropriate equation to use for planetary ranging experiments.

The derivation of the response of a two-way tracking system to a cosmic background of gravitational waves follows a derivation similar to that which led to equations (17.9). The Doppler response may be Fourier analyzed to give

$$y(t) = \frac{1}{16\pi^3}\int d\psi \int d\mu\, d\phi \int d\omega\, \cos 2\psi(1 - \mu)$$

$$\times \Re\{\bar{h}(\omega, \theta, \phi, \psi)[(1 - \mu) - 2\mu e^{-i\omega(1+\mu)l/c} - (1 + \mu)e^{-2i\omega l/c}]e^{i\omega t}\}. \tag{17.16}$$

With equation (17.16), we are led to

$$S_y(\omega) = R(\omega l)S_h(\omega) = \frac{16\pi G}{c^2}\frac{R(\omega l)}{\omega^2}\rho_g(\omega), \tag{17.17a}$$

with

$$R(u) = 1 - \frac{1}{3}\cos 2u - \frac{3}{u^2}\frac{\cos 2u}{u^2} + \frac{2\sin 2u}{u^3} \tag{17.17b}$$

where $u = \omega l/c$, as before.

17.3.3 Interferometers

An interferometer is formed when two Doppler signals with the same round-trip light-time are recorded simultaneously and then differenced. The interferometer signal would thus be

$$x(t, \theta_1, \theta_2, \psi_1, \psi_2) \equiv y_2(t, \theta_2, \psi_2) - y_1(t, \theta_1, \psi_1). \tag{17.18}$$

For most of the interferometers that have been proposed, the round-trip light-time is short compared to the period of the waves that are to be the targets of the detectors, so the long period condition applies. The signal in each arm of the interferometer will thus be given by equation (17.14), while the differenced signal, obtained by letting the signals from the two arms interfere, would be written:

$$\delta \equiv \Delta\phi_1 - \Delta\phi_2 = \Omega_0 \frac{l}{c}[\cos 2\psi_1 \sin^2 \theta_1 - \cos 2\psi_2 \sin^2 \theta_2]h(t). \tag{17.19}$$

The response of such an interferometer to a stochastic background of gravitational waves is obtained in a way similar to that which led to equations (17.9), though the simpler time dependence in equation (17.19) greatly simplifies the derivation. In the long-period limit consistent with equation (17.19), we have

$$S_\delta(\omega) = \Omega_0^2 \frac{l^2}{c^2} \frac{4}{5}(1 - \cos^2\gamma)S_h(\omega), \tag{17.20}$$

where γ is the angle between the two arms of the interferometer.

17.4 Pulsar timing analysis

In the last few years, several researchers have used timing data from pulsars to search for VLF gravitational waves, especially for the waves making up the stochastic cosmic background. Indeed these pulsar results have provided the only cosmologically significant limits on the possible energy density contained in such waves. In this section we discuss how these limits are obtained and point out several precautions that must be taken in the analysis of these data.

In pulsar timing, the times of arrival of pulses are measured and compared with a model. The proper (UTC) times of arrival τ are transformed to coordinate times of arrival t via well-known algorithms. The t's are related to the coordinate times of emission T by

$$ct = cT + \mathbf{k} \cdot (\mathbf{R} - \mathbf{r}) - (1 + \gamma) \sum_p \frac{GM_p}{c^2} \ln\left[\frac{\mathbf{k} \cdot \mathbf{r}_p + r_p}{\mathbf{k} \cdot \mathbf{R}_p + R_p}\right], \tag{17.21}$$

where $\mathbf{R} = \mathbf{R}_0 + \mathbf{V}T$ is the location of the pulsar at time T, $\mathbf{r}$ is the position of the radio observatory on the earth, $\mathbf{k}$ is a unit vector toward the pulsar, and $\mathbf{r}_p$ and $\mathbf{R}_p$ are the position of intervening body p relative to the earth and the pulsar, respectively, at the time when the signal passes closest to the body. The $(1 + \gamma)$ term is the Shapiro time delay, with PPN parameter γ parametrizing the curvature of space. The position of the observatory may be written as $\mathbf{r} = \mathbf{q} + \boldsymbol{\xi}$, where $\mathbf{q}$ is the position of the center of the earth, determined from numerically integrated planetary ephemerides, and $\boldsymbol{\xi}$ is the geocentric position vector of the observatory, determined from observatory coordinates and from a model of the physical ephemeris of the earth. Details of the pulsar timing modeling may be found in Backer and Hellings (1986).

The actual times of arrival may be compared with the predicted times of arrival to give timing residuals, Δt. Among the noise sources contributing to pulsar residuals may be the effects of a variation of the spacetime metric created by the passage of a gravitational wave. Equation (17.12) gives the induced Doppler shifts in a one-way tracking signal and is appropriate for describing the effect which a gravitational wave would have in a record of pulsar timing data. In a record of pulsar timing residuals $\Delta t(t)$, the apparent Doppler shift in the pulsar period (Ω) is found by taking the derivative of the residuals time series:

$$\frac{\Omega_2 - \Omega_1}{\Omega_1} = -\frac{d}{dt}(\Delta t) = \frac{1}{2}(1 - \cos\theta)\cos 2\psi[h(t) - h(t - l/c - l\cos\theta/c)]. \quad (17.22)$$

The effect is thus seen to depend on the value of the metric at the time and place of reception at the earth, $h(t)$, and at the time and place of emission at the pulsar, $h(t - l/c - l\cos\theta/c)$. For pulsars, the light-times are of the order of thousands of years while typical data spans are of the order of a decade, so only one of the two features would be seen from any particular gravitational wave. The features corresponding to the second term in equation (17.22) would thus be unique to each pulsar, carrying information about the gravitational wave at a place long ago and far away. However, the effect produced by the arrival of a gravitational wave at the earth, the first term of equation (17.22) will produce a common Doppler shift in the data from all pulsars in all directions, with an identical time behavior and with only the amplitude differing according to the line-of-sight to the pulsar through θ and ϕ.

By examining the noise spectrum of the pulsar residuals, one may use equations (17.9) to set limits on the spectral density of gravitational wave energy density passing through the galaxy. The long light-times and short data spans lead to the $\omega l/c \gg 1$ and $B \sim 1$ in equation (17.9b). If there is only a single pulsar being observed, then the spectral density of cosmic gravitational waves is simply less than or equal to the spectral density of the residuals. However, if there are several pulsars being observed over the same period of time, it is possible to dig into much larger noise to detect the gravitational wave noise source since it will be a common signal in the time series for each pulsar. The algorithm that allows this digging onto the noise is the cross-correlation function of the timing records of the two pulsars.

We write the frequency residuals from the ith pulsar as

$$\frac{\Delta\Omega_i}{\Omega} = \alpha_i h(t) + n_i(t), \tag{17.23}$$

where α_i contains geometrical factors resulting from the relation of the polarization and propagation vectors of the gravitational wave to the line-of-sight from the earth to the pulsar, and $n_i(t)$ is the independent noise in the data from each pulsar, including the independent second h-term in equation (17.22). Cross-correlating the data from pulsars i and j then gives

$$\Omega^{-2}\langle \Delta\Omega_i \Delta\Omega_j \rangle = \alpha_i \alpha_j \langle h^2 \rangle + \alpha_i \langle hn_j \rangle + \alpha_j \langle hn_i \rangle + \langle n_i n_j \rangle, \tag{17.24}$$

where the brackets indicate cross-section. Since n_i and n_j are independent of each other and of h, all of these terms will tend to zero as the square-root of the number of data points, except for $\langle h^2 \rangle$ which is the autocorrelation function of the gravitational wave amplitude. Thus, as data accumulate, the spectrum of the gravitational wave noise will begin to stand out from that of the other uncorrelated noise sources.

Detweiler (1979) was the first to point out that limits on the stochastic gravitational wave background could be set using the limits of residual noise in pulsar timing. Mashhoon, Carr and Hu (1981) and Bertotti, Carr and Rees (1983) examined the theoretical spectral response in more detail. Actual data analysis was performed by Romani and Taylor (1983) and by Hellings and Downs (1983), and by Davis *et al.* (1985). The most sensitive limit near 10^{-8} Hz was that of Hellings and Downs who analyzed the combined data from four quiet pulsars, in the manner indicated in equation (17.24). Near 10^{-7} Hz, the best results come from Davis *et al.* who used the residuals of the very quiet PSR 1937 + 21. These two limits are compared with other direct limits and with possible critical energy densities in figure 17.2 (see Zimmerman and Hellings, 1980, for details of this figure).

There is one caution that must be observed in analysis of pulsar data. That is that in order to reduce the timing residuals to the levels that appear in the literature, several deterministic signals have had to be subtracted away. These signals correspond to unknown (and therefore erroneous) values for the period, period derivatives, position, proper motion, and possible parallax of the pulsar and, as data accumulate for the most precise pulsars, the parameters of the earth's orbit and of the perturbing solar system. The point of this for gravitational wave analysis is that there may have been gravitational wave signals of unknown amplitude in the data originally which have now been subtracted away by adjusting one of the adjustable parameters of the model.

The method which must be used to take this process into account is to treat the parameter adjustment process as a data filter and to compute the transfer function of the filter. A transfer function is the function which multiplies the input spectrum, frequency by frequency, to produce the output spectrum. The

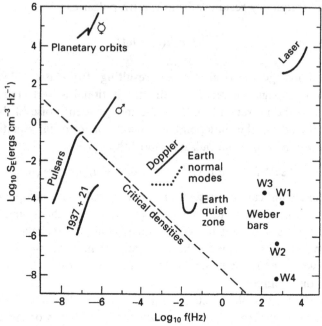

Figure 17.2. Limits on the spectrum of cosmic gravitational radiation energy density from several direct gravitational wave experiments. The line labeled 'critical densities' represents the locus of peaks of a set of broadband spectra, each of which would provide a critical energy density. The line labeled 'pulsars' is from the analysis of Hellings and Downs (1983). The line labeled '1937 + 21' comes from the results of Davis *et al.* (1985). The line labeled 'Doppler' is the result from analysis of the *Voyager* data (Hellings *et al.*, 1981). For further details see also Zimmerman and Hellings (1980).

spectrum of the post-fit residuals must therefore be divided by this transfer function in order to give the realistic limits that may be inferred on the original gravitational wave noise in the timing data records. The transfer function of the pulsar model, including phase, period, period derivative, position, proper motion, and parallax has been computed by Blandford, Narayan and Romani (1984). The results of their analysis are shown in figure 17.3(a). As may be seen, there is a strong absorption line at one year due to uncertainty in pulsar position and proper motion (the latter $t \sin t$ parameters acting to keep the line relatively broad) and strong absorption at the very lowest frequencies due to fitting of the pulsar period model. We have also recently worked out the transfer function for a *combined* adjustment of the pulsar parameters and of solar system parameters, consistent with the level at which the latter parameters are known from other solar system astrometric data. Since the solar system model is based on numerical integration, it was not possible to produce an analytical expression for this transfer function. Rather a Monte Carlo analysis was performed in which 20 years of pulsar timing data were simulated, one point per week, and these data added to the combined set of solar system data while all parameters, pulsar and solar system, were adjusted. Twenty such simulated data sets were analyzed and the pre- and post-fit power spectra were compared to get the transfer function for each set. A mean transfer function was found as an average of the 20 transfer functions. The results of this analysis are shown in figure 17.3(b). It should be noted that there has been a noticeable subtraction of power near Mars's orbital period (687 days) and that other longer-period planetary perturbations combine to subtract almost all power at periods longer than about five years.

The message here is that the rapid drop with decreasing frequency of the limit

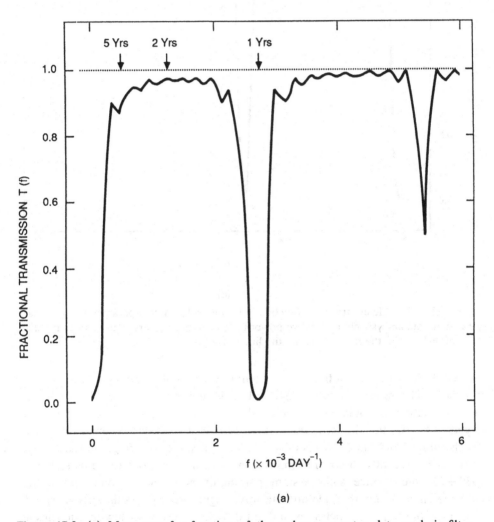

(a)

Figure 17.3. (a) Mean transfer function of the pulsar-parameter data analysis filter – relative power *vs.* frequency in inverse days. (From Blandford, Narayan and Romani, 1984.)   (continued overleaf)

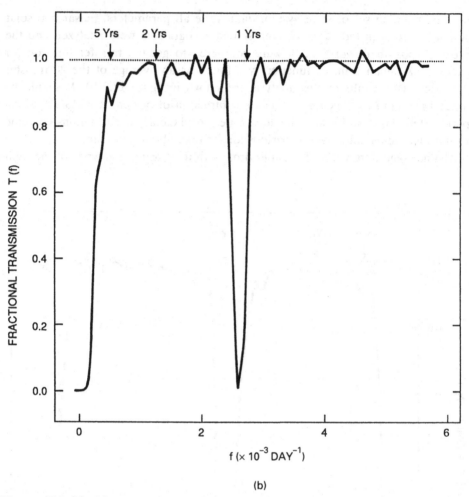

(b)

Figure 17.3 (b) Mean transfer function of the solar-system-parameter-plus-pulsar-parameter data analysis filter – relative power *vs* frequency in inverse days. Pulsar parallax was not included in the fit, eliminating the line at $t = \frac{1}{2}$yr.

labeled 'PSR 1937 + 21' in figure 17.2 cannot be continued very far beyond where it is now. The present cutoff in figure 17.2 is at about 1.7×10^{-7} Hz. With 20 years of data this could be extended down to 10^{-8} ergs cm^{-3} Hz^{-1} at a little beyond 10^{-8} Hz, but the limit would have to rise at frequencies lower than that. On the other hand, all of this is only true when applied to a single pulsar. If one of the other recently discovered millisecond pulsars turns out to be as stable as 1937 + 21, then, since solar system parameter adjustment would not be as effective in soaking up gravitational wave signatures for both pulsars simultaneously, the transfer function would not drop as severely for low frequencies as it does in figure 17.3(b).

17.5 Doppler spacecraft tracking

Gravitational wave detection using Doppler tracking of deep-space spacecraft is a concept pioneered by Dick Davies (1974) and Allen Anderson (1974, 1977). The three pulse response (equation 17.12) of the system to a gravitational wave was discovered by Estabrook and Wahlquist (1975). Early data analysis using the *Viking* spacecraft (Armstrong, Woo and Estabrook, 1979) investigated the noise sources present in the Doppler links and verified the dip in plasma scintillation noise that occurs at solar opposition, when the solar wind is streaming along the line-of-sight to the spacecraft. The first systematic search for gravitational waves with Doppler tracking used the *Voyager* spacecraft (Hellings *et al.*, 1981) and set what remains the current best limit over the band 2×10^{-2} Hz to 3×10^{-4} Hz ($h \leq 3 \times 10^{-14}$ at 10^{-3} Hz). A search at longer gravitational wave periods, down to 5×10^{-5} Hz, is being performed using the *Pioneer 10* spacecraft (Anderson and Mashoon, 1985). Finally, an order-of-magnitude improvement over the *Voyager* results is expected in the early 1990s when data from the *Galileo* spacecraft become available, and a two-spacecraft coincidence experiment will be a possibility using both the *Galileo* and the *Ulysses* spacecraft.

The Doppler spacecraft tracking system is probably most easily understood by reference to figure 17.4. The heart of the system is the hydrogen maser clock which both controls the frequency Ω of the low-noise transmitter and provides the reference frequency for producing the Doppler shift y. The uplink is a 220 kW S-band (2.1 GHz) signal radiated from one of the 64 m parabolic antennas in NASA's Deep Space Network (DSN). Both uplink and downlink are affected by troposphere, ionosphere, and interplanetary plasma scintillation. The signal is tracked in a phase-lock loop on board the spacecraft and coherently transponded at translated S-band (2.3 GHZ) and X-band (8.4 GHz) frequencies at powers of a few tens of watts. The downlink may be received by the same station which sent it (in which case it is called 'two-way tracking') or by some other station (in which case it is referred to as 'three-way tracking'). DSN receiving antennas and low-noise maser amplifiers operate at noise temperatures near 30K. In normal DSN operation, the received signal is tracked in a phase-lock loop, the hydrogen maser being used to beat the frequency down to the Doppler tone. This low-frequency tone is fed into a cycle counter which integrates the frequency, sampling the count at a specified interval and resolving the last partial cycle at each sampling to the nearest nanosecond. These phase counts are then differenced to calculate phase change and divided by the sample time to calculate average $\Delta\Omega$.

Each of the elements in figure 17.4 contributes noise to the system, and a comparison of noise levels to possible gravitational wave strengths would lead one to a pessimistic view for the prospects for such an experiment. However, we have

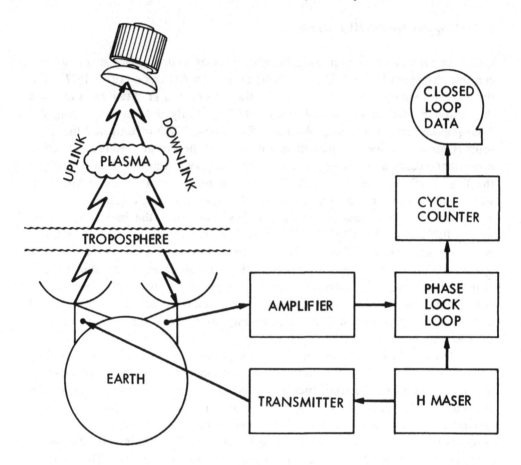

Figure 17.4. Schematic diagram of the Doppler spacecraft tracking system. The explanation is in the text.

as yet made no use of the unique Doppler signature of the gravitational waves (equation 17.12).

In searching for a solitary burst of gravitational waves, the best use of this unique signature will probably be to increase confidence that some event which protrudes strongly above the noise is in fact a gravitational wave. However, in a search for a stochastic background of cosmic gravitational radiation, it is possible to use this signature to actually dig into the noise to pull out its gravitational wave component. The key to the use of equation (17.12) in improving the signal-to-noise ratio is to notice that, of the three features in the Doppler record produced by each feature in h, the first and last will be separated by a round-trip light-time and will always be in inverse sense to each other with the intermediate feature turning up somewhere in between. A stochastic isotropic combination of waves will smear the intermediate pulse out, but the property that whenever $\Delta\Omega/\Omega$ is high now it will be low a round-trip light-time later will be preserved in the stochastic addition of waves. The existence of gravitational wave noise in the tracking record will therefore manifest itself as an anticorrelation in the autocovariance function $A(\tau)$ at a time lag equal to $2l/c$. The power in the anticorrelated feature will be one-sixth power in the zero-lag value of $A(\tau)$ (see Hellings, 1981).

The first results from a systematic search for gravitational waves in Doppler tracking data came from an experiment using the *Voyager* spacecraft in 1979 and 1980. In all, ten days of data were taken. The S-bank uplink was generated at the 64 m stations at Goldstone, California, or Madrid, Spain, and two-way S- and X-band data were then recorded back at each station. For part of the time, the 40 m antenna at Caltech's Owens Valley Radio Observatory (OVRO) was also used to receive the transponded X-band signal in a three-way tracking mode. The round-trip light-time in 1980 was about 6400 s. The experiments were timed to occur near solar opposition in order to take advantage of the drop in plasma noise when the solar wind is streaming along the line-of-sight. The experiments are discussed in more detail by Hellings *et al.* (1981).

As argued above, the preferred data analysis algorithm in a search for the stochastic background involves the autocovariance function of the Doppler record. Any gravitational wave spectrum that falls off faster at high frequency than the other noise sources will then be visible as a more-or-less sharp autocorrelated feature centered at a time lag $\tau = 2l/c$. As long as statistics give a standard deviation σ_A in the autocovariance function which is less than one-sixth the zero-lag autocovariance value, then the gravitational wave's one-sixth power feature at a round-trip light-time may be sought for. When the autocovariance statistics are not this good, then only an upper limit to the background may be found, given by the total zero-lag power.

Figure 17.5 shows autocovariance functions from the *Voyager* experiment on March 12, 1980, at sample times ranging from 100 to 800 s. The round-trip light-time of 6431 s is shown on each. In all cases but the 800 s case, statistics around $\tau = 2l$ produced the most sensitive limit. For the 800 s sample time, the upper limit came from the value of $A(0)$. The projected size of the background (determined by multiplying σ_A by six) was found to fit a spectral model given by

$$S_h = 10^{-27}f^{-1}, \qquad (17.25)$$

where it is assumed that the frequency band ran from a low frequency of 2×10^4 ($\approx 1/2l$) up to high frequency $f_H = 1/2\Delta t$, where Δt is the sample time. The equivalent energy density spectral density is

$$S_E = 5f \, (\mathrm{erg \, cm^{-3} \, Hz^{-1}}). \qquad (17.26)$$

In figure 17.6 the ability of the autocorrelation algorithm to dig into noise is made apparent by comparing S_h to the Doppler noise spectrum. At low frequencies (long sample times) there are so few data points that statistics do not dig below the rms Doppler noise, but, for high frequencies, the method gives

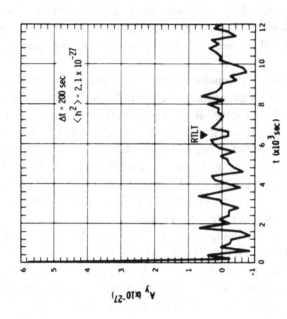

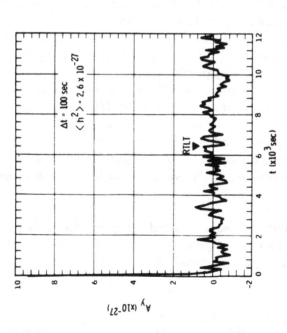

Figure 17.5. Autocovariance functions of data at different sample times from *Voyager* Doppler tracking on March 12, 1980. Non-detection of an anticorrelated feature at the round-trip light-time (RTLT) of 6431 s sets limits on the strength of the gravitational radiation background.

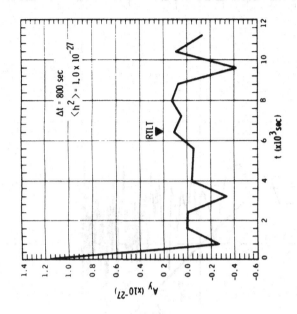

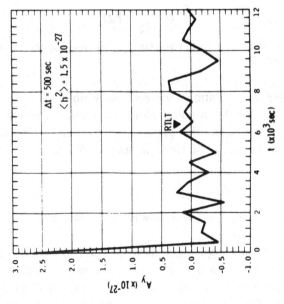

Figure 17.5. (*cont.*)

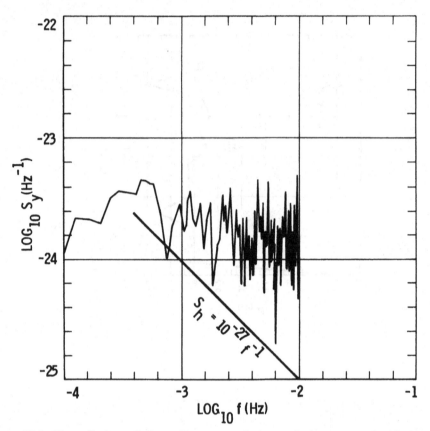

Figure 17.6. Upper limit on S_h from the autocorrelation analysis, compared with the total Doppler noise spectrum S_y.

over a 10 dB increase in gravitational wave signal-to-noise ratio. In figure 17.2, S_E is plotted as a line labeled 'Doppler' along with other existing limits on the background. As can be seen, the present results fail by about an order of magnitude in power (a factor of three in amplitude) to be at the interesting cosmological level.

17.6   Space interferometer gravitational wave experiments

The current sensitivity of Doppler gravitational wave detection is limited at the lower end of the LF spectrum by interplanetary plasma scintillation. Since this effect is radio frequency-dependent, one might suggest that tracking at higher radio frequencies would significantly improve the sensitivity; and this is indeed expected to be the case. A gravitational wave search using the *Galileo* X-band round-trip system is expected to improve the limits from the Voyager experiment by nearly an order of magnitude to a few parts in 10^{15}. However, it is not clear how to improve much beyond this level. Higher frequencies still, or dual frequencies (that allow one to calibrate the effect of the interplanetary plasma and remove it), might be tried, but it is expected that several noise sources lurk at about the 10^{-15} level, sealing off such improvements in sensitivity. Jitter in the best hydrogen maser clocks, for example, and difficult-to-model fluctuations in the troposphere density are known to occur at just this level. Therefore, with the *Galileo* experiment, it seems that earth-based Doppler tracking experiments have reached the limits of their sensitivity unless a significant technology development effort is launched on all three fronts (plasma scintillation, troposphere fluctuation, and clock jitter). Otherwise, a whole new approach to experiment design is what is required.

17.6.1 Microwave interferometer

Such a new approach could be provided by a microwave interferometer in space. One suggested design (Anderson, 1989, see figure 17.7) would use an existing element of the Tracking and Data Relay Satellite System (TDRSS) as a central free mass and would envision two simple and inexpensive Doppler and Ranging Transponders (DARTs) as end masses of an interferometer. This design would make use of the higher frequency K_u- and K_a-band links currently available for spacecraft-to-spacecraft communication on the TDRSS to reduce the effect of plasma scintillation; it would eliminate the troposphere fluctuation noise by going entirely above the atmosphere; and it would eliminate most of the clock jitter noise by canceling it via the interference in the two interferometer arms.

In more detail, the central spacecraft would generate a master frequency $F(t)$ which would be sent to the two end masses where it would be actively transponded in a phase-lock loop and reradiated to be received at the central mass a round-trip light-time later. Assuming the two end masses to be at constant radial distance (to simplify the discussion) and assuming that there are no gravitational waves present, we would find a Doppler residual in the ith arm of

$$D_i(t) = F(t) - F(t - 2l_i/c), \qquad (17.27)$$

where l_i is the length of each interferometer arm. In principle an interferometer signal could be formed by subtracting D_1 and D_2. However, we expect that the spacecraft clock will be noisy enough that the residual noise

$$I_{sc}(t) \equiv D_1(t) - D_2(t) = F(t - 2l_2/c) - F(t - 2l_1/c) \approx 2\dot{F}(t - 2l_1/c)(l_2 - l_1)/c \qquad (17.28)$$

would be too large without controlling the path length difference $(l_2 - l_1)$ very closely. To get around this requirement, the central spacecraft's master frequency could be transmitted to the earth where it would be compared with an earth-based hydrogen maser clock. This spacecraft clock calibration signal would provide

$$C(t) = F(t) + T(t) - f(t), \qquad (17.29)$$

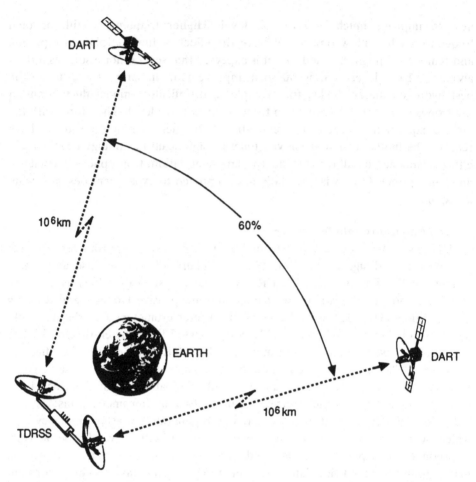

Figure 17.7. Conceptual design for the microwave interferometer. Central spacecraft is a TDRSS spacecraft in earth orbit. End spacecraft are small DART spacecraft in eccentric, inclined heliocentric orbits.

where $f(t)$ is the noise in the hydrogen maser and $T(t)$ is the troposphere fluctuation noise, both expected at the level of a few parts in 10^{15}. In addition, each interferometer arm's Doppler signal would be transmitted to the earth and received at the same station, with the same hydrogen maser reference that was used to monitor the master frequency. Each transmitted signal

$$D'_i(t) = D_i(t) + T(t) - f(t) \tag{17.30}$$

would then be corrected by $C(t) - C(t - 2l_i/c)$, and the two corrected signals would then be subtracted to produce the earth-based interferometer signal

$$I(t) = 2(\dot{T} - \dot{f})(l_2 - l_1)/c, \tag{17.31}$$

where the derivatives are both evaluated at $t - 2l_i/c$. The clock and troposphere noise remaining in this signal may thus be reduced to a few parts in 10^{18} by controlling the arm lengths to a few parts in 10^3.

Plasma noise at X-band (8.4 GHz) is currently known to be a few parts in 10^{15} over round-trip light-times of 1000 s. At K_u-band (15 GHz) that will be reduced to several parts in 10^{17} over 100 s round-trip light-times, and at K_a-band (28 GHz) the noise limit will be about 10^{-17}.

There still remain questions of other noise sources. Solar radiation pressure fluctuations, for example, are expected to produce noise at the 10^{-16} level and would have to be monitored with an accelerometer or a photodetector and subtracted out. Spacecraft and antenna buffeting would have to be investigated and engineered out if possible. Nevertheless, early studies of such a system lead to some optimism for a 10^{-17} end-to-end sensitivity gravitational wave antenna in the near future.

17.6.2 Laser interferometers

The major noise sources that remain in the interferometer system described above are residual interplanetary plasma scintillation, unmodeled radiation pressure fluctuations, and uncanceled clock jitter. If improvements are to be made over the hoped-for 10^{-17} sensitivity, then all three of these noise sources must be reduced.

There has been discussion in the recent literature (Stebbins *et al.*, 1989) of a design for a spaceborne laser interferometer which represents a forward leap in sensitivity to 10^{-21}. A system at this sensitivity will have the advantage that one can point to a particular binary star system, such as ι Boötes, and guarantee that the gravitational waves it produces will be seen in the instrument. As of this writing, the space laser is the only proposed experiment, space- or laboratory-based, that can make that claim.

The laser interferometer is similar in concept to the microwave interferometer discussed above. A single central mass spacecraft is equipped with a phase stabilized laser. The laser signal is received aboard two end-mass spacecraft and actively transponded by using the received signal as a reference for phase locking the return laser signals. The return signals are then recorded against the reference laser back on board the central spacecraft. All three spacecraft will need to be far from the earth to avoid the noisy local gravitational environment of the earth. One proposal would have a central spacecraft in 1 AU orbit, trailing the earth by about 30°. The other two end-mass spacecraft would likewise be in 1 AU orbits, but eccentric ($e \approx 0.0033$) and inclined ($i = 0.0058°$) so as to give roughly circular relative orbits around the central spacecraft (see figure 17.8).

The laser interferometer overcomes plasma scintillation by going to optical or near-optical frequencies. However, the laser system possesses technological challenges in developing sufficient laser power to obtain a good signal-to-noise ratio for the phase and in locking the normally phase-unstable laser to a more stable microwave cavity or other stabilising system. These technology goals,

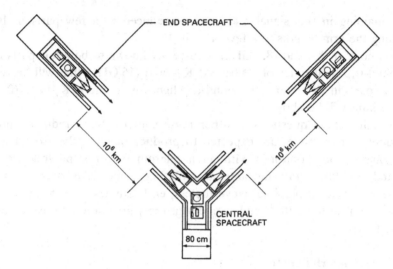

Figure 17.8. Conceptual design for the laser interferometer. All three spacecraft are in heliocentric orbit and are drag-free. Arm length is $\approx 10^6$ km.

however, are currently being pursued in several laboratories for a number of other applications, and the outlook is optimistic for the necessary technology being developed by the time it would be needed for the gravitational wave mission.

The real challenge in the plans to attain 10^{-21} sensitivity, however, is in the buffeting of the spacecraft by unwanted gravitational and non-gravitational forces. Reduction of spacecraft jitter to acceptable levels will require a drag-free system capable of $2 \times 10^{-15}\,\mathrm{cm\,s^{-2}\,Hz^{-0.5}}$ noise isolation over frequencies from 10^{-3} Hz to 10^{-5} Hz. Such a system will require, among other things, control of temperature gradients, magnetic fields, spacecraft position relative to the proof mass, and residual gas molecules in the test-mass cavity, and will require careful modeling of spacecraft mass inhomogeneity and the effect of solar radiation pressure fluctuations.

References

Adams, P. J., Hellings, R. W., Zimmerman, R. L., Farhoosh, H., Levine, D. I. and Zeldich, S. (1982). *Ap. J.* **253,** 1.

Anderson, A. J. (1974). JPL Engineering Memorandum 391–560, Pasadena: JPL Internal Document.

Anderson, A. J. (1977). *Atti dei Convegni Lincet* **34,** 235.

Anderson, A. J. (1989). Presented at the 12th International Congress on General Relativity and Gravitation, Boulder, Colorado.

Anderson, J. D. and Mashoon, B. (1985). *Ap. J.* **290,** 445.

Armstrong, J. W., Woo, R. and Estabrook, F. B. (1979). *Ap. J.* **230,** 560.

Backer, D. C. and Hellings, R. W. (1986). *Ann. Rev. Astron. Astrophys.* **24,** 537.

Barrow, J. and Carr, B. J. (1978). *MNRAS* **182,** 537.

Bertotti, B., Carr, B. J. and Rees, M. J. (1983). *M. Not. R. Ast.* **203,** 945.

Blandford, R., Narayan, R. and Romani, R. (1984). *J. Astrophys. Astron.* **5,** 369.

Carr, B. J. (1980). *Astron. Astrophys.* **89,** 6.

Carr, B. J. and Verdaguer, E. (1983). *Phys. Rev. D* **26,** 2995.

Davies, R. W. (1974). *Colloques Internationaux du CNRS* **220,** 33.

Davis, M. M., Taylor, J. H., Weisberg, J. M. and Backer, D. C. (1985). *Nature* **315,** 547.

Detweiler, S. (1979). *Ap. J.* **234,** 1100.

Epstein, R. and Clark, J. P. A. (1979). In *Sources of Gravitational Radiation,* ed. L. L. Smarr, p. 477, Cambridge University Press.

Estabrook, F. W. and Wahlquist, H. D. (1975). *Gen. Rev. Grav.* **6,** 439.

Grischuk, L. P. (1977). *Ann. NY Acad. Sci.* **302,** 439.

Hellings, R. W. (1981). *Phys. Rev. D* **23,** 832.

Hellings, R. W. (1983). In *Gravitational Radiation,* eds. N. Dereulle and T. Piran, p. 485, North Holland, Amsterdam.

Hellings, R. W., Anderson, J. D., Callahan, P. S. and Moffet, A. T. (1981). *Phys. Rev. D* **23,** 844.

Hellings, R. W. and Downs, G. S. (1983). *Ap. J. Lett.* **265,** L39.

Hogan, C. J. and Rees, M. J. (1984). *Nature* **311,** 109.

Krauss, L. M. (1985). *Nature* **313,** 32.

Mashoon, B., Carr, B. J. and Hu, B. L. (1981). *Astrophys. J.* **246,** 569.

Press, W. H. and Thorne, K. S. (1972). *Ann. Rev. Astron. Astrophys.* **10,** 335.

Romani, R. W. and Taylor, J. H. (1983). *Ap. J. Lett.* **265,** L35.

Ruffini, R. and Wheeler, J. A. (1969). In *Proc. ESRO Colloq.,* p. 45, ESA, Neuilly-sur-Seine, 1969.

Starobinskii, A. A. (1979). *JETP Lett.* **30,** 682.

Stebbins, R. T., Bender, P. L., Faller, J. E., Hall, J. L., Hils, D. and Vincent, M. A. (1989). In *Proceedings of the Fifth Marcel Grossman Conference,* ed. D. Blair, World Scientific, Singapore.

Vilenkin, A. (1981). *Phys. Lett. B* **107,** 47.

Vilenkin, A. (1985). *Phys. Rep.* **121,** 263.

Witten, E. (1984). *Phys. Rev. D* **30,** 272.

Zimmerman, R. L. and Hellings, R. W. (1980). *Ap. J.* **241,** 475.

Index